T0298520

PROOF THEORY

Sequent Calculi and Related Formalisms

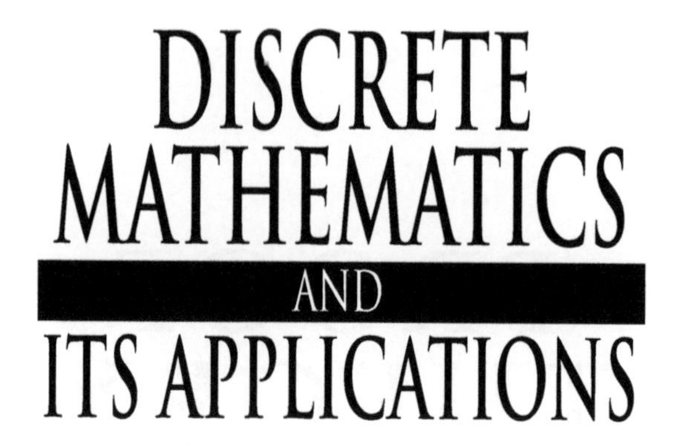

R. B. J. T. Allenby and Alan Slomson, How to Count: An Introduction to Combinatorics, Third Edition

Craig P. Bauer, Secret History: The Story of Cryptology

Juergen Bierbrauer, Introduction to Coding Theory

Katalin Bimbó, Combinatory Logic: Pure, Applied and Typed

Katalin Bimbó, Proof Theory: Sequent Calculi and Related Formalisms

Donald Bindner and Martin Erickson, A Student's Guide to the Study, Practice, and Tools of Modern Mathematics

Francine Blanchet-Sadri, Algorithmic Combinatorics on Partial Words

Miklós Bóna, Combinatorics of Permutations, Second Edition

Jason I. Brown, Discrete Structures and Their Interactions

Richard A. Brualdi and Dragoš Cvetković, A Combinatorial Approach to Matrix Theory and Its Applications

Kun-Mao Chao and Bang Ye Wu, Spanning Trees and Optimization Problems

Charalambos A. Charalambides, Enumerative Combinatorics

Gary Chartrand and Ping Zhang, Chromatic Graph Theory

Henri Cohen, Gerhard Frey, et al., Handbook of Elliptic and Hyperelliptic Curve Cryptography

Charles J. Colbourn and Jeffrey H. Dinitz, Handbook of Combinatorial Designs, Second Edition

Abhijit Das, Computational Number Theory

Martin Erickson, Pearls of Discrete Mathematics

Martin Erickson and Anthony Vazzana, Introduction to Number Theory

Steven Furino, Ying Miao, and Jianxing Yin, Frames and Resolvable Designs: Uses, Constructions, and Existence

Mark S. Gockenbach, Finite-Dimensional Linear Algebra

Titles (continued)

Randy Goldberg and Lance Riek, A Practical Handbook of Speech Coders

Jacob E. Goodman and Joseph O'Rourke, Handbook of Discrete and Computational Geometry, Second Edition

Jonathan L. Gross, Combinatorial Methods with Computer Applications

Jonathan L. Gross and Jay Yellen, Graph Theory and Its Applications, Second Edition

Jonathan L. Gross, Jay Yellen, and Ping Zhang Handbook of Graph Theory, Second Edition

David S. Gunderson, Handbook of Mathematical Induction: Theory and Applications

Richard Hammack, Wilfried Imrich, and Sandi Klavžar, Handbook of Product Graphs, Second Edition

Darrel R. Hankerson, Greg A. Harris, and Peter D. Johnson, Introduction to Information Theory and Data Compression, Second Edition

Darel W. Hardy, Fred Richman, and Carol L. Walker, Applied Algebra: Codes, Ciphers, and Discrete Algorithms, Second Edition

Daryl D. Harms, Miroslav Kraetzl, Charles J. Colbourn, and John S. Devitt, Network Reliability: Experiments with a Symbolic Algebra Environment

Silvia Heubach and Toufik Mansour, Combinatorics of Compositions and Words

Leslie Hogben, Handbook of Linear Algebra, Second Edition

Derek F. Holt with Bettina Eick and Eamonn A. O'Brien, Handbook of Computational Group Theory

David M. Jackson and Terry I. Visentin, An Atlas of Smaller Maps in Orientable and Nonorientable Surfaces

Richard E. Klima, Neil P. Sigmon, and Ernest L. Stitzinger, Applications of Abstract Algebra with Maple™ and MATLAB®, Second Edition

Richard E. Klima and Neil P. Sigmon, Cryptology: Classical and Modern with Maplets

Patrick Knupp and Kambiz Salari, Verification of Computer Codes in Computational Science and Engineering

William Kocay and Donald L. Kreher, Graphs, Algorithms, and Optimization

Donald L. Kreher and Douglas R. Stinson, Combinatorial Algorithms: Generation Enumeration and Search

Hang T. Lau, A Java Library of Graph Algorithms and Optimization

C. C. Lindner and C. A. Rodger, Design Theory, Second Edition

San Ling, Huaxiong Wang, and Chaoping Xing, Algebraic Curves in Cryptography

Nicholas A. Loehr, Bijective Combinatorics

Toufik Mansour, Combinatorics of Set Partitions

Alasdair McAndrew, Introduction to Cryptography with Open-Source Software

Elliott Mendelson, Introduction to Mathematical Logic, Fifth Edition

Titles (continued)

Alfred J. Menezes, Paul C. van Oorschot, and Scott A. Vanstone, Handbook of Applied Cryptography

Stig F. Mjølsnes, A Multidisciplinary Introduction to Information Security

Jason J. Molitierno, Applications of Combinatorial Matrix Theory to Laplacian Matrices of Graphs

Richard A. Mollin, Advanced Number Theory with Applications

Richard A. Mollin, Algebraic Number Theory, Second Edition

Richard A. Mollin, Codes: The Guide to Secrecy from Ancient to Modern Times

Richard A. Mollin, Fundamental Number Theory with Applications, Second Edition

Richard A. Mollin, An Introduction to Cryptography, Second Edition

Richard A. Mollin, Quadratics

Richard A. Mollin, RSA and Public-Key Cryptography

Carlos J. Moreno and Samuel S. Wagstaff, Jr., Sums of Squares of Integers

Gary L. Mullen and Daniel Panario, Handbook of Finite Fields

Goutam Paul and Subhamoy Maitra, RC4 Stream Cipher and Its Variants

Dingyi Pei, Authentication Codes and Combinatorial Designs

Kenneth H. Rosen, Handbook of Discrete and Combinatorial Mathematics

Douglas R. Shier and K.T. Wallenius, Applied Mathematical Modeling: A Multidisciplinary Approach

Alexander Stanoyevitch, Introduction to Cryptography with Mathematical Foundations and Computer Implementations

Jörn Steuding, Diophantine Analysis

Douglas R. Stinson, Cryptography: Theory and Practice, Third Edition

Roberto Tamassia, Handbook of Graph Drawing and Visualization

Roberto Togneri and Christopher J. deSilva, Fundamentals of Information Theory and Coding Design

W. D. Wallis, Introduction to Combinatorial Designs, Second Edition

W. D. Wallis and J. C. George, Introduction to Combinatorics

Jiacun Wang, Handbook of Finite State Based Models and Applications

Lawrence C. Washington, Elliptic Curves: Number Theory and Cryptography, Second Edition

DISCRETE MATHEMATICS AND ITS APPLICATIONS

PROOF THEORY

Sequent Calculi and Related Formalisms

Katalin Bimbó

University of Alberta
Edmonton, Canada

CRC Press is an imprint of the
Taylor & Francis Group, an informa business
A CHAPMAN & HALL BOOK

Front cover image by Katalin Bimbó

CRC Press
Taylor & Francis Group
6000 Broken Sound Parkway NW, Suite 300
Boca Raton, FL 33487-2742

Version Date: 20140714

International Standard Book Number-13: 978-1-4665-6466-4 (Hardback)

Library of Congress Cataloging-in-Publication Data

Bimbo, Katalin, 1963- author.
Proof theory : sequent calculi and related formalisms / Katalin Bimbo.
pages cm. -- (Discrete mathematics and its applications)
Includes bibliographical references and index.
ISBN 978-1-4665-6466-4 (hardback)
1. Proof theory. I. Title.

QA9.54.B55 2015
511.3'6--dc23 2014022998

Visit the Taylor & Francis Web site at
http://www.taylorandfrancis.com

and the CRC Press Web site at
http://www.crcpress.com

Contents

Preface

Sequent calculi constitute an interesting and important category of proof systems. They are much less known than axiomatic systems or natural deduction systems, and they are much less known than they should be. Sequent calculi were designed as a *theoretical framework* for investigations of logical consequence, and they live up to the expectations completely as an abundant source of meta-logical results.

The goal of this book is to provide a fairly comprehensive view of sequent calculi—including a wide range of variations. The focus is on sequent calculi for various *non-classical logics*, from intuitionistic logic to relevance logic, through linear and modal logics.

A particular version of sequent calculi, the so-called consecution calculi, has seen important new developments in the last decade or so. The invention of new consecution calculi for various relevance logics allowed the last major open problem in the area of relevance logic to be solved positively: *pure ticket entailment is decidable*. An exposition of this result from [42] is included in Chapter 9 together with further new decidability results (for less famous systems). A series of other results that were obtained by J. M. Dunn and me, or by me, in the last decade or so, are also presented in various places in the book. Some of these results are slightly improved in their current presentation. Obviously, many calculi and several important theorems are not new. They are included here to ensure the completeness of the picture; their original formulations may be found in the referenced publications.

This book contains little about semantics, in general, and about the semantics of non-classical logic in particular. It is impossible to combine the proof theoretical and the model theoretical results into a book of reasonable size. As an excuse, I should mention the book [38], which takes the dual approach to this book by focusing on *relational semantics of non-classical logics* while allots little attention to proof systems beyond specifying a logic formally.

The emphasis on calculi, which are inherently systems emphasizing syntax, is likely to appeal to computer scientists, who might also find intriguing the connections between sequent calculi and resolution, and between sequent calculi and typed systems. Those who are particularly interested in the constructive approach will find formalizations of intuitionistic logic as well as two calculi for linear logic in this book. The treatment of a wide range of variations on calculi for classical logic should appeal to those who favor that logic—whether they are mathematicians or philosophers. Philosophical logicians who do not shun more complicated logics will find charming the calculi for relevance logics. Lambek calculi and their extensions are given a rather detailed presentation, which may engage linguists (even without the analysis of concrete linguistic phenomena).

I hope that parts of this book are suitable as texts in various courses. Chapters 1, 2, 3 and 8 may be useful in an advanced undergraduate or beginning graduate course, where an emphasis is placed on classical logic and on a range of different proof calculi (mainly for classical logic). Chapters 4, 5 and 6 deal almost exclusively with non-classical logics. Chapters 7 and 9 are rich in meta-logical results, including results that have been obtained specifically using sequent calculus formalizations of various logics. These last five chapters might be used in a graduate course that embraces classical and non-classical logics together with their meta-theory.

To facilitate the use of the book as a text in a course (and to keep the size of the book under control), the text is peppered with exercises. In general, the starring indicates an increase in difficulty, however, sometimes an exercise is starred simply because it goes beyond the scope of the book or it is very lengthy. Solutions to selected exercises may be found on the web at the URL `www.ualberta.ca/~bimbo/ProofTheoryBook`.

In the winter term of 2013/14, I used a selection of chapters from this book in a combined upper-level undergraduate and graduate course at the University of Alberta. I would like to thank the students who spotted some typos, namely, Luke Aronyk, Jacob Denson, Nicole Deuchar, Mitchell Jenkins, Nicholas Juselius, Hassan Masoud, Alex Shulhan, Andrew Tedder, Alfred Ye and Lianghua Zhou.

Acknowledgments. I am grateful to J. Michael Dunn, who is a professor emeritus at Indiana University in Bloomington. I have had the privilege to learn a lot of logic from him, and then to work with him on a range of projects. Quite a few results in this book either directly come from or at least derive from our joint research and publications.

I would like to express my gratitude to Robert Stern, who, as acquiring editor, accepted my proposal to write this book. Robert Ross, who is the contact editor, provided useful advice to me. Jennifer Ahringer and Michele Dimont helped during the production process. I am grateful to all of them.

I am thankful for a grant from the Killam Research Fund awarded by the vice president (for research) of the University of Alberta, which allowed me to be excused from teaching a course that I had been scheduled to teach toward the final stages of the writing of this book.

The book was typeset using the program TEX, which was designed and written by Donald Knuth. Concretely, I used the LATEX format and I also utilized various packages and fonts that were developed under the auspices of the American Mathematical Society.

Edmonton, January 27, 2014 KB

Chapter 1

Proofs and proof theory

Proofs and proof theory may be understood in different ways. The aim of this short informal chapter is to sketch where proofs and proof theory come from and fit into logic and other disciplines.

Starting with a broad idea about what proofs are, then briefly connecting proofs with the three historical approaches to the foundations of mathematics in the early 20th century, we locate the place and role of proofs in logic now.

Sequent calculi were invented almost by accident, or at least, for purely theoretical reasons. Gentzen says the following.

> In order to be able to enunciate and prove the *Hauptsatz* in a convenient form, I had to provide a logical calculus especially suited to the purpose. For this the natural calculus proved unsuitable. For, although it already contains the properties essential to the validity of the *Hauptsatz*, it does so only with respect to its intuitionist form, in view of the fact that the law of excluded middle, as pointed out earlier, occupies a special position in relation to these properties.
>
> In Section III of this paper, therefore, I shall develop a new calculus of logical deduction containing all the desired properties in both their intuitionist and their classical form. [91, p. 289]

It should then be astounding that sequent calculi turned out to be not only a rich source of theoretical results, but their variants, such as consecution calculi inspired by structurally free logics, and their spin-offs, such as tableaux and resolution systems, turned out to be convenient, easy-to-use reasoning systems in practice.

1.1 Proofs of all kinds

Corroborations, verifications and confirmations of statements are *ubiquitous*. In everyday life or in a court of law, the process of demonstration is widely open-ended—both with respect to the sorts of claims that are to be shown and with respect to the methods that are permitted to be used. Philosophically speaking, most of those claims are likely to be empirical, and accordingly, what counts as a proof is evidence-based and experiential.

Proofs, in a sense pertinent to our enterprise, are crucial to *theoretical disciplines*. Scholars in empirical sciences, including the natural and the social sciences, seem to prefer to talk about theories, data, evidence and arguments, rather than proofs per se. Of course, empirical sciences also involve some deductive reasoning, however, quite often all of that remains fully unformalized—save bits and pieces of mathematics (which may include numbers or mathematical models). Some theoretical disciplines contain major applied components. We think of informatics and computer science, perhaps, theoretical physics from among the sciences, but also of exact philosophy from among the liberal arts. The areas of knowledge that are inherently tied up with utilizing proofs are *mathematics* and *logic*. Logicians not only prove theorems, but they prove theorems about proofs themselves, that is, the subject of their studies and investigations include proofs and the formal systems in which proofs exist or can be constructed rigorously.

Some philosophers of mathematics and some mathematicians think and argue that proofs are a *distinguishing feature*, or even a *defining feature* of mathematics.[1] This means that nothing can count as mathematics unless it contains proofs, but it also means that logic and some parts of computer science, philosophy, etc. are absorbed into mathematics. (The latter outcome might seem equally objectionable to some mathematicians and to some workers in those other fields.) We would like to leave some wiggle room so that proofs may occur and can be an integral part of other disciplines—outside of mathematics.

Logic is mainly concerned with reasoning, and the way to justify the correctness of an inference is to provide a proof. However, logicians do not craft custom-made proofs for whoever would like to prove something. Rather, logicians investigate when an inference is correct—for logical reasons. As it turns out, *logical reasons* are diverse and to some extent depend on the area one wants to reason about and on the methods that are deemed acceptable in that area. For example, quantum phenomena seem not to be amenable to reasoning according to classical logic.

Proofs in logic are of two kinds: proofs *within* some formalized system for a logic, and proofs *about* a formal system of logic. The latter belong to what is often called the *meta-theory* of a logic, or simply, *meta-logic*. This book is concerned with both kinds of proofs, but some of the more vital proofs such as *cut theorems* belong to meta-logic.

Mathematics is rightly perceived as a formal discipline—when compared to, let us say, art history. Nonetheless, most of mathematical practice and even most of published mathematical work remains somewhat informal. (Notable exceptions include computer-proofs of mathematical theorems, in which concepts, claims and the rules of inference must be formalized. We should be grateful to computers for their non-understanding of phrases like

[1]For example, J. Auslander puts forward this claim as the starting point of his paper "On the roles of proof in mathematics" in [103].

"The other case is obvious.") The not-quite-formal character of most mathematical proofs is shared with most proofs in the meta-theory of logics. Some mathematicians, though, are definitely interested in delineating the principles, axioms and methods, that is, a deductive system sufficient for proofs of their claims and theorems, which resembles setting up a formal system and proving theorems within it.

The following chapters contain some *completely formal proofs* carried out in one or another sequent calculus. However, we will also prove—often in a semi-formal fashion—*meta-theorems*, which concern the sequent calculi themselves or proofs in those sequent calculi.

1.2 Early history of proof theory in a nutshell

For centuries, it was believed that there is a certain collection of *laws of thought* that govern all correct reasoning. It is slightly ironic that thinkers were unwilling to give up this conviction—usually bordering on the status of faith—despite obvious failures.

Aristotle can be considered as the creator of the *first axiomatic system* of reasoning. The prototypical correct inference in syllogistic may be illustrated by concluding "All Greeks are mortal" from the premises "All men are mortal" and "All Greeks are men." Aristotle's syllogistic was innovative and advanced for his time, but some of the inferences he took to be logically valid turn out not to be correct, if we use modern formalizations of the statements involved in those inferences. Aristotle may also be credited with raising questions concerning reasoning about modal concepts and about the relationship of time and truth.

Shortly after Aristotle, the *philosophers of the Stoa* turned their attention to other components of their language that they believed to play a role is ensuring the correctness of reasoning. They are thought to have isolated some of the truth-functional connectives, most famously, what is called today, disjunction and material implication. Needless to say that contemporary analyses of "either ..., or ..." and of "if ..., then ..." are much more refined than what can be carried out in a two-valued truth-functional setting, which often leads to their replacement by other connectives.

It was not until the second half of the 19th century that the inadequacy of syllogistic as the general framework for correct reasoning had been widely accepted. However, when syllogistic finally had been supplanted by what is nowadays called *classical logic*, the latter soon assumed the same sacrosanct status that syllogistic enjoyed for centuries. Some of those who contributed to the invention of classical logic are Augustus De Morgan, George Boole, Charles S. Peirce and Gottlob Frege.

Frege, who gave an *axiomatic calculus* for classical first-order logic in his book entitled *Begriffsschrift*, and published in 1879, understood well that his "calculus of concepts" is not suitable to formalize a whole range of sentences. For example, statements that involve modalities (such as necessity) or tenses (as statements about the past or future do) cannot be formalized in classical logic. Further limitations of the expressive power of classical *first-order logic* and of *first-order theories*—with respect to their intended models—was made clear by the results of Leopold Löwenheim and Thoralf Skolem in the first decades of the past century.

Logic quickly developed into *multiple logics* in the early 20th century. Modal logics were introduced (first, as axiom systems of strict implication); higher-order logics and applied theories such as formal arithmetic and set theory were developed. Intuitionistic and many-valued logics appeared in the late 1920s–early 1930s. In this book, we take it to be manifest that there are many logics.[2] Undoubtedly, classical first-order logic has a special place in the landscape, because of its simplicity, and also because it has been investigated for so long by so many people. It is also clear to us that some logics may be better motivated than others or may have more applications than others. We will highlight various logics beyond classical logic starting in Chapter 3; indeed, we will allot much more attention to such logics than usual in books on proof systems.

The early 20th century saw not only the emergence of new logics, but it also witnessed new questions being asked about already existing logics. Questions about interpretations of classical logic led slowly to rigorous notions of a model of a logic and of a *formal semantics* for a logic. Another aspect, perhaps the defining aspect of a logic, is its *syntax* into which we include not only the *language* of a logic, but also a *formal specification* of how to generate theorems or to derive conclusions from premises.

Axiom systems are a great leap forward compared to informal reasoning, in which supporting principles and assumptions can be conjured up whenever that seems necessary or useful. Setting out an axiom system clarifies which principles and statements can be used and how. Traditionally, these basic principles are supposed to be self-evident (or at least, easily comprehensible) and few in their numbers. However, anybody who tried to prove even moderately complicated theorems of classical first-order logic in an axiomatic calculus knows how much heuristics, creativity and effort is required for that.

Let us step back to see clearly what the problem is. Given a *formal proof system*, such as an axiom system, proofs are completely formal objects and each step in a proof can be checked mechanically. It is conceivable that proofs could be churned out, so to speak, perhaps by a computer, because they comprise successive formal steps. A problem is that all those proofs may be proofs of something else than what we are looking or waiting for. Even if

[2] [17] provides some philosophical arguments in favor of certain logics beyond classical logic.

we know that the statement should be a theorem, obtaining a proof of it may take too long. In a sense, proofs in an axiomatic system are not and cannot be targeted easily toward proving a particular theorem.

It was the early 1930s when two new types of proof systems were introduced by Gerhard Gentzen. The *natural deduction systems* were intended to capture the way mathematicians actually reason when they prove theorems or lemmas. Rather than focusing exclusively on theorems, a natural deduction system typically permits *assumptions* to be made, which may be discharged, for instance, by conditioning the final conclusion on them. Using the rules, a reasoner can dismantle (or analyze) the given and derived statements, and assemble (or synthesize) new statements. This two-way procedure enables a more goal-directed proving process than is possible in axiomatic calculi.

Sequent calculi, the other sort of calculi introduced by Gentzen, are a step beyond the natural deduction systems. These calculi mainly focus on *consequence relations*, and the inference rules concern how the premises and the conclusions may be altered so that the inference remains correct. Sequent calculi—enjoying the cut theorem—provide unprecedented potential to analyze proofs, to discover properties of proof systems (hence, of logics) and to search for proofs of goal statements.

Since the publication of Gentzen's paper in 1935, new versions of old sequent calculi have been introduced, and new logics have been invented in the form of sequent calculi. Last but not least, new proofs of old theorems and altogether novel proof methods have been introduced. The topic of this book encompasses these kinds of calculi together with the central theorem concerning them—the cut theorem. We also look at certain consequences of the cut theorem as well as at some of the connections of these calculi to other proof systems.

1.2.1 Formalism and proof theory

An emphasis on proofs and on the theory of proofs is often associated with the name of David Hilbert, and with an approach to the foundations of mathematics known as formalism. Without disputing this connection, we intend to show that the main rivals of formalists in the early 20th century, that is, intuitionists and logicists, also have to be proponents of proofs.

A crucial difference between mathematics and rhetoric is that an argument in mathematics has to be *correct*—not merely convincing. From at least the time of Euclid and Aristotle, it has been understood that allowing arbitrary principles or arbitrary steps in a proof (even if the proof is informal), trivializes the proof. The idea of carefully choosing basic principles and permissible steps is embodied in the first axiomatic systems created by Euclid for geometry and by Aristotle for syllogistic. The axioms of geometry were supposed to be self-evident and the methods of construction were limited to certain operations that can be carried out by the straightedge and the compass. Axiomatization of a theory remained the ideal way to secure ultimate certainty for centuries afterward.

The proliferation of geometries in the 19th century questioned the self-evident character of the axioms. Doubtful moves in calculating with the infinitesimals, and difficulties in the reconstruction of reals from better-grounded numbers led to a search for improved methods. From today's point of view, all mathematics was at best semi-formal, if not completely informal at the time, and there was no clear separation between methods that would be specific to some area of mathematics (such as the differential calculus) and methods of reasoning in general.

New mathematical *concepts* were developed (such as sets and reals by cuts) and new *methods* for reasoning emerged (such as the algebra of thoughts and a calculus of concepts). A revived impulse for precision produced Frege's axiom system (for pure reasoning) and Hilbert's axiom system (for Euclidean geometry).

Hilbert, who had proved results in many areas of mathematics, and gained considerable prominence by 1900, became a leading figure in a quest for *rigorous mathematics*. Precision and rigor were thought to be guaranteed by axiomatization in a formal system—provided it had the additional property that something remained unprovable from the set of axioms.

As long as the axioms can be perceived as self-evident and the proof steps are sound, a mere axiomatization of a mathematical theory may be sufficient. However, the invention of hyperbolic geometry highlighted that neither the axioms nor their independence from each other is obvious—even to experts in the field. On the other hand, both Boole's and Frege's calculi canonized the pervasive character of contradictions. A contradiction in a theory infects the theory, because one contradiction brings all possible contradictions into the theory. The loss of perceptible basic truths and the adoption of the permeating character of contradictions naturally caused a desire for demonstrations of the consistency of mathematical theories.

Consistency may be proved in two different ways. One is to construct a *model* to show that the axioms are true and the rules preserve truth. (This is the so-called soundness theorem.) In the case of pure logic, there may be a modest number of assumptions about the objects and their interactions with each other that is needed for a model. However, in the case of a mathematical theory such as a geometry, the objects are assumed to have various (known) properties, which are used essentially in the construction of a model. (Beltrami's models are a simple illustration of this point.) Then, the question of consistency is propelled forward to another theory, which describes the models.

The other way to prove consistency is syntactically. Given a formal theory with a notion of theoremhood, the consistency proof shows that there is *no proof of a contradiction*, for example, of $\mathcal{A} \wedge \neg\mathcal{A}$ (that is, of both a sentence and its negation). At first, it may seem to be an easy task; eventually, proofs are typically defined to be finite objects such as sequences of formulas with some further properties. Even for an axiom system, let us say, an axiom system for classical first-order logic (without additional genuinely mathematical axioms), it is less than clear how to deal with this task. Perhaps, there is some ingenious (and exceedingly lengthy) proof going through the quantificational axioms and rules that produces a contradiction.

The problem is obviously challenging, but this is what Hilbert posed, specifically, for arithmetic in his famous lecture in 1900. Arguably, nobody, not even Hilbert himself, understood or knew exactly how challenging the problem was. At the time, crucial limitations of expressive capabilities of classical first-order logic have not yet been discovered; neither was the complexity of natural numbers or of their formal theory grasped. However, the importance of *proofs* was stressed, and a need for a development of a *theory of proofs* was clear.

Incidentally, we note that within Hilbert's *metamathematics* program the term "proof theory" denoted not the theory of proofs in general, but specifically, a theory aimed at proving the consistency of mathematical theories via syntactic means. Of course, when the only logic that had been started to be formally developed was classical logic, hence, a mathematical theory was—inescapably—embedded into classical logic, there was little difference between the theory of proofs and proof theory. However, in the last century, many new logics have been defined. Our concern in this book is with *proofs in sequent calculi*—including sequent calculi formalizing *non-classical logics*—and with the *meta-theory* of these calculi.

1.2.2 Proofs in intuitionism and constructivism

The invention of modern intuitionism is customarily attributed to Luitzen E. J. Brouwer. Intuitionism is connected to the philosophical notion of intuition, especially, in the philosophy of Immanuel Kant. Brouwer's philosophical ideas are caliginous, but he seems to have assigned only a secondary role to proofs in mathematics. A proof is merely an external representation and an attempt of communication of the internal thought processes of the mathematician who creates mathematical objects and discovers their properties in his mind.

Intuitionism acquired a more tangible form after Arendt Heyting's proposal for *intuitionistic logic* in 1930. The latter formal system instantaneously riveted attention. In parallel, Andreĭ N. Kolmogorov developed an axiomatic system and similar ideas appealing to mathematical *constructions*. Gentzen also paid a lot of attention to intuitionistic logic. His natural deduction system is apposite for intuitionistic logic, whereas classical logic requires the addition of a disparate rule (i.e., the double negation elimination rule). Gentzen designed the logistic sequent calculus *LK* for classical logic so that intuitionistic logic falls out via a simple restriction in the calculus.

Admittedly, intuitionism can be thought to focus on something else than proofs. However, proofs are mathematical objects themselves, which are effectively constructed. Moreover, intuitionistic logic can be given a *proof interpretation*, which is sometimes labeled as the BHK interpretation (after Brouwer, Heyting and Kolmogorov).[3]

The BHK interpretation takes the connectives and the quantifiers to be *operations on proofs* (which combine and change proofs)—rather than finitary and infinitary truth functions. The following examples illustrate the idea.

The proof of the conjunction of $\mathcal{A}$ and $\mathcal{B}$ (i.e., $\mathcal{A} \wedge \mathcal{B}$) is the proof that results from pairing the proofs that demonstrate $\mathcal{A}$ and $\mathcal{B}$, respectively. A disjunction of $\mathcal{A}$ and $\mathcal{B}$ (i.e., $\mathcal{A} \vee \mathcal{B}$) is proved when it is known which disjunct is proved and how. This means that a proof of a disjunction is a pair comprising a proof of a disjunct and a selection function, which assigns the proof to one or the other disjunct. Implication is not reducible to a disjunction (with a negation thrown in) as in classical logic. A proof of $\mathcal{A} \supset \mathcal{B}$ is a transformation such that given a proof of $\mathcal{A}$ it produces a proof of $\mathcal{B}$. The quantifiers must be interpreted independently from each other, because intuitionistically, they carry different meanings. A proof of a universally quantified formula such as $\forall x\, \mathcal{A}(x)$, generates a proof of every instance of the formula, that is, a proof for each $\mathcal{A}(a)$. The existentially quantified formula $\exists x\, \mathcal{A}(x)$ is similar to disjunction, that is, its proof consists of a proof and a selection of the instance of which the proof is a proof. These informally described ideas can be and have been made more precise by identifying proofs with functions.

Kurt Gödel made remarkable contributions to the meta-theory of intuitionistic logic and intuitionistic arithmetic. From the point of view of theorems, that is, from the point of view of provable sentences, he introduced a negative translation of classical logic into intuitionistic logic. (We briefly outline this in Section 3.7.) Gödel also introduced a logic with a unary operator, which is interpreted as provability, that allows an interpretation of propositional intuitionistic logic. This logic is the first modern presentation of a normal modal logic, namely, of the system $S4$, as an extension of classical propositional logic. This first link between intuitionistic logic and $S4$ has been further developed later on.

The connections between classical logic and classical arithmetic, on the one hand, and intuitionistic logic and intuitionistic arithmetic, on the other, may get confusing with the parallel treatments of these logics and their translations into each other. However, Gödel's results show that the intuitionistic understanding of proofs is not sufficient to make intuitionistic arithmetic unrecoverable from Peano arithmetic, which is based on classical logic. Gödel in his later work refined (or even replaced) the somewhat vague notion of intuitionistic proof.

Intuitionism greatly benefited from getting a formal remake in the form of proof systems. In turn, the insistence on proofs (without any use of excluded

[3]"HKG" would be a more appropriate label, given the considerable role Gödel played in developing such ideas versus Brouwer's qualms concerning non-subjective objects, in general.

middle or double negation elimination) aided the development of the theory of proofs for intuitionistic and for other non-classical logics.

1.2.3 Logicism and the theory of proofs

Finally, let us turn to logicism, which chronologically should be considered as preceding intuitionism, and possibly, formalism too. Usually Frege as well as Alfred N. Whitehead and Bertrand Russell are considered to be the leading figures in logicism in the late 19th–early 20th centuries.

As soon as logic had become formal and axiomatized, proofs turned out to be inevitable components of any theory. Of course, logicism places an emphasis on finding *sufficiently strong axioms* to deduce all the theorems of a mathematical theory, which is under consideration, rather than on studying proofs in a particular theory.

The fame of logicism and logicists as devotees of proof theory might have been somewhat damaged by Frege's inconsistent system that he proposed to deduce arithmetic from principles he took to be purely logical. The reaction, in the form of the monumental three-volume *Principia Mathematica*, seem to have gone to the other extreme by making excessive distinctions (in the form of ramified type theory), which then necessitated some more or less ad hoc axioms to be added.

In sum, adherents of logicism have to endorse and uphold proofs in one or another formalism—as long as logic is formal and symbolic.

1.3 Proofs as calculations

There is an undeniable similarity between building a proof and performing calculations—such as solving a system of equations. One aspect of this likeness is the *formal* character of the *steps*.

A deeper similarity stems from the *proof interpretation* of intuitionistic connectives. Gödel's system of computable functionals of finite type may be thought to concretize the intuitionistic notion of a proof. However, even without a whole calculus of computable functions such as combinatory logic or a λ-calculus, there are discernible connections between certain logics and classes of functions.

The correspondence between *implicational formulas* and *functions* strongly favors not only intuitionistic logic, but also some other non-classical logics (over classical logic). Among such non-classical logics are the logic of relevant implication and pure ticket entailment. Matching functions and proofs is more straightforward in an axiomatic setting than in sequent calculi; nonetheless, these connections inspired *structurally free logics*, which are sequent calculi that incorporate combinators as formulas into sequents. Some of the sequent calculi that were invented in the last decade or so also utilize insights from the connections between simple types and combinators. The constructive spirit permeates linear logic, which has been met with a lot of enthusiasm in computer science, and has been used in modeling aspects of computing.

Many of the latter developments are relatively new—compared to the first axiom system for classical logic or even the first sequent calculi. In order to provide a current view of the field of sequent calculi, we devote more chapters to non-classical logics than to some other more traditional themes.

Chapter 2

Classical first-order logic

The first sequent calculus, *LK* is the example after which many other calculi have been fashioned. We describe this calculus, prove its equivalence to the axiom system *K*, and provide a sound and complete interpretation too. A major part of this chapter presents the proof of the *cut theorem* using *triple induction*. We avoid the detour via the mix rule by using a suitable formulation of the *single cut rule*, and by adding a *new parameter* to the original double induction. Although we do not go through all the details of this proof, we hope that sufficient details are included so that the structure of the proof becomes completely clear. Later on, we will provide more condensed proofs of cut theorems or simply state cut theorems with reference to this proof.

2.1 The sequent calculus *LK*

The calculus *LK* was introduced by Gentzen in [89], where it is labeled *logistic* (*logistisch*, in German). We will simply use the general term *sequent calculus* until we introduce consecution calculi in Chapter 5. Our presentation of *LK* is faithful to the original in the shape of the axioms and rules, though we do not always use the same notation or symbols.

The formulas of *LK* belong to the following language. There are denumerably many *name constants*, and they are denoted by $a_0, a_1, a_2, \ldots$. There are *function symbols*, denoted by $f_0^{n_0}, f_1^{n_1}, f_2^{n_2}, \ldots$, where the subscript identifies the function symbol, whereas the superscript shows the arity of the function symbol. We define the language so that the *arity* of function symbols (and later on, of predicate symbols and of connectives) is always a *positive integer*. The language contains *predicate symbols* that are denoted by $P_0^{m_0}, P_1^{m_1}, P_2^{m_2}, \ldots$. Once again, the subscript identifies the predicate and the superscript indicates the number of arguments the predicate takes. The language includes denumerably many of both function symbols and of predicate symbols. We use $=$ as the symbol for the two-place predicate called *identity*. Identity is considered a *logical* component of the language; therefore, the interpretation of $=$ is fixed, as it will become obvious in Section 2.4. *Propositional variables* are denoted by $p_0, p_1, p_2, \ldots$, and we will stipulate that there are denumerably many of them.

Individual variables are a very important kind of basic expressions. Every first-order language must contain *denumerably many* individual variables, which are the only variables that can be quantified in a first-order language. We denote individual variables by $x_0, x_1, x_2, \ldots$. The set of logical components—beyond identity—includes *connectives*, a pair of *constant propositions* and *quantifiers*. The connectives are *negation* ($\neg$), *conjunction* ($\wedge$), *disjunction* ($\vee$) and *conditional* ($\supset$). The constant propositions are $\boldsymbol{T}$ and $\boldsymbol{F}$, which can be thought of as "the true proposition" and "the false proposition," respectively. The names of these constants clearly show the limited amount of information that is taken into account in classical logic. The *universal* and the *existential* quantifiers are denoted by $\forall$ and $\exists$, respectively.

Occasionally, we may simplify our notation—provided that no confusion is likely. For instance, we might also use a, b and c for name constants, f, g and h for function symbols, P, Q and R for predicate symbols, p, q and r for propositional variables and x, y and z for individual variables. Sometimes we will omit superscripts, when the arity of a symbol can be determined from the (implicit) assumption that an expression is well formed.

So far there is nothing in the set-up of the language that would be specific to a sequent calculus. The next two definitions are standard too.

Definition 2.1. (Terms) The *set of terms* is defined inductively by (1)–(2).

(1) Name constants and individual variables are *terms*;

(2) if $f_i^{n_i}$ is a function symbol and $t_1, \ldots, t_{n_i}$ are terms, then $f_i(t_1, \ldots, t_{n_i})$ is a *term*.

Having said that the definition is inductive, we assume that the set of terms is the *least set* generated from the set of *atomic terms* (i.e., the set specified in clause (1)) by finitely many applications of the inductive clause (i.e., clause (2)). Finitely many includes zero many; that is, all the atomic terms are terms—as their label intended to suggest.[1]

The set of terms could be defined using the *Backus–Naur form* (BNF) (as it is often done nowadays), if we stipulate that $\mathbb{A}$ is a non-terminal symbol that rewrites to an atomic term. Characterizing atomic terms by a *context-free grammar* (CFG) is not difficult, but it is not pretty. We give the following definition as an example, and later on we always assume that a similar definition can be given when there is a base set, the elements of which are indexed by the natural numbers.[2] (The subscript is concatenated to the previous letter and it is slightly lowered for aesthetic reasons only.)

[1] Inductive definitions and inductive proofs are explained in detail, for example, in [16] and in [107]. Here we only assume a basic understanding of inductive definitions.

[2] Knowledge of formal language theory is not essential for our purposes, though it may be useful. For example, Sipser [169] is a good introductory text on formal language theory.

Definition 2.2. (Atomic terms) Let A, I, S and N be non-terminal symbols. $\mathbb{A}$, the *set of atomic terms* is defined by the next context-free grammar.

1. $\mathbb{A} := A_I$
2. $A := a \mid x$
3. $I := 0 \mid N \mid NS$
4. $S := 0 \mid N \mid SS$
5. $N := 1 \mid 2 \mid 3 \mid 4 \mid 5 \mid 6 \mid 7 \mid 8 \mid 9$

The *start* symbol is $\mathbb{A}$, which is replaced by an "indexed atomic term" A_I. Since we do not want to have a 0 at the beginning of a string of digits, except when 0 is the whole string, we have to complicate the definition of I (the "index") with N (for "non-zero digit") and S (for "subsequent digit"). Permitting S to be replaced by SS ensures that indices greater than 99 can be generated too. Obviously, the complications in the above CFG stem from the steps that produce a natural number in decimal notation, with positive numbers not starting with 0's.

Definition 2.3. (Terms in BNF) The *set of terms* is defined as

$$T := \mathbb{A} \mid \mathbb{F}_i^{n_i}(\underbrace{T, \ldots, T}_{n_i}),$$

where $\mathbb{A}$ rewrites to an atomic term, and $\mathbb{F}_i^{n_i}$ rewrites to a function symbol.

Exercise 2.1.1. Give a definition for $\mathbb{F}_i^{n_i}$ that is similar to 2.2. [Hint: One has to make sure that for each i there is a unique positive integer n_i. Here is an idea that works, though it may not be the simplest solution (and it will not produce f_0, e.g.). Think about the subscript and superscript as the single string in_i, in which there is a substring of the form $\varrho(n_i)n_i$ (or $\varrho(n_i); n_i$ with $;$ to separate the subscript and the superscript), where $\varrho(n_i)$ is the reverse of the string n_i. Palindromes are easy to generate, though a slight complication emerges here, because we do not want n_i (hence, i either) to start with 0. Then i is $m\varrho(n_i)$, where m is any string that does not start with 0. Associating the superscript to the subscript in this way can also ensure that there are denumerably many function symbols of each arity.]

The following definition presupposes that we can refer not only to the set of terms but also to the set of variables as a category of expressions.

Definition 2.4. (Well-formed formulas) The *set of well-formed formulas* is inductively defined by (1)–(5).

(1) $\boldsymbol{T}$ and $\boldsymbol{F}$ are *well-formed formulas;*

(2) propositional variables are *well-formed formulas;*

(3) if $t_1, \ldots, t_{n_i}$ are terms, and $P_i^{n_i}$ is a predicate symbol, then $P_i^{n_i}(t_1, \ldots, t_{n_i})$ is a *well-formed formula;*

(4) if $\mathcal{A}$ and $\mathcal{B}$ are well-formed formulas, then $\neg\mathcal{A}$, $(\mathcal{A} \wedge \mathcal{B})$, $(\mathcal{A} \vee \mathcal{B})$ and $(\mathcal{A} \supset \mathcal{B})$ are *well-formed formulas;*

(5) if $\mathcal{A}$ is a well-formed formula and x_i is an individual variable, then $\forall x_i\, \mathcal{A}$ and $\exists x_i\, \mathcal{A}$ are *well-formed formulas*.

A well-formed formula is an *atomic formula*, if it is by clause (1), (2) or (3).

"Well-formed formula" is a lengthy expression; hence, we will often use instead the abbreviation *wff* or the shorter term "formula."

Note that we take the formation of an identity statement to be a special instance of (3). We introduced $=$ as the identity symbol, but one particular P_i could be fixed as the identity symbol. Then, by a notational convention, we would write $t_1 = t_2$ instead of $P_i(t_1, t_2)$.

Exercise 2.1.2. The expressions in (a) are terms, and those in (b) are wff's. Use the Definitions 2.1 and 2.4 to generate these expressions step by step.

(a) x_3, $f_1^2(a_4, x_6)$, $h(g(x,b), f(a,y,y,z))$

(b) $(p \supset P_1^1(x_0))$, $(\neg\forall x_0\, P_1^1(x_0) \wedge \forall x_1\, P_1^1(x_2))$, $\exists x_1 \forall x_2 \exists x_3 (P_1^2(x_1, x_2) \supset (P_2^4(x_1, a, b, x_2) \supset (P_3^1(x_3) \vee P_4^2(x_2, x_4))))$, $(\neg p \supset \neg\neg(Q(b,c,a) \wedge \neg r))$

Exercise 2.1.3. The following expressions are neither terms nor wff's. Explain for each why it does not belong to either category of expressions. (a) $\supset p$, (b) $a_2 p_1 \supset p_2 a_3$, (c) $\forall\neg(\neg P(x) \vee \exists Q(p,q))$. [Hint: You may discover that the notation has certain built-in assumptions that are usually left tacit.]

An occurrence of a variable in a formula is just what the informal meaning of the phrase suggests: one can look at the formula and find the variable in it. It is possible to assign unique labels to each occurrence of a variable, and then to define when an occurrence is free or bound. However, for our purposes it is sufficient to note this possibility without giving its details. Similarly, *a subformula* is a wff that occurs in a given wff.

Exercise 2.1.4. Design a schema that assigns a unique numerical label to occurrences of individual variables.

Definition 2.5. (Variable binding) An occurrence of x in $\mathcal{A}$ is *free* iff (i.e., if and only if) it is not within a subformula of $\mathcal{A}$ of the form $\forall x\, \mathcal{B}$ or $\exists x\, \mathcal{B}$. All occurrences of x in $\mathcal{A}$ that are not free are *bound*.

Each displayed quantifier in $\forall x\, \mathcal{A}$ and $\exists x\, \mathcal{A}$ *binds* the immediately following occurrence of x, as well as all the *free occurrences* of x in $\mathcal{A}$.

Exercise 2.1.5. Describe an informal procedure that allows one to decide if an occurrence of a variable is free or bound. Explain how to find the quantifier which binds a particular bound occurrence of a variable.

Every occurrence of every variable is either free or bound, but not both. It is also customary to talk about free and bound variables of a formula. x is a *free variable* of $\mathcal{A}$ iff there is a free occurrence of x in $\mathcal{A}$, and similarly, for a *bound variable*. Then, it is easy to see that, if x occurs in $\mathcal{A}$, then x is a free or a bound variable of $\mathcal{A}$, and possibly, both.

In sequent calculus rules, by $\mathcal{A}(y)$ we mean a formula $\mathcal{A}$ with finitely many free occurrences of y selected. Then, $\forall x\, \mathcal{A}(x)$ and $\exists x\, \mathcal{A}(x)$ are obtained by replacing the selected occurrences of y with x, which must be OK for all those occurrences of y in $\mathcal{A}(y)$ (as we explain on p. 22), and by attaching a quantifier prefix. That is, substituting y for x in $\mathcal{A}(x)$ yields $\mathcal{A}(y)$.

Exercise 2.1.6. List all the variables that occur in the following formulas. Then, for each occurrence determine whether it is free or bound. Lastly, for each formula, list its free and bound variables.

(a) $(\forall x_0\, (P_1^1(x_0) \supset P_2^2(x_0, x_1)) \wedge \exists x_3\, \neg P_1^1(x_3))$

(b) $\forall y\, (Q(z) \vee \neg\exists z\, R(z, y, a_3, z))$

(c) $\forall x_1\, \exists x_1\, (\forall x_2\, P_1^1(x_1) \supset \forall x_3\, \neg\exists x_1\, P_4^3(a_0, x_1, x_3))$

Exercise 2.1.7. Create formulas with the variable binding properties listed.

(a) x has no occurrence, y has one bound and z has one free occurrence.

(b) x has one free, y has two bound and z has two free and one bound occurrences.

A quintessential feature of sequent calculi is that they are designed to formalize reasoning *about inferences* rather than simply to be a framework to construct inferences. In the case of classical logic, there can be finitely many premises and finitely many conclusions. This idea is captured in the concept of *sequents*.

Definition 2.6. (Sequents) If $\langle \mathcal{A}_0, \ldots, \mathcal{A}_n \rangle$ and $\langle \mathcal{B}_0, \ldots, \mathcal{B}_m \rangle$ are (possibly empty) sequences of formulas, then $\mathcal{A}_0, \ldots, \mathcal{A}_n \vdash \mathcal{B}_0, \ldots, \mathcal{B}_m$ is a *sequent*. The $\mathcal{A}_0, \ldots, \mathcal{A}_n$ part is the *antecedent* and the $\mathcal{B}_0, \ldots, \mathcal{B}_m$ part is the *succedent* of the sequent. The comma (i.e., ,), which separates the formulas, is a *structural connective*.

The $\vdash$ symbol *does not* stand for a connective; it forms a sequent from two sequences of formulas. Either of these two sequences of formulas may be *empty*. The *turnstile* (i.e., $\vdash$) is often used to indicate that a formula is a theorem of an axiom system, and we will occasionally use $\vdash$ in that sense too. Thus, $\vdash$ can appear in connection with more than one system, but hopefully, this will not cause confusion, because the context determines which sense is meant. These uses of $\vdash$ are connected: in the sequent calculus LK, $\vdash \mathcal{A}$ will be the analog of $\vdash \mathcal{A}$ in the corresponding axiom system K.

Sequences of formulas will be denoted by capital Greek letters such as Γ and Δ. By $\Gamma, \mathcal{A}$ and $\mathcal{A}, \Gamma$, we mean the sequence of formulas obtained via appending the formula $\mathcal{A}$ to the end or to the beginning of the sequence Γ, respectively. Sequences are the *structures* in LK.

LK is a sequent calculus for *classical first-order logic*, FOL. It contains an *axiom* and *rules*. Some of the rules will only change the shape of a structure, and accordingly, they are called *structural rules*. The other rules are connected in an obvious way to connectives and quantifiers, hence, they are called *connective* and *quantifier rules*, or together, *operational rules*. To simplify our notation a bit further, we will omit outside parentheses from wff's. Note that we do not give any special rules involving $\boldsymbol{T}$, $\boldsymbol{F}$ or $=$ now; in effect, we temporarily exclude them from the language.

$$\mathcal{A} \vdash \mathcal{A} \;\; \text{id}$$

$$\frac{\mathcal{A}, \Gamma \vdash \Delta}{\mathcal{A} \wedge \mathcal{B}, \Gamma \vdash \Delta} \wedge\vdash_1 \qquad \frac{\mathcal{B}, \Gamma \vdash \Delta}{\mathcal{A} \wedge \mathcal{B}, \Gamma \vdash \Delta} \wedge\vdash_2 \qquad \frac{\Gamma \vdash \Delta, \mathcal{A} \quad \Gamma \vdash \Delta, \mathcal{B}}{\Gamma \vdash \Delta, \mathcal{A} \wedge \mathcal{B}} \vdash\wedge$$

$$\frac{\mathcal{A}, \Gamma \vdash \Delta \quad \mathcal{B}, \Gamma \vdash \Delta}{\mathcal{A} \vee \mathcal{B}, \Gamma \vdash \Delta} \vee\vdash \qquad \frac{\Gamma \vdash \Delta, \mathcal{A}}{\Gamma \vdash \Delta, \mathcal{A} \vee \mathcal{B}} \vdash\vee_1 \qquad \frac{\Gamma \vdash \Delta, \mathcal{B}}{\Gamma \vdash \Delta, \mathcal{A} \vee \mathcal{B}} \vdash\vee_2$$

$$\frac{\Gamma \vdash \Delta, \mathcal{A}}{\neg\mathcal{A}, \Gamma \vdash \Delta} \neg\vdash \qquad \frac{\mathcal{A}, \Gamma \vdash \Delta}{\Gamma \vdash \Delta, \neg\mathcal{A}} \vdash\neg$$

$$\frac{\Gamma \vdash \Delta, \mathcal{A} \quad \mathcal{B}, \Theta \vdash \Lambda}{\mathcal{A} \supset \mathcal{B}, \Gamma, \Theta \vdash \Delta, \Lambda} \supset\vdash \qquad \frac{\mathcal{A}, \Gamma \vdash \Delta, \mathcal{B}}{\Gamma \vdash \Delta, \mathcal{A} \supset \mathcal{B}} \vdash\supset$$

$$\frac{\mathcal{A}(y), \Gamma \vdash \Delta}{\forall x\, \mathcal{A}(x), \Gamma \vdash \Delta} \forall\vdash \qquad \frac{\Gamma \vdash \Delta, \mathcal{A}(y)}{\Gamma \vdash \Delta, \forall x\, \mathcal{A}(x)} \vdash\forall^{\oslash}$$

$$\frac{\mathcal{A}(y), \Gamma \vdash \Delta}{\exists x\, \mathcal{A}(x), \Gamma \vdash \Delta} \exists\vdash^{\oslash} \qquad \frac{\Gamma \vdash \Delta, \mathcal{A}(y)}{\Gamma \vdash \Delta, \exists x\, \mathcal{A}(x)} \vdash\exists$$

The rules $(\vdash\forall)$ and $(\exists\vdash)$, which are marked by $^{\oslash}$, are restricted to be applicable when the lower sequent *does not contain* the variable y free. An obvious consequence of this stipulation is that there can be no free occurrences of y in the elements of Γ or Δ.

The rules above are the *operational rules* in *LK*, whereas the next six rules are the *structural rules*.

$$\frac{\Gamma \vdash \Delta}{\mathcal{A}, \Gamma \vdash \Delta} K\vdash \qquad \frac{\Gamma \vdash \Delta}{\Gamma \vdash \Delta, \mathcal{A}} \vdash K$$

$$\frac{\mathcal{A}, \mathcal{A}, \Gamma \vdash \Delta}{\mathcal{A}, \Gamma \vdash \Delta} W\vdash \qquad \frac{\Gamma \vdash \Delta, \mathcal{A}, \mathcal{A}}{\Gamma \vdash \Delta, \mathcal{A}} \vdash W$$

$$\frac{\Gamma, \mathcal{A}, \mathcal{B}, \Delta \vdash \Theta}{\Gamma, \mathcal{B}, \mathcal{A}, \Delta \vdash \Theta} C\vdash \qquad \frac{\Theta \vdash \Gamma, \mathcal{A}, \mathcal{B}, \Delta}{\Theta \vdash \Gamma, \mathcal{B}, \mathcal{A}, \Delta} \vdash C$$

The labels for the structural rules allude to the Curry–Howard correspondence between implicational formulas and combinators. We will have and will explain a more rigorous version of the correspondence between structural rules and combinators in Chapter 5.

The $(K\vdash)$ and $(\vdash K)$ rules are called *thinning* or *weakening*. If $\Gamma \vdash \Delta$ expresses that from the premises in Γ, at least one of the conclusions in Δ can be inferred, then adding a premise or a conclusion weakens the original claim about the inference. The $(W\vdash)$ and $(\vdash W)$ rules are called *contraction*, because two occurrences of a formula are shrunk to one occurrence. The $(C\vdash)$ and $(\vdash C)$ rules are called *permutation, interchange* or *exchange*. These two rules allow the switching of two formulas anywhere within a sequent. Although any of Γ, Δ and Θ may be empty, if they are not, then the shape of the permutation rules guarantees that $\mathcal{A}$ and $\mathcal{B}$ still can be swapped.

Keeping the structural rules in mind, we could (informally) view sequences of wff's as sets, where the order and multiplicity of occurrences of wff's does not matter. Replacing sequences of wff's by other structures will be important in other sequent calculi (e.g., in Chapters 3, 4 and 5). However, in this chapter, we consider LK in its original form, where sequents are built from sequences of wff's.

Sequences and sequents imply a linear ordering or a linear succession of formulas; proofs are more complex.

Definition 2.7. (Proofs) A *proof* is a finite tree comprising occurrences of sequents, where each leaf of the tree is an instance of the axiom and all the other nodes are justified by applications of the rules.[3]

We say that a sequent $\Gamma \vdash \Delta$ is *provable* when there is a proof in which $\Gamma \vdash \Delta$ is the end sequent. To make the definition of proofs more palpable we give two proofs as examples followed by detailed explanations.

Example 2.8. Our first example is an instance of exchanging different quantifiers.

$$\dfrac{\dfrac{\dfrac{\dfrac{\dfrac{R(z,v) \vdash R(z,v)}{\forall y\, R(z,y) \vdash R(z,v)}\,\forall\vdash}{\forall y\, R(z,y) \vdash \exists x\, R(x,v)}\,\vdash\exists}{\forall y\, R(z,y) \vdash \forall y\, \exists x\, R(x,y)}\,\vdash\forall}{\exists x\, \forall y\, R(x,y) \vdash \forall y\, \exists x\, R(x,y)}\,\exists\vdash}{\vdash \exists x\, \forall y\, R(x,y) \supset \forall y\, \exists x\, R(x,y)}\,\vdash\supset$$

The first line is an instance of the axiom, from which we can get the next sequent by $(\forall\vdash)$. There is no restriction on this rule, and there is no restriction on the $(\vdash\exists)$ rule either, which yields the next sequent. Each of the variables z and v has exactly one occurrence in the sequent, which means that the rules $(\vdash\forall)$ and $(\exists\vdash)$ can be applied in the next two steps. The last move

[3]The notion of trees as we use it here and elsewhere in this book is explained in Section A.1 of Appendix A. (Our usage is the standard one in logic and parts of computer science, however, the term is used with a wider meaning in some other areas.)

is to introduce $\supset$. The proof also shows that Γ and Δ can be empty, for instance, in the last step they are both empty. (Incidentally, note that the labels are not part of the proof, they are added to help to see the proof as a proof.)

Definition 2.9. (Theoremhood) A formula $\mathcal{A}$ is a *theorem* of LK, if there is a proof ending in the sequent $\vdash \mathcal{A}$.

From the previous example, we know that $\exists x \forall y\, R(x,y) \supset \forall y \exists x\, R(x,y)$ is a theorem of LK.

Example 2.10. Now we prove a theorem that involves connective and structural rules.

$$\dfrac{\mathcal{C} \vdash \mathcal{C}}{\mathcal{A},\mathcal{C} \vdash \mathcal{C}}\ K\vdash \qquad \dfrac{\mathcal{A},\mathcal{C} \vdash \mathcal{C}}{\mathcal{C},\mathcal{A} \vdash \mathcal{C}}\ C\vdash$$

$$\dfrac{\mathcal{A} \vdash \mathcal{A}}{\mathcal{A} \vdash \mathcal{A},\mathcal{C}}\ \vdash K \qquad \dfrac{\mathcal{A} \vdash \mathcal{A},\mathcal{C}}{\mathcal{A} \vdash \mathcal{C},\mathcal{A}}\ \vdash C \qquad \dfrac{\mathcal{A} \vdash \mathcal{C},\mathcal{A}}{\neg\mathcal{A},\mathcal{A} \vdash \mathcal{C}}\ \neg\vdash$$

$$\dfrac{\mathcal{C},\mathcal{A} \vdash \mathcal{C} \qquad \neg\mathcal{A},\mathcal{A} \vdash \mathcal{C}}{\mathcal{C} \vee \neg\mathcal{A},\mathcal{A} \vdash \mathcal{C}}\ \vee\vdash$$

$$\dfrac{\mathcal{C} \vee \neg\mathcal{A},\mathcal{A} \vdash \mathcal{C}}{\mathcal{A} \vdash \mathcal{C},\neg(\mathcal{C} \vee \neg\mathcal{A})}\ \vdash\neg \qquad \dfrac{\mathcal{A} \vdash \mathcal{C},\neg(\mathcal{C} \vee \neg\mathcal{A})}{\mathcal{A} \vdash \mathcal{C},\mathcal{B} \vee \neg(\mathcal{C} \vee \neg\mathcal{A})}\ \vdash\vee$$

$$\dfrac{\mathcal{A} \vdash \mathcal{C},\mathcal{B} \vee \neg(\mathcal{C} \vee \neg\mathcal{A})}{\mathcal{A} \vdash \mathcal{B} \vee \neg(\mathcal{C} \vee \neg\mathcal{A}),\mathcal{C}}\ \vdash C$$

$$\dfrac{\mathcal{A} \vdash \mathcal{B} \vee \neg(\mathcal{C} \vee \neg\mathcal{A}),\mathcal{C} \qquad \mathcal{B} \vdash \mathcal{B}}{\mathcal{C} \supset \mathcal{B},\mathcal{A} \vdash \mathcal{B} \vee \neg(\mathcal{C} \vee \neg\mathcal{A}),\mathcal{B}}\ \supset\vdash$$

$$\dfrac{\mathcal{C} \supset \mathcal{B},\mathcal{A} \vdash \mathcal{B} \vee \neg(\mathcal{C} \vee \neg\mathcal{A}),\mathcal{B}}{\mathcal{C} \supset \mathcal{B},\mathcal{A} \vdash \mathcal{B} \vee \neg(\mathcal{C} \vee \neg\mathcal{A}),\mathcal{B} \vee \neg(\mathcal{C} \vee \neg\mathcal{A})}\ \vdash\vee$$

$$\dfrac{\mathcal{C} \supset \mathcal{B},\mathcal{A} \vdash \mathcal{B} \vee \neg(\mathcal{C} \vee \neg\mathcal{A}),\mathcal{B} \vee \neg(\mathcal{C} \vee \neg\mathcal{A})}{\mathcal{A} \wedge (\mathcal{C} \supset \mathcal{B}),\mathcal{A} \vdash \mathcal{B} \vee \neg(\mathcal{C} \vee \neg\mathcal{A}),\mathcal{B} \vee \neg(\mathcal{C} \vee \neg\mathcal{A})}\ \wedge\vdash$$

$$\dfrac{\mathcal{A} \wedge (\mathcal{C} \supset \mathcal{B}),\mathcal{A} \vdash \mathcal{B} \vee \neg(\mathcal{C} \vee \neg\mathcal{A}),\mathcal{B} \vee \neg(\mathcal{C} \vee \neg\mathcal{A})}{\mathcal{A},\mathcal{A} \wedge (\mathcal{C} \supset \mathcal{B}) \vdash \mathcal{B} \vee \neg(\mathcal{C} \vee \neg\mathcal{A}),\mathcal{B} \vee \neg(\mathcal{C} \vee \neg\mathcal{A})}\ C\vdash$$

$$\dfrac{\mathcal{A},\mathcal{A} \wedge (\mathcal{C} \supset \mathcal{B}) \vdash \mathcal{B} \vee \neg(\mathcal{C} \vee \neg\mathcal{A}),\mathcal{B} \vee \neg(\mathcal{C} \vee \neg\mathcal{A})}{\mathcal{A} \wedge (\mathcal{C} \supset \mathcal{B}),\mathcal{A} \wedge (\mathcal{C} \supset \mathcal{B}) \vdash \mathcal{B} \vee \neg(\mathcal{C} \vee \neg\mathcal{A}),\mathcal{B} \vee \neg(\mathcal{C} \vee \neg\mathcal{A})}\ \wedge\vdash$$

$$\dfrac{\mathcal{A} \wedge (\mathcal{C} \supset \mathcal{B}),\mathcal{A} \wedge (\mathcal{C} \supset \mathcal{B}) \vdash \mathcal{B} \vee \neg(\mathcal{C} \vee \neg\mathcal{A}),\mathcal{B} \vee \neg(\mathcal{C} \vee \neg\mathcal{A})}{\mathcal{A} \wedge (\mathcal{C} \supset \mathcal{B}),\mathcal{A} \wedge (\mathcal{C} \supset \mathcal{B}) \vdash \mathcal{B} \vee \neg(\mathcal{C} \vee \neg\mathcal{A})}\ \vdash W$$

$$\dfrac{\mathcal{A} \wedge (\mathcal{C} \supset \mathcal{B}),\mathcal{A} \wedge (\mathcal{C} \supset \mathcal{B}) \vdash \mathcal{B} \vee \neg(\mathcal{C} \vee \neg\mathcal{A})}{\mathcal{A} \wedge (\mathcal{C} \supset \mathcal{B}) \vdash \mathcal{B} \vee \neg(\mathcal{C} \vee \neg\mathcal{A})}\ W\vdash$$

$$\dfrac{\mathcal{A} \wedge (\mathcal{C} \supset \mathcal{B}) \vdash \mathcal{B} \vee \neg(\mathcal{C} \vee \neg\mathcal{A})}{\vdash (\mathcal{A} \wedge (\mathcal{C} \supset \mathcal{B})) \supset (\mathcal{B} \vee \neg(\mathcal{C} \vee \neg\mathcal{A}))}\ \vdash\supset$$

The proof is a tree with the *root* being the bottom sequent. There are *three leaves* in the tree, and each leaf is an instance of the axiom with wff's $\mathcal{C}$, $\mathcal{A}$ and $\mathcal{B}$, respectively. This is not the only possible proof of the end sequent $\vdash (\mathcal{A} \wedge (\mathcal{C} \supset \mathcal{B})) \supset (\mathcal{B} \vee \neg(\mathcal{C} \vee \neg\mathcal{A}))$.

Exercise 2.1.8. Take a look at the proof above. Create a couple of other proofs ending in the sequent $\vdash (\mathcal{A} \wedge (\mathcal{C} \supset \mathcal{B})) \supset (\mathcal{B} \vee \neg(\mathcal{C} \vee \neg\mathcal{A}))$. How do the proofs differ from each other?

The nodes in the tree are *occurrences of sequents*, and in this case it so happens that all the occurrences of sequents are occurrences of distinct sequents. However, this does not need to be so, in general. Here is a simple example.

$$\dfrac{\mathcal{A} \vdash \mathcal{A} \qquad \mathcal{A} \vdash \mathcal{A}}{\mathcal{A} \supset \mathcal{A},\mathcal{A} \vdash \mathcal{A}}\ \supset\vdash$$

$$\dfrac{\mathcal{A} \supset \mathcal{A},\mathcal{A} \vdash \mathcal{A}}{\mathcal{A},\mathcal{A} \supset \mathcal{A} \vdash \mathcal{A}}\ C\vdash \qquad \dfrac{\mathcal{A},\mathcal{A} \supset \mathcal{A} \vdash \mathcal{A}}{\mathcal{A} \supset \mathcal{A} \vdash \mathcal{A} \supset \mathcal{A}}\ \vdash\supset \qquad \dfrac{\mathcal{A} \supset \mathcal{A} \vdash \mathcal{A} \supset \mathcal{A}}{\vdash (\mathcal{A} \supset \mathcal{A}) \supset (\mathcal{A} \supset \mathcal{A})}\ \vdash\supset$$

The two leaves are the same instances of the axiom, and if we would not distinguish between occurrences of sequents, then the tree would have only five nodes. In that situation, the two top lines would look like the following.

$$\frac{\mathcal{A} \vdash \mathcal{A}}{\mathcal{A} \supset \mathcal{A}, \mathcal{A} \vdash \mathcal{A}}$$

This is *not* an instance of the $(\supset\vdash)$ rule; hence, the lower sequent is unjustified. Having clarified and exemplified when the same sequent appears repeatedly in a proof tree at distinct nodes, now we introduce the *convention* of omitting the "occurrence of" from "occurrence of a sequent," when it is unlikely to cause a confusion.

Exercise 2.1.9. Take a look at the proof of $\vdash (\mathcal{A} \land (\mathcal{C} \supset \mathcal{B})) \supset (\mathcal{B} \lor \neg(\mathcal{C} \lor \neg\mathcal{A}))$ above. Suppose that all the occurrences of $\mathcal{A}$ and $\mathcal{B}$ are replaced by $\mathcal{D}$ and that the occurrences of sequents are not distinguished. How would the graph underlying the modified proof tree look? [Hint: Draw the tree corresponding to the proof tree above as an unlabeled graph and modify the graph by collapsing the nodes that stand for the same sequent.]

All the nodes in a proof tree that are not leaves are justified by the rules shown in the *annotation*. As we already noted, the annotation is *not* part of the proof, but it is often useful. Also, once a (purported) proof is given, it is decidable if the tree is indeed a proof. One can imagine how to go through the nodes of a finite tree while verifying that they are justified to be where they are. Annotations are normally intended to be correct, and if they are, then only one rule needs to be checked. In general, it is neither required nor true in all sequent calculi that a sequent in a proof has a *unique* justification.

Exercise* 2.1.10. Determine whether proofs in *LK* always have a unique justification or not. [Hint: Either construct a proof containing a sequent that can be annotated with more than one rule, or sketch an argument that shows that no such proof can be constructed.]

For instance, in the above simple example $\mathcal{A} \supset \mathcal{A} \vdash \mathcal{A} \supset \mathcal{A}$ is an interior node, hence, it must be justified by a rule, and it cannot be justified as an instance of the axiom (though it is an instance of the axiom). Reflecting on this fact, we can see that the same end sequent has a much shorter proof.

Exercise 2.1.11. Prove that the formulas given in (a) and (b) are theorems of *LK*. [Hint: The formula in (a) is an instance of the principal type schema of the combinator S'. That is, not all structural rules will be needed for a proof of the formula. There are various ways to prove the formula in (b). Try to find a proof in which $(\mathrm{C}\vdash)$ is the only structural rule used.]

(a) $(\mathcal{A} \supset \mathcal{D}) \supset ((\mathcal{A} \supset (\mathcal{D} \supset (\mathcal{B} \supset \mathcal{D}))) \supset (\mathcal{A} \supset (\mathcal{B} \supset \mathcal{D})))$

(b) $(\neg(\mathcal{A} \land \mathcal{B}) \supset (\neg\mathcal{A} \lor \neg\mathcal{B})) \land ((\neg\mathcal{A} \lor \neg\mathcal{B}) \supset \neg(\mathcal{A} \land \mathcal{B}))$

You might have discovered while trying to prove the two formulas in the previous exercise that it is often helpful to construct a proof (when there is a concrete formula to be proved) from the bottom to the top. Building the proof tree from its root, of course, means that the rules are "applied backward," so to speak.

Exercise 2.1.12. Prove the formulas given in (a) and (b). [Hint: The restrictions on some of the quantifier rules mean that occasionally the order in which quantifier rules are applied does matter.]

(a) $\forall x\,(P(x) \supset \forall x\,Q(x,x)) \supset \forall x\,(\exists x\,P(x) \supset Q(x,x))$

(b) $\forall x\,\forall y\,(R(x,y) \supset \forall z\,(R(x,z) \supset R(y,z))) \supset \forall x\,(\exists y\,R(y,x) \supset R(x,x))$

The formula in (b) expresses that if R is *Euclidean*, then it is *end-reflexive* (i.e., a point with an R-predecessor is reflexive). (Such properties are of interest, e.g., in the meta-theory of modal logics.)

Now that we have a sense of how LK as a formal system works, let us turn back to exploring Gentzen's original idea that his logistic calculus is a calculus of *logical deduction*.

A system of logic has a notion of proof and deduction associated to it. However, the *properties* of those notions are investigated in the so-called *meta-theory* of the logic. LK is intended to be a calculus that allows not simply proofs and derivations, but also reasoning about logical inferences.

Sequents in LK have a natural informal interpretation: $\mathcal{A}_1,\ldots,\mathcal{A}_n \vdash \mathcal{B}_1, \ldots,\mathcal{B}_m$ is to be thought of as the inference from *all* the premises $\mathcal{A}_1,\ldots,\mathcal{A}_n$ to *some* of the conclusions $\mathcal{B}_1,\ldots,\mathcal{B}_m$. Using the language of LK, we can rephrase this as $\mathcal{A}_1 \wedge \ldots \wedge \mathcal{A}_n$ implies $\mathcal{B}_1 \vee \ldots \vee \mathcal{B}_m$. (To ensure that these expressions are formulas, we may assume that the parentheses are to be restored by association to the left.)

This way of looking at sequences imparts a new interpretation of the rules. Let us start with the three pairs of structural rules.

The left rules, $(K\vdash)$, $(W\vdash)$ and $(C\vdash)$ allow, respectively, the *addition* of a new premise, the *omission* of extra copies of a premise and the *reordering* of the premises. If we think about the premises as joined by $\wedge$, then $(W\vdash)$ and $(C\vdash)$ can be viewed to express half of the *idempotence* of $\wedge$ and its *commutativity*, respectively. The left thinning rule can be seen to express that a conjunction implies its right conjunct. (Sometimes the wff $(\mathcal{A} \wedge \mathcal{B}) \supset \mathcal{B}$ and the wff $(\mathcal{A} \wedge \mathcal{B}) \supset \mathcal{A}$ are called *simplification*.)

There is a tendency in everyday reasoning, and even in certain areas of logic, to prefer *single-conclusion* inferences. For instance, syllogisms, no matter if in their ancient or medieval form, contain more than one premise, but only one conclusion. Thus, perhaps, we are less familiar with how sets of conclusions can be manipulated without retracting a conclusion and reexamining the inference. Still, it is not difficult to see that the right structural rules, $(\vdash K)$, $(\vdash W)$ and $(\vdash C)$ are informally justifiable. *Adding* a new conclusion is unproblematic, because if we already have some formulas at least one of which follows, then throwing in one more formula maintains that property, that is, at least one formula still follows. It is even easier to see that *omitting repeated occurrences* of a conclusion or *changing the order* of the conclusions does not affect the correctness of an inference.

Exercise 2.1.13. Prove that $\wedge$ and $\vee$ are idempotent, commutative and associative. Which structural rules are used in the proofs?

In Section 2.3 (on p. 26), we introduce formally the *cut rule*, which is perhaps the most important rule to modify inferences. The cut rule allows us to *intertwine* two deductions. First, let us consider what happens when Δ_1 and Δ_2 are empty in the sequent $\Gamma \vdash \Delta_1, \mathcal{C}, \Delta_2$. Then, in the deduction from $\Theta_1, \mathcal{C}, \Theta_2$ to Λ, we can replace $\mathcal{C}$ by Γ, provided that $\mathcal{C}$ follows from Γ, that is, $\Gamma \vdash \mathcal{C}$. This move is very much like how lemmas can be eliminated in favor of their whole proof in mathematical and in real-life reasoning. If the Δ's are not empty, then they must be retained, because we cannot say, on the basis of $\Gamma \vdash \Delta_1, \mathcal{C}, \Delta_2$, that $\mathcal{C}$ is *the* consequence of Γ. However, we can still say that if all the wff's in $\Theta_1, \Gamma, \Theta_2$ are available as premises, then at least one of the wff's in $\Delta_1, \Lambda, \Delta_2$ is derivable, when $\Theta_1, \mathcal{C}, \Theta_2 \vdash \Lambda$.

As it turns out, the most important uses of the cut rule, such as the proof of the *replacement theorem* or the emulation of the *detachment* rule, do not require the full power of the cut rule. However, it is easier to prove a more general form of the cut rule to be admissible.

Lastly, it should be emphasized that all sequents are *finite*, that is, they contain finitely many formulas. The informal interpretation turns a sequent into an inference; but if inferences between infinite sets of formulas are allowed (and they often are), then some of the latter do not have formal equivalents among the sequents.

2.2 An axiom system for FOL

FOL may be axiomatized in more than one way; indeed, it has been axiomatized in many ways.[4] For different purposes, such as to *prove theorems* or to *prove meta-theorems* more easily, various axiom systems are useful.

The language is defined as before, and for the sake of easy translation, we use the same symbols. Then we may assume the identity translation between the formulas that belong to LK and those that belong to K, that is, the axiom system defined below.

[4]We present the axiom system K, which is the one given in Mendelson [132, p. 62]. We harmonize the notation with ours.

Definition 2.11. (Axiom system K) The *axioms* and *rules* of K are (A1)–(A5), (MP) and (UG).

(A1) $(\mathcal{A} \supset (\mathcal{B} \supset \mathcal{A}))$

(A2) $((\mathcal{A} \supset (\mathcal{B} \supset \mathcal{C})) \supset ((\mathcal{A} \supset \mathcal{B}) \supset (\mathcal{A} \supset \mathcal{C})))$

(A3) $((\neg\mathcal{A} \supset \neg\mathcal{B}) \supset ((\neg\mathcal{A} \supset \mathcal{B}) \supset \mathcal{A}))$

(A4) $(\forall x\, \mathcal{A}(x) \supset \mathcal{A}(y))$, where y is OK for substitution for x in $\mathcal{A}(x)$

(A5) $(\forall x\, (\mathcal{A} \supset \mathcal{B}) \supset (\mathcal{A} \supset \forall x\, \mathcal{B}))$, where x is not free in $\mathcal{A}$

(MP) $\mathcal{A}$ and $(\mathcal{A} \supset \mathcal{B})$ imply $\mathcal{B}$

(UG) $\mathcal{A}$ implies $\forall x\, \mathcal{A}$

Axiom (A4) requires some explanation. The wff $\mathcal{A}(x)$ may have some occurrences of x, and some of those may be free in $\mathcal{A}(x)$, hence, bound in $\forall x\, \mathcal{A}(x)$. By $\mathcal{A}(y)$, we mean the formula that is obtained from $\mathcal{A}(x)$ by replacing all the free occurrences of x by y. There is a catch though, because we want all the newly inserted y's to be free in $\mathcal{A}(y)$. The situation in which this does happen is what we call "y is OK for substitution for x in $\mathcal{A}(x)$," and then we say that $\mathcal{A}(y)$ is the result of the *substitution* of y for x in $\mathcal{A}(x)$.

A slightly more complicated description of when y is OK for substitution for x can be given by characterizing where x occurs in $\mathcal{A}(x)$. x may have both free and bound occurrences in $\mathcal{A}(x)$, and we can forget about all the bound occurrences for now. A free occurrence of x cannot be in the scope of $\forall x$, however, it may be in a subformula of the form $\forall y\, \mathcal{B}$, that is, somewhere within $\mathcal{B}$. Since we assume that x and y are distinct variables, $\forall y$ leaves the free occurrences of x free. However, should such an x be replaced by a y, that y would become bound by $\forall y$. To sum up in a sentence, *y is OK for substitution for x in $\mathcal{A}(x)$* iff no free occurrence of x in $\mathcal{A}(x)$ is within a subformula of the form $\forall y\, \mathcal{B}$.

We do not allow substitutions when a variable is not OK for substitution, which makes substitution a partial operation (i.e., it is not always defined) for a wff and a pair of variables. There are other ways to deal with the problem of this "clash of variables," as it is sometimes called. For instance, by renaming the bound variables that would cause the new occurrences of the substituted variable to become bound. However, the problem can always be circumvented by choosing a variable, which has *no occurrences* in $\mathcal{A}(x)$ for the role of y.

Exercise 2.2.1. Decide if the "OK for substitution" conditions are satisfied in cases (a)–(d).

(a) y is OK for substitution for x in $(\forall x\, P(x, y) \supset \neg(\forall y\, P(y, x) \supset Q(x)))$

(b) x is OK for substitution for y in $(\forall x\, P(x, y) \supset \neg(\forall y\, P(y, x) \supset Q(x)))$

(c) z is OK for substitution for x in $\forall z\,(P(x,z) \supset \neg\forall x\,P(z,x))$

(d) z is OK for substitution for y in $\forall z\,(P(x,z) \supset \neg\forall x\,P(z,x))$

The language of K contains five other logical constants that have not figured into the axioms and rules. They are viewed as contextually defined by certain formulas.

Definition 2.12. The following *abbreviations* are adopted in K.

(D1) $(\mathcal{A} \vee \mathcal{B})$ for $(\neg\mathcal{A} \supset \mathcal{B})$

(D2) $(\mathcal{A} \wedge \mathcal{B})$ for $\neg(\mathcal{A} \supset \neg\mathcal{B})$

(D3) $\exists x\,\mathcal{A}$ for $\neg\forall x\,\neg\mathcal{A}$

(D4) $\boldsymbol{F}$ for $\neg(\mathcal{A} \supset \mathcal{A})$

(D5) $\boldsymbol{T}$ for $(\mathcal{A} \supset \mathcal{A})$

Definition 2.13. (Proofs, theorems) A *proof* is a finite sequence of wff's, in which each formula satisfies (at least) one of (1)–(3).

(1) The wff is an instance of an axiom;

(2) the wff is obtained from preceding elements of the sequence by an application of the rule (MP);

(3) the wff is obtained from a previous wff in the sequence by an application of the rule (UG).

The last formula in a proof is called a *theorem*. $\vdash \mathcal{A}$ indicates that $\mathcal{A}$ is a theorem of K.

Definition 2.14. (Derivations, consequence) A *derivation* of the wff $\mathcal{A}$ from the set of wff's Γ is a finite sequence of wff's, in which each formula satisfies (at least) one of (1)–(3).

(1) The wff is a theorem;

(2) the wff is an element of Γ;

(3) the wff is obtained from preceding elements of the sequence by an application of the rule (MP).

The wff $\mathcal{A}$ is a *logical consequence* of the set of wff's Γ, which is denoted by $\Gamma \vdash \mathcal{A}$, when there is a derivation of $\mathcal{A}$ from Γ.

Notice that although both proofs and derivations are *finite* sequences of wff's, the premise set Γ itself does not need to be finite. That is, the consequence relation can hold between an infinite set of wff's and a wff.

Axiom systems have a certain *elegance*, because typically, a few self-evident principles and rules suffice to generate an infinite set of less obvious, or in some sense, more complex theorems. As a way to organize knowledge, axiom systems have had a long and successful history since the 4th century BCE. However, quite frequently, it is not easy to find a proof of a theorem in an axiom system.

For example, in K, an application of the rule (MP) leads to a loss of a subformula, so to speak: $\mathcal{B}$ is obtained from $\mathcal{A} \supset \mathcal{B}$ and $\mathcal{A}$. That is, if we try to construct a proof of $\mathcal{B}$, we have to find an $\mathcal{A}$ such that both the major premise (i.e., the implicational wff) and the minor premise of the detachment rule are theorems. Since the theorems themselves could have been obtained by detachment, and we may not know, to start with, how many steps are required, the process is challenging. Solving the following exercises may give a flavor of this difficulty.

Exercise 2.2.2. Prove the following wff's in the axiom system K. (a) $(\mathcal{A} \supset \mathcal{A})$, (b) $((\mathcal{C} \supset \mathcal{A}) \supset ((\mathcal{A} \supset \mathcal{B}) \supset (\mathcal{C} \supset \mathcal{B})))$, (c) $((\mathcal{A} \supset \mathcal{B}) \supset ((\mathcal{A} \supset (\mathcal{B} \supset \mathcal{C})) \supset (\mathcal{A} \supset \mathcal{C})))$ and (d) $(((\mathcal{B} \supset \mathcal{C}) \supset \mathcal{B}) \supset \mathcal{B})$. [Hint: The first wff is the principal type schema of the combinator I. The wff's in (b) and (c) are similarly related to B' and S'. The last formula is usually termed as Peirce's law.]

Exercise 2.2.3. Prove that the wff's in (a)–(c) are theorems of K, and that the logical consequence stated in (d) obtains. (a) $(\forall x\, \forall y\, \mathcal{A}(x,y) \supset \forall x\, \mathcal{A}(x,x))$, (b) $(\forall x\, \mathcal{A} \supset \forall x\, (\mathcal{A} \vee \mathcal{B}))$, (c) $\exists x\, (\mathcal{A}(x) \supset \forall x\, \mathcal{A}(x))$, (d) $\forall x\, \forall y\, (R(x,y) \supset \forall z\, (R(y,z) \supset R(x,z)))$, $\forall x\, \forall y\, (R(y,x) \supset R(x,y)) \vdash \forall x\, \forall y\, (R(x,y) \supset R(x,x))$.

Exercise* 2.2.4. Prove that the next wff's are theorems of K. (a) $(\exists x\, (\mathcal{A}(x) \supset \mathcal{B}(x)) \supset (\forall x\, \mathcal{A}(x) \supset \exists x\, \mathcal{B}(x)))$, (b) $(\exists x\, (\mathcal{A}(x) \wedge \mathcal{B}(x)) \supset \exists x\, \mathcal{A}(x))$, (c) $(\exists x\, \forall y\, R(x,y) \supset \forall x\, \exists y\, R(y,x))$.

"Axiom-chopping," as it is sometimes called, helps to develop an understanding of the interactions and relationships between the axioms, rules and theorems. Notwithstanding, our focus is on another type of proof system in this book. There are classic texts that contain excellent examples of how to build up a set of useful theorems from an axiom system.[5] We will freely assume that known wff's (that are needed for our purposes) are provable, and we leave it to the interested reader to find their proofs in K.

2.3 Equivalence of *LK* and *K*

The two proof systems, K and LK, are equivalent, which means that if $\mathcal{A}$ is a theorem of K, then the sequent $\vdash \mathcal{A}$ is provable in LK, and vice versa.

[5] See, for example, Church [64] and Kleene [113], and of course, Mendelson [132].

It is straightforward to prove that every axiom of K is provable in LK. As an example, we give a proof of axiom (A1).

$$\dfrac{\dfrac{\dfrac{\mathcal{A} \vdash \mathcal{A}}{\mathcal{B}, \mathcal{A} \vdash \mathcal{A}}\; K\vdash}{\mathcal{A} \vdash \mathcal{B} \supset \mathcal{A}}\; \vdash\supset}{\vdash \mathcal{A} \supset (\mathcal{B} \supset \mathcal{A})}\; \vdash\supset$$

Exercise 2.3.1. Construct proofs in LK of axioms (A2) $(\mathcal{A} \supset (\mathcal{B} \supset \mathcal{C})) \supset ((\mathcal{A} \supset \mathcal{B}) \supset (\mathcal{A} \supset \mathcal{C}))$ and (A3) $(\neg\mathcal{A} \supset \neg\mathcal{B}) \supset ((\neg\mathcal{A} \supset \mathcal{B}) \supset \mathcal{A})$.

Exercise 2.3.2. Prove that the quantificational axioms (A4) and (A5) of K are theorems of LK. [Hint: See Definition 2.11 on page 22.]

K did not contain all the logical constants of LK as primitives. Therefore, we have to prove that certain wff's that are counterparts of the Definitions (D1)–(D3) in K are provable in LK. Namely, we have to prove the following six wff's: (1) $(\mathcal{A} \vee \mathcal{B}) \supset (\neg\mathcal{A} \supset \mathcal{B})$, (2) $(\neg\mathcal{A} \supset \mathcal{B}) \supset (\mathcal{A} \vee \mathcal{B})$, (3) $(\mathcal{A} \wedge \mathcal{B}) \supset \neg(\mathcal{A} \supset \neg\mathcal{B})$, (4) $\neg(\mathcal{A} \supset \neg\mathcal{B}) \supset (\mathcal{A} \wedge \mathcal{B})$, (5) $\exists x\, \mathcal{A} \supset \neg\forall x\, \neg\mathcal{A}$ and (6) $\neg\forall x\, \neg\mathcal{A} \supset \exists x\, \mathcal{A}$. We prove (4) and leave the other formulas for an exercise.

$$\dfrac{\dfrac{\neg\vdash\;\dfrac{\vdash\supset\;\dfrac{\vdash K\;\dfrac{\mathcal{A} \vdash \mathcal{A}}{\mathcal{A} \vdash \mathcal{A}, \neg\mathcal{B}}}{\vdash \mathcal{A}, \mathcal{A} \supset \neg\mathcal{B}}}{\neg(\mathcal{A} \supset \neg\mathcal{B}) \vdash \mathcal{A}} \qquad \dfrac{\dfrac{\dfrac{\dfrac{\mathcal{B} \vdash \mathcal{B}}{\vdash \mathcal{B}, \neg\mathcal{B}}\; \vdash\neg}{\mathcal{A} \vdash \mathcal{B}, \neg\mathcal{B}}\; K\vdash}{\vdash \mathcal{B}, \mathcal{A} \supset \neg\mathcal{B}}\; \vdash\supset}{\neg(\mathcal{A} \supset \neg\mathcal{B}) \vdash \mathcal{B}}\; \neg\vdash}{\neg(\mathcal{A} \supset \neg\mathcal{B}) \vdash \mathcal{A} \wedge \mathcal{B}}\; \vdash\wedge}{\vdash \neg(\mathcal{A} \supset \neg\mathcal{B}) \supset (\mathcal{A} \wedge \mathcal{B})}\; \vdash\supset$$

Exercise 2.3.3. Prove that the wff's listed above as (1), (2), (3), (5) and (6) are also theorems of LK.

The rules of K are detachment, (i.e., (MP)) and universal generalization, (i.e., (UG)). We have to show that if the premises of these rules are provable in LK, then so are their respective conclusions.

For modus ponens, we may assume that there are some proof trees rooted in $\mathcal{A}$ and $\mathcal{A} \supset \mathcal{B}$. We would like to be able to have the following proof tree.

$$\dfrac{\begin{matrix} \vdots \\ \vdash \mathcal{A} \end{matrix} \qquad \begin{matrix} \vdots \\ \vdash \mathcal{A} \supset \mathcal{B} \end{matrix}}{\vdash \mathcal{B}}\; ?$$

The ? indicates that there is no rule (or combination of rules) that we could apply to the premises to obtain the lower sequent. (We made the horizontal line a bit thicker to suggest the possibility that several rules had been applied.)

Exercise 2.3.4. The claim we just made may or may not be obvious. Convince yourself that chaining together several rules of LK will not help.

We could add the above pattern as a rule to LK, and then try to show that it does not increase the set of provable sequents. Instead, we add a more general rule, which is useful for various other purposes too.[6]

$$\frac{\Gamma \vdash \Delta_1, \mathcal{C}, \Delta_2 \qquad \Theta_1, \mathcal{C}, \Theta_2 \vdash \Lambda}{\Theta_1, \Gamma, \Theta_2 \vdash \Delta_1, \Lambda, \Delta_2} \text{ cut}$$

Sometimes, this rule is called the *single cut* rule, because the premises are only required to contain an occurrence of $\mathcal{C}$, the *cut formula*, exactly one of which is affected. The left premise of the rule has to have an occurrence of $\mathcal{C}$ in the succedent, whereas the right premise has to have an occurrence of $\mathcal{C}$ in the antecedent for the rule to be applicable. ($\mathcal{C}$ may have further occurrences, but those will be ignored from the point of view of an application of this rule; they are merely components in the Δ's and Θ's.) The occurrences of $\mathcal{C}$, which are singled out, are replaced by Γ and Λ, respectively, each of which might comprise several formulas.

If we have the cut rule, then we can show that modus ponens can be emulated in the following way.

$$\text{cut}\ \frac{\begin{matrix}\vdots \\ \vdash \mathcal{A}\end{matrix} \qquad \text{cut}\ \dfrac{\begin{matrix}\vdots \\ \vdash \mathcal{A} \supset \mathcal{B}\end{matrix} \qquad \dfrac{\mathcal{A} \vdash \mathcal{A} \qquad \mathcal{B} \vdash \mathcal{B}}{\mathcal{A} \supset \mathcal{B}, \mathcal{A} \vdash \mathcal{B}} \supset\vdash}{\mathcal{A} \vdash \mathcal{B}}}{\vdash \mathcal{B}}$$

This chunk of the proof shows that if cut would be a rule, then (MP) would be a *derived* rule in LK. Notice that the first application of the cut rule suggests more uses for the cut rule beyond emulating detachment. If the formula $\mathcal{A} \supset \mathcal{B}$ is a theorem, then the corresponding sequent $\vdash \mathcal{A} \supset \mathcal{B}$ could be obtained only by one of the following two rules: $(\vdash K)$ and $(\vdash \supset)$.

We could exclude the first, if we would know that $\vdash$ is not a provable sequent. Notice that there is a huge difference between this sequent and the sequent $\mathcal{A} \vdash \mathcal{A}$, which is the axiom of LK. If the sequent with empty antecedent and empty succedent would be provable, then using the $(\vdash K)$ rule *all wff's* would be theorems. The provability of $\vdash$ would have a catastrophic effect on the calculus. We will show that LK is consistent, hence, this sequent is not provable.

Then the last rule must have been $(\vdash \supset)$. An application of the cut allows us "to take back" this step. Cut is a very powerful rule, and we look at it in connection to LK in the next section, where we also prove its *admissibility* in LK. Later, in Chapter 7, we consider further versions and other useful consequences of this rule. For now, we only record that the single cut rule suffices to mimic in LK the (MP) rule of K.

[6] I gave this formulation of the single cut for LK in [33, p. 564].

The other rule of K is (UG). We have to justify the following step in a proof.

$$\frac{\begin{matrix}\vdots\\ \vdash \mathcal{A}(x)\end{matrix}}{\vdash \forall x\, \mathcal{A}(x)}\ ?$$

It might seem that we can simply say that we have the rule that we need in place of ?, namely, $(\vdash\forall)$. However, this rule has a side condition, which requires that there are no free occurrences of y anywhere in the lower sequent, which means that the formula $\forall x\, \mathcal{A}(x)$ may not contain such occurrences of y. This may be guaranteed by requiring that all the free occurrences of y are selected—provided that x is OK for all those occurrences. However, x is always OK for x in any $\mathcal{A}$, and the $(\vdash\forall)$ rule does not require—though it permits—changing the y in $\mathcal{A}(y)$ (in the upper sequent) to x in $\forall x\, \mathcal{A}(x)$.

However, if $\forall x\, \mathcal{A}(x)$ results by (UG) from $\mathcal{A}(x)$, then the substitution of y for x in $\mathcal{A}(x)$ yields $\mathcal{A}(y)$, which means that in the sequent calculus proof, all the free occurrences of y may be selected.

In the context of axiomatic calculi, it is usual to talk about *renaming bound variables*. In the context of sequent calculi, the renaming of the *free* y's in and above the $(\vdash\forall)$ and $(\exists\vdash)$ rules is essential.

Lemma 2.15. *A proof of the sequent* $\Gamma \vdash \Delta$ *may be transformed into a proof of the same end sequent in which* y *in the upper sequent of an application of* $(\exists\vdash)$ *or* $(\vdash\forall)$ *occurs* only in the subtree *rooted in the upper sequent.*

Proof: There are finitely many occurrences of applications of these two rules in any proof, hence, iterating the following step proves the lemma. We select an application of the two rules that either has no application of $(\exists\vdash)$ or $(\vdash\forall)$ above it, or if it has, then the variables of all those rules have already been renamed. We replace all the occurrences of y in the subtree rooted in the upper sequent by z, which occurs nowhere in the proof. The process yields a proof from a proof by the following lemma. qed

Lemma 2.16. *If* a free variable, *which is not the* y *in* $(\exists\vdash)$ *or in* $(\vdash\forall)$ *is* replaced *everywhere in the axiom or in a rule by a variable, which is not the* y *of* $(\exists\vdash)$ *or* $(\vdash\forall)$, *then the result is an instance of the axiom or of the same inference rule, respectively.*

Proof: For most of the rules, the claim is obvious, since nothing depends on the concrete shape of the $\mathcal{A}$'s and $\mathcal{B}$'s. However, the quantifier rules are either not restricted, or by the condition of the lemma, the variable entering the restriction is not renamed. qed

Theorem 2.17. (From K to LK) *If* $\mathcal{A}$ *is a theorem of* K, *then* $\mathcal{A}$ *is a* theorem *of* LK.

Proof: We only have to put together the components that already have been proved. The axioms of K are theorems of LK, and its rules can be simulated in LK. The connectives that were defined in K behave in LK in accordance with those definitions. qed

To view the axiom system and the sequent calculus as equivalent, we need to be able to claim that the converse of the if–then statement in the theorem is true too.

The interpretation of sequents, which we introduced somewhat informally earlier, comes handy.

Definition 2.18. For $\Gamma \vdash \Delta$, we define $\tau(\Gamma \vdash \Delta)$ to be $\bigwedge(\Gamma) \supset \bigvee(\Delta)$, where $\bigwedge$ and $\bigvee$ are as follows.

(1) $\bigwedge(\ ) := (\mathcal{A} \supset \mathcal{A})$

(2) $\bigwedge(\mathcal{A}) := \mathcal{A}$

(3) $\bigwedge(\mathcal{A}_{n+1}, \mathcal{A}_n, \ldots, \mathcal{A}_1) := (\mathcal{A}_{n+1} \wedge \bigwedge(\mathcal{A}_n, \ldots, \mathcal{A}_1))$

(4) $\bigvee(\ ) := \neg(\mathcal{A} \supset \mathcal{A})$

(5) $\bigvee(\mathcal{B}) := \mathcal{B}$

(6) $\bigvee(\mathcal{B}_1, \ldots, \mathcal{B}_m, \mathcal{B}_{m+1}) := (\bigvee(\mathcal{B}_1, \ldots, \mathcal{B}_m) \vee \mathcal{B}_{m+1})$

Obviously, $\tau(\Gamma \vdash \Delta)$ is a wff, and $\bigwedge(\Gamma)$ is a conjunction when Γ is not empty, whereas $\bigvee(\Delta)$ is a disjunction when Δ is not empty.

The connectives $\wedge$ and $\vee$ are defined in K, and they may be proved *associative*. This means that we could omit all but the outside parentheses obtained from the wff's $\bigwedge(\Gamma)$ and $\bigvee(\Delta)$. The wff in (1) and (4) can be chosen arbitrarily, or it could be stipulated to be a fixed one. Alternatively, the empty antecedent of a sequent could be thought of as $\boldsymbol{T}$ and the empty succedent could be taken to be $\boldsymbol{F}$, in accordance with (D4) and (D5).

Theorem 2.19. (From LK to K) *If $\mathcal{A}$ is a theorem of LK, then $\mathcal{A}$ is a* theorem *of K.*

Proof: What we show is that τ takes the axiom of LK into a theorem, and further, if the upper sequent(s) of a rule are theorems, then so is the lower sequent. (The proof is by induction on the height of the proof tree, and we include here only a small selection of the cases.)

1. $\tau(\mathcal{A} \vdash \mathcal{A})$ is $\mathcal{A} \supset \mathcal{A}$, which is a theorem of K.

2.1. If the last rule is $(\wedge\vdash_1)$, then $\tau(\mathcal{A} \wedge \mathcal{B}, \Gamma \vdash \Delta)$ should follow from $\tau(\mathcal{A}, \Gamma \vdash \Delta)$ (or from $\tau(\mathcal{B}, \Gamma \vdash \Delta)$, by $(\wedge\vdash_2)$). The formula $((\mathcal{A} \wedge \mathcal{C}) \supset \mathcal{D}) \supset ((\mathcal{A} \wedge \mathcal{B} \wedge \mathcal{C}) \supset \mathcal{D})$ is a theorem of K; hence, if its antecedent is provable, then so is its consequent. (Finding proofs of this and of other formulas in K is Exercise 2.3.6.)

2.2. Let us assume that the last rule applied is $(\vdash\neg)$. Then $\tau(\mathcal{A}, \Gamma \vdash \Delta)$ has to imply $\tau(\Gamma \vdash \Delta, \neg\mathcal{A})$. The wff $((\mathcal{A} \wedge \mathcal{C}) \supset \mathcal{D}) \supset (\mathcal{C} \supset (\mathcal{D} \vee \neg\mathcal{A}))$ is a

theorem of K, thus, if $\tau(\mathcal{A}, \Gamma \vdash \Delta)$ is a theorem of K, so is $\tau(\Gamma \vdash \Delta, \neg\mathcal{A})$, as needed.
2.3. If the last rule is $(\forall\vdash)$, then the formula that should be shown to be a theorem of K is $((\mathcal{A}(y) \wedge \mathcal{C}) \supset \mathcal{D}) \supset ((\forall x\, \mathcal{A}(x) \wedge \mathcal{C}) \supset \mathcal{D})$. The antecedent of this wff is $\tau(\mathcal{A}(y), \Gamma \vdash \Delta)$, whereas the consequent is $\tau(\forall x\, \mathcal{A}(x), \Gamma \vdash \Delta)$. Again, the hypothesis of the induction together with the theorem of K give the desired conclusion.
2.4. Let us consider $(\vdash\forall)$. $\tau(\Gamma \vdash \Delta, \mathcal{A}(y))$ is $\mathcal{C} \supset (\mathcal{D} \vee \mathcal{A}(y))$ and $\tau(\Gamma \vdash \Delta, \forall x\, \mathcal{A}(x))$ is $\mathcal{C} \supset (\mathcal{D} \vee \forall x\, \mathcal{A}(x))$. There is no reason why the first formula would imply the second, but the $(\vdash\forall)$ rule comes with a restriction that saves the implication. y cannot occur free in the second formula at all, hence, it certainly cannot occur free in $\mathcal{C}$ or $\mathcal{D}$, the antecedent or the disjunct in the wff. With this restriction in place, $\forall y\, (\mathcal{C} \supset (\mathcal{D} \vee \mathcal{A}(y))) \supset (\mathcal{C} \supset (\mathcal{D} \vee \forall x\, \mathcal{A}(x)))$ is a theorem of K.
3. The theoremhood of $\mathcal{B}$ in LK is defined as the provability of $\vdash \mathcal{B}$. τ gives $(\mathcal{A} \supset \mathcal{A}) \supset \mathcal{B}$, from which we immediately get $\mathcal{B}$, by detachment, because $\mathcal{A} \supset \mathcal{A}$ is a theorem of K. qed

Exercise 2.3.5. Complete the proof of the previous theorem. [Hint: You may assume that the wff's that turn out to be necessary for the proof are provable in K.]

Exercise* 2.3.6. Prove in K the wff's that have to be shown to be theorems in order for the above proof to work. [Hint: Remember that proving theorems in an axiom system sometimes can get tricky and lengthy.]

2.3.1 Cut rules

We used the single cut rule in the previous section with the promise of showing it admissible later. The cut rule was introduced in [89], in fact it was included into LK as a structural rule. To be more precise, Gentzen introduced *a cut rule*, which is not exactly the same rule that we gave above. Gentzen's cut rule is cut_G.

$$\frac{\Gamma \vdash \Delta, \mathcal{C} \qquad \mathcal{C}, \Theta \vdash \Lambda}{\Gamma, \Theta \vdash \Delta, \Lambda}\ \text{cut}_G$$

$\mathcal{C}$ must have an occurrence in the premises, just as we stated for the cut rule introduced in the previous section. As before, only a single occurrence of $\mathcal{C}$ is omitted from $\Delta, \mathcal{C}$ and $\mathcal{C}, \Theta$, respectively.

The left and right permutation rules—together with the absence of grouping—guarantee that wff's can be "moved around" both in the antecedent and in the succedent of a sequent. Thus it may appear at first that it is absolutely unimportant where a formula, for example, the cut formula is placed within the antecedent and the succedent. Gentzen seems to have preferred the edges: all the formulas affected by his rules—except those in $(C\vdash)$ and

$(\vdash \mathsf{C})$—are either first in the antecedent or last in the succedent. (We defined τ accordingly so that τ yields formulas with a transparent structure.) In the case of the cut rule, however, this positioning of the cut formulas is clearly problematic, because we want to prove the admissibility of the cut. Gentzen missed this observation, though he obviously realized that the permutation rules must have a more general form. An *insight* gained from applications of the single cut rule and from proofs of the admissibility of cut in sequent calculi for non-classical logics (especially, in consecution calculi where the structural connective is not commutative) is that the *appropriate formulation* of the single cut rule is the cut we gave earlier (and not cut_{G}).

[89] introduced *LK* starting from scratch, so to speak. At the same time, when we acknowledge the novelty and the cleverness of the original sequent calculus *LK*, it may be worthwhile to mention other problematic points too.

Pertaining to the edge positioning of formulas is a peculiarity that became well understood only after the work of Curry [69] (that was later reinforced by Howard [110]). Consider the following proof.

$$\dfrac{\dfrac{\dfrac{\mathcal{A} \vdash \mathcal{A} \qquad \mathcal{B} \vdash \mathcal{B}}{\mathcal{A} \supset \mathcal{B}, \mathcal{A} \vdash \mathcal{B}} \supset\vdash}{\mathcal{A} \vdash (\mathcal{A} \supset \mathcal{B}) \supset \mathcal{B}} \vdash\supset}{\vdash \mathcal{A} \supset ((\mathcal{A} \supset \mathcal{B}) \supset \mathcal{B})} \vdash\supset$$

The formula $\mathcal{A} \supset ((\mathcal{A} \supset \mathcal{B}) \supset \mathcal{B})$ is a theorem (and a valid wff) of classical logic; it is called *assertion*. However, it is not an instance of self-implication $(\mathcal{A} \supset \mathcal{A})$, which is provable from the axiom of *LK* by one application of the $(\vdash \supset)$ rule. Assertion is the principal type schema of the combinator CI (also denoted by T), with axiom $\mathsf{CI}xy \rhd yx$.[7] Obviously, some permutation is happening in the combinatory axiom, whereas the $(\mathsf{C}\vdash)$ rule has not been applied in the proof displayed above. In other words, the $(\supset\vdash)$ and $(\vdash\supset)$ rules are mismatched, when we look beyond mere provability of wff's in classical logic.

Another potential complaint is that Gentzen considered cut to be a structural rule. It turns out that cut is a very useful and important rule, but has very different properties than the six other structural rules, which allow the manipulation of the formulas in the antecedent or in the succedent of *one* sequent. Cut combines *two* sequents—while it drops an occurrence of a formula from each.

The appropriately formulated cut rule can be shown to be admissible by a straightforward induction. We give some details of this proof in this section. On the other hand, Gentzen had to devise a roundabout proof to show the admissibility (really, from his point of view, the *eliminability*) of cut_{G}. The *mix rule* can be shown to be equivalent to cut_{G} in *LK* (but not in some other sequent calculi).

[7] We will say more about combinators and types in Chapters 5 and 9.

$$\frac{\Gamma \vdash \Delta \qquad \Theta \vdash \Lambda}{\Gamma, \Theta^* \vdash \Delta^*, \Lambda} \text{ mix}$$

The rule is applicable if there is a wff C such that it occurs both in Δ and Θ. The starred versions of Δ and Θ stand for the same sequences of formulas, Δ and Θ, respectively, but with *all the occurrences of* C omitted.

It is also interesting that this rule is formulated somewhat informally—there is no notation introduced to *locate* the occurrences of C. The rule even may appear not to require the existence of a suitable C (though it does). Mix is decidedly a strange rule. Instead of the idea of locating one occurrence of the cut formula within Δ and Θ, Gentzen opted for replacing *all* occurrences with *one copy* of Γ and Λ. Mix is equivalent to cut_G within LK (and also in LJ), but this equivalence is peculiar to classical (and to intuitionistic) logic.

Lemma 2.20. *The cut*$_G$ *rule and the mix rule are* equivalent, *that is, a sequent is provable in LK using cut iff it is provable in LK using mix.*

Proof: Let us assume that $\Gamma, \Theta \vdash \Delta, \Lambda$ is by cut_G from the premises $\Gamma \vdash \Delta, C$ and $C, \Theta \vdash \Lambda$. If Δ, C and C, Θ have no other occurrences of C, then an application of mix using the same premises yields the same lower sequent. In general, let there be n occurrences of C in the succedent of the left premise, and m occurrences of C in the antecedent of the right premise (where $n, m \geq 1$). Then the following proof segment yields the derived sequent $\Gamma, \Theta \vdash \Delta, \Lambda$. (The thicker horizontal lines indicate that the sequents above and below may be separated by more than one step.) If $m = 1$ or $n = 1$, then the corresponding thinning and permutation steps are altogether omitted.

$$\cfrac{\cfrac{\cfrac{\Gamma \vdash \Delta, C \qquad C, \Theta \vdash \Lambda}{\Gamma, \Theta^* \vdash \Delta^*, \Lambda} \text{ mix}}{\Gamma, \Theta \vdash \Delta^*, \Lambda} \; K\vdash (m-1\times), C\vdash\text{'s}}{\Gamma, \Theta \vdash \Delta, \Lambda} \; \vdash K \; (n-1\times), \vdash C\text{'s}$$

We cannot specify the exact numbers of the applications of the $(C\vdash)$ and $(\vdash C)$ rules, which are necessary to restore the sequence of formulas Γ, Θ^* to Γ, Θ (and similarly, to derive Δ, Λ from Δ^*, Λ). (This is unlike the number of applications of the thinning rules, which is determined by the number of occurrences of C in Θ and Δ, respectively.) However, the size of the sequences of wff's Γ, Θ^* and Δ^*, Λ clearly induces an upper bound. If the numbers of wff's in Γ, Θ^* and in Δ^*, Λ are i and j, respectively, then the bounds are $i \cdot (m-1)$ and $j \cdot (n-1)$.

Now let cut_G be given, and let us assume that in the premises $\Gamma \vdash \Delta$ and $\Theta \vdash \Lambda$, there are n and m occurrences of C (with $n, m \geq 1$). If $m = n = 1$, then one application of cut_G yields the same sequent—provided that the cut formulas are in the right positions in the premises. Finitely many applications of $(\vdash C)$ to the left premise and finitely many applications of $(C\vdash)$ to the right premise can ensure that the C's are on the edges.

Let us consider the more complicated situation when $m \neq 1$ or $n \neq 1$.

$$
\dfrac{
  \vdash W\text{'s } (m{-}1\times)\;
  \dfrac{\vdash C\text{'s}\;\dfrac{\Gamma \vdash \Delta}{\Gamma \vdash \Delta^*, \mathcal{C}^m}}{\Gamma \vdash \Delta^*, \mathcal{C}}
  \qquad
  \dfrac{\dfrac{\Theta \vdash \Lambda}{\mathcal{C}^n, \Theta^* \vdash \Lambda}\; C{\vdash}\text{'s}}{\mathcal{C}, \Theta^* \vdash \Lambda}\; W{\vdash}\text{'s } (n{-}1\times)
}{\Gamma, \Theta^* \vdash \Delta^*, \Lambda}\;\text{cut}_G
$$

Applications of the permutation rules at the top may or may not be needed, depending on the concrete shape of Δ and Θ. If $m = 1$ or $n = 1$, then no contraction is applied on the left or on the right branch. qed

The proof shows not only how powerful the mix rule is, but also that the equivalence of this rule and the cut_G rule relies on *all* the structural rules.

Exercise 2.3.7. Consider the second half of the proof above. It is possible to postpone the applications of the contraction rules. What does the new proof segment look like? [Hint: You may want to work out a small concrete example first, for instance, assuming that the premises are $\mathcal{A} \vdash \mathcal{C}, \mathcal{C}, \mathcal{B}$ and $\mathcal{D}, \mathcal{C}, \mathcal{C}, \mathcal{C} \vdash \mathcal{E}$.]

There are other parts of LK that could be (and later on, were) sharpened. The axiom $\mathcal{A} \vdash \mathcal{A}$ could be stated in the form $p \vdash p$, where p is an atomic formula (that is, a propositional variable or a predicate followed by sufficiently many terms). There is a proof of the sequent $\mathcal{A} \supset \mathcal{A} \vdash \mathcal{A} \supset \mathcal{A}$ on page 18, which is an instance of the original axiom. If we restrict the axiom to atomic formulas, then it is natural to ask if the modified system (let us say, LK_{at}) can prove all the theorems that LK can.

Exercise 2.3.8. Either prove that LK and LK_{at} have the same set of theorems, or give an example, i.e., a wff, that is provable in one but not in the other system. [Hint: The inclusion is obvious in one direction.]

As a last remark concerning certain peculiarities of LK, let us note a discrepancy in the formulation of the two-premise rules. The $(\vdash \wedge)$ and $(\vee \vdash)$ rules assume that the two premises are identical except $\mathcal{A}$ and $\mathcal{B}$. On the other hand, the $(\supset \vdash)$ rule does not prescribe that Γ and Θ, or Δ and Λ are the same. The rule could have been formulated instead as follows.

$$
\dfrac{\Gamma \vdash \Delta, \mathcal{A} \qquad \mathcal{B}, \Gamma \vdash \Delta}{\Gamma, \mathcal{A} \supset \mathcal{B} \vdash \Delta}\;{\supset}{\vdash}_{\text{e}}
$$

We switched $\mathcal{A} \supset \mathcal{B}$ to the other edge in the antecedent (in view of our previous complaint related to typing), but the main difference is that now $\mathcal{A}$ and $\mathcal{B}$ must be proved in the same sequents, though on different sides of the turnstile. (The subscript $_{\text{e}}$ is to indicate that this is the genuinely *extensional* version of the $(\supset \vdash)$ rule. The distinction between intensional and extensional connectives, which are otherwise alike or similar, will be explained in more detail in Chapter 5.)

Classical logic cannot distinguish between the connectives that are introduced on the left by the two rules $(\supset\vdash)$ and $(\supset\vdash_e)$. However, Gentzen defined LK with an eye toward LJ (which we will look at in Chapter 3).[8] The rule $(\supset\vdash)$ has an advantage over $(\supset\vdash_e)$, if one wants to define LJ by simply restricting the number of wff's on the right-hand side of the $\vdash$ to at most one.

Exercise 2.3.9. Show that $(\supset\vdash_e)$ is a derivable rule in LK. Conversely, prove that if the left introduction rule for $\supset$ is $(\supset\vdash_e)$, then $(\supset\vdash)$ is derivable (when the rest of LK is kept unchanged).

We listed some features of the original formulation of LK that later turned out to be puzzling, undesirable or suboptimal.

To further motivate our formulation of the cut rule, let us consider the sequent calculus again as a system to reason about inferences.

The right premise of the cut rule says that Λ can be derived from $\Theta_1, \mathcal{C}, \Theta_2$. If from Γ the formula $\mathcal{C}$ is derivable within Δ_1 and Δ_2 (as given by the left premise), then placing Γ in the spot where $\mathcal{C}$ is (that is, starting with $\Theta_1, \Gamma, \Theta_2$), should suffice for the derivation of Λ within Δ_1 and Δ_2 (that is, $\Delta_1, \Lambda, \Delta_2$).

The two cut rules, cut and cut_G are, obviously, equivalent in LK, and from now on we always mean our single cut rule (rather than cut_G). In order to prove the admissibility of cut, we introduce a series of new concepts. Some of them characterize the *roles* various formulas play in a rule; others pertain to proofs in which there is an application of the cut rule.

Definition 2.21. (Degree of a wff) The *degree* of a formula $\mathcal{A}$ is denoted by $\delta(\mathcal{A})$, and it is defined inductively by (1)–(7).

(1) $\delta(p) = 0$, where p is an atomic formula;

(2) $\delta(\neg\mathcal{A}) = \delta(\mathcal{A}) + 1$;

(3) $\delta(\mathcal{A} \wedge \mathcal{B}) = \max(\delta(\mathcal{A}), \delta(\mathcal{B})) + 1$;

(4) $\delta(\mathcal{A} \vee \mathcal{B}) = \max(\delta(\mathcal{A}), \delta(\mathcal{B})) + 1$;

(5) $\delta(\mathcal{A} \supset \mathcal{B}) = \max(\delta(\mathcal{A}), \delta(\mathcal{B})) + 1$;

(6) $\delta(\forall x\, \mathcal{A}) = \delta(\mathcal{A}) + 1$;

(7) $\delta(\exists x\, \mathcal{A}) = \delta(\mathcal{A}) + 1$.

The degree of a formula is a natural number assigned to the formula, which aligns with the complexity of the formula. (This is not the only possible way though to arithmetize the complexity of a formula.) Formulas can be represented as trees with atomic formula occurrences labeling the leaves and

[8] See [151] for more historical considerations.

the intermediate nodes labeled by occurrences of connectives or quantifiers. Then the degree of a formula corresponds to the height of the formation tree of the formula minus one.

Next we provide an *analysis* (in the sense of Curry [68]) for the rules of the calculus. Clearly, some formulas are simply copied from the upper sequent(s) to the lower sequent, whereas others are combined, modified, or at least, moved around. We introduce three categories of formulas, *principal formulas, subalterns* and *parametric formulas.*

Informally, the categories capture certain types of behaviors for formulas that we describe briefly. A principal formula is a formula in the lower sequent of a rule, which is in the center of attention or it is the formula that is affected by the rule. A subaltern is a formula in an upper sequent in a rule that is intimately connected to the principal formula, most frequently, because it is a proper subformula of the principal formula, but sometimes it is an occurrence of the same formula as the principal formula. A parametric formula is either in the lower or in the upper sequents, but it is neither a principal formula nor a subaltern. Parametric formulas are in the sequents, because the sequent calculus is formulated to reason about consequences, which may involve more than two or three wff's; hence, some formulas are not in the spotlight, so to speak, in some proof steps.

Definition 2.22. (Analysis) We consider each rule, and identify the *principal* formulas and the *subalterns*. All the other formulas are *parametric*.
$(\wedge\vdash_1)$, $(\wedge\vdash_2)$ and $(\vdash\wedge)$. The principal formula is $\mathcal{A}\wedge\mathcal{B}$. There is one subaltern in the left introduction rules, depending on whether $\mathcal{A}$ or $\mathcal{B}$ occurs at the edge of the upper sequent. In $(\vdash\wedge)$, both $\mathcal{A}$ and $\mathcal{B}$ are subalterns.
$(\vee\vdash)$, $(\vdash\vee_1)$ and $(\vdash\vee_2)$. The principal formula is $\mathcal{A}\vee\mathcal{B}$. $\mathcal{A}$ and $\mathcal{B}$ are subalterns in $(\vee\vdash)$, and whichever of them occurs at the far right in the succedent is the subaltern in $(\vdash\vee_1)$ and $(\vdash\vee_2)$.
$(\neg\vdash)$ and $(\vdash\neg)$. The principal formula is $\neg\mathcal{A}$, and the subaltern is the displayed $\mathcal{A}$ in both rules.
$(\supset\vdash)$ and $(\vdash\supset)$. The principal formula is $\mathcal{A}\supset\mathcal{B}$ with the subalterns being the two immediate proper subformulas, $\mathcal{A}$ and $\mathcal{B}$, which are displayed in the rules.
$(\forall\vdash)$ and $(\vdash\forall)$. The principal formula is $\forall x\,\mathcal{A}(x)$, whereas the subaltern is the $\mathcal{A}(y)$ at the edge.
$(\exists\vdash)$ and $(\vdash\exists)$. The principal formula is $\exists x\,\mathcal{A}(x)$, and the subaltern is the displayed $\mathcal{A}(y)$.
$(K\vdash)$ and $(\vdash K)$. The principal formula is the newly introduced $\mathcal{A}$. These rules are exceptional in the sense that there is no subaltern in either of them.
$(W\vdash)$ and $(\vdash W)$. The principal formula is $\mathcal{A}$ in the lower sequent. The two displayed occurrences of the same formula in the upper sequent are the subalterns.
$(C\vdash)$ and $(\vdash C)$. The principal formulas are the $\mathcal{B}$ and $\mathcal{A}$ in the lower sequent, and the subalterns are $\mathcal{A}$ and $\mathcal{B}$ in the upper sequent.

The descriptions should make it clear, but it is perhaps useful to emphasize that we label as principal formulas or subalterns only the formula occurrences that are explicit in the formulation of the rules. From the point of view of easy identification of subalterns and principal formulas, it is helpful that the formulas affected by the rules are at the edge of the antecedent or succedent—except in the permutation rules.

Definition 2.23. (Parametric ancestors) A formula $\mathcal{A}$ (occurring in a sequent higher in the proof tree) is a *parametric ancestor* of a formula $\mathcal{A}$ (occurring in a sequent lower in the proof tree) iff the two formulas are related via the transitive closure of the relation specified in (1)–(3).

(1) A parametric formula $\mathcal{A}$ in an upper sequent of a rule is a parametric ancestor of the same $\mathcal{A}$ in the lower sequent of the rule;

(2) the subalterns in $(C\vdash)$ and $(\vdash C)$ are parametric ancestors of the corresponding principal formulas in the lower sequent of the rules;

(3) the subalterns in the $(W\vdash)$ and $(\vdash W)$ rules are parametric ancestors of the principal formula in the lower sequent of the rules.

Parametric ancestors allow us to trace formulas upward in the proof tree, which provides a more refined view of proofs than if we would rely on occurrences of formulas.

Let us assume that we are given a proof containing an application of the cut rule. We define the three parameters that the inductive proof relies on.

Definition 2.24. (Degree of cut) The *degree* of (an application of) the cut rule, δ, is the degree of the cut formula.

Definition 2.25. (Contraction measure of cut) The *contraction measure* of (an application of) the cut rule, μ, is the number of applications of $(W\vdash)$ and $(\vdash W)$ rules to parametric ancestors of the cut formula.[9]

Definition 2.26. (Rank of cut) The *left rank* of (an application of) the cut rule, ϱ_l, is the maximal number of consecutive sequents (from the left premise upward) in the subtree rooted in the left premise in which the cut formula occurs in the succedent.

The *right rank* of (an application of) the cut rule, ϱ_r, is the maximal number of consecutive sequents (from the right premise upward) in the subtree rooted in the right premise in which the cut formula occurs in the antecedent.

The *rank* of (an application of) the cut rule, ϱ, is the *sum* of the left and right ranks.

Obviously, δ, μ and ϱ are all natural numbers; moreover, ϱ is positive, and at least 2, because $\varrho_l \geq 1$ and $\varrho_r \geq 1$.

[9] The notion of contraction measures was introduced in [33, p. 559], first, for $LE^t_{\rightarrow}$. The idea of having this new parameter in an inductive proof comes from certain observations recorded in [31].

Theorem 2.27. (Cut theorem) *The cut rule is* admissible *in* LK.

Proof: We prove that if a sequent is provable with applications of the cut rule, then it is provable without using the cut rule. The cut rule is *not* a derived rule, that is, there is no way, in general, to obtain the lower sequent from the premises by applications of the rules of LK.

Exercise 2.3.10. Prove the claim in the last sentence. [Hint: Construct a concrete proof with an application of the cut rule.]

The proof we give is *constructive,* that is, it is effective, and it is *proof-theoretical,* that is, it does not appeal to the interpretation of the sequents. Effectiveness means that given a proof containing applications of the cut rule, one can produce a proof of the end sequent without cut using appropriate cases from the proof of the cut theorem.

In the steps of the proof, we will specify *local* changes to be made to the given proof tree—unlike in some other proofs, where global modifications of the proof tree are stipulated. In other words, the steps will not require replacement of formulas throughout a subtree of the proof tree, though we sometimes will assume that it is possible to copy a subtree.

The proof is by *complete triple induction* on δ, μ and ϱ. An inductive proof comprises a base step and an inductive step (or several of each). A triple induction involves a complete induction on the third parameter both within the base step and within the inductive step of the induction on the second parameter and similarly for the first parameter.[10] However, when we use the term "triple" (or "double") rather than "three" (or "two"), we imply that it should be understood that the inductions are intertwined. Indeed, they cannot be separated from each other, because of the shape of the inductive hypotheses.

Inside the triple induction, steps are split into cases, because there are 20 rules in LK. This means that with all the cases detailed, the proof turns out to be enormous. At the same time, steps in the proof, and even some segments of the induction, look alike. This is one of the reasons why proofs of cut theorems are rarely (if ever) given in full. Sometimes these proofs may appear to be a jumbled collection of cases.

Our strategy is to make clear the *overall structure of the proof,* and indicate all the cases that need to be considered. However, when we reach a point where a pattern in the proof may be discerned, we will insert exercises, and omit the writing out of all the details.

1.1.1 $\varrho = 2$, $\delta = 0$, $\mu = 0$. The cut formula is atomic, let us say p, and it does not occur above the left or the right premise. The latter severely constrains the shape of the proof tree: either both premises of the cut are instances of the axiom, or one or the other premise is by a thinning rule with

[10] This way of looking at a double induction may be found, for example, in [113, p. 454]. See Section A.2 in the *Appendix* for more details.

the principal formula p. Accordingly, there are *four* subcases. (We characterize the cases by a pair of labels.)

(a) $\langle(\mathrm{id}),(\mathrm{id})\rangle$. The proof tree is of the form given on the left below. We can transform the proof by retaining one of the premises, as shown on the right.

$$\text{cut}\ \frac{p \vdash p \qquad p \vdash p}{p \vdash p} \quad \rightsquigarrow \quad p \vdash p$$

(b) $\langle(\mathrm{id}),(K\vdash)\rangle$. The proof given and the proof after the transformation are as follows. The justification is exactly the same as for the proof tree rooted in the right premise of the cut.

$$\text{cut}\ \frac{p \vdash p \qquad \dfrac{\vdots \atop \Gamma \vdash \Delta}{p,\Gamma \vdash \Delta}\ K\vdash}{p,\Gamma \vdash \Delta} \quad \rightsquigarrow \quad \frac{\vdots \atop \Gamma \vdash \Delta}{p,\Gamma \vdash \Delta}\ K\vdash$$

(c) $\langle(\vdash K),(\mathrm{id})\rangle$. This case is symmetric with respect to the previous one. This time the right premise and the conclusion of the cut rule are omitted, because the end sequent and the left premise are the same.

$$\text{cut}\ \frac{\vdash K\ \dfrac{\vdots \atop \Gamma \vdash \Delta}{\Gamma \vdash \Delta, p} \qquad p \vdash p}{\Gamma \vdash \Delta, p} \quad \rightsquigarrow \quad \frac{\vdots \atop \Gamma \vdash \Delta}{\Gamma \vdash \Delta, p}\ \text{cut}$$

(d) $\langle(\vdash K),(K\vdash)\rangle$. Both premises are by thinning, which leads to a transformed proof in which there are $(K\vdash)$ and $(C\vdash)$ steps (if Θ is non-empty), and there are $(\vdash K)$ steps (if Λ is non-empty). (If we would start with the right premise, then the $(\vdash C)$ rule would be needed, if Δ is not empty.)

$$\text{cut}\ \frac{\vdash K\ \dfrac{\vdots \atop \Gamma \vdash \Delta}{\Gamma \vdash \Delta, p} \qquad \dfrac{\vdots \atop \Theta \vdash \Lambda}{p,\Theta \vdash \Lambda}\ K\vdash}{\Gamma,\Theta \vdash \Delta,\Lambda} \quad \rightsquigarrow \quad \dfrac{\dfrac{\vdots \atop \Gamma \vdash \Delta}{\Gamma,\Theta \vdash \Delta}\ K\vdash\text{'s, } C\vdash\text{'s}}{\Gamma,\Theta \vdash \Delta,\Lambda}\ \vdash K\text{'s}$$

Notice that in the first three cases, the result of the cut is identical to a premise of the cut, hence, the subtree rooted in that premise suffices. This observation does not depend on the shape of the formula in the axiom, which is the reason why some presentations of proofs of the cut theorem collate such cases no matter what δ is.

1.1.2 $\varrho = 2,\ \delta = 0,\ \mu > 0$. Now we know that there is at least one contraction on p above the cut in the proof tree. However, the principal formula of the contraction cannot be p, because $\varrho = 2$. These two requirements contradict each other. That is, there is no proof in which $\varrho = 2$ and $\mu > 0$.

1.2.1 $\varrho = 2,\ \delta > 0,\ \mu = 0$. Since the cut formula is no longer atomic, it may have been introduced in one or in both premises by a connective or by a quantifier rule. Now an axiom (as the left or the right premise) may be combined by 7 rules, plus two axioms can be paired—totaling *15 cases*. If either premise is by a thinning rule, then the other premise may be by one of 7 rules, which gives *13 cases*. Lastly, each of the two premises may be by one of 6 operational rules, which yields another *6 cases*.

This is the first step in the proof, where the number of possibilities sort of "blows up," because of the number of rules and their combinations. We consider a small selection of cases from each of the three groups.

(a) $\langle(\vdash\supset),(\mathrm{id})\rangle$. The segment of the proof given and of the transformed proof are as follows.

$$\dfrac{\dfrac{\overset{\vdots}{\mathcal{A},\Gamma\vdash\Delta,\mathcal{B}}}{\Gamma\vdash\Delta,\mathcal{A}\supset\mathcal{B}}{\scriptstyle\vdash\supset}\quad \mathcal{A}\supset\mathcal{B}\vdash\mathcal{A}\supset\mathcal{B}}{\Gamma\vdash\Delta,\mathcal{A}\supset\mathcal{B}}{\scriptstyle\mathrm{cut}} \quad\rightsquigarrow\quad \dfrac{\overset{\vdots}{\mathcal{A},\Gamma\vdash\Delta,\mathcal{B}}}{\Gamma\vdash\Delta,\mathcal{A}\supset\mathcal{B}}{\scriptstyle\vdash\supset}$$

Exercise 2.3.11. Go through the other 14 cases from the first group that we outlined above.

(b) $\langle(\vdash K),(\vee\vdash)\rangle$. The proof trees with cut and without cut are given below.

$$\dfrac{\dfrac{\overset{\vdots}{\Gamma\vdash\Delta}}{\Gamma\vdash\Delta,\mathcal{A}\vee\mathcal{B}}{\scriptstyle\vdash K}\quad \dfrac{\overset{\vdots}{\mathcal{A},\Theta\vdash\Lambda}\quad \overset{\vdots}{\mathcal{B},\Theta\vdash\Lambda}}{\mathcal{A}\vee\mathcal{B},\Theta\vdash\Lambda}{\scriptstyle\vee\vdash}}{\Gamma,\Theta\vdash\Delta,\Lambda}{\scriptstyle\mathrm{cut}} \quad\rightsquigarrow\quad \dfrac{\dfrac{\overset{\vdots}{\Gamma\vdash\Delta}}{\Gamma,\Theta\vdash\Delta}{\scriptstyle K\vdash\text{'s},\ C\vdash\text{'s}}}{\Gamma,\Theta\vdash\Delta,\Lambda}{\scriptstyle\vdash K\text{'s}}$$

Exercise 2.3.12. There are 12 cases remaining in the second group. Can you give a general description of the transformation so that it is applicable in all of these cases?

The gist of Case **1.2.1** is when the cut formula is introduced by matching connective or quantifier rules in the premises.

(c) $\langle(\vdash\wedge),(\wedge\vdash_1)\rangle$. There are two versions of the $(\wedge\vdash)$ rule. We consider the case when the left conjunct occurs in the upper premise. (The other rule yields a transformation that is symmetric to the one we give.)

$$\dfrac{\dfrac{\overset{\vdots}{\Gamma\vdash\Delta,\mathcal{A}}\quad \overset{\vdots}{\Gamma\vdash\Delta,\mathcal{B}}}{\Gamma\vdash\Delta,\mathcal{A}\wedge\mathcal{B}}{\scriptstyle\vdash\wedge}\quad \dfrac{\overset{\vdots}{\mathcal{A},\Theta\vdash\Lambda}}{\mathcal{A}\wedge\mathcal{B},\Theta\vdash\Lambda}{\scriptstyle\wedge\vdash_1}}{\Gamma,\Theta\vdash\Delta,\Lambda}{\scriptstyle\mathrm{cut}} \quad\rightsquigarrow\quad \dfrac{\overset{\vdots}{\Gamma\vdash\Delta,\mathcal{A}}\quad \overset{\vdots}{\mathcal{A},\Theta\vdash\Lambda}}{\Gamma,\Theta\vdash\Delta,\Lambda}{\scriptstyle\mathrm{cut}}$$

This is the first case in the proof where an application of the cut rule occurs in the transformed proof. It appears that we moved the cut upward in the proof, so to speak. This idea is right in spirit. However, the formal justification of this move is not a reduction of the rank of the cut formula, because the cut formula has changed. ($\mathcal{A}$ very well may have a left or right rank greater than 1.) $\delta(\mathcal{A}\wedge\mathcal{B})>\delta(\mathcal{A})$, that is, the *degree of the cut formula* is reduced.

(d) $\langle(\vdash\supset),(\supset\vdash)\rangle$. The original and the transformed proof look like the following.

$$\text{cut}\ \dfrac{\vdash\supset\ \dfrac{\mathcal{A},\Gamma\vdash\Delta,\mathcal{B}}{\Gamma\vdash\Delta,\mathcal{A}\supset\mathcal{B}} \qquad \dfrac{\Theta_1\vdash\Lambda_1,\mathcal{A}\quad \mathcal{B},\Theta_2\vdash\Lambda_2}{\mathcal{A}\supset\mathcal{B},\Theta_1,\Theta_2\vdash\Lambda_1,\Lambda_2}\ \supset\vdash}{\Gamma,\Theta_1,\Theta_2\vdash\Delta,\Lambda_1,\Lambda_2} \quad\rightsquigarrow$$

$$\rightsquigarrow\quad \dfrac{\text{cut}\ \dfrac{\text{cut}\ \dfrac{\Theta_1\vdash\Lambda_1,\mathcal{A}\quad \mathcal{A},\Gamma\vdash\Delta,\mathcal{B}}{\Theta_1,\Gamma\vdash\Lambda_1,\Delta,\mathcal{B}}\qquad \mathcal{B},\Theta_2\vdash\Lambda_2}{\Theta_1,\Gamma,\Theta_2\vdash\Lambda_1,\Delta,\Lambda_2}}{\Gamma,\Theta_1,\Theta_2\vdash\Delta,\Lambda_1,\Lambda_2}\ C\vdash\text{'s},\vdash C\text{'s}$$

It might seem that the transformation is a step backward, because we have two applications of the cut rule instead of one. However, both cuts have (strictly) lower degree.

(e) $\langle(\vdash\neg),(\neg\vdash)\rangle$. If the cut formula is $\neg\mathcal{A}$, then the given and the transformed proofs are as follows.

$$\text{cut}\ \dfrac{\vdash\neg\ \dfrac{\mathcal{A},\Gamma\vdash\Delta}{\Gamma\vdash\Delta,\neg\mathcal{A}}\qquad \dfrac{\Theta\vdash\Lambda,\mathcal{A}}{\neg\mathcal{A},\Theta\vdash\Lambda}\ \neg\vdash}{\Gamma,\Theta\vdash\Delta,\Lambda}\quad\rightsquigarrow\quad \dfrac{\dfrac{\Theta\vdash\Lambda,\mathcal{A}\quad \mathcal{A},\Gamma\vdash\Delta}{\Theta,\Gamma\vdash\Lambda,\Delta}\ \text{cut}}{\Gamma,\Theta\vdash\Delta,\Lambda}\ C\vdash\text{'s},\vdash C\text{'s}$$

(f) $\langle(\vdash\forall),(\forall\vdash)\rangle$. The cut formula is $\forall x\,\mathcal{A}(x)$, and the segments of the proofs are as follows.

$$\text{cut}\ \dfrac{\vdash\forall\ \dfrac{\Gamma\vdash\Delta,\mathcal{A}(z)}{\Gamma\vdash\Delta,\forall x\,\mathcal{A}(x)}\qquad \dfrac{\mathcal{A}(y),\Theta\vdash\Lambda}{\forall x\,\mathcal{A}(x),\Theta\vdash\Lambda}\ \forall\vdash}{\Gamma,\Theta\vdash\Delta,\Lambda}\quad\rightsquigarrow\quad \dfrac{\Gamma\vdash\Delta,\mathcal{A}(y)\quad \mathcal{A}(y),\Theta\vdash\Lambda}{\Gamma,\Theta\vdash\Delta,\Lambda}\ \text{cut}$$

The new cut formula is of lower degree, but we have to justify the change of the upper sequent in the left premise of this cut. The $(\vdash\forall)$ rule comes with a restriction that implies that z does not occur free in any formulas in Γ or Δ. We can use Lemma 2.15 about renaming of free variables. Since the $(\vdash\forall)$ rule is not applied in the transformed proof, we do not need to worry about whether y does (or does not) occur free in formulas in Γ or Δ.

(g) $\langle(\vdash\exists),(\exists\vdash)\rangle$. Dually to the previous case, the cut formula may be $\exists x\,\mathcal{A}(x)$. Here are the given and the modified proofs.

$$\text{cut}\ \dfrac{\vdash\exists\ \dfrac{\Gamma\vdash\Delta,\mathcal{A}(y)}{\Gamma\vdash\Delta,\exists x\,\mathcal{A}(x)}\qquad \dfrac{\mathcal{A}(z),\Theta\vdash\Lambda}{\exists x\,\mathcal{A}(x),\Theta\vdash\Lambda}\ \exists\vdash}{\Gamma,\Theta\vdash\Delta,\Lambda}\quad\rightsquigarrow\quad \dfrac{\Gamma\vdash\Delta,\mathcal{A}(y)\quad \mathcal{A}(y),\Theta\vdash\Lambda}{\Gamma,\Theta\vdash\Delta,\Lambda}\ \text{cut}$$

Exercise 2.3.13. We omitted one case from the third group. Identify the logical constant involved and complete the step.

1.2.2 $\varrho = 2,\ \delta > 0,\ \mu > 0.$ Once again, there is no proof that satisfies the constraints.

2.1.1 $\varrho > 2,\ \delta = 0,\ \mu = 0.$ First of all, since $\delta = 0$, we know that the cut formula is atomic, let us say, p. It is useful to distinguish between two possibilities: either $\varrho_l > 1$ or $\varrho_r > 1$. Let us look at the shape of the chunk of the proof when $\varrho_l > 1$.

$$\dfrac{\dfrac{\stackrel{\vdots}{\Gamma' \vdash \Delta_1', p, \Delta_2'}}{\Gamma \vdash \Delta_1, p, \Delta_2} \qquad \stackrel{\vdots}{\Theta_1, p, \Theta_2 \vdash \Lambda}}{\Theta_1, \Gamma, \Theta_2 \vdash \Delta_1, \Lambda, \Delta_2}$$

The priming indicates that some rule has been applied to obtain the left premise of the cut rule, and the rule may have affected the antecedent or the succedent. We have already seen in Case **1.2.1** how the number of subcases can suddenly increase while many of them can be dealt with in a uniform way. Here all left rules and right rules have to be considered as leading to the left premise. However, p cannot be the principal formula of the $(\vdash W)$ rule. The subcases really add up now.

We consider three sample subcases from among those in which a left rule is applied (in the left premise), and leave the verification of the rest for the reader.

(11) $\langle(\neg\vdash),\ \_\rangle$. The shape of the given and transformed proofs are as follows.

$$\dfrac{\dfrac{\stackrel{\vdots}{\Gamma \vdash \Delta_1, p, \Delta_2, \mathcal{A}}}{\neg\mathcal{A}, \Gamma \vdash \Delta_1, p, \Delta_2}\,\neg\vdash \qquad \stackrel{\vdots}{\Theta_1, p, \Theta_2 \vdash \Lambda}}{\Theta_1, \neg\mathcal{A}, \Gamma, \Theta_2 \vdash \Delta_1, \Lambda, \Delta_2}\,\text{cut} \quad \rightsquigarrow$$

$$\rightsquigarrow \quad \dfrac{\dfrac{\dfrac{\stackrel{\vdots}{\Gamma \vdash \Delta_1, p, \Delta_2, \mathcal{A}} \qquad \stackrel{\vdots}{\Theta_1, p, \Theta_2 \vdash \Lambda}}{\Theta_1, \Gamma, \Theta_2 \vdash \Delta_1, \Lambda, \Delta_2, \mathcal{A}}\,\text{cut}}{\neg\mathcal{A}, \Theta_1, \Gamma, \Theta_2 \vdash \Delta_1, \Lambda, \Delta_2}\,\neg\vdash}{\Theta_1, \neg\mathcal{A}, \Gamma, \Theta_2 \vdash \Delta_1, \Lambda, \Delta_2}\,\text{C}\vdash\text{'s}$$

The transformation is justified by a reduction in the left rank of the cut. (The $(\text{C}\vdash)$ steps may or may not be necessary depending on whether Θ_1 contains any formulas or it does not.)

(12) $\langle(\forall\vdash),\ \_\rangle$. This case is very easy, because there is no restriction for the $(\forall\vdash)$ rule. The proof segments are as follows.

$$\dfrac{\dfrac{\stackrel{\vdots}{\mathcal{A}(y), \Gamma \vdash \Delta_1, p, \Delta_2}}{\forall x\, \mathcal{A}(x), \Gamma \vdash \Delta_1, p, \Delta_2}\,\forall\vdash \qquad \stackrel{\vdots}{\Theta_1, p, \Theta_2 \vdash \Lambda}}{\Theta_1, \forall x\, \mathcal{A}(x), \Gamma, \Theta_2 \vdash \Delta_1, \Lambda, \Delta_2}\,\text{cut} \quad \rightsquigarrow$$

$$\rightsquigarrow\quad \dfrac{\dfrac{\dfrac{\dfrac{\mathcal{A}(y),\Gamma\vdash\Delta_1,p,\Delta_2 \qquad \Theta_1,p,\Theta_2\vdash\Lambda}{\Theta_1,\mathcal{A}(y),\Gamma,\Theta_2\vdash\Delta_1,\Lambda,\Delta_2}\,\text{cut}}{\mathcal{A}(y),\Theta_1,\Gamma,\Theta_2\vdash\Delta_1,\Lambda,\Delta_2}\,C\vdash\text{'s}}{\forall x\,\mathcal{A}(x),\Theta_1,\Gamma,\Theta_2\vdash\Delta_1,\Lambda,\Delta_2}\,\forall\vdash}{\Theta_1,\forall x\,\mathcal{A}(x),\Gamma,\Theta_2\vdash\Delta_1,\Lambda,\Delta_2}\,C\vdash\text{'s}$$

The justification is as in the previous case. In a sense, the transformation may be viewed as swapping the cut rule and the left rule that produces the left premise—except that to ensure the applicability of the $(\forall\vdash)$ rule additional permutations may have to be performed in the antecedent.

(13) $\langle (K\vdash),\ \_\, \rangle$. There is no complication in this case either as the proof segments show.

$$\text{cut}\ \dfrac{K\vdash\ \dfrac{\Gamma\vdash\Delta_1,p,\Delta_2}{\mathcal{A},\Gamma\vdash\Delta_1,p,\Delta_2} \qquad \Theta_1,p,\Theta_2\vdash\Lambda}{\Theta_1,\mathcal{A},\Gamma,\Theta_2\vdash\Delta_1,\Lambda,\Delta_2}\quad\rightsquigarrow$$

$$\rightsquigarrow\quad \dfrac{\dfrac{\dfrac{\Gamma\vdash\Delta_1,p,\Delta_2 \qquad \Theta_1,p,\Theta_2\vdash\Lambda}{\Theta_1,\Gamma,\Theta_2\vdash\Delta_1,\Lambda,\Delta_2}\,\text{cut}}{\mathcal{A},\Theta_1,\Gamma,\Theta_2\vdash\Delta_1,\Lambda,\Delta_2}\,K\vdash}{\Theta_1,\mathcal{A},\Gamma,\Theta_2\vdash\Delta_1,\Lambda,\Delta_2}\,C\vdash\text{'s}$$

The transformation reduces the left rank of the cut.

Exercise 2.3.14. Other left rules could have been applied here to obtain the left premise. Verify those subcases. [Hint: There are 7 other subcases when we do not count the two $(\wedge\vdash)$ rules separately.]

Again, we include three subcases as examples from among the cases in which a right rule is applied (in the left premise).

(r1) $\langle (\vdash\wedge),\ \_\, \rangle$. We consider the following proof and its transformation.

$$\text{cut}\ \dfrac{\vdash\wedge\ \dfrac{\Gamma\vdash\Delta_1,p,\Delta_2,\mathcal{A} \qquad \Gamma\vdash\Delta_1,p,\Delta_2,\mathcal{B}}{\Gamma\vdash\Delta_1,p,\Delta_2,\mathcal{A}\wedge\mathcal{B}} \qquad \Theta_1,p,\Theta_2\vdash\Lambda}{\Theta_1,\Gamma,\Theta_2\vdash\Delta_1,\Lambda,\Delta_2,\mathcal{A}\wedge\mathcal{B}}\quad\rightsquigarrow$$

$$\rightsquigarrow\quad \dfrac{\text{cut}\ \dfrac{\Gamma\vdash\Delta_1,p,\Delta_2,\mathcal{A} \qquad \Theta_1,p,\Theta_2\vdash\Lambda}{\Theta_1,\Gamma,\Theta_2\vdash\Delta_1,\Lambda,\Delta_2,\mathcal{A}} \qquad \dfrac{\Gamma\vdash\Delta_1,p,\Delta_2,\mathcal{B} \qquad \Theta_1,p,\Theta_2\vdash\Lambda}{\Theta_1,\Gamma,\Theta_2\vdash\Delta_1,\Lambda,\Delta_2,\mathcal{B}}\,\text{cut}}{\Theta_1,\Gamma,\Theta_2\vdash\Delta_1,\Lambda,\Delta_2,\mathcal{A}\wedge\mathcal{B}}\,\vdash\wedge$$

The transformation results in two cuts, and the proof tree rooted in the right premise of the original cut is duplicated. But both cuts have a left rank lower than the previous one. Notice that the subalterns of the $(\vdash\wedge)$ rule are on the right edges of the sequents, whereas the cut formula in not; therefore, the $(\vdash\wedge)$ rule is immediately applicable after the two cuts without additional permutations.

(r2) $\langle(\vdash\forall),\,\_\rangle$. The proof segments are as follows.

$$\dfrac{\dfrac{\Gamma\vdash\Delta_1,p,\Delta_2,\mathcal{A}(z)}{\Gamma\vdash\Delta_1,p,\Delta_2,\forall x\,\mathcal{A}(x)}\vdash\forall \qquad \Theta_1,p,\Theta_2\vdash\Lambda}{\Theta_1,\Gamma,\Theta_2\vdash\Delta_1,\Lambda,\Delta_2,\forall x\,\mathcal{A}(x)}\text{cut} \quad\rightsquigarrow$$

$$\rightsquigarrow\quad \dfrac{\dfrac{\Gamma\vdash\Delta_1,p,\Delta_2,\mathcal{A}(y)\qquad \Theta_1,p,\Theta_2\vdash\Lambda}{\Theta_1,\Gamma,\Theta_2\vdash\Delta_1,\Lambda,\Delta_2,\mathcal{A}(y)}\text{cut}}{\Theta_1,\Gamma,\Theta_2\vdash\Delta_1,\Lambda,\Delta_2,\forall x\mathcal{A}(x)}\vdash\forall$$

The left premise of the cut in the transformed proof contains $\mathcal{A}(y)$, which indicates that y may have to be relabeled to guarantee that the side condition of the $(\vdash\forall)$ rule is satisfied after the cut. Lemma 2.15 ensures that the renaming always can be carried out. Again, the left rank of the new cut is strictly less than that of the earlier one.

(r3) $\langle(\vdash\text{C}),\,\_\rangle$. The last subcase that we consider involves permutation.

$$\dfrac{\dfrac{\Gamma\vdash\Delta_1,p,\Delta_2,\mathcal{A},\mathcal{B},\Delta_3}{\Gamma\vdash\Delta_1,p,\Delta_2,\mathcal{B},\mathcal{A},\Delta_3}\vdash\text{C} \qquad \Theta_1,p,\Theta_2\vdash\Lambda}{\Theta_1,\Gamma,\Theta_2\vdash\Delta_1,\Lambda,\Delta_2,\mathcal{B},\mathcal{A},\Delta_3}\text{cut} \quad\rightsquigarrow$$

$$\rightsquigarrow\quad \dfrac{\dfrac{\Gamma\vdash\Delta_1,p,\Delta_2,\mathcal{A},\mathcal{B},\Delta_3\qquad \Theta_1,p,\Theta_2\vdash\Lambda}{\Theta_1,\Gamma,\Theta_2\vdash\Delta_1,\Lambda,\Delta_2,\mathcal{A},\mathcal{B},\Delta_3}\text{cut}}{\Theta_1,\Gamma,\Theta_2\vdash\Delta_1,\Lambda,\Delta_2,\mathcal{B},\mathcal{A},\Delta_3}\vdash\text{C}$$

The justification for the transformation is as before. The chunk of the proof depicted above is not the only possibility. We do not have a sufficiently general notation (at this point) to indicate that a formula is anywhere inside the succedent. First of all, p could be inside of Δ_3. It is easy to see that swapping the application of the cut rule with the structural rule works then too. Further, the $(\vdash\text{C})$ rule—like the other structural rules—does not increase the degree of any formula; hence, one of its principal formulas could be p. The transformation will switch the rules again, but if Λ comprises more than a single formula, then several applications of the $(\vdash\text{C})$ rule are required.

Exercise 2.3.15. Complete the verification for the subcases in which a right rule is applied to obtain the left premise. This will complete the case for $\varrho_l > 1$. [Hint: We omitted 7 subcases.]

Of course, $\varrho_r > 1$ is also possible (independently of $\varrho_l > 1$). Then the general shape of the proof segment is the following.

$$\dfrac{\begin{matrix}\vdots\\ \Gamma\vdash\Delta_1,p,\Delta_2\end{matrix}\qquad \dfrac{\begin{matrix}\vdots\\ \Theta'_1,p,\Theta'_2\vdash\Lambda'\end{matrix}}{\Theta_1,p,\Theta_2\vdash\Lambda}}{\Theta_1,\Gamma,\Theta_2\vdash\Delta_1,\Lambda,\Delta_2}$$

The priming shows that the respective sequences of wff's may differ depending on the concrete rule that has been applied. There is an abundance of rules that could have been applied either on the left or on the right to obtain the right premise. First, we consider some of the right rules, and then some of the left rules.

(r1) $\langle\,\_\,,(\vdash\neg)\rangle$. The relevant part of the proof and its transformation are as follows.

$$\text{cut}\;\dfrac{\begin{matrix}\vdots\\ \Gamma\vdash\Delta_1,p,\Delta_2\end{matrix}\qquad \dfrac{\begin{matrix}\vdots\\ \mathcal{A},\Theta_1,p,\Theta_2\vdash\Lambda\end{matrix}}{\Theta_1,p,\Theta_2\vdash\Lambda,\neg\mathcal{A}}\;{\scriptstyle\vdash\neg}}{\Theta_1,\Gamma,\Theta_2\vdash\Delta_1,\Lambda,\neg\mathcal{A},\Delta_2}\quad\rightsquigarrow$$

$$\rightsquigarrow\quad\dfrac{\dfrac{\dfrac{\begin{matrix}\vdots\\ \Gamma\vdash\Delta_1,p,\Delta_2\end{matrix}\qquad\begin{matrix}\vdots\\ \mathcal{A},\Theta_1,p,\Theta_2\vdash\Lambda\end{matrix}}{\mathcal{A},\Theta_1,\Gamma,\Theta_2\vdash\Delta_1,\Lambda,\Delta_2}\;\text{cut}}{\Theta_1,\Gamma,\Theta_2\vdash\Delta_1,\Lambda,\Delta_2,\neg\mathcal{A}}\;{\scriptstyle\vdash\neg}}{\Theta_1,\Gamma,\Theta_2\vdash\Delta_1,\Lambda,\neg\mathcal{A},\Delta_2}\;{\scriptstyle\vdash C\text{'s}}$$

The cut in the transformed proof has been moved a step up in the right premise, that is, ϱ_r is reduced.

(r2) $\langle\,\_\,,(\vdash\exists)\rangle$. This subcase is completely straightforward—as the proof segments below show.

$$\text{cut}\;\dfrac{\begin{matrix}\vdots\\ \Gamma\vdash\Delta_1,p,\Delta_2\end{matrix}\qquad \dfrac{\begin{matrix}\vdots\\ \Theta_1,p,\Theta_2\vdash\Lambda,\mathcal{A}(y)\end{matrix}}{\Theta_1,p,\Theta_2\vdash\Lambda,\exists x\,\mathcal{A}(x)}\;{\scriptstyle\vdash\exists}}{\Theta_1,\Gamma,\Theta_2\vdash\Delta_1,\Lambda,\exists x\,\mathcal{A}(x),\Delta_2}\quad\rightsquigarrow$$

$$\rightsquigarrow\quad\dfrac{\dfrac{\dfrac{\dfrac{\begin{matrix}\vdots\\ \Gamma\vdash\Delta_1,p,\Delta_2\end{matrix}\qquad\begin{matrix}\vdots\\ \Theta_1,p,\Theta_2\vdash\Lambda,\mathcal{A}(y)\end{matrix}}{\Theta_1,\Gamma,\Theta_2\vdash\Delta_1,\Lambda,\mathcal{A}(y),\Delta_2}\;\text{cut}}{\Theta_1,\Gamma,\Theta_2\vdash\Delta_1,\Lambda,\Delta_2,\mathcal{A}(y)}\;{\scriptstyle\vdash C\text{'s}}}{\Theta_1,\Gamma,\Theta_2\vdash\Delta_1,\Lambda,\Delta_2,\exists x\,\mathcal{A}(x)}\;{\scriptstyle\vdash\exists}}{\Theta_1,\Gamma,\Theta_2\vdash\Delta_1,\Lambda,\exists x\,\mathcal{A}(x),\Delta_2}\;{\scriptstyle\vdash C\text{'s}}$$

The application of the cut in the modified proof (on the right-hand side) has a lower right rank, hence, a lower rank than the cut in the given proof (on the left-hand side).

(r3) $\langle\, \_\, , ({\vdash} K) \rangle$. This subcase is easy too. Notice that the principal formula of the thinning rule does not have a subaltern, which means that the cut and the thinning rules are simply switched in the modified proof.

$$\text{cut}\ \dfrac{\begin{matrix}\vdots\\ \Gamma \vdash \Delta_1, p, \Delta_2\end{matrix} \qquad \dfrac{\begin{matrix}\vdots\\ \Theta_1, p, \Theta_2 \vdash \Lambda\end{matrix}}{\Theta_1, p, \Theta_2 \vdash \Lambda, \mathcal{A}}\ {\vdash}K}{\Theta_1, \Gamma, \Theta_2 \vdash \Delta_1, \Lambda, \mathcal{A}, \Delta_2} \quad \rightsquigarrow$$

$$\rightsquigarrow \quad \dfrac{\dfrac{\dfrac{\begin{matrix}\vdots\\ \Gamma \vdash \Delta_1, p, \Delta_2\end{matrix} \qquad \begin{matrix}\vdots\\ \Theta_1, p, \Theta_2 \vdash \Lambda\end{matrix}}{\Theta_1, \Gamma, \Theta_2 \vdash \Delta_1, \Lambda, \Delta_2}\ \text{cut}}{\Theta_1, \Gamma, \Theta_2 \vdash \Delta_1, \Lambda, \Delta_2, \mathcal{A}}\ {\vdash}K}{\Theta_1, \Gamma, \Theta_2 \vdash \Delta_1, \Lambda, \mathcal{A}, \Delta_2}\ {\vdash}C\text{'s}$$

Once again, the new ϱ_r is one less that the old ϱ_r.

Exercise 2.3.16. Consider the other right introduction rules that could have yielded the right premise in this case. [Hint: There are 7 other possibilities if you do not count two versions of the $\vdash \vee$ rule as two rules.]

(l1) $\langle\, \_\, , ({\supset}{\vdash}) \rangle$. The $({\supset}{\vdash})$ rule has two premises, and we have to distinguish two subcases depending on the location of the cut formula. The given proof segments and their modifications look sufficiently different to make it worthwhile to consider both of them.

$$\text{cut}\ \dfrac{\begin{matrix}\vdots\\ \Gamma \vdash \Delta_1, p, \Delta_2\end{matrix} \qquad \dfrac{\begin{matrix}\vdots\\ \Theta_1, p, \Theta_2' \vdash \Lambda_1, \mathcal{A}\end{matrix} \qquad \begin{matrix}\vdots\\ \mathcal{B}, \Theta_2'' \vdash \Lambda_2\end{matrix}}{\mathcal{A} \supset \mathcal{B}, \Theta_1, p, \Theta_2 \vdash \Lambda_1, \Lambda_2}\ {\supset}{\vdash}}{\mathcal{A} \supset \mathcal{B}, \Theta_1, \Gamma, \Theta_2 \vdash \Delta_1, \Lambda_1, \Lambda_2, \Delta_2} \quad \rightsquigarrow$$

$$\rightsquigarrow \quad \dfrac{\dfrac{\dfrac{\dfrac{\begin{matrix}\vdots\\ \Gamma \vdash \Delta_1, p, \Delta_2\end{matrix} \qquad \begin{matrix}\vdots\\ \Theta_1, p, \Theta_2' \vdash \Lambda_1, \mathcal{A}\end{matrix}}{\Theta_1, \Gamma, \Theta_2' \vdash \Delta_1, \Lambda_1, \mathcal{A}, \Delta_2}\ \text{cut}}{\Theta_1, \Gamma, \Theta_2' \vdash \Delta_1, \Lambda_1, \Delta_2, \mathcal{A}}\ {\vdash}C\text{'s} \qquad \begin{matrix}\vdots\\ \mathcal{B}, \Theta_2'' \vdash \Lambda_2\end{matrix}}{\mathcal{A} \supset \mathcal{B}, \Theta_1, \Gamma, \Theta_2', \Theta_2'' \vdash \Delta_1, \Lambda_1, \Delta_2, \Lambda_2}\ {\supset}{\vdash}}{\mathcal{A} \supset \mathcal{B}, \Theta_1, \Gamma, \Theta_2 \vdash \Delta_1, \Lambda_1, \Lambda_2, \Delta_2}\ {\vdash}C\text{'s}$$

We assume that Θ_2', Θ_2'' is Θ_2, and similarly, Θ_1', Θ_1'' is Θ_1.

$$\text{cut}\ \dfrac{\begin{matrix}\vdots\\ \Gamma \vdash \Delta_1, p, \Delta_2\end{matrix} \qquad \dfrac{\begin{matrix}\vdots\\ \Theta_1' \vdash \Lambda_1, \mathcal{A}\end{matrix} \qquad \begin{matrix}\vdots\\ \mathcal{B}, \Theta_1'', p, \Theta_2 \vdash \Lambda_2\end{matrix}}{\mathcal{A} \supset \mathcal{B}, \Theta_1, p, \Theta_2 \vdash \Lambda_1, \Lambda_2}\ {\supset}{\vdash}}{\mathcal{A} \supset \mathcal{B}, \Theta_1, \Gamma, \Theta_2 \vdash \Delta_1, \Lambda_1, \Lambda_2, \Delta_2} \quad \rightsquigarrow$$

$$\rightsquigarrow \quad \dfrac{\Theta_1' \vdash \Lambda_1, \mathcal{A} \qquad \dfrac{\Gamma \vdash \Delta_1, p, \Delta_2 \qquad \mathcal{B}, \Theta_1'', p, \Theta_2 \vdash \Lambda_2}{\mathcal{B}, \Theta_1'', \Gamma, \Theta_2 \vdash \Delta_1, \Lambda_2, \Delta_2}\,\text{cut}}{\dfrac{\mathcal{A} \supset \mathcal{B}, \Theta_1', \Theta_1'', \Gamma, \Theta_2 \vdash \Lambda_1, \Delta_1, \Lambda_2, \Delta_2}{\mathcal{A} \supset \mathcal{B}, \Theta_1, \Gamma, \Theta_2 \vdash \Delta_1, \Lambda_1, \Lambda_2, \Delta_2}\,\vdash C\text{'s}}\,\supset\vdash$$

In both modified proofs, the cut is performed on a premise of the $(\supset\vdash)$ rule, which reduces ϱ_r and justifies the step.

(12) $\langle\, \_, (\exists\vdash) \rangle$. The given and the modified proof segments look like the following.

$$\text{cut}\ \dfrac{\Gamma \vdash \Delta_1, p, \Delta_2 \qquad \dfrac{\mathcal{A}(y), \Theta_1, p, \Theta_2 \vdash \Lambda}{\exists x\, \mathcal{A}(x), \Theta_1, p, \Theta_2 \vdash \Lambda}\,\exists\vdash}{\exists x\, \mathcal{A}(x), \Theta_1, \Gamma, \Theta_2 \vdash \Delta_1, \Lambda, \Delta_2} \quad \rightsquigarrow$$

$$\rightsquigarrow \quad \dfrac{\dfrac{\Gamma \vdash \Delta_1, p, \Delta_2 \qquad \mathcal{A}(z), \Theta_1, p, \Theta_2 \vdash \Lambda}{\mathcal{A}(z), \Theta_1, \Gamma, \Theta_2 \vdash \Delta_1, \Lambda, \Delta_2}\,\text{cut}}{\exists x\, \mathcal{A}(x), \Theta_1, \Gamma, \Theta_2 \vdash \Delta_1, \Lambda, \Delta_2}\,\exists\vdash$$

At first, it may seem that only an exchange of the cut and of the $(\exists\vdash)$ rule has happened. However, notice that y may have to be relabeled in order to ensure that the variable that becomes existentially quantified does not occur free in the whole sequent, which now contains the Θ's and Λ too. Again, the transformation has decreased ϱ_r by 1.

(13) $\langle\, \_, (C\vdash) \rangle$. This is the neatest case, because it shows that it is possible that rules, which are formulated carefully with respect to the location of the principal and subaltern formulas, lead to a transformation, which means nothing more than swapping two rules. Several earlier cases required applications of $(C\vdash)$ or $(\vdash C)$ to reposition formulas.

$$\text{cut}\ \dfrac{\Gamma \vdash \Delta_1, p, \Delta_2 \qquad \dfrac{\Theta_1, \mathcal{A}, \mathcal{B}, \Theta_2, p, \Theta_3 \vdash \Lambda}{\Theta_1, \mathcal{B}, \mathcal{A}, \Theta_2, p, \Theta_3 \vdash \Lambda}\,C\vdash}{\Theta_1, \mathcal{B}, \mathcal{A}, \Theta_2, \Gamma, \Theta_3 \vdash \Delta_1, \Lambda, \Delta_2} \quad \rightsquigarrow$$

$$\rightsquigarrow \quad \dfrac{\dfrac{\Gamma \vdash \Delta_1, p, \Delta_2 \qquad \Theta_1, \mathcal{A}, \mathcal{B}, \Theta_2, p, \Theta_3 \vdash \Lambda}{\Theta_1, \mathcal{A}, \mathcal{B}, \Theta_2, \Gamma, \Theta_3 \vdash \Delta_1, \Lambda, \Delta_2}\,\text{cut}}{\Theta_1, \mathcal{B}, \mathcal{A}, \Theta_2, \Gamma, \Theta_3 \vdash \Delta_1, \Lambda, \Delta_2}\,C\vdash$$

The notation we use does not allow us to place $\mathcal{A}$ and $\mathcal{B}$ in a completely general way within the antecedent of the right premise, but it is not difficult to see that if p is in Θ_3, then switching $(C\vdash)$ and the cut gives the desired sequent, as in the above situation.

There is another potential difficulty, though, that still may result in more than one application of the permutation rule after the cut. Namely, if $\mathcal{A}$ or $\mathcal{B}$ is p, the cut formula, and Γ contains more than one formula, then $\mathcal{B}$ or $\mathcal{A}$ must be permuted several times to get it across Γ. In Chapter 5, we will see that structural rules have to be *structuralized* (i.e., to be applicable to structures)—in addition to being *contextualized* (i.e., their subalterns and principal wff's located appropriately within sequents).

Exercise 2.3.17. Finish the proof of Case 2.1.1. [Hint: There are 7 more cases if we do not consider the two variants of the $(\wedge\vdash)$ as separate.]

2.1.2 $\varrho > 2,\ \delta = 0,\ \mu > 0$. This case is very similar to the previous one—as to be expected—because the only difference is with respect to μ. The positive contraction measure means that the contraction rule $(\vdash W)$ applied to p must be considered too. We give details of two subcases.

(1) $\langle (W\vdash),\ \_\ \rangle$. Let us assume that $\varrho_l > 1$ and the left premise is by $(W\vdash)$. The given segment of the proof and its transformation have the shape below.

$$\dfrac{\dfrac{\mathcal{A},\mathcal{A},\Gamma\vdash\Delta_1,p,\Delta_2}{\mathcal{A},\Gamma\vdash\Delta_1,p,\Delta_2}\,W\vdash \qquad \Theta_1,p,\Theta_2\vdash\Lambda}{\Theta_1,\mathcal{A},\Gamma,\Theta_2\vdash\Delta_1,\Lambda,\Delta_2}\,\text{cut} \quad \rightsquigarrow$$

$$\rightsquigarrow \quad \dfrac{\dfrac{\dfrac{\dfrac{\mathcal{A},\mathcal{A},\Gamma\vdash\Delta_1,p,\Delta_2 \qquad \Theta_1,p,\Theta_2\vdash\Lambda}{\Theta_1,\mathcal{A},\mathcal{A},\Gamma,\Theta_2\vdash\Delta_1,\Lambda,\Delta_2}\,\text{cut}}{\mathcal{A},\mathcal{A},\Theta_1,\Gamma,\Theta_2\vdash\Delta_1,\Lambda,\Delta_2}\,C\vdash\text{'s}}{\mathcal{A},\Theta_1,\Gamma,\Theta_2\vdash\Delta_1,\Lambda,\Delta_2}\,W\vdash}{\Theta_1,\mathcal{A},\Gamma,\Theta_2\vdash\Delta_1,\Lambda,\Delta_2}\,C\vdash\text{'s}$$

The left rank of the cut has been reduced by 1.

(2) $\langle (\vdash W),\ \_\ \rangle$. The right contraction rule can yield the left premise. Contraction is a structural rule, which implies that the principal formula can be of any shape; hence, it can be an atomic formula too. Then, it can be the cut formula too. The following proof and its transformation show the situation when there is no such coincidence.

$$\dfrac{\dfrac{\Gamma\vdash\Delta_1,p,\Delta_2,\mathcal{A},\mathcal{A}}{\Gamma\vdash\Delta_1,p,\Delta_2,\mathcal{A}}\,\vdash W \qquad \Theta_1,p,\Theta_2\vdash\Lambda}{\Theta_1,\Gamma,\Theta_2\vdash\Delta_1,\Lambda,\Delta_2,\mathcal{A}}\,\text{cut} \quad \rightsquigarrow$$

$$\rightsquigarrow \quad \dfrac{\dfrac{\Gamma\vdash\Delta_1,p,\Delta_2,\mathcal{A},\mathcal{A} \qquad \Theta_1,p,\Theta_2\vdash\Lambda}{\Theta_1,\Gamma,\Theta_2\vdash\Delta_1,\Lambda,\Delta_2,\mathcal{A},\mathcal{A}}\,\text{cut}}{\Theta_1,\Gamma,\Theta_2\vdash\Delta_1,\Lambda,\Delta_2,\mathcal{A}}\,\vdash W$$

If p happens to be $\mathcal{A}$, then the given proof segment looks like the one on the left.

$$\text{cut}\ \frac{\vdash W\ \dfrac{\begin{matrix}\vdots\\ \Gamma \vdash \Delta, p, p\end{matrix}}{\Gamma \vdash \Delta, p} \qquad \begin{matrix}\vdots\\ \Theta_1, p, \Theta_2 \vdash \Lambda\end{matrix}}{\Theta_1, \Gamma, \Theta_2 \vdash \Delta, \Lambda} \quad \rightsquigarrow$$

$$\rightsquigarrow \quad \cfrac{\cfrac{\cfrac{\text{cut}\ \dfrac{\begin{matrix}\vdots\\ \Gamma \vdash \Delta, p, p\end{matrix} \qquad \begin{matrix}\vdots\\ \Theta_1, p, \Theta_2 \vdash \Lambda\end{matrix}}{\Theta_1, \Gamma, \Theta_2 \vdash \Delta, \Lambda, p} \qquad \begin{matrix}\vdots\\ \Theta_1, p, \Theta_2 \vdash \Lambda\end{matrix}}{\Theta_1, \Theta_1, \Gamma, \Theta_2, \Theta_2 \vdash \Delta, \Lambda, \Lambda}\ \text{cut}}{\Theta_1, \Gamma, \Theta_2 \vdash \Delta, \Lambda, \Lambda}\ C\vdash\text{'s},\ W\vdash\text{'s}}{\Theta_1, \Gamma, \Theta_2 \vdash \Delta, \Lambda}\ \vdash C\text{'s},\ \vdash W\text{'s}$$

The proof segment on the right-hand side illustrates again that if structural rules apply to formulas (rather than to sequences of formulas or structures), then after a cut, several applications of structural rules may be required. In the proof chunk above, the two Λ's could be contracted to one Λ if $(\vdash W)$ would be applicable not only to a single formula. Depending on the length and concrete shape of Θ_1, Θ_2 and Λ, several alternating permutation and contraction steps may be needed.

Each of the two cuts in the transformed proofs are justified as having a lower μ. The number of contractions on parametric ancestors of p in the antecedent of the right premise is unchanged. However, both occurrences of p in the succedent of the upper sequent of $(\vdash W)$ have fewer contractions applied to their parametric ancestors than the p in the lower sequent has. The latter has one more contraction applied to its parametric ancestors than the sum of the contractions applied to parametric ancestors of its two subalterns.

Exercise 2.3.18. There are two other subcases involving a contraction rule. Describe those cases—together with the transformations and their justifications.

Exercise 2.3.19. Tally all the subcases in **2.1.2**. [Hint: Do not forget that some rules have slightly different versions, and occasionally the cut formula may coincide with a principal formula of a rule.]

2.2.1 $\varrho > 2,\ \delta > 0,\ \mu = 0$. This case has a strong similarity to **2.1.1**, though now the cut formula is not atomic. The subcases may be grouped again by considering when $\varrho_l > 1$ or $\varrho_r > 1$, and then taking all the left and right rules into account.

An important new possibility to notice is that when the left rank is greater than 1, and the last rule is a right rule in the proof of the left premise, then the cut formula may coincide with the principal formula in almost all subcases (not only for structural rules). The cut formula cannot coincide with the principal formula of $(\vdash W)$ though. We give details of a handful of subcases illustrating the interplay between the principal and the cut formula.

(r1) $\langle(\vdash\neg),\_\rangle$. Here and in the next two subcases we assume $\varrho_l > 1$. The left premise of the cut is by the negation introduction on the right and the cut formula is the newly introduced occurrence of $\neg\mathcal{A}$.

$$\dfrac{\dfrac{\mathcal{A},\Gamma\vdash\Delta_1,\neg\mathcal{A},\Delta_2}{\Gamma\vdash\Delta_1,\neg\mathcal{A},\Delta_2,\neg\mathcal{A}}\,{\vdash\neg}\qquad \Theta_1,\neg\mathcal{A},\Theta_2\vdash\Lambda}{\Theta_1,\Gamma,\Theta_2\vdash\Delta_1,\neg\mathcal{A},\Delta_2,\Lambda}\,\text{cut}\quad\rightsquigarrow$$

$$\rightsquigarrow\quad\dfrac{\dfrac{\dfrac{\dfrac{\mathcal{A},\Gamma\vdash\Delta_1,\neg\mathcal{A},\Delta_2\qquad \Theta_1,\neg\mathcal{A},\Theta_2\vdash\Lambda}{\Theta_1,\mathcal{A},\Gamma,\Theta_2\vdash\Delta_1,\Lambda,\Delta_2}\,\text{cut}}{\mathcal{A},\Theta_1,\Gamma,\Theta_2\vdash\Delta_1,\Lambda,\Delta_2}\,\text{C}\vdash\text{'s}}{\Theta_1,\Gamma,\Theta_2\vdash\Delta_1,\Lambda,\Delta_2,\neg\mathcal{A}}\,{\vdash\neg}}{\Theta_1,\Gamma,\Theta_2\vdash\Delta_1,\neg\mathcal{A},\Delta_2,\Lambda}\,{\vdash\text{C's}}$$

We made explicit that there is another occurrence of the principal formula of the $(\vdash\neg)$ rule in the succedent of the left premise. Then the cut can be performed on that occurrence in the transformed proof. The new cut has a lower left rank, which justifies the transformation. What remains is to apply the $(\vdash\neg)$ rule; which may have to be preceded and followed by some permutations to obtain exactly the same succedent as before.

(r2) $\langle(\vdash\supset),\_\rangle$. Let us consider the subcase when the $(\vdash\supset)$ rule has been applied, and the newly introduced $\mathcal{A}\supset\mathcal{B}$ is the cut formula.

$$\dfrac{\dfrac{\mathcal{A},\Gamma\vdash\Delta_1,\mathcal{A}\supset\mathcal{B},\Delta_2,\mathcal{B}}{\Gamma\vdash\Delta_1,\mathcal{A}\supset\mathcal{B},\Delta_2,\mathcal{A}\supset\mathcal{B}}\,{\vdash\supset}\qquad \Theta_1,\mathcal{A}\supset\mathcal{B},\Theta_2\vdash\Lambda}{\Theta_1,\Gamma,\Theta_2\vdash\Delta_1,\mathcal{A}\supset\mathcal{B},\Delta_2,\Lambda}\,\text{cut}\quad\rightsquigarrow$$

$$\rightsquigarrow\quad\dfrac{\dfrac{\dfrac{\dfrac{\mathcal{A},\Gamma\vdash\Delta_1,\mathcal{A}\supset\mathcal{B},\Delta_2,\mathcal{B}\qquad \Theta_1,\mathcal{A}\supset\mathcal{B},\Theta_2\vdash\Lambda}{\Theta_1,\mathcal{A},\Gamma,\Theta_2\vdash\Delta_1,\Lambda,\Delta_2,\mathcal{B}}\,\text{cut}}{\mathcal{A},\Theta_1,\Gamma,\Theta_2\vdash\Delta_1,\Lambda,\Delta_2,\mathcal{B}}\,\text{C}\vdash\text{'s}}{\Theta_1,\Gamma,\Theta_2\vdash\Delta_1,\Lambda,\Delta_2,\mathcal{A}\supset\mathcal{B}}\,{\vdash\supset}}{\Theta_1,\Gamma,\Theta_2\vdash\Delta_1,\mathcal{A}\supset\mathcal{B},\Delta_2,\Lambda}\,{\vdash\text{C's}}$$

In the modified proof the cut has lower ϱ_l, hence, a lower ϱ.

(r3) $\langle(\vdash\exists),\_\rangle$. Let the left premise be obtained by the right introduction rule for the existential quantifier. The given proof and the transformed one are the following.

$$\dfrac{\dfrac{\Gamma\vdash\Delta_1,\exists x\,\mathcal{A}(x),\Delta_2,\mathcal{A}(y)}{\Gamma\vdash\Delta_1,\exists x\,\mathcal{A}(x),\Delta_2,\exists x\,\mathcal{A}(x)}\,{\vdash\exists}\qquad \Theta_1,\exists x\,\mathcal{A}(x),\Theta_2\vdash\Lambda}{\Theta_1,\Gamma,\Theta_2\vdash\Delta_1,\exists x\,\mathcal{A}(x),\Delta_2,\Lambda}\,\text{cut}\quad\rightsquigarrow$$

$$\rightsquigarrow\quad \dfrac{\dfrac{\dfrac{\begin{array}{c}\vdots\\ \Gamma \vdash \Delta_1, \exists x\, \mathcal{A}(x), \Delta_2, \mathcal{A}(y)\end{array} \qquad \begin{array}{c}\vdots\\ \Theta_1, \exists x\, \mathcal{A}(x), \Theta_2 \vdash \Lambda\end{array}}{\Theta_1, \Gamma, \Theta_2 \vdash \Delta_1, \Lambda, \Delta_2, \mathcal{A}(y)}\,\text{cut}}{\Theta_1, \Gamma, \Theta_2 \vdash \Delta_1, \Lambda, \Delta_2, \exists x\, \mathcal{A}(x)}\,\vdash\exists}{\Theta_1, \Gamma, \Theta_2 \vdash \Delta_1, \exists x\, \mathcal{A}(x), \Delta_2, \Lambda}\,\vdash C\text{'s}$$

The justification is exactly the same as in the previous case.

Now we consider some symmetric cases. The right premise is by a left rule, and the cut is performed on the principal formula of the rule.

(l1) $\langle\, \_, (\wedge\vdash_1)\rangle$. The $(\wedge\vdash)$ rule has two versions, depending on which conjunct is in the upper sequent, but the difference does not affect the transformation. Let the proof segment on the left be given. Its transformation is on the right.

$$\dfrac{\begin{array}{c}\vdots\\ \Gamma \vdash \Delta_1, \mathcal{A}\wedge\mathcal{B}, \Delta_2\end{array} \qquad \dfrac{\begin{array}{c}\vdots\\ \mathcal{A}, \Theta_1, \mathcal{A}\wedge\mathcal{B}, \Theta_2 \vdash \Lambda\end{array}}{\mathcal{A}\wedge\mathcal{B}, \Theta_1, \mathcal{A}\wedge\mathcal{B}, \Theta_2 \vdash \Lambda}\,\wedge\vdash_1}{\Gamma, \Theta_1, \mathcal{A}\wedge\mathcal{B}, \Theta_2 \vdash \Delta_1, \Lambda, \Delta_2}\,\text{cut}\quad\rightsquigarrow$$

$$\rightsquigarrow\quad \dfrac{\dfrac{\dfrac{\begin{array}{c}\vdots\\ \Gamma \vdash \Delta_1, \mathcal{A}\wedge\mathcal{B}, \Delta_2\end{array} \qquad \begin{array}{c}\vdots\\ \mathcal{A}, \Theta_1, \mathcal{A}\wedge\mathcal{B}, \Theta_2 \vdash \Lambda\end{array}}{\mathcal{A}, \Theta_1, \Gamma, \Theta_2 \vdash \Delta_1, \Lambda, \Delta_2}\,\text{cut}}{\mathcal{A}\wedge\mathcal{B}, \Theta_1, \Gamma, \Theta_2 \vdash \Delta_1, \Lambda, \Delta_2}\,\wedge\vdash_1}{\Theta_1, \mathcal{A}\wedge\mathcal{B}, \Gamma, \Theta_2 \vdash \Delta_1, \Lambda, \Delta_2}\,C\vdash\text{'s}$$

The cut and the $(\wedge\vdash)$ rules are swapped in the new proof, and this move subtracts 1 from the right rank of the cut.

(l2) $\langle\, \_, (\forall\vdash)\rangle$. This case is the dual of (r3) above. Given the proof chunk on the left-hand side, we get the modified proof on the right-hand side.

$$\dfrac{\begin{array}{c}\vdots\\ \Gamma \vdash \Delta_1, \forall x\, \mathcal{A}(x), \Delta_2\end{array} \qquad \dfrac{\begin{array}{c}\vdots\\ \mathcal{A}(y), \Theta_1, \forall x\, \mathcal{A}(x), \Theta_2 \vdash \Lambda\end{array}}{\forall x\, \mathcal{A}(x), \Theta_1, \forall x\, \mathcal{A}(x), \Theta_2 \vdash \Lambda}\,\forall\vdash}{\Gamma, \Theta_1, \forall x\, \mathcal{A}(x), \Theta_2 \vdash \Delta_1, \Lambda, \Delta_2}\,\text{cut}\quad\rightsquigarrow$$

$$\rightsquigarrow\quad \dfrac{\dfrac{\dfrac{\begin{array}{c}\vdots\\ \Gamma \vdash \Delta_1, \forall x\, \mathcal{A}(x), \Delta_2\end{array} \qquad \begin{array}{c}\vdots\\ \mathcal{A}(y), \Theta_1, \forall x\, \mathcal{A}(x), \Theta_2 \vdash \Lambda\end{array}}{\mathcal{A}(y), \Theta_1, \Gamma, \Theta_2 \vdash \Delta_1, \Lambda, \Delta_2}\,\text{cut}}{\forall x\, \mathcal{A}(x), \Theta_1, \Gamma, \Theta_2 \vdash \Delta_1, \Lambda, \Delta_2}\,\forall\vdash}{\Gamma, \Theta_1, \forall x\, \mathcal{A}(x), \Theta_2 \vdash \Delta_1, \Lambda, \Delta_2}\,C\vdash\text{'s}$$

There is a reduction in the right rank by 1, which justifies the modification.

(l3) $\langle\, \_, (C\vdash)\rangle$. This turns out to be the simplest case.

$$\dfrac{\begin{array}{c}\vdots\\ \Gamma \vdash \Delta_1, \mathcal{A}, \Delta_2\end{array} \qquad \dfrac{\begin{array}{c}\vdots\\ \Theta_1, \mathcal{A}, \mathcal{B}, \Theta_2, \mathcal{A}, \Theta_3 \vdash \Lambda\end{array}}{\Theta_1, \mathcal{B}, \mathcal{A}, \Theta_2, \mathcal{A}, \Theta_3 \vdash \Lambda}\,C\vdash}{\Theta_1, \mathcal{B}, \Gamma, \Theta_2, \mathcal{A}, \Theta_3 \vdash \Delta_1, \Lambda, \Delta_2}\,\text{cut}\quad\rightsquigarrow$$

$$\rightsquigarrow \quad \dfrac{\dfrac{\begin{array}{c}\vdots\\ \Gamma \vdash \Delta_1, \mathcal{A}, \Delta_2\end{array} \qquad \begin{array}{c}\vdots\\ \Theta_1, \mathcal{A}, \mathcal{B}, \Theta_2, \mathcal{A}, \Theta_3 \vdash \Lambda\end{array}}{\Theta_1, \Gamma, \mathcal{B}, \Theta_2, \mathcal{A}, \Theta_3 \vdash \Delta_1, \Lambda, \Delta_2}\,\text{cut}}{\Theta_1, \mathcal{B}, \Gamma, \Theta_2, \mathcal{A}, \Theta_3 \vdash \Delta_1, \Lambda, \Delta_2}\,\text{C}\vdash\text{'s}$$

Our notation does not allow us to capture this case in its full generality. However, it is not difficult to see that placing $\mathcal{A}, \mathcal{B}$ somewhere else in the antecedent of the right premise, or identifying the cut formula with the other principal formula (i.e., with $\mathcal{B}$) of the $(\text{C}\vdash)$ rule, can be dealt with similarly. A reduction in the right rank of the cut justifies the transformation.

Exercise 2.3.20. Convince yourself, by working out the details, that in the other subcases where the principal formula of a rule coincides with the cut formula, that there is a suitable transformation of the proof. [Hint: There are over 15 subcases to check.]

Exercise 2.3.21. Write out the subcases in which the cut formula cannot be the principal formula of the last rule. [Hint: For $\varrho_l > 1$, consider when the left premise is by a left rule, and dually for $\varrho_r > 1$.]

2.2.2 $\varrho > 2,\ \delta > 0,\ \mu > 0.$ This case is similar to **2.1.2** with the additional possibility that the principal formula of an operational rule occasionally may coincide with the cut formula. We consider two cases, which involve contraction with the principal formula being the cut formula.

(r1) $\langle(\vdash W),\ \_\,\rangle.$ If the principal formula of the contraction rule is distinct from the cut formula, then swapping the rules is all that is required. However, if the cut formula is the principal formula of $(\vdash W)$, then the given and the transformed proof segments are as follows.

$$\text{cut}\ \dfrac{\vdash W\ \dfrac{\begin{array}{c}\vdots\\ \Gamma \vdash \Delta_1, \mathcal{A}, \Delta_2, \mathcal{A}, \mathcal{A}\end{array}}{\Gamma \vdash \Delta_1, \mathcal{A}, \Delta_2, \mathcal{A}} \qquad \begin{array}{c}\vdots\\ \Theta_1, \mathcal{A}, \Theta_2 \vdash \Lambda\end{array}}{\Theta_1, \Gamma, \Theta_2 \vdash \Delta_1, \mathcal{A}, \Delta_2, \Lambda} \quad \rightsquigarrow$$

$$\rightsquigarrow \quad \dfrac{\dfrac{\text{cut}\ \dfrac{\begin{array}{c}\vdots\\ \Gamma \vdash \Delta_1, \mathcal{A}, \Delta_2, \mathcal{A}, \mathcal{A}\end{array} \qquad \begin{array}{c}\vdots\\ \Theta_1, \mathcal{A}, \Theta_2 \vdash \Lambda\end{array}}{\Theta_1, \Gamma, \Theta_2 \vdash \Delta_1, \mathcal{A}, \Delta_2, \Lambda, \mathcal{A}} \qquad \begin{array}{c}\vdots\\ \Theta_1, \mathcal{A}, \Theta_2 \vdash \Lambda\end{array}}{\Theta_1, \Theta_1, \Gamma, \Theta_2, \Theta_2 \vdash \Delta_1, \mathcal{A}, \Delta_2, \Lambda, \Lambda}\,\text{cut}}{\Theta_1, \Gamma, \Theta_2 \vdash \Delta_1, \mathcal{A}, \Delta_2, \Lambda}\,\text{C}\vdash\text{'s, W}\vdash\text{'s, }\vdash\text{C's, }\vdash\text{W's}$$

Exercise 2.3.22. The above proof segment on the left shows an occurrence of $\mathcal{A}$ that is not affected by the contraction rule. If there is such an $\mathcal{A}$ in the left premise, then there is a different transformation that also works, because it can be justified by a reduction in ϱ. [Hint: It does not require left contractions and includes one application of the $(\vdash W)$ rule.]

(11) $\langle\, \_, (W\vdash) \rangle$. Let us assume that the principal formula of the contraction rule is the same as the cut formula.

$$\dfrac{\begin{matrix}\vdots\\ \Gamma \vdash \Delta_1, \mathcal{A}, \Delta_2\end{matrix} \qquad \dfrac{\begin{matrix}\vdots\\ \mathcal{A}, \mathcal{A}, \Theta_1, \mathcal{A}, \Theta_2 \vdash \Lambda\end{matrix}}{\mathcal{A}, \Theta_1, \mathcal{A}, \Theta_2 \vdash \Lambda}\, W\vdash}{\Gamma, \Theta_1, \mathcal{A}, \Theta_2 \vdash \Delta_1, \Lambda, \Delta_2}\,\text{cut} \quad \rightsquigarrow$$

$$\rightsquigarrow \quad \dfrac{\dfrac{\begin{matrix}\vdots\\ \Gamma \vdash \Delta_1, \mathcal{A}, \Delta_2\end{matrix} \qquad \dfrac{\begin{matrix}\vdots\\ \Gamma \vdash \Delta_1, \mathcal{A}, \Delta_2\end{matrix} \qquad \begin{matrix}\vdots\\ \mathcal{A}, \mathcal{A}, \Theta_1, \mathcal{A}, \Theta_2 \vdash \Lambda\end{matrix}}{\Gamma, \mathcal{A}, \Theta_1, \mathcal{A}, \Theta_2 \vdash \Delta_1, \Lambda, \Delta_2}\,\text{cut}}{\Gamma, \Gamma, \Theta_1, \mathcal{A}, \Theta_2 \vdash \Delta_1, \Delta_1, \Lambda, \Delta_2, \Delta_2}\,\text{cut}}{\Gamma, \Theta_1, \mathcal{A}, \Theta_2 \vdash \Delta_1, \Lambda, \Delta_2}\, C\vdash\text{'s}, W\vdash\text{'s}, \vdash C\text{'s}, \vdash W\text{'s}$$

Both cuts have lower contraction measure than the original cut.

Exercise 2.3.23. Scrutinize Case 2.1.2, and convince yourself that viewing the atomic formula p as the compound wff $\mathcal{A}$ does not affect the suitability of the transformations. Then, take another look at the cases, and work out what happens when the cut formula is principal (if that possibility is not excluded).

This completes the proof. qed

Exercise 2.3.24. Having the details of the triple inductive proof of cut, explain why a double induction (without μ) works for mix but not for cut_G.

The transformations in the proof can be viewed as a non-deterministic algorithm, because if $\varrho_l > 1$ and $\varrho_r > 1$, then no order is specified on applicable subcases. This is not how Gentzen specified his double induction in the proof of the admissibility of the mix rule. In the case of LK, we could require that ϱ_r must be reduced first, and then the assumption $\varrho_l > 1$ may be strengthened with $\varrho_r = 1$. The latter approach is advantageous in some other sequent calculi, which have exactly one formula on the right-hand side of the $\vdash$ and have a richer language than classical logic.

2.4 Interpretations, soundness and completeness

The previous sections have not given a rigorous interpretation of the formulas and sequents. We used notation that may be familiar from other presentations of classical logic or suggestive of the intended meaning of the formal expressions, but it is time to have a formal interpretation in place. The interpretation that we present follows the approach that is usual nowadays:

it defines inductively the notion of a structure (also called a *model*) *satisfying* (or making true) a formula. Then we extend this notion to sequents in a straightforward way, which is in accordance with viewing the comma as $\wedge$ or $\vee$—as in Definition 2.18 or in the informal sense we introduced earlier on page 20.

Definition 2.28. (Classical models) A classical *model* $\mathfrak{M}$ is a triple $\langle D, I, v\rangle$, where D is a non-empty set of objects, the *domain of interpretation*; I is a function, the *interpretation function* satisfying conditions (1)–(4) below; and v is a function, the *valuation function* that interprets individual variables into D.

(1) If t is a name constant, then $I(t) \in D$;

(2) if f^n is an n-ary function symbol, then $I(f) \in D^{D^n}$;

(3) if p is a propositional variable, then $I(p) \in \{T, F\}$ (where $T, F \notin D$ and they stand for the two truth values);

(4) if P^n is an n-place predicate, then $I(P) \subseteq D^n$.

This definition does not specify the cardinality of the domain of interpretation D, which is sometimes called the *universe of discourse*. Nor does it define a unique structure; in other words, there is more than one, indeed, there are infinitely many models—though some wff's are true only in finitely many models. The variability of the interpretations of non-logical components is in contrast with the interpretation of the logical particles, which are fixed by Definition 2.30 below.

In order to make the definition of satisfaction smooth, we introduce a function, denoted by i, which gives the *denotation* of a term. An atomic term can be a constant or a variable, and a function symbol can take either kind of atomic term as its argument, as well as it can have a complex term as its argument. Clearly, I and v are sufficient to compute the denotation of a term in a bottom-up fashion, thus i's role is merely to make the presentation easier rather than to add something new to an interpretation.

Definition 2.29. (Denotation of terms) If t is a term, then the *denotation of* t, denoted by $\mathrm{i}(t)$, is determined by a classical model $\mathfrak{M} = \langle D, I, v\rangle$ in accordance with (1)–(3).

(1) If t is a name constant, then $\mathrm{i}(t) = I(t)$;

(2) if t is a variable, then $\mathrm{i}(t) = v(t)$;

(3) if t is of the form $f^n(t_1, \ldots, t_n)$, where $t_1, \ldots, t_n$ are terms and f^n is a function symbol, then $\mathrm{i}(f(t_1, \ldots, t_n)) = I(f)(\mathrm{i}(t_1), \ldots, \mathrm{i}(t_n))$.

It is an easy induction on the structure of terms to prove that i is well-defined, that is, i is a total function on the set of terms, and maps each term into an element of D.

We introduce a further notational device to indicate the *pointwise modification* (including null or no modification) for valuation functions. By $v[x : d]$ (where $d \in D$) we mean a function such that for any y that is (syntactically) not identical to the variable x, $v(y) = v[x : d](y)$, whereas $v[x : d](x) = d$. Obviously, if $v(x) = d$, then $v[x : d]$ does not literally modify v at all, and we labeled this as null modification. Otherwise (i.e., when $v(x) \neq d$), $v[x : d]$ differs from v in exactly one place, in the value for x.

Definition 2.30. (Truth in classical models) Let $\mathfrak{M} = \langle D, I, v\rangle$ be a classical model. The relation $\mathfrak{M} \vDash_v \mathcal{A}$ is defined recursively by (1)–(10). (We omit $\mathfrak{M}$, as well as $_v$, when it is unaltered in a clause.)

(1) $\vDash \boldsymbol{T}$ and $\nvDash \boldsymbol{F}$ (i.e., $\boldsymbol{F}$ is never satisfied);

(2) $\vDash p$ iff $I(p) = T$;

(3) $\vDash t_1 = t_2$ iff $\mathrm{i}(t_1) = \mathrm{i}(t_2)$;

(4) if P^n is an n-place predicate (other than $=$) and $t_1, \ldots, t_n$ are terms, then $\vDash P(t_1, \ldots, t_n)$ iff $\langle \mathrm{i}(t_1), \ldots, \mathrm{i}(t_n)\rangle \in I(P)$;

(5) $\vDash \neg\mathcal{A}$ iff $\nvDash \mathcal{A}$, that is, not $\vDash \mathcal{A}$;

(6) $\vDash \mathcal{A} \wedge \mathcal{B}$ iff $\vDash \mathcal{A}$ and $\vDash \mathcal{B}$;

(7) $\vDash \mathcal{A} \vee \mathcal{B}$ iff $\vDash \mathcal{A}$ or $\vDash \mathcal{B}$;

(8) $\vDash \mathcal{A} \supset \mathcal{B}$ iff $\nvDash \mathcal{A}$ or $\vDash \mathcal{B}$;

(9) $\vDash_v \forall x\, \mathcal{A}(x)$ iff for any $v[x : d]$, $\vDash_{v[x:d]} \mathcal{A}(x)$;

(10) $\vDash_v \exists x\, \mathcal{A}(x)$ iff there is a $v[x : d]$ such that $\vDash_{v[x:d]} \mathcal{A}(x)$.

$\mathfrak{M} \vDash_v \mathcal{A}$ is read as "the model $\mathfrak{M}$ with the valuation v *satisfies* (or makes *true*) the wff $\mathcal{A}$." In most of the clauses, $_v$ may be left implicit, because $\mathfrak{M}$ includes v. The definition makes clear that given a model $\mathfrak{M}$, the truth of a formula $\mathcal{A}$ is completely determined by $\mathfrak{M}$ itself. Although v's pointwise modification must be considered in (9) and (10), which yield models of the form $\mathfrak{M}' = \langle D, I, v[x : d]\rangle$, the d's are chosen from D, which is a component of $\mathfrak{M}$. (9) and (10) differ from the rest of the clauses, because they show that quantification is not compositional. In the case of quantificational formulas, the semantic values of the component expressions are not sufficient to compute the semantic value of the complex formula.

Exercise 2.4.1. Are the next four formulas satisfiable or not? (a) $(\exists x\, \forall y\, R(x, y) \supset \forall y\, \exists x\, R(x, y))$, (b) $(\forall x\, (P(x) \supset \exists y\, P(y)) \supset (\exists x\, P(x) \supset \exists y\, P(y)))$, (c) $(\forall x\, (\forall x\, \neg P(x) \supset Q(x)) \supset \forall y\, (\neg Q(y) \supset P(y)))$, (d) $(\forall x\, \forall y\, \forall z\, ((R(x, y) \wedge R(y, z)) \supset R(x, z)) \wedge (\forall x\, \exists y\, R(x, y) \wedge \forall x\, \forall y\, (R(x, y) \supset \neg R(y, x))))$. [Hint: Use the semantic notions introduced so far to substantiate your claims.]

Definition 2.31. (Validity in classical models) $\mathcal{A}$ is *logically valid* (or *valid*, for short) in classical models iff $\mathfrak{M} \models_v \mathcal{A}$ for all $\mathfrak{M}$, that is, the wff $\mathcal{A}$ is true in all classical models.

The definition can be rephrased—equivalently—to state that $\mathcal{A}$ is valid when given any domain D and any interpretation function I, all valuations v make $\mathcal{A}$ true.

Exercise 2.4.2. Determine if the following wff's are logically valid or not. (a) $(\exists x\,(P(x) \wedge \forall y\,(Q(y) \supset R(x,y))) \supset \forall x\,(Q(x) \supset \exists y\,(P(y) \wedge R(y,x))))$, (b) $(\forall x\,(P(x) \supset Q(x)) \supset \exists y\,(P(y) \wedge Q(y)))$, (c) $((\forall x\,P(x) \vee \forall y\,Q(y)) \supset \forall z\,(\neg P(z) \supset Q(z)))$, (d) $\exists x\,\exists y\,(x = y \wedge (P(x) \supset P(y)))$. [Hint: The negation of a valid wff is not satisfiable.]

Definition 2.18 allows us to view a sequent as a wff. If we think of $\mathcal{A}$ in clauses (1) and (4) of the definition as a fixed wff, then τ is a *function*, that is, every sequent is mapped into a unique formula.

Exercise 2.4.3. Prove that τ is a well-defined function, that is, for each sequent $\Gamma \vdash \Delta$, $\tau(\Gamma \vdash \Delta)$ is a wff, and if $\tau(\Gamma \vdash \Delta)$ is the same wff as $\tau(\Gamma' \vdash \Delta')$, then $\Gamma \vdash \Delta$ is $\Gamma' \vdash \Delta'$.

The soundness and completeness theorems guarantee that the proof system is adequate with respect to the intended interpretation. Our focus in this book is on proof systems; hence, we only outline the proofs and do not give many details.

Theorem 2.32. (Soundness, 1) *If $\Gamma \vdash \Delta$ is a provable sequent, then $\tau(\Gamma \vdash \Delta)$ is a* logically valid *formula.*

Proof: We outline the structure of the proof and detail two steps. The rest of the proof is relegated to an exercise.

The proof is by induction on the height of the proof tree. If the height is 1, then the sequent is provable, because it is an instance of the axiom. Otherwise, the bottom sequent in the proof is by a rule, and one has to consider the possible cases rule by rule.

1. If the sequent is of the form $\mathcal{A} \vdash \mathcal{A}$, then $\tau(\mathcal{A} \vdash \mathcal{A})$ is $\mathcal{A} \supset \mathcal{A}$. For this wff not to be logically valid, there should be a model $\mathfrak{M}$ such that $\mathfrak{M} \not\models_v \mathcal{A} \supset \mathcal{A}$. The latter, by clause (8) from Definition 2.30, means that $\mathfrak{M} \models_v \mathcal{A}$ and $\mathfrak{M} \not\models_v \mathcal{A}$. This is obviously impossible.
2. Let us assume that the last rule applied in the proof is $(K\vdash)$, in particular, let the shape of the lower sequent be $\mathcal{A}, \Gamma \vdash \Delta$. The hypothesis of the induction is that $\tau(\Gamma \vdash \Delta)$ is valid. By Definition 2.18, $\tau(\Gamma \vdash \Delta)$ is $\mathcal{C} \supset \mathcal{D}$, and $\tau(\mathcal{A}, \Gamma \vdash \Delta)$ is $(\mathcal{A} \wedge \mathcal{C}) \supset \mathcal{D}$. If $\mathfrak{M} \not\models_v (\mathcal{A} \wedge \mathcal{C}) \supset \mathcal{D}$, for some $\mathfrak{M}$, then $\mathfrak{M} \models_v \mathcal{A} \wedge \mathcal{C}$ but $\mathfrak{M} \not\models_v \mathcal{D}$. However, then also $\mathfrak{M} \models_v \mathcal{C}$, which is a contradiction. qed

Exercise 2.4.4. There are 19 other rules to consider (if we count $(\wedge\vdash)$ and $(\vdash\vee)$ each as two rules). [Hint: Each step is straightforward.]

Informally, the idea behind the soundness proof is that an instance of the axiom is *logically valid*, and the rules—whether operational or structural—*preserve validity*.

The soundness theorem is often stated by saying that theorems are valid. Definition 2.9 introduced the notion of theoremhood into LK.

Theorem 2.33. (Soundness, 2) *If $\mathcal{A}$ is a theorem of LK, then $\mathcal{A}$ is* valid.

Exercise* 2.4.5. Prove the previous theorem. [Hint: Notice that no instance of the axiom—without further applications of rules—yields a theorem.]

The converse of the soundness theorem is the completeness theorem.

Theorem 2.34. (Completeness) *If $\mathcal{A}$ is a logically valid formula, then $\vdash \mathcal{A}$ is a* provable sequent.

Proof: We sketch the proof of the contrapositive of the claim, that is, if $\mathcal{A}$ is not a theorem of LK, then there is a model $\mathfrak{M}$ such that $\mathfrak{M} \nvDash \mathcal{A}$.

We outline some steps, and leave filling in the details as exercises. If there are free variables in $\mathcal{A}$, then we change each of them to a new name constant, and from now on, we take $\mathcal{A}$ to be this closed wff. We also exclude function symbols and $=$ from the language in order to provide a brief general description of the construction. We take $\forall$ as a defined logical constant, that is, $\forall x\, \mathcal{B}(x)$ is $\neg\exists x\, \neg\mathcal{B}(x)$. We consider sets of wff's starting with $\{\neg\mathcal{A}\}$. This is obviously a finite set, hence there are finitely many non-logical symbols that occur in $\neg\mathcal{A}$. Nonetheless, there are denumerably many closed formulas that can be generated using those non-logical components. We fix an enumeration of those wff's. Originally, we assumed that we have a denumerable sequence of name constants in the language; now, we take a "fresh" sequence of name constants, let us say, $b_0, b_1, b_2, \ldots$. These name constants enter into terms and wff's exactly as the name constants, which are in the original language, do.

The following construction is usually called *Lindenbaumizing*.[11] Starting with a set of wff's Γ_0 (i.e., $\{\neg\mathcal{A}\}$), we consider the wff's that we stipulated to be enumerated, and at each stage we expand the set if we can do so without introducing a contradiction.

Some of the enumerated wff's are of the form $\exists x\, \mathcal{C}(x)$, and we want to make sure that each of these formulas is instantiated. We define $\mathcal{C}^{b/x}$ to be $\mathcal{C}$, if $\mathcal{C}$ is not an existentially quantified formula. Otherwise, $\mathcal{C}^{b/x}$ is the formula obtained from $\mathcal{C}$ by omitting the quantifier prefix and substituting for the previously quantified variable the next unused name constant from the sequence of b's.

[11]The verb is derived from the name of A. Lindenbaum.

1. $\Gamma_{n+1} = \begin{cases} \Gamma_n \cup \{\mathcal{B}_{n+1}, \mathcal{B}^{b/x}_{n+1}\}, & \text{if } \Gamma_n \vdash \neg\mathcal{B}_{n+1} \text{ is not provable;} \\ \Gamma_n, & \text{otherwise.} \end{cases}$

2. $\Gamma = \bigcup_{n \in \omega} \Gamma_n$

The construction *preserves consistency*, that is, if there was a wff $\mathcal{E}$ such that $\Gamma_0 \vdash \mathcal{E}$ was not provable, then there is no $\Gamma' \subseteq \Gamma$ such that $\Gamma' \vdash \mathcal{E}'$ is provable, for all $\mathcal{E}'$. The starting set $\{\neg\mathcal{A}\}$ also has this property, because if $\neg\mathcal{A} \vdash \mathcal{B}$, for any $\mathcal{B}$, then $\vdash \mathcal{A}$ is provable.

The next step is to use the set Γ to *define a model* $\mathfrak{M}$ with the property that—due to the definition of $\mathfrak{M}$ itself—$\mathfrak{M} \vDash \mathcal{C}$, for all $\mathcal{C} \in \Gamma$. The wff's in Γ (may) contain some name constants, and if they do, then we take D to be the set of those name constants. If no wff in Γ contains a name constant, then we take an object d to be the only element of D.

Lastly, it remains to be shown that $\mathfrak{M}$ indeed has the property that it makes all the elements of Γ *true*. qed

The next exercises ask you to fill in the details in the above proof sketch.

Exercise 2.4.6. Prove that if $\mathcal{A}$ is not a theorem of LK, then $\{\neg\mathcal{A}\}$ is consistent. [Hint: Find a suitable wff in place of $\mathcal{B}$, and show that $\vdash \neg\neg\mathcal{A}$ is provable, hence, by cut, $\vdash \mathcal{A}$ is provable.]

Exercise 2.4.7. Show that Γ, obtained by Lindenbaumizing, is consistent, provided that Γ_0 is consistent.

Exercise 2.4.8. Make the definition of $\mathfrak{M}$ precise. [Hint: The atomic formulas determine the model, because Γ contains all the formulas that it can, in the restricted fragment of the language.]

Exercise 2.4.9. Prove that the model $\mathfrak{M}$ defined in the previous exercise has the desired property, that is, for all $\mathcal{C} \in \Gamma$, $\mathfrak{M} \vDash \mathcal{C}$.

Chapter 3

Variants of the first sequent calculi

Classical and intuitionistic logics have a certain simplicity, which makes them amenable to variations. For instance, the algebra of classical propositional logic may be presented in different signatures, even with a single binary operator. Similarly, the *sequent calculi* LK and LJ permit a lot of variations—omissions and additions alike.

The shape of the connective rules in Gentzen's calculi is somewhat arbitrary, if we drop the idea of establishing a close connection between LK and LJ. The rules can be reshaped in several ways: to make them to adhere to some general pattern or to have desirable properties such as invertibility.

Originally, sequents are pairs of sequences of wff's, however, this can be changed too. This can be also viewed as a way of *hiding* some *structural rules*, which can also be incorporated into the axiom or the operational rules.

The negation in classical logic obeys the De Morgan principles. This leads to the well-known duality between some pairs of connectives and between the quantifiers, but this also means that we could accumulate all the formulas on one side of the $\vdash$.

This chapter overviews a range of these *variations*, and will also serve as a preparation for the next several chapters in terms of exploring possibilities and their repercussions. To conclude our more or less tandem consideration of classical and intuitionistic logics, we conclude this chapter with the *disjunction property* and some *translations* between classical logic and intuitionistic logic.

3.1 Intuitionistic logic and other modifications

The first modification of LK that we consider is the sequent calculus LJ for *intuitionistic logic*. Intuitionistic logic—as the logic behind intuitionism or Brouwer's approach to the foundations of mathematics—was formalized by Heyting by omitting some classical theorems. Gentzen aimed at formalizing classical and intuitionistic logics at once, and in fact, designed LK so that LJ can be obtained from it by a straightforward restriction.

The *sequent calculus* LJ is defined by the same axiom and rules—save

$(\vdash W)$ and $(\vdash C)$—as LK, but with the restriction that each sequent may contain *at most one formula* on the right-hand side of the $\vdash$. Right contraction and permutation is excluded simply because, by their form, the rules violate the restriction concerning the number of formulas in the succedent.

The *single cut rule* takes the following form, where X is a wff or the empty sequence of wff's, and $[\mathcal{C}]$ is a particular occurrence of $\mathcal{C}$.

$$\frac{\Gamma \vdash \mathcal{C} \quad \Delta[\mathcal{C}] \vdash X}{\Delta[\Gamma] \vdash X}\ \text{cut}$$

Theorem 3.1. *The single cut rule is* admissible *in LJ.*

Exercise 3.1.1. Prove the theorem. [Hint: The triple inductive proof in Chapter 2 may be adapted to LJ.]

Intuitionistic logic has been thoroughly investigated since its invention in 1930, and we cannot and do not intend to give a comprehensive study of intuitionistic logic. A natural deduction formulation of intuitionistic logic is presented in Section 8.1. We give an axiomatic formulation here to give a sense of how much more complex intuitionistic logic is (than classical logic).

Classical logic may be formalized very concisely, for example, as in Section 2.2. Intuitionistic logic may be obtained by omitting certain axioms from an axiomatization of classical logic, however, the axiom system has to contain a more versatile collection of axioms than K does.

The *axiomatic calculus* J for intuitionistic logic comprises the following axioms and rules. (Parentheses are omitted from right-associated implications.)

(A1) $\mathcal{A} \supset \mathcal{B} \supset \mathcal{A}$

(A2) $(\mathcal{A} \supset \mathcal{B} \supset \mathcal{C}) \supset (\mathcal{A} \supset \mathcal{B}) \supset \mathcal{A} \supset \mathcal{C}$

(A3) $(\mathcal{A} \wedge \mathcal{B}) \supset \mathcal{A}$

(A4) $(\mathcal{A} \wedge \mathcal{B}) \supset \mathcal{B}$

(A5) $(\mathcal{A} \supset \mathcal{B}) \supset (\mathcal{A} \supset \mathcal{C}) \supset \mathcal{A} \supset (\mathcal{B} \wedge \mathcal{C})$

(A6) $\mathcal{A} \supset (\mathcal{A} \vee \mathcal{B})$

(A7) $\mathcal{B} \supset (\mathcal{A} \vee \mathcal{B})$

(A8) $(\mathcal{A} \supset \mathcal{B}) \supset (\mathcal{C} \supset \mathcal{B}) \supset (\mathcal{A} \vee \mathcal{C}) \supset \mathcal{B}$

(A9) $(\mathcal{A} \supset \neg\mathcal{B}) \supset \mathcal{B} \supset \neg\mathcal{A}$

(A10) $\mathcal{A} \supset \neg\mathcal{A} \supset \mathcal{B}$

(A11) $\forall x\, \mathcal{A}(x) \supset \mathcal{A}(y)$, where y is OK for substitution for x in $\mathcal{A}(x)$

(A12) $\forall x\, (\mathcal{A} \supset \mathcal{B}(x)) \supset \mathcal{A} \supset \forall x\, \mathcal{B}(x)$, where x is not free in $\mathcal{A}$

(A13) $\mathcal{A}(y) \supset \exists x\, \mathcal{A}(x)$, where y is OK for substitution for x in $\mathcal{A}(x)$

(A14) $\forall x\, (\mathcal{A}(x) \supset \mathcal{B}) \supset \exists x\, \mathcal{A}(x) \supset \mathcal{B}$, where x is not free in $\mathcal{B}$

(R1) $\mathcal{A} \supset \mathcal{B}$ and $\mathcal{A}$ imply $\mathcal{B}$

(R2) $\vdash \mathcal{A}$ implies $\vdash \forall x\, \mathcal{A}$

The equivalence of the calculus J and LJ can be established like the equivalence of K and LK. We will not go into those details, or into interpretations of intuitionistic logic (of which there is a wide variety).

Exercise 3.1.2. Prove that all the axioms of J are theorems of LJ, and that (R1) and (R2) preserve theoremhood. [Hint: The cut rule is admissible in LJ, by the previous exercise.]

Exercise* 3.1.3. Prove that if $\vdash \mathcal{A}$ is provable in LJ, then $\mathcal{A}$ is a theorem of J.

Intuitionistic logic, obviously, differs from classical logic, which is a well-known fact nowadays. This also means that the connectives and quantifiers in the two logics are *not the same* either—despite that following Gentzen and many other authors, we used the same set of symbols in the languages of the two logics. A pair of prototypical theorems of classical logic, (the analogs of) which are not theorems of LJ are $\mathcal{A} \vee \neg\mathcal{A}$ and $\neg\neg\mathcal{A} \supset \mathcal{A}$.

Exercise* 3.1.4. Prove that $\mathcal{A} \vee \neg\mathcal{A}$ and $\neg\neg\mathcal{A} \supset \mathcal{A}$ are not provable in LJ. [Hint: There are many ways to prove this. You may use the results of Section 3.6 for the former, and those of Section 9.1.1 for the latter.]

Not all modifications to a sequent calculus change the logic, that is, the set of provable theorems. Given that sequent calculi have to be pieced together carefully, there is considerable interest in modifications that support the same set of theorems, and possibly, bestow certain desirable properties on the system. First, we overview some modifications of the classical and of the intuitionistic sequent calculi, which can be attributed to O. Ketonen, H. B. Curry and S. C. Kleene.

Ketonen in his thesis of 1944 introduced what later on became known as the "Ketonen rules" for conjunction on the left $(\wedge\vdash)$ and for disjunction on the right $(\vdash\vee)$. He also modified the left horseshoe rule $(\supset\vdash)$ to have shared contexts. The formulation of the $(\supset\vdash)$ rule in Gentzen's LK was accommodating LJ as a simple restriction of LK. Here are the three new rules.

$$\frac{\mathcal{A}, \mathcal{B}, \Gamma \vdash \Delta}{\mathcal{A} \wedge \mathcal{B}, \Gamma \vdash \Delta}\wedge\vdash \qquad \frac{\Gamma \vdash \Delta, \mathcal{A}, \mathcal{B}}{\Gamma \vdash \Delta, \mathcal{A} \vee \mathcal{B}}\vdash\vee \qquad \frac{\Gamma \vdash \Delta, \mathcal{A} \qquad \mathcal{B}, \Gamma \vdash \Delta}{\mathcal{A} \supset \mathcal{B}, \Gamma \vdash \Delta}\supset\vdash$$

Exercise 3.1.5. Prove that $(\wedge\vdash)$ is equivalent to the two original left introduction rules; similarly, prove the appropriate equivalences for the two other new rules. [Hint: You can establish that all the rules are derivable, when their counterparts are taken to be primitive.]

An advantage of this formulation is that all the rules for the connectives ($\neg$, $\wedge$, $\vee$ and $\supset$) are invertible. A rule is *invertible* iff the upper sequent(s) of the rule can be derived from the lower sequent together with (1) instances of the axiom, (2) applications of the structural rules and of the cut rule, and (3) the other introduction rule(s) for the same connective.

Example 3.2. We give a proof segment that demonstrates the invertibility of the $(\wedge\vdash)$ rule.

$$\dfrac{\dfrac{\dfrac{\dfrac{\mathcal{A}\vdash\mathcal{A}}{\mathcal{B},\mathcal{A}\vdash\mathcal{A}}\,K\vdash}{\mathcal{A},\mathcal{B}\vdash\mathcal{A}}\,C\vdash \qquad \dfrac{\mathcal{B}\vdash\mathcal{B}}{\mathcal{A},\mathcal{B}\vdash\mathcal{A}}\,K\vdash}{\mathcal{A},\mathcal{B}\vdash\mathcal{A}\wedge\mathcal{B}}\,\vdash\wedge \qquad \begin{array}{c}\vdots\\ \mathcal{A}\wedge\mathcal{B},\Gamma\vdash\Delta\end{array}}{\mathcal{A},\mathcal{B},\Gamma\vdash\Delta}\,\text{cut}$$

Exercise 3.1.6. Prove that the other connective rules are invertible too.

Curry noted that in the negation-free propositional fragment of LJ, it is sufficient to impose a restriction on the right-hand side of sequents in $(\vdash\supset)$ in order to obtain LJ_+. If negation would be included too, then the right introduction rule have to be restricted similarly. The rules are as follows.

$$\dfrac{\Gamma\vdash\Delta,\mathcal{A} \quad \mathcal{B},\Gamma\vdash\Delta}{\mathcal{A}\supset\mathcal{B},\Gamma\vdash\Delta}\,\supset\vdash \qquad\qquad \dfrac{\mathcal{A},\Gamma\vdash\mathcal{B}}{\Gamma\vdash\mathcal{A}\supset\mathcal{B}}\,\vdash\supset$$

$$\dfrac{\Gamma\vdash\Delta,\mathcal{A}}{\neg\mathcal{A},\Gamma\vdash\Delta}\,\neg\vdash \qquad\qquad \dfrac{\mathcal{A},\Gamma\vdash}{\Gamma\vdash\neg\mathcal{A}}\,\vdash\neg$$

Exercise 3.1.7. Scrutinize where the proofs (in LK) of $((\mathcal{A}\supset\mathcal{B})\supset\mathcal{A})\supset\mathcal{A}$, $\neg\neg\mathcal{A}\supset\mathcal{A}$ and $\mathcal{A}\vee\neg\mathcal{A}$ fail, if the rules are modified as above.

Let a sequent calculus for the propositional fragment of intuitionistic logic be formulated by the LK rules for $\wedge$ and $\vee$, with the above rules for $\neg$ and $\supset$—together with the three pairs of structural rules. We denote this calculus as LJ'.

Exercise* 3.1.8. Investigate whether the cut rule is admissible in LJ', and if it is, then prove that every theorem of propositional J is a theorem of LJ'.

Kleene, in [113], introduced the first sequent calculi for classical and intuitionistic logics that do not contain any structural rules. His calculi, which we denote by LK_3 and LJ_3, differ from those in Section 3.3 below.[1] In particular, LK_3 and LJ_3 build *thinning into the axiom*.

Definition 3.3. (Cognate sequents) Let Γ, Γ', Δ and Δ' be sequences of wff's. $\Gamma\vdash\Delta$ and $\Gamma'\vdash\Delta'$ are *cognate sequents* iff exactly the same formulas occur in Γ and Γ', as well as in Δ and Δ'. (The number of occurrences of a formula in Γ may differ from that in Γ'; similarly, for Δ and Δ'.)

[1] Kleene labeled both calculi as $G3$, because his numbering was to indicate the way the calculus is formulated, rather than the logic it formalizes. We do not follow his labeling or notation.

Kleene's calculi include the principal formulas in the premises (whenever possible), which leads to operational rules that might even appear pointless at first. Kleene also stipulated that in an application of a rule, a sequent may be replaced by another sequent as long as the sequents are cognate.

The axiom and the rules for the *sequent calculus* LK_3 are as follows.

$$\mathcal{A}, \Gamma \vdash \Delta, \mathcal{A} \text{ id}$$

$$\frac{\mathcal{A}, \mathcal{A} \wedge \mathcal{B}, \Gamma \vdash \Delta}{\mathcal{A} \wedge \mathcal{B}, \Gamma \vdash \Delta} \wedge\vdash_1 \qquad \frac{\mathcal{B}, \mathcal{A} \wedge \mathcal{B}, \Gamma \vdash \Delta}{\mathcal{A} \wedge \mathcal{B}, \Gamma \vdash \Delta} \wedge\vdash_2$$

$$\frac{\mathcal{A}, \mathcal{A} \vee \mathcal{B}, \Gamma \vdash \Delta \quad \mathcal{B}, \mathcal{A} \vee \mathcal{B} \vdash \Delta}{\mathcal{A} \vee \mathcal{B}, \Gamma \vdash \Delta} \vee\vdash \qquad \frac{\Gamma \vdash \Delta, \mathcal{A} \wedge \mathcal{B}, \mathcal{A} \quad \Gamma \vdash \Delta, \mathcal{A} \wedge \mathcal{B}, \mathcal{B}}{\Gamma \vdash \Delta, \mathcal{A} \wedge \mathcal{B}} \vdash\wedge$$

$$\frac{\Gamma \vdash \Delta, \mathcal{A} \vee \mathcal{B}, \mathcal{A}}{\Gamma \vdash \Delta, \mathcal{A} \vee \mathcal{B}} \vdash\vee_1 \qquad \frac{\Gamma \vdash \Delta, \mathcal{A} \vee \mathcal{B}, \mathcal{B}}{\Gamma \vdash \Delta, \mathcal{A} \vee \mathcal{B}} \vdash\vee_2$$

$$\frac{\neg\mathcal{A}, \Gamma \vdash \Delta, \mathcal{A}}{\neg\mathcal{A}, \Gamma \vdash \Delta} \neg\vdash \qquad \frac{\mathcal{A}, \Gamma \vdash \Delta, \neg\mathcal{A}}{\Gamma \vdash \Delta, \neg\mathcal{A}} \vdash\neg$$

$$\frac{\mathcal{A} \supset \mathcal{B}, \Gamma \vdash \Delta, \mathcal{A} \quad \mathcal{B}, \mathcal{A} \supset \mathcal{B}, \Gamma \vdash \Delta}{\mathcal{A} \supset \mathcal{B}, \Gamma \vdash \Delta} \supset\vdash \qquad \frac{\mathcal{A}, \Gamma \vdash \Delta, \mathcal{A} \supset \mathcal{B}, \mathcal{B}}{\Gamma \vdash \Delta, \mathcal{A} \supset \mathcal{B}} \vdash\supset$$

$$\frac{\mathcal{A}(y), \forall x\, \mathcal{A}(x), \Gamma \vdash \Delta}{\forall x\, \mathcal{A}(x), \Gamma \vdash \Delta} \forall\vdash \qquad \frac{\Gamma \vdash \Delta, \forall x\, \mathcal{A}(x), \mathcal{A}(y)}{\Gamma \vdash \Delta, \forall x\, \mathcal{A}(x)} \vdash\forall^{\oslash}$$

$$\frac{\mathcal{A}(y), \exists x\, \mathcal{A}(x), \Gamma \vdash \Delta}{\exists x\, \mathcal{A}(x), \Gamma \vdash \Delta} \exists\vdash^{\oslash} \qquad \frac{\Gamma \vdash \Delta, \exists x\, \mathcal{A}(x), \mathcal{A}(y)}{\Gamma \vdash \Delta, \exists x\, \mathcal{A}(x)} \vdash\exists$$

The superscript $^{\oslash}$ indicates the same variable restriction as in LK, namely, y cannot occur free in the lower sequent. The notions of a *proof* and of a *theorem* are unchanged.

The motif in the rules is the presence of an occurrence of a formula that is the principal formula of the rule in the premises. This clearly cannot be transposed onto all rules in the intuitionistic calculus.

The axiom and rules for the *sequent calculus* LJ_3 are the following.

$$\mathcal{A}, \Gamma \vdash \mathcal{A} \text{ id}$$

$$\frac{\mathcal{A}, \mathcal{A} \wedge \mathcal{B}, \Gamma \vdash X}{\mathcal{A} \wedge \mathcal{B}, \Gamma \vdash X} \wedge\vdash_1 \qquad \frac{\mathcal{B}, \mathcal{A} \wedge \mathcal{B}, \Gamma \vdash X}{\mathcal{A} \wedge \mathcal{B}, \Gamma \vdash X} \wedge\vdash_2 \qquad \frac{\Gamma \vdash \mathcal{A} \quad \Gamma \vdash \mathcal{B}}{\Gamma \vdash \mathcal{A} \wedge \mathcal{B}} \vdash\wedge$$

$$\frac{\mathcal{A}, \mathcal{A} \vee \mathcal{B}, \Gamma \vdash X \quad \mathcal{B}, \mathcal{A} \vee \mathcal{B}, \Gamma \vdash X}{\mathcal{A} \vee \mathcal{B}, \Gamma \vdash X} \vee\vdash \qquad \frac{\Gamma \vdash \mathcal{A}}{\Gamma \vdash \mathcal{A} \vee \mathcal{B}} \vdash\vee_1 \qquad \frac{\Gamma \vdash \mathcal{B}}{\Gamma \vdash \mathcal{A} \vee \mathcal{B}} \vdash\vee_2$$

$$\frac{\neg\mathcal{A}, \Gamma \vdash \mathcal{A}}{\neg\mathcal{A}, \Gamma \vdash X} \neg\vdash \qquad \frac{\mathcal{A}, \Gamma \vdash}{\Gamma \vdash \neg\mathcal{A}} \vdash\neg$$

$$\frac{\mathcal{A}\supset\mathcal{B},\Gamma\vdash\mathcal{A}\quad \mathcal{B},\mathcal{A}\supset\mathcal{B},\Gamma\vdash X}{\mathcal{A}\supset\mathcal{B},\Gamma\vdash X}\supset\vdash \qquad \frac{\mathcal{A},\Gamma\vdash\mathcal{B}}{\Gamma\vdash\mathcal{A}\supset\mathcal{B}}\vdash\supset$$

$$\frac{\mathcal{A}(y),\forall x\,\mathcal{A}(x),\Gamma\vdash X}{\forall x\,\mathcal{A}(x),\Gamma\vdash X}\forall\vdash \qquad \frac{\Gamma\vdash\mathcal{A}(y)}{\Gamma\vdash\forall x\,\mathcal{A}(x)}\vdash\forall^{\oslash}$$

$$\frac{\mathcal{A}(y),\exists x\,\mathcal{A}(x),\Gamma\vdash X}{\exists x\,\mathcal{A}(x),\Gamma\vdash X}\exists\vdash^{\oslash} \qquad \frac{\Gamma\vdash\mathcal{A}(y)}{\Gamma\vdash\exists x\,\mathcal{A}(x)}\vdash\exists$$

X stands for either a single formula or for the empty sequence of formulas. The variable restrictions, the notions of *proof* and *theorem* are as before.

Exercise* 3.1.9. Prove that if $\Gamma\vdash\Delta$ is provable in LK, then it is provable in LK_3, and vice versa. [Hint: Use inductions on the height of the proofs.]

Exercise* 3.1.10. Prove the equivalence of LJ and LJ_3. [Hint: Prove similar claims as in the previous exercise.]

3.2 Sequent calculi with multisets and sets

A sequence of formulas may seem a very general arrangement of multiple formulas. However, at least since the late 1950s, it has been clear that a sequent calculus—an example of which is the non-associative Lambek calculus—can be formulated so that the antecedent and the succedent are not sequences of formulas. At the same time, it was also realized, for example by Curry, that the permutation rules are not quite like the other rules, not even completely like contraction or thinning. Indeed, omitting the permutation rule results in certain simplifications in the analysis of proofs, because the effect of this rule can always be undone by the same rule. This may be the reason why sometimes applications of the permutation rules are left tacit in concrete examples of proofs—with a remark pointing this out.

If we do not wish to distinguish between sequences of wff's that can be obtained from each other by permutations, then we are interested in multisets of formulas. There are denumerably many wff's in all the languages that we consider, hence we define multisets fine-tuned to that situation.

Definition 3.4. (Multisets) Let A be a denumerable set with a fixed injective enumeration e. f is a *multiset* (with respect to A and e) when $f\colon\mathbb{N}\longrightarrow\mathbb{N}$ and $\{\,a\in A\colon f(e^{-1}(a))\neq 0\,\}$ is a finite subset of A.

Informally, a multiset is like a set except that the *multiplicity* of the elements counts. Alternatively, a multiset is like a sequence except that the *order* of the elements is unimportant. Multisets are sandwiched between sets and sequences. A notation that we may use occasionally is a list of the elements as many times as they occur in the multiset, enclosed into square brackets. For instance, $[a_0, a_1, a_1, a_5, a_7, a_7]$ is a multiset.

Exercise 3.2.1. Consider the case when A is finite, and simplify the above definition accordingly.

Given e, composing f with e^{-1} we get the number which shows how many times an element occurs in a multiset. The restriction that there are finitely many non-zeros in the codomain of f means that multisets are *finite*; this is in harmony with strings (or words) and sequences of formulas in sequent calculi being finite.[2] (This restriction may be behind the name that has been used earlier for multisets, namely, *firesets*, which seems to be an elision of *fi*nite *re*peating *sets*.)

An advantage of defining multisets as functions is that analogs of set operations can be defined by *pointwise operations* on the functions. We assume that A and e are the same for the multisets to which the operations apply.

Definition 3.5. The *union* of the multisets Γ and Δ is the *sum* of f_Γ and f_Δ. The *intersection* of the multisets Γ and Δ is the *minimum* of f_Γ and f_Δ. The *relative complement* of Δ with respect to Γ (i.e., the set difference $\Gamma \setminus \Delta$) is the *truncated subtraction* of f_Δ from f_Γ.

We will use $\Gamma, \Delta, \ldots$ as meta-variables for multisets of formulas, and Γ, Δ indicates the union of the multisets Γ and Δ; similarly, $\Gamma, \mathcal{A}$ is the multiset that is the union of Γ with the multiset that contains one copy of $\mathcal{A}$. The empty multiset (which is unique, like the empty set is) is denoted by $[\ ]$ or by (space).

It is convenient to assume that although there is a fixed enumeration of the formulas, we do not need to list the formulas in the order derived from the enumeration. That is, even if $\mathcal{A}$, $\mathcal{B}$ and $\mathcal{C}$ appear in this order in the enumeration, it does not matter whether we actually list the elements of a multiset as $[\mathcal{A}, \mathcal{B}, \mathcal{C}, \mathcal{C}]$ or $[\mathcal{C}, \mathcal{B}, \mathcal{C}, \mathcal{A}]$, because matching the wff's by their shape, we can always determine whether the multisets are the same.

We define a *sequent* to be a pair of multisets of formulas. In the case of intuitionistic logic, we impose the earlier restriction, namely, that on the right-hand side of the turnstile there is at most one formula, that is, that multiset is the empty multiset or a multiset containing one copy of one formula.

The *sequent calculi* $LK^{[\,]}$ and $LJ^{[\,]}$ for classical and for intuitionistic logic, respectively, with the modified notion of sequents may be easily described with respect to the original LK and LJ. $LK^{[\,]}$ and $LJ^{[\,]}$ are like LK and LJ, except that $(C\vdash)$ and $(\vdash C)$ are omitted. The notions of a *proof* and of a *theorem* are unchanged.

[2]Multisets are not always defined to be finite in the literature. Later on, we occasionally add "finite" for emphasis.

We do not repeat the rules, because they look like the rules in LK and LJ. However, we would like to emphasize that it is the result of our notational conventions that the rules are look-alikes.

The two formulations of the cut rule (i.e., cut and cut$_G$)), which we gave in Chapter 2, are indistinguishable now. Also, the exact location of $\mathcal{C}$ in the listing of the wff's in the cut rule from Section 3.1 is unimportant. It is not difficult to see that grounding the notion of sequents in multisets instead of sequences of formulas does not ruin the admissibility of the cut rule either for $LK^{[\,]}$ or for $LJ^{[\,]}$.

Exercise 3.2.2. Scrutinize the earlier proofs of the cut theorems to see that they remain proofs—provided that we omit the cases for the permutation rules and some applications of those rules.

The next step toward concealing structural rules in the data type that underlies sequents is to disregard the *number of occurrences* of formulas. Of course, in the case of intuitionistic logic, we want to keep the numerical restriction on the right-hand side of the $\vdash$.

A *sequent* is a pair of sets of formulas, for classical logic, and a set of formulas and a formula or the empty set, for intuitionistic logic. We will now use $\Gamma, \Delta, \Theta, \ldots$ as meta-variables for sets of formulas, including the empty set. $\Gamma, \mathcal{A}$ or $\mathcal{A}, \Gamma$ is the union of Γ with $\{\mathcal{A}\}$, and Γ, Δ is $\Gamma \cup \Delta$. We do not change the notions of a *proof* or of a *theorem*.

The *sequent calculus* $LK^{\{\,\}}$ is defined by the same axiom and set of rules as LK, except that the permutation rules ($(C\vdash)$ and $(\vdash C)$) and the contraction rules ($(W\vdash)$ and $(\vdash W)$) are omitted.

The *sequent calculus* $LJ^{\{\,\}}$ is defined by the same axiom and set of rules as LJ, except that the $(C\vdash)$ and $(W\vdash)$ rules are omitted.

Exercise 3.2.3. Prove that $LK^{\{\,\}}$ and LK, as well as $LJ^{\{\,\}}$ and LJ are equivalent with respect to their theorems.

We might wonder whether the Ketonen rules are still equivalent to the original conjunction and disjunction rules.[3] If $\mathcal{A}$ and $\mathcal{B}$, the subalterns of the conjunction $\mathcal{A} \wedge \mathcal{B}$, are distinct, then it appears that from the premise of $(\wedge\vdash_i)$ (with $i = 1$ or $i = 2$), with a $(K\vdash)$ step, we get $\mathcal{A} \wedge \mathcal{B}$ with the Ketonen rule. The other direction is obvious too; successive applications of $(\wedge\vdash_1)$ and of $(\wedge\vdash_2)$ to $\mathcal{A}, \mathcal{B}$ simply result in a set containing $\mathcal{A} \wedge \mathcal{B}$.

The question is, what happens if $\mathcal{A}$ and $\mathcal{B}$ are one and the same formulas, let us say they are both $\mathcal{D}$? First of all, we note that none of the connective rules have side conditions. (In $(\vdash\forall)$ and $(\exists\vdash)$, a certain distinctness of free variables is required, but no other rules have any similar restrictions.) The upper sequent of the $(\wedge\vdash)$ rule is $\mathcal{D}, \mathcal{D}, \Gamma \vdash \Delta$, and this is the same sequent as $\mathcal{D}, \Gamma \vdash \Delta$, because the antecedent is a set of wff's. The difference

[3]See [144, Ch. 6] for a different view than what we present here.

between the Ketonen and the other rules is that the former allows us to get the conjunction $\mathcal{D} \wedge \mathcal{D}$ only, whereas the two others can yield $\mathcal{D} \wedge \mathcal{E}$ and $\mathcal{E} \wedge \mathcal{D}$, where $\mathcal{E}$ is any formula, possibly, $\mathcal{D}$ itself. However, we are considering the case when the two subalterns are the same, and hence, $(\wedge\vdash_1)$ and $(\wedge\vdash_2)$ look the same.

There is *no problem* with defining sequent calculi for classical or intuitionistic logics using sets of formulas. Some logics—as we will see in later chapters—are formalizable using stricter data types, for instance, structures built by binary structural connectives. It may even happen that for a particular logic one can pick the *best suited* data type. For classical and intuitionistic logics, the data types can be easily varied, because of the abundance of features that $\wedge$ and $\vee$ have. The trend of allowing a structural connective to absorb certain features (hence, relaxing strictures of data types) goes back to Gentzen. He could have chosen $,$ to closely mimic $\wedge$ and $\vee$, which are *binary* connectives, however, he made his $,$ polyadic or associative.

The use of sets of formulas in sequents proved to be a useful simplification in several cases as illustrated in Sections 3.4, 3.5 and 8.3.

3.3 Sequent calculi with no structural rules

The separation of the operational and structural rules may be viewed as advantageous when we are interested in characterizing the meaning of the connectives and quantifiers. However, there is a certain *arbitrariness* to this idea, because the structural connectives such as $,$ play a role in some of the operational rules. Also, the interpretation of sequents relies on certain connectives standing in for a structural connective, perhaps, in a context-sensitive way. (This sort of connection will become even more apparent in Chapter 6, where in most cases there are several structural connectives unlike in LK and LJ.) It is also debatable what is and what is not part of the meaning of, let us say, conjunction in classical logic. Does it include the commutativity of conjunction or its distributivity over disjunction?

There are two completely different ways to understand what the phrase "no structural rules" means. One may wonder what happens if we retain all the operational rules of LK, for instance, but we omit all six structural rules. This is not the project in this section. Another way to understand the phrase is that the structural rules are not postulated explicitly, but may be admissible, or at least they are admissible to the extent that every theorem remains a theorem. *Hiding contraction in connective rules* is a crucial step in the calculi we consider in Section 9.1, where we deal with decidability.

We have already considered the possibility of omitting some structural rules (without changing the logic itself) in the previous section. This section presents sequent calculi in which thinning and contraction are integrated into operational rules, whereas permutation is dropped in favor of using multisets.[4]

Classical logic and intuitionistic logic permit the definition of negation from $\boldsymbol{F}$ and $\supset$. The two calculi in this section omit $\neg$, but they include $\boldsymbol{F}$.

The *sequent calculus* with *no* structural rules for classical logic is denoted by LK^F. A *sequent* is a pair of multisets of wff's separated by $\vdash$, where either multiset may be empty. $\Gamma, \Delta, \Theta, \ldots$ are *multisets*. Γ, Δ is the union of the multisets Γ and Δ. Similarly, if a formula $\mathcal{A}$ is joined by a comma with a multiset Γ, then that is the union of $[\mathcal{A}]$ and Γ. The union of multisets is a commutative and associative operation, which justifies omitting parentheses. Superscripts on formulas such as $\mathcal{A}^m$ indicate that $\mathcal{A}$ occurs m times, where $m \in \mathbb{N}$.

The *axiom and the rules* of LK^F are the following.

$$\mathcal{A} \vdash \mathcal{A}$$

$$\frac{\mathcal{A}^m, \mathcal{B}^n, \Gamma \vdash \Delta}{\mathcal{A} \wedge \mathcal{B}, \Gamma \vdash \Delta} \wedge\vdash \qquad \frac{\Gamma \vdash \Delta, \mathcal{A}^m \quad \Theta \vdash \Lambda, \mathcal{B}^n}{\Gamma, \Theta \vdash \Delta, \Lambda, \mathcal{A} \wedge \mathcal{B}} \vdash\wedge$$

$$\frac{\mathcal{A}^m, \Gamma \vdash \Delta \quad \mathcal{B}^n, \Theta \vdash \Lambda}{\mathcal{A} \vee \mathcal{B}, \Gamma, \Theta \vdash \Delta, \Lambda} \vee\vdash \qquad \frac{\Gamma \vdash \Delta, \mathcal{A}^m, \mathcal{B}^n}{\Gamma \vdash \Delta, \mathcal{A} \vee \mathcal{B}} \vdash\vee$$

$$\frac{\Gamma \vdash \Delta, \mathcal{A}^m \quad \mathcal{B}^n, \Theta \vdash \Lambda}{\mathcal{A} \supset \mathcal{B}, \Gamma, \Theta \vdash \Delta, \Lambda} \supset\vdash \qquad \frac{\mathcal{A}^m, \Gamma \vdash \Delta, \mathcal{B}^n}{\Gamma \vdash \Delta, \mathcal{A} \supset \mathcal{B}} \vdash\supset$$

$$\frac{}{\boldsymbol{F} \vdash \Delta} \boldsymbol{F}\vdash$$

$$\frac{\mathcal{A}(y)^m, \Gamma \vdash \Delta}{\forall x\, \mathcal{A}(x), \Gamma \vdash \Delta} \forall\vdash \qquad \frac{\Gamma \vdash \Delta, \mathcal{A}(y)^m}{\Gamma \vdash \Delta, \forall x\, \mathcal{A}(x)} \vdash\forall^{\oslash}$$

$$\frac{\mathcal{A}(y)^m, \Gamma \vdash \Delta}{\exists x\, \mathcal{A}(x), \Gamma \vdash \Delta} \exists\vdash^{\oslash} \qquad \frac{\Gamma \vdash \Delta, \mathcal{A}(y)^m}{\Gamma \vdash \Delta, \exists x\, \mathcal{A}(x)} \vdash\exists$$

The $^{\oslash}$ notation indicates that a rule is applicable when y is not free in the resulting sequent. *Proofs* and *theorems* are defined as usual.

Some choices in this calculus may seem unwarranted at first. The vocabulary contains $\boldsymbol{F}$, which is a zero-ary connective. This is motivated by a goal in [142], which is to provide a 1–1 correspondence between normal natural deduction proofs and cut-free sequent calculus proofs. ($\neg\mathcal{A}$ is defined as $\mathcal{A} \supset \boldsymbol{F}$.) Furthermore, $\boldsymbol{F}\vdash$ is a rule, which has no premise. The motivation for this arrangement—beyond aesthetic reasons such as that a connective should be introduced by a rule—is that $\boldsymbol{F}$ in the rule counts as principal,

[4]The following calculi LK^F and LJ^F are from [143], where they are called GM and GN. See also [142, pp. 105–108]. We do not follow their notation.

which makes a difference in the presentation of the proof of the cut theorem in [142].

This calculus is more like LK than LK_3 is, though the two-premise conjunction and disjunction rules (i.e., $(\vdash\wedge)$ and $(\vee\vdash)$) do not have the same parametric formulas. The subalterns and the principal formulas in the rules are listed on the "edges" of a sequent, but of course, the antecedent and the succedent are multisets now.

A difference between LK and LK^F is that the latter has *infinitely many rule schemas*, which is essential for not having contraction in the calculus. This also allows us to have the identity axiom for formulas with no parametric formulas on either side—without weakening rules. Let us consider how one rule with superscripts subsumes several rules. (For the sake of easy comparison, we will ignore applications of the permutation rules in LK in the example.)

Example 3.6. (1) $(\vdash\wedge)$ may have $m = n = 0$. Then this rule yields $\Gamma, \Theta \vdash \Delta, \Lambda, \mathcal{A} \wedge \mathcal{B}$ from the premises $\Gamma \vdash \Delta$ and $\Theta \vdash \Lambda$. Given a proof in LK of the latter two sequents, we could proceed in a couple of different ways to obtain the lower sequent of the rule. First all, we could discard the proof of $\Gamma \vdash \Delta$ or $\Theta \vdash \Lambda$ altogether, and then use $(K\vdash)$ (if Γ and Θ are not empty) and $(\vdash K)$, possibly repeatedly, to obtain a proof of the lower sequent. Alternatively, we could use $(K\vdash)$ and $(\vdash K)$ to obtain $\Gamma, \Theta \vdash \Delta, \Lambda, \mathcal{A}$ and $\Gamma, \Theta \vdash \Delta, \Lambda, \mathcal{B}$ and then apply $(\vdash\wedge)$ from LK. Notice that weakening must be used either way.

(2) If $m = 1$ or $n = 1$, then deleting the proof of one of the premises may not be an option, because there is no reason to assume that $\mathcal{A}$ or $\mathcal{B}$ occurs among the parametric formulas on the right-hand side of the lower sequent (i.e., in Δ, Λ). Even if $m = 1 = n$, and Γ is Θ, as well as Δ is Λ, in general, thinning would be required after an application of the $(\vdash\wedge)$ rule of LK, because there are two copies of Γ and Δ in the lower sequent. Once again, the rule $(\vdash\wedge)$ in LK^F, obviously, includes thinning.

(3) Lastly, if $m > 1$ or $n > 1$, then the right contraction rule $(\vdash W)$ has to be applied too. For instance, if $m = 3$ and $n = 5$ and Δ and Λ have no occurrences of $\mathcal{A}$ and of $\mathcal{B}$, respectively, then beyond ensuring that the parametric formula sets are Γ, Θ and Δ, Λ, the number of $\mathcal{A}$'s and $\mathcal{B}$'s must be reduced to one before an application of LK's $(\vdash\wedge)$ rule.

The calculi LK and LK^F are *not equivalent* with respect to *provable sequents*. The absence of structural rules means that contractions and weakenings are tied to subalterns and principal formulas—as we saw above in the case of $(\vdash\wedge)$. This constrains applications of thinning, in particular, *when* they can happen. In the $(\vdash\wedge)$ rule of LK^F, Γ, Δ, Θ and Λ may contain formulas of any shape (and in this sense $(\vdash\wedge)$ is not restricted); however, the rule does not permit the addition of a parametric wff on the left that is not in Γ or Θ.

Example 3.7. None of the following sequents is provable in LK^F, though they are all provable in LK. (a) $\forall x\, \mathcal{A}(x), p \vdash q, \exists y\, \mathcal{A}(y)$, (b) $\mathcal{A} \vdash \mathcal{A} \wedge \mathcal{A}$, (c) $\mathcal{A} \vdash \mathcal{B} \supset \mathcal{B}$.

Exercise 3.3.1. Scrutinize the $(\wedge\vdash)$ and $(\vdash\vee)$ rules. How are some previous rules recovered? How are the structural rules incorporated into these rules?

Exercise 3.3.2. Prove that the following wff's are theorems of LK^F. (a) $\mathcal{A} \supset \mathcal{B} \supset \mathcal{A}$, (b) $(\mathcal{A} \supset \mathcal{B} \supset \mathcal{C}) \supset (\mathcal{A} \supset \mathcal{B}) \supset \mathcal{A} \supset \mathcal{C}$, (c) $((\mathcal{A} \supset \mathcal{B}) \supset \mathcal{A}) \supset \mathcal{A}$. The proofs of these formulas illustrate how some of the typical uses of contraction and thinning are emulated in LK^F.

Exercise 3.3.3. Prove theorems in LK^F that express the distributivity of $\wedge$ over $\vee$, and vice versa.

The intuitionistic sequent calculus LJ can be given a similar remake. The resulting calculus is denoted by LJ^F. A *sequent* is a pair of a multiset of formulas and a singleton or empty multiset of formulas.

The *axiom* and *rules* of LJ^F are the following. (Again, $m, n \in \mathbb{N}$.)

$$\mathcal{A} \vdash \mathcal{A} \ \text{id}$$

$$\frac{\mathcal{A}^m, \mathcal{B}^n, \Gamma \vdash \mathcal{C}}{\mathcal{A} \wedge \mathcal{B}, \Gamma \vdash \mathcal{C}} \wedge\vdash \qquad \frac{\Gamma \vdash \mathcal{A} \quad \Theta \vdash \mathcal{B}}{\Gamma, \Theta \vdash \mathcal{A} \wedge \mathcal{B}} \vdash\wedge$$

$$\frac{\mathcal{A}^m, \Gamma \vdash \mathcal{C} \quad \mathcal{B}^n, \Theta \vdash \mathcal{C}}{\mathcal{A} \vee \mathcal{B}, \Gamma, \Theta \vdash \mathcal{C}} \vee\vdash \qquad \frac{\Gamma \vdash \mathcal{A}}{\Gamma \vdash \mathcal{A} \vee \mathcal{B}} \qquad \frac{\Gamma \vdash \mathcal{B}}{\Gamma \vdash \mathcal{A} \vee \mathcal{B}} \vdash\vee$$

$$\frac{\Gamma \vdash \mathcal{A} \quad \mathcal{B}^n, \Theta \vdash \mathcal{C}}{\mathcal{A} \supset \mathcal{B}, \Gamma, \Theta \vdash \mathcal{C}} \supset\vdash \qquad \frac{\mathcal{A}^m, \Gamma \vdash \mathcal{B}}{\Gamma \vdash \mathcal{A} \supset \mathcal{B}} \vdash\supset$$

$$\frac{}{\boldsymbol{F} \vdash \mathcal{C}} \boldsymbol{F}\vdash$$

$$\frac{\mathcal{A}(y)^m, \Gamma \vdash \mathcal{C}}{\forall x\, \mathcal{A}(x), \Gamma \vdash \mathcal{C}} \forall\vdash \qquad \frac{\Gamma \vdash \mathcal{A}(y)}{\Gamma \vdash \forall x\, \mathcal{A}(x)} \vdash\forall^{\oslash}$$

$$\frac{\mathcal{A}(y)^m, \Gamma \vdash \mathcal{C}}{\exists x\, \mathcal{A}(x), \Gamma \vdash \mathcal{C}} \exists\vdash^{\oslash} \qquad \frac{\Gamma \vdash \mathcal{A}(y)}{\Gamma \vdash \exists x\, \mathcal{A}(x)} \vdash\exists$$

The $^{\oslash}$ superscripts on two of the quantifier rules indicates the same restriction as before. The notion of a *proof* and of a *theorem* is unchanged.

The cut rule is almost admissible in LK^F and LJ^F. The "almost" refers to a potential change in the multiplicities of formulas in the end sequent in the cut-free proof. Nevertheless, the subformula property holds.[5]

Exercise 3.3.4.** Prove that $\mathcal{A}$ is a theorem of LJ iff it is a theorem of LJ^F. [Hint: $\boldsymbol{F}$ is definable as $\mathcal{A} \wedge \neg\mathcal{A}$.]

[5] Further details about these calculi may be found in [142, Ch. 5].

3.4 One-sided sequent calculi

The principle of *excluded middle* is clearly paramount to classical logic; it also distinguishes classical logic from intuitionistic logic. If we add that the two connectives, namely, $\neg$ and $\vee$, are sufficient to express all two-valued truth functions, and double negation is fully available in classical logic, then the possibility of another change to the language and to the notion of a sequent becomes salient.[6]

The language contains $\neg$ (negation) and $\vee$ (disjunction) as connectives, together with $\exists$ (the existential quantifier). The set of formulas is defined as in Chapter 2, but with the other logical components omitted. A *sequent* is a sequence of formulas.

The axiom and rules of the *sequent calculus* $LK^{\curlyvee}$ are as follows.

$$\mathcal{A}, \neg\mathcal{A} \ \text{id}$$

$$\frac{\Gamma, \mathcal{A}, \mathcal{B}}{\Gamma, \mathcal{A} \vee \mathcal{B}} \vee \qquad \frac{\Gamma, \neg\mathcal{A} \qquad \Gamma, \neg\mathcal{B}}{\Gamma, \neg(\mathcal{A} \vee \mathcal{B})} \text{DM} \qquad \frac{\Gamma, \mathcal{A}}{\Gamma, \neg\neg\mathcal{A}} \text{DN}$$

$$\frac{\Gamma, \mathcal{A}(y)}{\Gamma, \exists x\, \mathcal{A}(x)} \text{EG} \qquad \frac{\Gamma, \neg\mathcal{A}(y)}{\Gamma, \neg\exists x\, \mathcal{A}(x)} \text{UG}^{\oslash}$$

$$\frac{\Gamma, \mathcal{A}, \mathcal{B}, \Delta}{\Gamma, \mathcal{B}, \mathcal{A}, \Delta} \text{C} \qquad \frac{\Gamma, \mathcal{A}, \mathcal{A}}{\Gamma, \mathcal{A}} \text{W} \qquad \frac{\Gamma}{\Gamma, \mathcal{A}} \text{K}$$

The $^{\oslash}$ indicates the usual variable restriction, that is, y may not occur free in $\Gamma, \neg\exists x\, \mathcal{A}(x)$. The notion of a *proof* is the earlier one. If Γ is the root of a proof, then $\bigvee \Gamma$ is a *theorem*, where $\bigvee$ is as in (5)–(6) in Definition 2.18.

Informally, each sequent could be thought of as being prefixed by $\vdash$, that is, the left-hand side of $\vdash$ is always empty. Most labels for the operational rules allude to often-used names for classical principles. (DM) abbreviates De Morgan, (DN) is double negation, (EG) and (UG) are existential and universal generalization, respectively.

The *cut rule* is of the following form.

$$\frac{\Gamma, \mathcal{A} \qquad \Delta, \neg\mathcal{A}}{\Gamma, \Delta} \text{cut}$$

The cut rule is admissible in $LK^{\curlyvee}$, and we give proofs of the cut theorem in Section 7.6. We point out though that the cut-free proofs do not quite have the subformula property, because occasionally, a negation interferes.

This calculus could be *dualized*. The notion of a sequent remains the same, however, the language and the rules change. Instead of $\vee$, we use $\wedge$, and $\forall$ replaces $\exists$. The axiom and rules for the *sequent calculus* $LK^{\curlywedge}$ are as follows.

[6]The calculus $\mathbf{K}_1$ was originally proposed by Kurt Schütte, but we more or less follow [8, §39.3].

$$\mathcal{A}, \neg\mathcal{A} \ \text{id}$$

$$\frac{\Gamma, \mathcal{A}, \mathcal{B}}{\Gamma, \mathcal{A} \wedge \mathcal{B}} \wedge \qquad \frac{\Gamma, \neg\mathcal{A} \qquad \Gamma, \neg\mathcal{B}}{\Gamma, \neg(\mathcal{A} \wedge \mathcal{B})} \text{DM} \qquad \frac{\Gamma, \mathcal{A}}{\Gamma, \neg\neg\mathcal{A}} \text{DN}$$

$$\frac{\Gamma, \mathcal{A}(y)}{\Gamma, \forall x\, \mathcal{A}(x)} \text{UG} \qquad \frac{\Gamma, \neg\mathcal{A}(y)}{\Gamma, \neg\forall x\, \mathcal{A}(x)} \text{EG}^{\oslash}$$

$$\frac{\Gamma, \mathcal{A}, \mathcal{B}, \Delta}{\Gamma, \mathcal{B}, \mathcal{A}, \Delta} \text{C} \qquad \frac{\Gamma, \mathcal{A}, \mathcal{A}}{\Gamma, \mathcal{A}} \text{W} \qquad \frac{\Gamma}{\Gamma, \mathcal{A}} \text{K}$$

The notion of a *proof* is unchanged. $\mathcal{A}$ is a *theorem* of LK^{τ}, if $\neg\mathcal{A}$ is the root of a proof.

Exercise*3.4.1. Determine whether LK and LK^{τ} (with the language restricted to that of the latter calculus) are equivalent.

At first, the one-sided calculi may seem to be playful variations on a well-motivated sequent calculus such as LK. However, one-sided calculi can and has been defined for logics beyond classical logic, and the one-sided calculi for classical logic are straightforwardly connected to the uniform calculi in the next section, as well as to calculi in Sections 8.2 and 8.3.

3.5 Uniform sequent calculi

We have seen various sequent calculi that capture K and J. Clearly, a calculus may be more or less suitable for certain purposes even if it formalizes a given logic. It may seem baffling to some people that logicians prove more theorems *about* logics than *within* logics. In Chapter 2, we introduced LK and K, and proved various properties of these calculi, including their equivalence. We already mentioned there that in an axiomatic calculus it is advantageous to have as few logical constants as possible, because the proofs of the *meta-theorems* tend to become shorter.

The *uniform sequent calculi* for classical logic, which we are about to describe, are motivated by the idea that if a new level of abstraction is introduced into the definition of a calculus, then one can retain a range of logical constants, and at the same time provide concise proofs of meta-theorems, which are not obscured by a spate of details.[7]

[7] These uniform calculi were introduced by Raymond M. Smullyan, and they may be found in [171] and [170]. We harmonize the notation with ours.

We are not concerned here with the different ways in which the completeness (and soundness) of classical logic is proved—though we think that is an interesting and important topic too. Thus our motivation to describe the uniform sequent calculi for classical logic is that it provides a link to another kind of proof system, namely, to analytic tableaux, which we will look at in Section 8.2. Of course, the aim to provide a comprehensive view of sequent calculi would also be a sufficient reason to include these calculi too.

We denote the first uniform sequent calculus by LU_1. The logical constants include $\neg$, $\wedge$, $\vee$, $\supset$, $\forall$ and $\exists$. We also include *name constants* in the language. Formulas are as in Definition 2.4 with the listed logical particles and name constants. Furthermore, we restrict our consideration to wff's that are *closed*, that is, they have no free variables.

Definition 3.8. (Signed formulas) If $\mathcal{A}$ is a formula, then $\mathfrak{t}\mathcal{A}$ and $\mathfrak{f}\mathcal{A}$ are *signed formulas.*

$\mathfrak{t}$ and $\mathfrak{f}$ are merely two symbols that may be prefixed to a wff. (They are not $\boldsymbol{t}$ or $\boldsymbol{f}$ that we use elsewhere.) Nonetheless, they are intended to suggest a connection to a potential interpretation of wff's as "true" or "false." Note that signed formulas cannot be signed.

The role of the signs in the sequent calculus will be to distinguish between formulas that would appear on the left and on the right of the $\vdash$. $\Gamma \vdash \Delta$, where Γ and Δ are sets of wff's (like in Section 3.2) is to be viewed as $\mathfrak{t}\Gamma \cup \mathfrak{f}\Delta$. That is, if Γ is $\{ \mathcal{A}_1, \ldots, \mathcal{A}_n \}$ and Δ is $\{ \mathcal{B}_1, \ldots, \mathcal{B}_m \}$, then $\Gamma \vdash \Delta$ is turned into a set of signed formulas $\{ \mathfrak{t}\mathcal{A}_1, \ldots, \mathfrak{t}\mathcal{A}_n, \mathfrak{f}\mathcal{B}_1, \ldots, \mathfrak{f}\mathcal{B}_m \}$. Of course, the $\mathfrak{t}$'s and $\mathfrak{f}$'s are not meant to suggest that when the $\mathcal{A}$'s are true, false $\mathcal{B}$'s follow from them. Rather, if the sequent is provable, then an interpretation that makes all the $\mathcal{A}$'s true, cannot make all the $\mathcal{B}$'s false, that is, the above set of signed formulas is unsatisfiable.

An observation that might look banal at this point is that some rules for $\wedge$, $\vee$ and $\supset$ have *two premises*, whereas some other rules for these connectives have *one premise*. Moreover, despite variations on the rules themselves (as in Section 3.2), the number of premises is the same across versions of calculi. Similarly, $(\exists\vdash)$ and $(\vdash\forall)$ are alike, and so are $(\forall\vdash)$ and $(\vdash\exists)$. The rules for $\neg$ do not seem to fit into these categories. However, if we are not concerned with the size of proofs, then we could easily view them just as we view the other connectives by repeating the premise in one of the $(\neg\vdash)$ and $(\vdash\neg)$ rules.

From the above observations, we obtain a *fourfold classification* of the rules. There are α, β, γ and δ type rules. Γ is a set of signed formulas, whereas the α's, β's, etc. are signed formulas. $\gamma(a)$ and $\delta(a)$ are γ and δ, respectively, with their quantifier prefix omitted and a substituted for the free occurrences of x in the rest of the formula.

The *axioms and rules* of LU_1 are the following.

$$\Gamma, \mathfrak{t}\mathcal{A}, \mathfrak{f}\mathcal{A} \ \text{id}$$

$$\frac{\Gamma, \alpha_1, \alpha_2}{\Gamma, \alpha}\ \alpha \qquad \frac{\Gamma, \beta_1 \qquad \Gamma, \beta_2}{\Gamma, \beta}\ \beta$$

$$\frac{\Gamma, \gamma(a)}{\Gamma, \gamma}\ \gamma \qquad \frac{\Gamma, \delta(a)}{\Gamma, \delta}\ \delta^{\oslash}$$

The $^{\oslash}$ indicates that the rule requires that a does not occur in the set Γ, δ. Of course, we also have to specify which formulas stand in for the lower-case Greek letters. We summarize them in the following tables. On the right edge, we place labels of concrete rules in usual notation.

α	α_1	α_2	rule	β	β_1	β_2	rule
$\mathfrak{t}(\mathcal{A} \wedge \mathcal{B})$	$\mathfrak{t}\mathcal{A}$	$\mathfrak{t}\mathcal{B}$	$(\wedge\vdash)$	$\mathfrak{f}(\mathcal{A} \wedge \mathcal{B})$	$\mathfrak{f}\mathcal{A}$	$\mathfrak{f}\mathcal{B}$	$(\vdash\wedge)$
$\mathfrak{f}(\mathcal{A} \vee \mathcal{B})$	$\mathfrak{f}\mathcal{A}$	$\mathfrak{f}\mathcal{B}$	$(\vdash\vee)$	$\mathfrak{t}(\mathcal{A} \vee \mathcal{B})$	$\mathfrak{t}\mathcal{A}$	$\mathfrak{t}\mathcal{B}$	$(\vee\vdash)$
$\mathfrak{f}(\mathcal{A} \supset \mathcal{B})$	$\mathfrak{t}\mathcal{A}$	$\mathfrak{f}\mathcal{B}$	$(\vdash\supset)$	$\mathfrak{t}(\mathcal{A} \supset \mathcal{B})$	$\mathfrak{f}\mathcal{A}$	$\mathfrak{t}\mathcal{B}$	$(\supset\vdash)$
$\mathfrak{t}\neg\mathcal{A}$	$\mathfrak{f}\mathcal{A}$	$\mathfrak{f}\mathcal{A}$	$(\neg\vdash)$	$\mathfrak{f}\neg\mathcal{A}$	$\mathfrak{t}\mathcal{A}$	$\mathfrak{t}\mathcal{A}$	$(\vdash\neg)$

Table 3.1. α–β constituents.

We folded the negation rules into the α–β classification by taking the right introduction rule for $\neg$ to be a two-premise rule.

γ	$\gamma(a)$	rule	δ	$\delta(a)$	rule
$\mathfrak{t}\,\forall x\, \mathcal{A}(x)$	$\mathfrak{t}\mathcal{A}(a)$	$(\forall\vdash)$	$\mathfrak{f}\,\forall x\, \mathcal{A}(x)$	$\mathfrak{f}\mathcal{A}(a)$	$(\vdash\forall)^{\oslash}$
$\mathfrak{f}\,\exists x\, \mathcal{A}(x)$	$\mathfrak{f}\mathcal{A}(a)$	$(\vdash\exists)$	$\mathfrak{t}\,\exists x\, \mathcal{A}(x)$	$\mathfrak{t}\mathcal{A}(a)$	$(\exists\vdash)^{\oslash}$

Table 3.2. γ–δ constituents.

The notion of a *proof* is similar to the notion of a proof in one-sided sequent systems; that is, the nodes in the proof tree are occurrences of sets of signed wff's. A wff $\mathcal{A}$ is a *theorem* iff there is a proof of $\mathfrak{f}\mathcal{A}$.

Exercise 3.5.1. Write out the rules of LU_1 with signed formulas, given the two tables above.

Exercise 3.5.2. Assuming a correspondence between sets of signed formulas and sequents, which version of LK would result from these rules?

Exercise 3.5.3. Prove that the axioms of K are theorems. [Hint: The axioms are in Definition 2.11 on p. 22. Since we excluded open formulas, you may assume that every free variable is replaced by a name constant that otherwise does not occur in the formula.]

Exercise 3.5.4. In elementary courses on classical logic, other connectives are mentioned too. Here are the truth tables for four of them (with T for "true" and F for "false"). Would it be possible to fit them into the α–β classification? [Hint: If they fit in, then incorporate them; if they do not fit in, then explain the reason and create a category they would need.]

$\leftrightarrow$	T	F
T	T	F
F	F	T

∇	T	F
T	F	T
F	T	F

$\downarrow$	T	F
T	F	T
F	T	T

$\mid$	T	F
T	F	F
F	F	T

Common names for these connectives are *biconditional* (or equivalence) for $\leftrightarrow$, *Xor* (or exclusive or) for ∇, *Peirce's joint denial* (or Nor) for $\downarrow$ and *Sheffer's stroke* (or Nand) for $\mid$.

One might wonder whether the uniform notation depends on certain specifics of the above system. First of all, let us note that it may be somewhat subjective how uniform one wishes a system to be. We could have a classification of the rules so that each category contains exactly one rule and no duality connects the categories. It would, probably, be frivolous to call the system uniform; still that would not be impossible.

What we are interested in is whether the four categories, which are well-motivated and useful, can be retained even though some concrete details of the calculus are changed. The answer is "yes," which means that uniformity characterizes the coherence or harmony of the rules with respect to each other, rather than their concrete shape.

The *uniform sequent calculus* LU_2 comprises the following axioms and rules. $\Gamma, \Delta, \ldots$ range over sequences of signed wff's (including the empty sequence).

$$\Gamma, \mathfrak{t}\mathcal{A}, \Delta, \mathfrak{f}\mathcal{A}, \Theta \ \text{id}_1 \qquad\qquad \Gamma, \mathfrak{f}\mathcal{A}, \Delta, \mathfrak{t}\mathcal{A}, \Theta \ \text{id}_2$$

$$\frac{\Gamma, \alpha, \Delta, \alpha_1, \Theta, \alpha_2}{\Gamma, \alpha, \Delta, \Theta}\ \alpha \qquad\qquad \frac{\Gamma, \beta, \Delta, \beta_1 \qquad \Gamma, \beta, \Delta, \beta_2}{\Gamma, \beta, \Delta}\ \beta$$

$$\frac{\Gamma, \gamma, \Delta, \gamma(a)}{\Gamma, \gamma, \Delta}\ \gamma \qquad\qquad \frac{\Gamma, \delta, \Delta, \delta(a)}{\Gamma, \delta, \Delta}\ \delta^{\oslash}$$

Proofs and *theorems* are defined as before. We can specify the rules with signed formulas by using the components in Tables 3.1 and 3.2 above.

Exercise 3.5.5. Spell out the rules of LU_2—both as sequences of signed formulas and as rules in a two-sided sequent calculus (if possible).

It may seem that the rules are nothing like the rules in Gentzen's original sequent calculus, because every premise already contains the principal formula of the rule—like the rules in LK_3 do. It may be also suspicious that we have sequences of formulas, yet we do not have structural rules.

<u>Example</u> 3.9. We prove a formula, namely, $(\mathcal{A} \supset \mathcal{B}) \supset (\mathcal{B} \supset \mathcal{C}) \supset \mathcal{A} \supset \mathcal{C}$. To keep the proof to a reasonable size, we abbreviate this formula as $X \supset Y$, and the sequence $\langle \mathfrak{f}(X \supset Y), \mathfrak{t}X, \mathfrak{f}Y, \mathfrak{t}(\mathcal{B} \supset \mathcal{C}), \mathfrak{f}(\mathcal{A} \supset \mathcal{C})\rangle$ as Γ.

$$\dfrac{\dfrac{\dfrac{\Gamma, \mathfrak{t}\mathcal{A}, \mathfrak{f}\mathcal{C}, \mathfrak{f}\mathcal{A} \qquad \dfrac{\Gamma, \mathfrak{t}\mathcal{A}, \mathfrak{f}\mathcal{C}, \mathfrak{t}\mathcal{B}, \mathfrak{f}\mathcal{B} \qquad \Gamma, \mathfrak{t}\mathcal{A}, \mathfrak{f}\mathcal{C}, \mathfrak{t}\mathcal{B}, \mathfrak{t}\mathcal{C}}{\Gamma, \mathfrak{t}\mathcal{A}, \mathfrak{f}\mathcal{C}, \mathfrak{t}\mathcal{B}}\supset\vdash}{\mathfrak{f}(X \supset Y), \mathfrak{t}X, \mathfrak{f}Y, \mathfrak{t}(\mathcal{B} \supset \mathcal{C}), \mathfrak{f}(\mathcal{A} \supset \mathcal{C}), \mathfrak{t}\mathcal{A}, \mathfrak{f}\mathcal{C}}\supset\vdash}{\vdash\supset\quad \mathfrak{f}(X \supset Y), \mathfrak{t}X, \mathfrak{f}Y, \mathfrak{t}(\mathcal{B} \supset \mathcal{C}), \mathfrak{f}(\mathcal{A} \supset \mathcal{C}) \quad = \Gamma}}{\vdash\supset\quad \dfrac{\mathfrak{f}(X \supset Y), \mathfrak{t}X, \mathfrak{f}Y}{\vdash\supset\quad \mathfrak{f}((\mathcal{A} \supset \mathcal{B}) \supset (\mathcal{B} \supset \mathcal{C}) \supset \mathcal{A} \supset \mathcal{C}) \quad = \mathfrak{f}(X \supset Y)}}$$

The example reveals that working with sequences without structural rules is not a problem, because the subformulas are absorbed by the principal formula of the rule. The signs assure that the disappearing formulas are *exact subformulas*—rather than subformulas or negations of subformulas. Of course, anybody familiar with *analytic tableaux* could see this proof tree as an upside-down tableau in the style, where all the formulas are copied from node to node. (We introduce tableaux in Section 8.2.)

We consider yet another uniform calculus. Once again, we change the underlying data type, and we also change the rules. Uniformity forces us to consider certain rules, that perhaps would not have been contemplated otherwise.

The *sequent calculus* LU_3 is based on multisets of signed wff's. X is a signed wff. The axiom and rules are as follows.

$$\mathfrak{t}\mathcal{A}, \mathfrak{f}\mathcal{A}\ \text{id} \qquad \frac{\Gamma, \alpha_1}{\Gamma, \alpha}\ \alpha_1 \qquad \frac{\Gamma, \alpha_2}{\Gamma, \alpha}\ \alpha_2 \qquad \frac{\Gamma, \beta_1 \qquad \Gamma, \beta_2}{\Gamma, \beta}\ \beta$$

$$\frac{\Gamma, \gamma(a)}{\Gamma, \gamma}\ \gamma \qquad \frac{\Gamma, \delta(a)}{\Gamma, \delta}\ \delta^{\oslash} \qquad \frac{\Gamma}{\Gamma, X}\ K \qquad \frac{\Gamma, X, X}{\Gamma, X}\ W$$

Proofs are defined as earlier; $\mathcal{A}$ is a *theorem* iff $\mathfrak{f}\mathcal{A}$ is provable.

The most noticeable change is the presence of two α rules, instead of just one. Of course, *LK* has two $(\wedge\vdash)$ and two $(\vdash\vee)$ rules too. But we have not seen two $(\vdash\supset)$ rules.[8] Still using Table 3.1, we obtain the following rules (in usual $\vdash$ notation). (Γ_1 and Γ_2 add up to a multiset of wff's, and $\mathfrak{t}\Gamma_1 \cup \mathfrak{f}\Gamma_2 = \Gamma$ in the (α_1) and (α_2) rules.)

$$\frac{\Gamma_1, \mathcal{A} \vdash \Gamma_2}{\Gamma_1 \vdash \Gamma_2, \mathcal{A} \supset \mathcal{B}}\ \vdash\supset_1 \qquad \frac{\Gamma_1 \vdash \Gamma_2, \mathcal{B}}{\Gamma_1 \vdash \Gamma_2, \mathcal{A} \supset \mathcal{B}}\ \vdash\supset_2$$

A relevance logician may note that these rules make much more sense for material implication than the original $(\vdash\supset)$ does, and they do not leave doubts or uncertainty about what $\supset$ really is.

[8]Smullyan credits uniformity as the source of his discovery of the two versions of the right introduction rule for the horseshoe; see [171, p. 550].

3.6 Disjunction property

We started off this chapter by introducing LJ as a result of syntactical restrictions on LK. In the last two sections we gravitated to variations of LK, which were possible due to properties of $\neg$ in classical logic (and, because of the ensuing interdefinability of $\wedge$ and $\vee$). Before we turn to other non-classical logics, we mention a prominent property of intuitionistic logic, which is easily proved using LJ.

Definition 3.10. (Disjunction property) A logic has the *disjunction property* iff from the provability of $\mathcal{A} \vee \mathcal{B}$ it follows that $\mathcal{A}$ is provable or $\mathcal{B}$ is provable.

Obviously, classical logic does not have the disjunction property, because $p \vee \neg p$ is a theorem, however, neither p nor $\neg p$ is provable. For any logic, which includes excluded middle as its theorem, and permits uniform substitution, the disjunction property would be disastrous (or nearly so). On the other hand, this property accords well with interpretations of intuitionistic logic—including informal ones.

Theorem 3.11. *Intuitionistic logic has the* disjunction property.

Proof: The first proof was given by Gentzen in [92, §1], and the disjunction property can be proved in a similar fashion for a range of logics that have a single right-handed sequent calculus formalization. (Some variations may be necessary, depending on the definition of theoremhood.)

Let us assume that $\mathcal{A} \vee \mathcal{B}$ is a theorem of J. Then $\vdash \mathcal{A} \vee \mathcal{B}$ is provable in LJ. Furthermore, there is a cut-free proof of this sequent. The sequent is not an axiom, hence, the proof ends with an application of a rule. There are two possibilities; the first one is that the rule is a $\vee$-introduction on the right. Either version of the rule guarantees the truth of the theorem. The second possibility is that the rule is $(\vdash K)$. However, then $\vdash$ would have a cut-free proof, which is impossible. qed

The disjunction property is sometimes stated by saying that a logic is *prime*. This terminology originates from algebraic considerations of logics. If a logic, which contains $\vee$ and $\wedge$, is algebraized, and the set of the equivalence classes of its theorems forms a *prime filter*, then the logic is prime. (Under this view, it is obvious again that classical logic does not have the disjunction property, because the set of its theorems is not an ultrafilter on its algebra.) Alternatively, if *theories*, that is, sets of formulas closed under the logic's consequence relation are considered, then the set of theorems of a logic that possesses the disjunction property is a *prime theory*.

3.7 Translations between classical and intuitionistic logics

Intuitionistic logic emerged as a leading alternative to classical logic early in the 20th century. J differs from K, but it does not expand the language of K with $\Box$ or $\prec$, as modal logics do. Indeed, Gentzen's treatment of J and K in parallel, especially, in the sequent calculus where LJ is obtained from LK by a restriction, encourages one to overlook that LK and LJ are two (distinct) logics in two (distinct) languages. If we continue to neglect the difference between the languages, then intuitionistic logic turns out to be a *subsystem* of classical logic.[9]

It is well known that $\mathcal{A} \vee \neg\mathcal{A}$ is not a theorem of J, that is, not all instances of this schema are provable. This easily follows from the disjunction property for LJ that we proved in the previous section. On the other hand, $\neg\neg(\mathcal{A} \vee \neg\mathcal{A})$ is provable, and so is $\neg\neg(\neg\neg\mathcal{A} \supset \mathcal{A})$.

Exercise 3.7.1. Prove in LJ the formulas $\neg\neg(\mathcal{A} \vee \neg\mathcal{A})$ and $\neg\neg(\neg\neg\mathcal{A} \supset \mathcal{A})$. [Hint: The double negation in front of the formulas allows one to move the formulas out of the way, so to speak, in the proof.]

In the early 1930s, in the absence of a semantics for intuitionistic, and even for classical logic, a tool to compare logics was via *translations*.[10] The results of those investigations have been further refined later on, and they remain of interest, especially, in thinking about the foundations of mathematics.

The view of intuitionistic logic as "classical logic limited" can be made rigorous via a translation. It is tempting to call this translation the identity translation (as we do), but we tweak the claim so that it supports both the view that the logics are in different languages and that their connectives are (almost) the same.

Lemma 3.12. (Identity translation) *Let $\neg_K, \wedge_K, \vee_K, \supset_K, \forall_K$ and $\exists_K$ be the logical components in the language of K; similarly, let $\neg_J, \wedge_J, \vee_J, \supset_J, \forall_J$ and $\exists_J$ be the logical components in the language of J. The* identity translation, *ι is defined by (1)–(7), and it guarantees that* $\vdash_J \mathcal{A}$ implies $\vdash_K \iota(\mathcal{A})$.

(1) *If p is an atomic formula, then $\iota(p)$ is p;*

(2) $\iota(\neg_J\mathcal{A})$ *is* $\neg_K\iota(\mathcal{A})$;

(3) $\iota(\mathcal{A} \wedge_J \mathcal{B})$ *is* $\iota(\mathcal{A}) \wedge_K \iota(\mathcal{B})$;

(4) $\iota(\mathcal{A} \vee_J \mathcal{B})$ *is* $\iota(\mathcal{A}) \vee_K \iota(\mathcal{B})$;

[9] Arend Heyting arrived at a set of intuitionistically acceptable formulas by scrutinizing theorems of classical logic, which would impart a similar view.

[10] We mean here rigorous semantics, such as Saul A. Kripke's semantics for J in [118], which was invented some 30 years after the axiomatic presentation of J.

(5) $\iota(\mathcal{A} \supset_J \mathcal{B})$ *is* $\iota(\mathcal{A}) \supset_K \iota(\mathcal{B})$;

(6) $\iota(\forall_J x\, \mathcal{A})$ *is* $\forall_K x\, \iota(\mathcal{A})$;

(7) $\iota(\exists_J x\, \mathcal{A})$ *is* $\exists_K x\, \iota(\mathcal{A})$.

Exercise 3.7.2. Outline two proofs for the lemma, one using LK and LJ, and another using a pair of other proof systems such as the axiomatic formulations, K and J. [Hint: The claim is practically obvious, thus, focus on the structure of the proofs.]

One of the earliest translations was provided by Valeriĭ Glivenko in [98]. Given the provability of the doubly negated versions of the typical non-theorems of J above, it could be conjectured that placing $\neg\neg$ in front of a classical propositional theorem is sufficient to obtain an intuitionistic theorem.

Definition 3.13. Let Γ be a set of formulas. Then $\neg\neg\Gamma$ denotes the set of formulas obtained by prefixing each element of Γ by $\neg\neg$, that is, $\neg\neg\Gamma = \{\neg\neg\mathcal{A} \colon \mathcal{A} \in \Gamma\}$. Similarly, $\neg\Gamma = \{\neg\mathcal{A} \colon \mathcal{A} \in \Gamma\}$.

Theorem 3.14. (Glivenko's theorem) *If* $\Gamma \vdash_K \mathcal{A}$, *then* $\neg\neg\Gamma \vdash_J \neg\neg\mathcal{A}$, *and if* $\neg\Gamma, \Delta \vdash_K \neg\mathcal{A}$, *then* $\neg\Gamma, \neg\neg\Delta \vdash_J \neg\mathcal{A}$, *where all formulas are propositional.*

A proof of this theorem using axiom systems may be found in [113, §81]. Glivenko's theorem does not extend to quantified K and J. Kleene delineated the subsystem of his axiom system for classical logic to which the theorem is applicable. The restriction, essentially, is to omit the universal generalization rule.

The non-extendability of the Glivenko translation was noted by Gödel in [102, fn. 4.], who was interested in the relationship between formal arithmetic based on classical versus on intuitionistic logics. Gödel's insight is that usual formulations of classical logic are vastly redundant. It may be convenient to use a relatively large set of connectives as well as both quantifiers, but a more economical approach is to use just $\neg$, $\wedge$ and $\forall$.

Definition 3.15. (Gödel translation) The *Gödel translation* of a formula of $\mathcal{A}$ of K is denoted by $\mathfrak{g}(\mathcal{A})$, and it is defined by (1)–(6).

(1) If p is an atomic formula, then $\mathfrak{g}(p)$ is $\neg_J\neg_J p$;

(2) $\mathfrak{g}(\neg_K\mathcal{A})$ is $\neg_J\mathfrak{g}(\mathcal{A})$;

(3) $\mathfrak{g}(\mathcal{A} \wedge_K \mathcal{B})$ is $\mathfrak{g}(\mathcal{A}) \wedge_J \mathfrak{g}(\mathcal{B})$;

(4) $\mathfrak{g}(\mathcal{A} \vee_K \mathcal{B})$ is $\neg_J(\neg_J\mathfrak{g}(\mathcal{A}) \wedge_J \neg_J\mathfrak{g}(\mathcal{B}))$;

(5) $\mathfrak{g}(\mathcal{A} \supset_K \mathcal{B})$ is $\neg_J(\mathfrak{g}(\mathcal{A}) \wedge_J \neg_J\mathfrak{g}(\mathcal{B}))$;

(6) $\mathfrak{g}(\forall_K x\, \mathcal{A})$ is $\forall_J x\, \mathfrak{g}(\mathcal{A})$.

Gödel's translation as well as other closely related translations are also called *negative translations*, because the resulting wff's in J do not contain $\vee$ and $\exists$. These logical particles, under a certain interpretation of J, require a *positive demonstration* of the provability of a wff, hence, they are positive logical particles. The next theorem is related to what Gödel proved in [102] about arithmetical theories.

Theorem 3.16. *If* $\vdash_K \mathcal{A}$, *then* $\vdash_J \mathfrak{g}(\mathcal{A})$.

Exercise 3.7.3. Prove the previous theorem. [Hint: Use induction on the structure of wff's.]

According to the theorem, intuitionistic logic can be thought to coincide with classical logic in its negation–conjunction–universal quantifier fragment, which, however, brings additional components to classical logic. In J, $\vee$ and $\exists$ are not definable from the other connectives and $\forall$. The function $\mathfrak{g}$ maps formulas of K into a proper subset of wff's of J. Moreover, this proper subset of formulas in the co-domain of $\mathfrak{g}$ is included in the set of wff's, which is delineated by the absence of $\vee$ and $\exists$. The possibility to segregate potential classical theorems (within the set of wff's of J) by looking at the connectives and quantifiers contained in formulas means that intuitionistic logic has some genuinely new content—compared to classical logic.

Gentzen and Paul Bernays also discovered a translation that is very similar to Gödel's translation.[11]

Definition 3.17. (Gentzen–Bernays translation) The *Gentzen–Bernays translation* of a formula $\mathcal{A}$ of K is denoted by $\mathfrak{gb}(\mathcal{A})$, and it is defined by (1)–(4) and (6) from Definition 3.15 (with $\mathfrak{gb}$ for $\mathfrak{g}$ everywhere), together with (5′).

(5′) $\mathfrak{gb}(\mathcal{A} \supset_K \mathcal{B})$ is $\mathfrak{gb}(\mathcal{A}) \supset_J \mathfrak{gb}(\mathcal{B})$.

Theorem 3.18. *If* $\vdash_K \mathcal{A}$, *then* $\vdash_J \mathfrak{gb}(\mathcal{A})$.

A proof of this theorem—using axiomatic formulations of K and J—may be found in [113]; Gentzen's proof, which extends the claim to arithmetical theories, also uses axiomatic systems.

Andreĭ N. Kolmogorov provided a translation of classical logic into intuitionistic logic in which *double negations* are prominent.

Definition 3.19. (Kolmogorov translation) The *Kolmogorov translation* of a formula $\mathcal{A}$ of K is denoted by $\mathfrak{n}(\mathcal{A})$, and it is defined by (1)–(5).

(1) If p is an atomic formula, then $\mathfrak{n}(p)$ is $\neg_J\neg_J p$;

(2) $\mathfrak{n}(\neg_K\mathcal{A})$ is $\neg_J\neg_J\neg_J\mathfrak{n}(\mathcal{A})$;

(3) $\mathfrak{n}(\mathcal{A} \supset_K \mathcal{B})$ is $\neg_J\neg_J(\mathfrak{n}(\mathcal{A}) \supset_J \mathfrak{n}(\mathcal{B}))$;

[11] Gentzen's paper [93] was published in [175], some 30 years after the paper was written. However, he mentioned the translation and its slight difference from Gödel's translation in his [90].

(4) $\mathfrak{n}(\forall_K x\, \mathcal{A}(x))$ is $\neg_J \neg_J \forall_J x\, \mathfrak{n}(\mathcal{A}(x))$;

(5) $\mathfrak{n}(\exists_K x\, \mathcal{A}(x))$ is $\neg_J \neg_J \exists_J x\, \mathfrak{n}(\mathcal{A}(x))$.

The profusion of $\neg_J$'s may seem an overkill. In fact, Kolmogorov's translation maps K into a (proper) subsystem of J. If Γ is a set of wff's, then let $\mathfrak{n}(\Gamma)$ mean the set comprising the $\mathfrak{n}$-translations of each wff in Γ.[12]

Theorem 3.20. *If* $\Gamma \vdash_K \mathcal{A}$, *then* $\mathfrak{n}(\Gamma) \vdash_J \mathfrak{n}(\mathcal{A})$.

The relationship between classical and intuitionistic logics, as well as between mathematical theories based on these logics, has been extensively investigated since the 1930s. With this brief overview of some of the translations between K and J, we hope to have conveyed a sense that these are distinct logics—despite LK and LJ being so closely related.

[12]The following theorem is a slight extrapolation by Hao Wang of Kolmogorov's results in the preface to the English translation of Kolmogorov's original paper from 1925.

Chapter 4

Sequent calculi for non-classical logics

The title of this chapter may give you a pause at first. Have not we looked at several sequent calculi for a non-classical logic already in the previous chapter? Of course, we looked at calculi for intuitionistic logic, which is a non-classical logic. Moreover, adherents of intuitionism in mathematics, as we mentioned in Chapter 1, often mount a fierce polemic against the classical approach, and vice versa. In sum, *intuitionistic logic* is definitely a non-classical logic, but there are many others. In this and in the next chapter, we will focus on some non-classical logics that are less closely related to classical logic. It is a historical accident that sequent calculi were invented at once for classical logic and for intuitionistic logic. That is the reason why we looked at LK and LJ, as well as their versions together.

Although Gentzen himself did not invent other sequent calculi beyond LK, LJ and some of their extensions, his idea of obtaining LJ from LK via a restriction on the notion of a sequent and omitting some of the structural rules created a prototype method to design new logical calculi. Once we start to modify LK or LJ, we realize that there are many ways to go about adding and excluding connectives and structural features. The main dividing line between the calculi in this chapter and in the next one is whether the antecedent of a sequent is a *sequence of formulas* or it has more structure, because it is a *binary tree of formulas*.

4.1 Associative Lambek calculus

We have seen in Section 3.3 that it is possible to drop all the structural rules from LK or LJ—provided that the notion of a sequent, and also the rules themselves are changed. A different project is to omit all the structural rules, while leaving the notion of a sequent intact and then to investigate what logic results with a certain set of rules. Lambek in [120] did just this. He introduced a calculus that is nowadays called the *associative Lambek calculus*.

The *connectives* of LA are $\circ$, $\leftarrow$ and $\rightarrow$; each of them is binary. The set of well-formed formulas is inductively generated from a denumerable set of propositional variables.

Definition 4.1. (Well-formed formulas) The set of *well-formed formulas*, denoted by *wff*, is generated according to (1)–(2).

(1) If p is a propositional variable, then $p \in \text{wff}$;

(2) if $\mathcal{A}, \mathcal{B} \in \text{wff}$ and $\div \in \{\leftarrow, \circ, \rightarrow\}$, then $(\mathcal{A} \div \mathcal{B}) \in \text{wff}$.

We will use this definition as a template in other logics too—with some obvious modifications, when necessary. (E.g., if we add further binary connectives, then we assume that the set in (2) is expanded or modified.) Note that we changed the notation from the original $\backslash$, $\cdot$ and $/$ notation to bring it closer to the notation in classical and intuitionistic logics, as well as other logics to be introduced. Each logic has its *own language*. By re-using symbols, we aim at highlighting *abstract similarities* between proof systems rather than erasing their differences. We hope that context disambiguates which language a symbol belongs to—without us decorating the symbols with multiple subscripts and superscripts.

The *sequent calculus* LA is defined as a single right-handed sequent calculus. A *sequent* is a non-empty sequence of wff's followed by $\vdash$ and a wff. We use $\Gamma, \Delta, \ldots$ as meta-variables for sequences of wff's with the usual assumption about finiteness, but now, we also require that they include at least one formula. These kinds of sequences are easily definable in BNF.

$$\Gamma ::= \text{wff} \mid \Gamma; \Gamma$$

We use "wff" above as a non-terminal symbol in a CFG (context-free grammar) and we assume that there is an accompanying CFG that produces a concrete formula by rewriting the symbol wff. We use "wff" primarily as an abbreviation for well-formed formulas, as we have done so far. The use of this abbreviation as a non-terminal symbol in a CFG should not cause confusion; its role is similar to that of $\mathbb{A}$ in Definition 2.3. But we reiterate that $\Gamma, \Delta, \ldots$ *cannot* be instantiated by the empty sequence.[1]

Notice that we have put $;$ between the Γ's (not a $,$). The *semi-colon* still behaves very much like the comma did, at least in the sense that it is an *associative structural connective*—if viewed as a binary connective. However, in LA, $;$ is not connected to $\wedge$ or $\vee$ (which are not even in the language to start with). $;$ is the structural peer of $\circ$ (fusion), which does not have all the properties such as idempotence that $\wedge$ has in LK or LJ.

To indicate a particular *occurrence* of a wff or of a sequence in a sequence we employ a bracket notation. For example, $\Gamma[\mathcal{A}]$ indicates that $\mathcal{A}$ occurs in Γ and we have fixed one of its occurrences. Similarly, $\Gamma[\Delta]$ shows that Δ is a *subsequence* of Γ, and if there is more than one subsequence of Γ that is identical to Δ, then we have decided to focus on one of them. This sounds rather informal (especially, our talk about what we focus on), hence, we hasten to

[1]Given Definition 4.1, it is straightforward to define a CFG to generate wff's. We have discussed briefly the use of CFG's and of BNF in Section 2.1.

point out that it is possible to design a notation to make occurrences completely formal. We do not do that here, because we use the bracket notation for two limited purposes only. First, we sometimes want to make sure that a formula *does occur* in a sequence. Second, within a rule, we want to *replace an occurrence* of a sequence (possibly, of a wff) by a sequence. A usual notation for the replacement of a subformula $\mathcal{A}$ in $\mathcal{C}$ by the wff $\mathcal{B}$ is $\mathcal{C}[\mathcal{A}/\mathcal{B}]$, which we could abbreviate in a context, where $\mathcal{A}$ is understood, by $\mathcal{C}[\mathcal{B}]$. Now $\Gamma[\Delta]$ can be such an abbreviation for the replacement of a subsequent of Γ by Δ (understood from the context). A difference between these notations is that $\mathcal{C}[\mathcal{A}/\mathcal{B}]$ does not require or assume that $\mathcal{A}$ has any occurrences in $\mathcal{C}$.

Example 4.2. Let Γ be $\langle p; q \to p; p \to r; p; p \rangle$ (where p, q and r are propositional variables). p has three occurrences, and so we may use $\Gamma[p]$ to single out one of those occurrences. Suppose that $\Gamma[p]$ occurs in the upper sequent of a rule, and $\Gamma[\Delta]$ occurs in its lower sequent. The latter is the result of taking out that particular occurrence of p and inserting Δ in the hole. The bracket notation is not overly specific, and depending on which occurrence "we focused on," so to speak, $\Gamma[\Delta]$ can be $\langle \Delta; q \to p; p \to r; p; p \rangle$, $\langle p; q \to p; p \to r; \Delta; p \rangle$ or $\langle p; q \to p; p \to r; p; \Delta \rangle$. The three sequences are not the same (provided that Δ is not p), but it is an advantage conferred by the notation that we can refer to all these sequents at once, while excluding from consideration $\langle p; q \to p; \Delta; p; p \rangle$, for instance.

The *axiom* and the *rules* of LA are as follows.

$$\mathcal{A} \vdash \mathcal{A} \ \text{id}$$

$$\frac{\Gamma \vdash \mathcal{A} \qquad \Delta[\mathcal{B}] \vdash \mathcal{C}}{\Delta[\mathcal{A} \to \mathcal{B}; \Gamma] \vdash \mathcal{C}} \to\vdash \qquad\qquad \frac{\Gamma; \mathcal{A} \vdash \mathcal{B}}{\Gamma \vdash \mathcal{A} \to \mathcal{B}} \vdash\to$$

$$\frac{\Gamma[\mathcal{A}; \mathcal{B}] \vdash \mathcal{C}}{\Gamma[\mathcal{A} \circ \mathcal{B}] \vdash \mathcal{C}} \circ\vdash \qquad\qquad \frac{\Gamma \vdash \mathcal{A} \qquad \Delta \vdash \mathcal{B}}{\Gamma; \Delta \vdash \mathcal{A} \circ \mathcal{B}} \vdash\circ$$

$$\frac{\Gamma \vdash \mathcal{A} \qquad \Delta[\mathcal{B}] \vdash \mathcal{C}}{\Delta[\Gamma; \mathcal{B} \leftarrow \mathcal{A}] \vdash \mathcal{C}} \leftarrow\vdash \qquad\qquad \frac{\mathcal{A}; \Gamma \vdash \mathcal{B}}{\Gamma \vdash \mathcal{B} \leftarrow \mathcal{A}} \vdash\leftarrow$$

Exercise 4.1.1. Sequences are flat data structures—exactly as it is suggested by the way we write them. Perhaps, instead of using the bracket notation $\Gamma[\mathcal{A}]$, the notation exploited in LK, such as $\Gamma, \mathcal{A}$ or $\Gamma_1, \mathcal{A}, \Gamma_2$ could be used here. Formulate the rules of LA without the bracket notation. [Hint: A balance should be struck between ensuring that the antecedent of a sequent is not empty and that the rules are sufficiently general.]

The sequent calculus LA is considerably different from all the sequent calculi in Chapters 2 and 3. Nonetheless, the notion of a proof is fairly similar. A *proof* is a tree, in which each leaf is an instance of the axiom, and all other nodes are justified by applications of the rules of LA, that is, a node and its child or children together form an instance of one of the rules. The *root* of such a tree is the (bottom) sequent that is proved.

In LK and LJ, it is easy to define when a formula is provable, because the antecedent of a sequent can be empty. That cannot happen in LA, but we could define theoremhood in a circumvented way.

First, we need a cut rule. The *single cut rule* is of the following form.

$$\frac{\Gamma \vdash \mathcal{C} \qquad \Delta[\mathcal{C}] \vdash \mathcal{B}}{\Delta[\Gamma] \vdash \mathcal{B}}\ \text{cut}$$

This cut rule is very much like the single cut rule that we introduced for LK on page 26—except, of course, that now we cannot have more than one wff on the right. A more important difference between the two cuts is the use of brackets here. We want to avoid cumbersome circumscriptions (or "side conditions") specifying which occurrences of $\Gamma, \Delta, \ldots$ can or cannot be empty. If we maintain that these meta-variables range over non-empty sequences and the bracket notation singles out an occurrence, then $\Delta[\mathcal{C}]$ guarantees that $\mathcal{C}$ has an occurrence in the antecedent of the right premise. In the left premise, $\mathcal{C}$ has at least one occurrence, indeed, $\mathcal{C}$ is the whole succedent of the left premise.

Theorem 4.3. *The single cut rule is* admissible *in* LA.

Proof: The proof of this theorem is very simple, because LA itself is very simple in some sense. There are few rules, and none of them is a structural rule. Therefore, there is no potential problem caused by contraction-like rules. There are no zero-ary logical constants, that could require a special treatment. Also, the rules are not formulated with the principal formulas located on the edge of the antecedent of their lower sequents. (We leave filling in the details as an exercise.) qed

Exercise 4.1.2. Prove Theorem 4.3 taking into account the considerations above. [Hint: The proofs of cut theorems in Chapter 2 and Section 7.2 may be useful too.]

Lemma 4.4. *If* $\mathcal{A}_1; \mathcal{A}_2; \ldots; \mathcal{A}_n \vdash \mathcal{B}$ *is a provable sequent, then* $(\ldots(\mathcal{A}_1 \circ \mathcal{A}_2) \circ \ldots \circ \mathcal{A}_n) \vdash \mathcal{B}$ *is a* provable sequent, *and the other way around.*

Proof: **1.** For the "if–then" claim, we note that applying $(\circ\vdash)$ as many times as there are ; in the antecedent is sufficient to obtain the sequent $(\ldots(\mathcal{A}_1 \circ \mathcal{A}_2) \circ \ldots \circ \mathcal{A}_n) \vdash \mathcal{B}$.

2. For the converse we note that $(\circ\vdash)$ is an *invertible* rule, because the cut rule is admissible. Given the result of an application of the $(\circ\vdash)$ rule, we can restore the upper sequent of the rule by cut.

$$\frac{\vdash\circ\ \dfrac{\mathcal{A} \vdash \mathcal{A} \qquad \mathcal{B} \vdash \mathcal{B}}{\mathcal{A}; \mathcal{B} \vdash \mathcal{A} \circ \mathcal{B}} \qquad \begin{array}{c}\vdots \\ \Gamma[\mathcal{A} \circ \mathcal{B}] \vdash \mathcal{C}\end{array}}{\Gamma[\mathcal{A}; \mathcal{B}] \vdash \mathcal{C}}\ \text{cut}$$

Clearly, finitely many similar steps show that the second half of the statement holds too. qed

The lemma allows us to see the antecedent of a provable sequent as a wff. Then, we can think that if $\mathcal{A} \vdash \mathcal{B}$ is provable, then $\mathcal{A} \to \mathcal{B}$ is a *theorem* of LA. Although it is good to have a notion of a theoremhood, in this particular case, its usefulness seems to be limited to the statement of non-axiomatizability by a certain type of calculus.[2]

Equational and inequational logics are well-understood.[3] Let us consider a weakened notion of inequational logic, where the order, $\leq$ is a pre-order. (The relation could be lifted to a partial order on equivalence classes, if we would identify $\mathcal{A}$ and $\mathcal{B}$ whenever $\mathcal{A} \leq \mathcal{B}$ and $\mathcal{B} \leq \mathcal{A}$.)

The *C-calculus* that we consider consists of three parts, which characterize (a) the pre-order relation, (b) fusion and its interaction with implication and (c) the properly implicational components.[4]

The first part comprises an axiom and a rule.

$$\mathcal{A} \leq \mathcal{A} \qquad\qquad \frac{\mathcal{A} \leq \mathcal{B} \qquad \mathcal{B} \leq \mathcal{C}}{\mathcal{A} \leq \mathcal{C}}$$

The second component consists of two axioms and two rules. The axioms relate fusion and the implications. The rules state the monotonicity of fusion in both of its argument places.

$$\mathcal{A} \circ (\mathcal{B} \leftarrow \mathcal{A}) \leq \mathcal{B} \qquad\qquad (\mathcal{A} \to \mathcal{B}) \circ \mathcal{A} \leq \mathcal{B}$$

$$\frac{\mathcal{A} \leq \mathcal{B}}{\mathcal{A} \circ \mathcal{C} \leq \mathcal{B} \circ \mathcal{C}} \qquad\qquad \frac{\mathcal{A} \leq \mathcal{B}}{\mathcal{C} \circ \mathcal{A} \leq \mathcal{C} \circ \mathcal{B}}$$

Lastly, as the third part, four axioms and four rules are added. This C-calculus is equivalent to LA. (Incidentally, we use the same convention for $\leftarrow$ as for $\to$ to omit parentheses.)

$$\mathcal{A} \leq \mathcal{B} \leftarrow \mathcal{A} \to \mathcal{B} \qquad\qquad \mathcal{A} \leq (\mathcal{B} \leftarrow \mathcal{A}) \to \mathcal{B}$$

$$\mathcal{A} \to \mathcal{B} \leq (\mathcal{C} \to \mathcal{A}) \to \mathcal{C} \to \mathcal{B} \qquad\qquad \mathcal{B} \leftarrow \mathcal{A} \leq (\mathcal{B} \leftarrow \mathcal{C}) \leftarrow \mathcal{A} \leftarrow \mathcal{C}$$

$$\frac{\mathcal{A} \leq \mathcal{B}}{\mathcal{C} \to \mathcal{A} \leq \mathcal{C} \to \mathcal{B}}\ \mathsf{B}{\to} \qquad\qquad \frac{\mathcal{A} \leq \mathcal{B}}{\mathcal{B} \to \mathcal{C} \leq \mathcal{A} \to \mathcal{C}}\ \mathsf{B}'{\to}$$

$$\frac{\mathcal{A} \leq \mathcal{B}}{\mathcal{A} \leftarrow \mathcal{C} \leq \mathcal{B} \leftarrow \mathcal{C}}\ \mathsf{b}{\leftarrow} \qquad\qquad \frac{\mathcal{A} \leq \mathcal{B}}{\mathcal{C} \leftarrow \mathcal{B} \leq \mathcal{C} \leftarrow \mathcal{A}}\ \mathsf{b}'{\leftarrow}$$

[2] This result is essentially due to Zielonka in [191], though we do not follow his terminology or notation.

[3] See [79, §2.11 and §3.9] for detailed and accessible expositions of these kinds of calculi.

[4] In [191], a closely related calculus is called "categorial calculus," which we abbreviate here as C-calculus in order to avoid confusion with categorial grammars or with sequent calculi.

The labels for the last four rules are to intimate that the lower inequation in a rule is related to the principal type schema of the combinator or dual combinator in the label. Alternatively, the rules may be viewed to express the tonicity of $\leftarrow$ and $\rightarrow$.

Lemma 4.5. *If the rules* $(\mathsf{B}\rightarrow)$, $(\mathsf{B}'\rightarrow)$, $(\mathsf{b}\leftarrow)$ *and* $(\mathsf{b}'\leftarrow)$ *are omitted from the above calculus, then* an infinite set of implicational axioms *is required to formalize* LA.

The lemma is, essentially, the main result in [191]. We do not reproduce its proof here, because the C-calculus is closer to categorial grammars and inequational presentations of certain classes of algebras than to sequent calculi. Another way to look at the result is to say that LA cannot be formalized by finitely many axiom schemas, if the only rule for implicational formulas is modus ponens. A moral that we would like to draw from the lemma is that LA has a certain *complexity*, despite its simple formulation and its obvious decidability. (The decidability of LA follows easily from the cut theorem, which is, as we claimed, easy to prove.) We will see in Section 5.3, that the need for prefixing and suffixing rules (like the last four rules above), has been understood in the case of some other logics too, for example, in the axiomatization of the relevance logic $B_{\rightarrow}$.

It is useful to note, and it may be obvious from the C-calculus, that the notion of theoremhood is not needed to algebraize LA, because we may define $\mathcal{A}$ being equivalent to $\mathcal{B}$ by both $\mathcal{A}\vdash\mathcal{B}$ and $\mathcal{B}\vdash\mathcal{A}$ being provable.[5] This relation is reflexive, transitive and symmetric. (Transitivity follows from the cut Theorem 4.3, whereas symmetry and reflexivity are obvious.) Then, the elements of the equivalence classes of wff's under this relation are indistinguishable by the connectives of LA.

Lemma 4.6. (Replacement) *If* $\mathcal{A}\vdash\mathcal{B}$ *and* $\mathcal{B}\vdash\mathcal{A}$ *are provable, then for any wff* $\mathcal{C}$, *the sequents* $\mathcal{A}\rightarrow\mathcal{C}\vdash\mathcal{B}\rightarrow\mathcal{C}$, $\mathcal{C}\rightarrow\mathcal{A}\vdash\mathcal{C}\rightarrow\mathcal{B}$, $\mathcal{A}\circ\mathcal{C}\vdash\mathcal{B}\circ\mathcal{C}$, $\mathcal{C}\circ\mathcal{A}\vdash\mathcal{C}\circ\mathcal{B}$, $\mathcal{C}\leftarrow\mathcal{A}\vdash\mathcal{C}\leftarrow\mathcal{B}$ *and* $\mathcal{A}\leftarrow\mathcal{C}\vdash\mathcal{B}\leftarrow\mathcal{C}$ *are* provable.

Proof: First of all, we note that $\mathcal{A}$ and $\mathcal{B}$ may be interchanged in the sequents, which are proved, because each proof uses one of the two assumptions $\mathcal{A}\vdash\mathcal{B}$ and $\mathcal{B}\vdash\mathcal{A}$. This gives that $\mathcal{A}\rightarrow\mathcal{C}$ and $\mathcal{B}\rightarrow\mathcal{C}$, etc. are equivalent when so are $\mathcal{A}$ and $\mathcal{B}$. We include some details of two cases and leave the other cases for Exercise 4.1.3.

1. Let us consider $\rightarrow$.

$$\dfrac{\dfrac{\begin{matrix}\vdots\\ \mathcal{B}\vdash\mathcal{A}\end{matrix}\qquad \mathcal{C}\vdash\mathcal{C}}{\mathcal{A}\rightarrow\mathcal{C};\mathcal{B}\vdash\mathcal{C}}\;{\scriptstyle\rightarrow\vdash}}{\mathcal{A}\rightarrow\mathcal{C}\vdash\mathcal{B}\rightarrow\mathcal{C}}\;{\scriptstyle\vdash\rightarrow}
\qquad
\dfrac{\dfrac{\mathcal{C}\vdash\mathcal{C}\qquad \begin{matrix}\vdots\\ \mathcal{A}\vdash\mathcal{B}\end{matrix}}{\mathcal{C}\rightarrow\mathcal{A};\mathcal{C}\vdash\mathcal{B}}\;{\scriptstyle\rightarrow\vdash}}{\mathcal{C}\rightarrow\mathcal{A}\vdash\mathcal{C}\rightarrow\mathcal{B}}\;{\scriptstyle\vdash\rightarrow}$$

qed

[5]The term "algebraization" is used here in its usual sense rather than in the sense of [48], which is used in some of the recent literature.

Exercise 4.1.3. Finish the proof of the lemma above. [Hint: Notice that no application of the cut rule is required, because of the way the equivalence relation is defined.]

We are not going to go into the details of the algebraization of LA or of its semantics, both of which may be found in [38, §5.1].

In Lemma 4.4, we chose a wff—from among several wff's, in general—to correspond to the antecedent of a provable sequent. If there is more than one ; in the antecedent, then the wff we picked is created by introducing fusions starting from the left side. This is an arbitrary choice.

Claim 4.7. *If* $\mathcal{A}_1; \ldots; \mathcal{A}_n \vdash \mathcal{B}$ *is a provable sequent (with* $n \geq 3$*), then so is* $(\mathcal{A}_1 \circ \ldots \circ \mathcal{A}_n) \vdash \mathcal{B}$, *where the antecedent is a wff with the parentheses inserted in any one of the possible ways that make* $(\mathcal{A}_1 \circ \ldots \circ \mathcal{A}_n)$ *a wff.*

For $n = 3$, the *associativity* of the fusion operation (chopped into two provable sequents) suffices for a proof—provided the cut rule is available.

$$\circ\vdash\ \dfrac{\circ\vdash\ \dfrac{\vdash\circ\ \dfrac{\vdash\circ\ \dfrac{\mathcal{A}\vdash\mathcal{A} \qquad \mathcal{B}\vdash\mathcal{B}}{\mathcal{A};\mathcal{B}\vdash\mathcal{A}\circ\mathcal{B}} \qquad \mathcal{C}\vdash\mathcal{C}}{\mathcal{A};\mathcal{B};\mathcal{C}\vdash(\mathcal{A}\circ\mathcal{B})\circ\mathcal{C}}}{\mathcal{A};\mathcal{B}\circ\mathcal{C}\vdash(\mathcal{A}\circ\mathcal{B})\circ\mathcal{C}}}{\mathcal{A}\circ(\mathcal{B}\circ\mathcal{C})\vdash(\mathcal{A}\circ\mathcal{B})\circ\mathcal{C}}
\qquad
\dfrac{\dfrac{\dfrac{\mathcal{A}\vdash\mathcal{A} \qquad \dfrac{\mathcal{B}\vdash\mathcal{B} \qquad \mathcal{C}\vdash\mathcal{C}}{\mathcal{B};\mathcal{C}\vdash\mathcal{B}\circ\mathcal{C}}\ \vdash\circ}{\mathcal{A};\mathcal{B};\mathcal{C}\vdash\mathcal{A}\circ(\mathcal{B}\circ\mathcal{C})}\ \vdash\circ}{\mathcal{A}\circ\mathcal{B};\mathcal{C}\vdash\mathcal{A}\circ(\mathcal{B}\circ\mathcal{C})}\ \circ\vdash}{(\mathcal{A}\circ\mathcal{B})\circ\mathcal{C}\vdash\mathcal{A}\circ(\mathcal{B}\circ\mathcal{C})}\ \circ\vdash$$

If n is greater than 3, then there are more than 2 possibilities as to how to group the formulas.

Exercise 4.1.4. Prove the claim when $n = 4$. How many different formulas are there for other n's? [Hint: There are 5 possible ways to parenthesize a sequence of four formulas such as $\mathcal{A}_1; \mathcal{A}_2; \mathcal{A}_3; \mathcal{A}_4$.]

Exercise 4.1.5. Prove the general claim (for $n > 3$).

4.2 Extensions of the associative Lambek calculus

The associative Lambek calculus is sometimes called *directed* because $\mathcal{A} \to \mathcal{B}; \mathcal{A} \vdash \mathcal{B}$ and $\mathcal{A}; \mathcal{B} \leftarrow \mathcal{A} \vdash \mathcal{B}$ are provable for any $\mathcal{A}$ and $\mathcal{B}$, whereas $\mathcal{A}; \mathcal{A} \to \mathcal{B} \vdash \mathcal{B}$ and $\mathcal{B} \leftarrow \mathcal{A}; \mathcal{A} \vdash \mathcal{B}$ are not. It is very easy to establish that the two sequents are provable, but the demonstration of the non-provabilities is not difficult either. The deeper meaning for the label "directed," which may be thought to be accurately reflected by the arrows pointing into opposite directions, is that a function takes its argument from *exactly one* of its two sides. This functional view is explored in more detail in Sections 5.2 and 9.3.

One idea then is to add back into LA some of the structural rules (such as permutation) that were omitted from LJ to get LA and explore the emerging logics in the gulf between LA and LJ (or a fragment of LJ). Another idea is to take some of the rules (or their variants) from LJ to be rules for new connectives and add them to LA to see whether an interesting (or at least, a reasonable system) results.

The so-called *non-directed Lambek calculus* has been discovered and re-discovered—sometimes as a fragment of a logic that is richer in connectives. The idea is to make ; *commutative* so that the two unprovable sequents mentioned above become provable. Then there will be no reason to have both $\rightarrow$ and $\leftarrow$, because $\mathcal{A} \rightarrow \mathcal{B} \vdash \mathcal{B} \leftarrow \mathcal{A}$ and $\mathcal{A} \leftarrow \mathcal{B} \vdash \mathcal{B} \rightarrow \mathcal{A}$ are also provable. There are various, but equivalent, ways to achieve this effect.

We define the *sequent calculus* $LL^{\circ}_{\rightarrow}$ as LA with the $(C\vdash)$ rule added.

$$\frac{\Gamma[\mathcal{A};\mathcal{B}] \vdash \mathcal{C}}{\Gamma[\mathcal{B};\mathcal{A}] \vdash \mathcal{C}}\ C\vdash$$

To be precise, we omit $\leftarrow$ from the language and the $(\leftarrow\vdash)$ and $(\vdash\leftarrow)$ rules from LA before we add $(C\vdash)$. Although we typically *do keep* around logically equivalent formulas in logics, having two symbols for what is essentially one connective strikes us as silly. It seems at least misleading (or even simply a typo) to have both $\leftarrow$ and $\rightarrow$, when it is already understood that the two symbols are freely interchangeable in all contexts.

The second L in the $LL^{\circ}_{\rightarrow}$ is to allude to *linear logic*, which includes $LL^{\circ}_{\rightarrow}$, as a fragment. Linear logic attracted a lot of attention in the past decades, mainly because its wide applicability is computer science. Section 4.5 is a relatively short exploration of some of the proof-theoretic aspects of that logic.

Exercise 4.2.1. Prove that the single cut rule is admissible in $LL^{\circ}_{\rightarrow}$.

We may take the two other left structural rules from LK or LJ, rewrite them into our bracket notation and add them, one at a time, to LA.

Let the sequent calculi $LA + (W\vdash)$ and LA_K be defined by adding $(W\vdash)$ and $(K\vdash)$ to LA, in the following form. (There is a dissimilarity in our notation for the calculi, because we will use the label LA_W for another sequent calculus soon.)

$$\frac{\Gamma[\mathcal{A};\mathcal{A}] \vdash \mathcal{B}}{\Gamma[\mathcal{A}] \vdash \mathcal{B}}\ W\vdash \qquad\qquad \frac{\Gamma[\mathcal{A}] \vdash \mathcal{C}}{\Gamma[\mathcal{A};\mathcal{B}] \vdash \mathcal{C}}\ K\vdash$$

Exercise 4.2.2. Prove that the single cut rule is admissible in LA_K, but not in $LA + (W\vdash)$. [Hint: Case **2.1.2 (2)** in the proof of the cut theorem for LK on p. 47 may be helpful.]

These logics result as straightforward extensions of LA. However, there seems to be no clear-cut linguistic motivation for these structural rules. LA is very rigorous about not allowing permutations or repeated uses of one occurrence of a formula, but it can be contended that natural languages that have a (more or less) fixed word order (such as English) are described by LA (perhaps, a bit too restrictively). But at the other extreme, $LL^{\circ}_{\rightarrow}$ allows arbitrary permutations of formulas, and even languages with free word order, typically, allow all permutations only in very simple sentences. Inserting a word randomly, so to speak, as by an application of the $(K\vdash)$ rule, seems to be out of question in all languages.

$LL^{\circ}_{\rightarrow}$, though, has non-linguistic motivations readily available. Abelian semi-groups are a well-known algebraic abstraction, and one can think concretely of $\circ$ as multiplication (e.g., on $\mathbb{R}$). It is not so easy to find concrete algebraic examples with the analogs of $(W\vdash)$ or $(K\vdash)$.

Lastly, if we look at $LL^{\circ}_{\rightarrow}$, $LA + (W\vdash)$ and LA_K, and we view $\circ$ as an operation that combines premises and $\rightarrow$ as logical implication, then $LL^{\circ}_{\rightarrow}$ appears to be a much more restricted yet more sensible logic than LK or LJ are. $LL^{\circ}_{\rightarrow}$ permits rearranging the premises arbitrarily, but allows each of them to be used exactly once. Similarly interpreted, LA_K allows inserting arbitrary premises anywhere after the first one. Maybe, this is a sort of quintessential classical reasoning where the consequence relation's most important property is its monotonicity. $LA + (W\vdash)$, on the other hand, can be thought of as a logic that overlooks that a premise has been used more than once. Incidentally, LA_K and $LA + (W\vdash)$ are in need of an interpretation for $\leftarrow$ too. In general, both calculi would have to be connected to some concrete or abstract examples.

The three original left structural rules in LK are *independent* from each other in the sense that having chosen two of them, we cannot derive the third.

This means that we may consider *three* further extensions of LA before recouping a notational variant of the implication-conjunction fragment of LJ. One of these logics, *relevant implication,* is very well known; indeed, it is so well known that we devote to it and to its extensions three sections in this chapter, in Sections 4.3, 4.4 and 4.6. The combination of $LL^{\circ}_{\rightarrow}$ and $LA + (W\vdash)$ is known as *implicational affine logic*. But first, we take a quick look at $LA + (W\vdash)$ and LA_K blended together. This calculus takes order seriously, but suffers from the same problem as $LA + (W\vdash)$ did—the cut rule is not admissible in this calculus. A brief exploration of this problem and its remedy is quite worthwhile.

First of all, we substantiate our claim that the single cut rule is *not admissible.* We construct a counterexample.

The idea is to use contraction and cut to create a provable formula, and then to show that the formula is not provable without an application of the cut rule.

<u>*Example*</u> 4.8.

$$\dfrac{\dfrac{\dfrac{\dfrac{\mathcal{C}\vdash\mathcal{C}\quad \dfrac{\mathcal{D}\vdash\mathcal{D}\quad \mathcal{A}\vdash\mathcal{A}}{\mathcal{D}\to\mathcal{A};\mathcal{D}\vdash\mathcal{A}}}{\mathcal{D}\to\mathcal{A};\mathcal{C}\to\mathcal{D};\mathcal{C}\vdash\mathcal{A}}}{\mathcal{D}\to\mathcal{A};\mathcal{C}\to\mathcal{D}\vdash\mathcal{C}\to\mathcal{A}}\quad \dfrac{\dfrac{\mathcal{C}\to\mathcal{A}\vdash\mathcal{C}\to\mathcal{A}\quad \dfrac{\mathcal{C}\to\mathcal{A}\vdash\mathcal{C}\to\mathcal{A}\quad \mathcal{B}\vdash\mathcal{B}}{(\mathcal{C}\to\mathcal{A})\to\mathcal{B};\mathcal{C}\to\mathcal{A}\vdash\mathcal{B}}}{(\mathcal{C}\to\mathcal{A})\to(\mathcal{C}\to\mathcal{A})\to\mathcal{B};\mathcal{C}\to\mathcal{A};\mathcal{C}\to\mathcal{A}\vdash\mathcal{B}}}{(\mathcal{C}\to\mathcal{A})\to(\mathcal{C}\to\mathcal{A})\to\mathcal{B};\mathcal{C}\to\mathcal{A}\vdash\mathcal{B}}}{(\mathcal{C}\to\mathcal{A})\to(\mathcal{C}\to\mathcal{A})\to\mathcal{B};\mathcal{D}\to\mathcal{A};\mathcal{C}\to\mathcal{D}\vdash\mathcal{B}}\,\text{cut}}{(\mathcal{C}\to\mathcal{A})\to(\mathcal{C}\to\mathcal{A})\to\mathcal{B}\vdash(\mathcal{D}\to\mathcal{A})\to(\mathcal{C}\to\mathcal{D})\to\mathcal{B}}\,{\vdash\to}\text{'s}$$

Let us assume that $\mathcal{A}$, $\mathcal{B}$, $\mathcal{C}$ and $\mathcal{D}$ are distinct formulas; they may be propositional variables, for instance. If we try to build a proof tree from the bottom to the top, then $(\to\vdash)$ is a possible, but hopeless move, because $\mathcal{C}\to\mathcal{A}$ is, obviously, not a theorem. The other possibility is to think that $(W\vdash)$ has been applied, but there is no proof in sight after that step either. The most promising avenue seems to be to assume that $(\vdash\to)$ has been applied—once or twice. After a little thought, we see that moving all the subformulas possible to the left of $\vdash$ does not prevent us from contemplating that a particular rule was applied.

Thus, we continue with $(\mathcal{C}\to\mathcal{A})\to(\mathcal{C}\to\mathcal{A})\to\mathcal{B};\mathcal{D}\to\mathcal{A};\mathcal{C}\to\mathcal{D}\vdash\mathcal{B}$ as the bottom sequent. We can duplicate one or each of the three formulas, or we can assume that one or each but the first has been thinned into the sequent. Yet another possibility is that $(\to\vdash)$ has been applied to obtain one of the three formulas. Here are three possible ways how the first or the second implicational formula could have been introduced into the sequent.

$$\dfrac{\mathcal{D}\to\mathcal{A};\mathcal{C}\to\mathcal{D}\overset{\vdots}{\vdash}\mathcal{C}\to\mathcal{A}\qquad (\mathcal{C}\to\mathcal{A})\to\mathcal{B}\overset{?}{\vdash}\mathcal{B}}{(\mathcal{C}\to\mathcal{A})\to(\mathcal{C}\to\mathcal{A})\to\mathcal{B};\mathcal{D}\to\mathcal{A};\mathcal{C}\to\mathcal{D}\vdash\mathcal{B}}\,{\to\vdash}$$

The left branch can be turned into a proof (just as in the proof with an application of the cut rule), but the sequent in the right premise is not provable.

$$\dfrac{\mathcal{D}\to\mathcal{A}\overset{?}{\vdash}\mathcal{C}\to\mathcal{A}\qquad (\mathcal{C}\to\mathcal{A})\to\mathcal{B};\mathcal{C}\to\mathcal{D}\overset{?}{\vdash}\mathcal{B}}{(\mathcal{C}\to\mathcal{A})\to(\mathcal{C}\to\mathcal{A})\to\mathcal{B};\mathcal{D}\to\mathcal{A};\mathcal{C}\to\mathcal{D}\vdash\mathcal{B}}\,{\to\vdash}$$

In this attempt, we assumed the first formula in the sequent to be principal, but we apportioned the rest of the sequent between the two premises differently than before. If we suppose, as we do now, that $\mathcal{D}$ is not the same as $\mathcal{C}$, and $\mathcal{D}$ is also distinct from $\mathcal{A}$, then there is no reason for the sequents in either of the premises to be provable. Lastly, let $\mathcal{D}\to\mathcal{A}$ be the principal wff.

$$\dfrac{\mathcal{C}\to\mathcal{D}\overset{?}{\vdash}\mathcal{D}\qquad (\mathcal{C}\to\mathcal{A})\to(\mathcal{C}\to\mathcal{A})\to\mathcal{B};\mathcal{A}\overset{?}{\vdash}\mathcal{B}}{(\mathcal{C}\to\mathcal{A})\to(\mathcal{C}\to\mathcal{A})\to\mathcal{B};\mathcal{D}\to\mathcal{A};\mathcal{C}\to\mathcal{D}\vdash\mathcal{B}}\,{\to\vdash}$$

Perhaps, unsurprisingly, neither branch can be turned into a proof.

The following chunk shows what would be the right idea, and at the same time, we see what becomes a problem.

$$\dfrac{\dfrac{(\mathcal{C} \to \mathcal{A}) \to (\mathcal{C} \to \mathcal{A}) \to \mathcal{B}; \mathcal{D} \to \mathcal{A}; \mathcal{D} \to \mathcal{A}; \mathcal{C} \to \mathcal{D}; \mathcal{C} \to \mathcal{D} \overset{?}{\vdash} \mathcal{B}}{(\mathcal{C} \to \mathcal{A}) \to (\mathcal{C} \to \mathcal{A}) \to \mathcal{B}; \mathcal{D} \to \mathcal{A}; \mathcal{D} \to \mathcal{A}; \mathcal{C} \to \mathcal{D} \vdash \mathcal{B}}\ W\vdash}{(\mathcal{C} \to \mathcal{A}) \to (\mathcal{C} \to \mathcal{A}) \to \mathcal{B}; \mathcal{D} \to \mathcal{A}; \mathcal{C} \to \mathcal{D} \vdash \mathcal{B}}\ W\vdash$$

In the proof with an application of the cut rule, the cut came after $(W\vdash)$. The cut replaced one wff with a sequence comprising two wff's. This is not a problem in general, but in this particular case, we would like to have not just two copies of each of $\mathcal{D} \to \mathcal{A}$ and $\mathcal{C} \to \mathcal{D}$, but we want to be able to dissect the sequence of those four formulas into the two subsequences so that each premise gets one copy of each wff. Since there is no permutation, either the left premise ends up not having $\mathcal{C} \to \mathcal{D}$, or the right premise lacks $\mathcal{D} \to \mathcal{A}$. Incidentally, having one too many copies of a wff in either premise would not be a problem, because we could rid the sequent of the superfluous wff by $(K\vdash)$. (In $LA + (W\vdash)$, that would be an additional problem, which shows that the single cut rule is "even less" admissible in $LA + (W\vdash)$.)

Exercise 4.2.3. Write out all the possible ways in which the sequent $(\mathcal{C} \to \mathcal{A}) \to (\mathcal{C} \to \mathcal{A}) \to \mathcal{B}; \mathcal{D} \to \mathcal{A}; \mathcal{C} \to \mathcal{D} \vdash \mathcal{B}$ could have been obtained. [Hint: We considered in detail only a few cases.]

The presence of $(W\vdash)$ among the rules means that there is no stage in the construction of the search trees (from the bottom to the top), at which the search would stop, simply because no rule is applicable. In the concrete example above, one could argue—having scrutinized the search tree—that the multiplicities of wff's cannot turn a sequent that is unprovable as it is into a provable one. For instance, if $\mathcal{C} \to \mathcal{D} \vdash \mathcal{D}$ is unprovable, so is $\mathcal{C} \to \mathcal{D}; \mathcal{C} \to \mathcal{D} \vdash \mathcal{D}$.

Exercise 4.2.4. We claimed above that $(\mathcal{C} \to \mathcal{A}) \to \mathcal{B} \vdash \mathcal{B}$ is not a provable sequent. Why is $(W\vdash)$ not useful in a purported proof of this sequent?

Having shown that the cut rule is not admissible in $LA_K + (W\vdash)$, we might wonder whether the calculus can be modified so that the same sequents remain provable (as in $LA_K + (W\vdash)$ with cut), but the cut rule is admissible. The solution is quite obvious if one is familiar with LR_+, which dates back to 1969. We describe LR_+ in Section 4.6. As we will see, LR_+ contains two kinds of sequences of formulas, and the structural rules operate on *sequences of wff's* and *sequences of ... sequences of wff's*, not merely on individual wff's. This is a key insight as to how to formulate more refined, but well-behaved sequent and consecution calculi.

First, let us consider informally why the adoption of *properly structuralized structural rules* should be unproblematic from the point of view of what can be proved. The sequent $(\mathcal{C} \to \mathcal{A}) \to (\mathcal{C} \to \mathcal{A}) \to \mathcal{B} \vdash ((\mathcal{D} \to \mathcal{A}) \circ (\mathcal{C} \to$

$\mathcal{D})) \to \mathcal{B}$ could not have served as our counterexample above, because this sequent has a cut-free proof. Once $(\mathcal{D} \to \mathcal{A}) \circ (\mathcal{C} \to \mathcal{D})$ is on the left-hand side of $\vdash$ (in our backward proof-search tree), we can take the wff as the result of an application of $(W\vdash)$, because it is a wff. Thus, if we could somehow introduce $\circ$ in the proof of the purely implicational wff, but then change $\circ$ back to $;$, that would solve the problem. Of course, we do not want to allow such moves, because it would spoil some of the goodies delivered by the admissibility of the cut rule. But allowing $(W\vdash)$ to operate on a sequence of wff's rather than merely on one wff saves us the burden of having to fuse and then disassemble formulas.

We formulate a *sequent calculus* that adds $(W'\vdash)$ (rather than $(W\vdash)$) to LA_K. The rule $(W'\vdash)$ is the *structuralized contraction rule.*

$$\frac{\Gamma\,[\Delta;\Delta] \vdash \mathcal{A}}{\Gamma\,[\Delta] \vdash \mathcal{A}}\ W'\vdash$$

We denote this calculus by LA_{KW}. Similarly, LA with $(W'\vdash)$ added is denoted by LA_W.

$(W'\vdash)$ is a more powerful rule than $(W\vdash)$ is, but that is exactly what we need to ensure that the cut rule cannot increase the set of provable sequents. A harsher opinion about $LA + (W\vdash)$ and $LA_K + (W\vdash)$ is that they are ill-conceived. They compare to defining Boolean algebras in the signature $\langle \wedge, \vee, - \rangle$, then omitting some of the equations (such as $a \vee -a = b \vee -b$), but still hoping to have a bounded lattice.

We have not modified the $(K\vdash)$ rule to operate on sequences. First of all, in LA_K, the cut rule is admissible, hence, we may suspect that if we add further structural rules, then $(K\vdash)$ will remain unproblematic. However, before considering the admissibility of the cut rule in LA_{KW}, let us compare $(W\vdash)$ and $(K\vdash)$.

We have seen that if $\mathcal{A}$, the principal formula of the contraction rule, is replaced by $\mathcal{B}_1, \ldots, \mathcal{B}_n$, then the lower sequent of $(W'\vdash)$ does not result by $(W\vdash)$, given arbitrary $\mathcal{B}$'s. If $\mathcal{A}$ is introduced by $(K\vdash)$, then it is part of a subsequence of the form $\mathcal{B}, \mathcal{A}$—according to the formulation of the rule. Then the question is whether $\mathcal{B}, \mathcal{C}_1, \ldots, \mathcal{C}_n$ can be obtained by $(K\vdash)$, or perhaps we have to introduce a more general $(K'\vdash)$ rule too. Obviously, n applications of $(K\vdash)$ suffice. If $n > 1$, then this means that one application of $(K\vdash)$ followed by an application of the cut rule is traded in for more than one application of $(K\vdash)$. However, this affects only the length of the proof, not its provability.

Notwithstanding, we can consider the *structuralized* form of the *thinning* rule, which is $(K'\vdash)$.

$$\frac{\Gamma\,[\Delta] \vdash \mathcal{A}}{\Gamma\,[\Delta;\Theta] \vdash \mathcal{A}}\ K'\vdash$$

Exercise 4.2.5. Prove that $(K'\vdash)$ is a derived rule in LA_K, therefore, also in its extensions with either of the contraction rules that we considered.

The structural rule that operates on a single formula is trivially an *instance* of its structuralized version. Even if it is not forced upon us, we may structuralize thinning too—for aesthetic reasons, or to keep the length of a proof under control when applications of the cut rule are eliminated.

Theorem 4.9. *The single cut rule is* admissible *in* LA_W*, as well as in the extensions of* LA_W *with* $(K\vdash)$ *or* $(K'\vdash)$*.*

Exercise* 4.2.6. Prove the three claims contained in the theorem. [Hint: A straightforward modification of the proof for LK does the job here.]

The other combination of a pair of structural rules with LA yields BCK°, that is, implicational *BCK logic* with fusion, which is also a fragment of *affine logic*. In other words, we add $(K\vdash)$ to $LL^\circ_\rightarrow$—once again, forgetting about $\leftarrow$ and the two rules that introduce that connective. The label for BCK logic will become clearer in Section 5.3, where we give a different formulation motivated by insights from structurally free logics.

Theorem 4.10. *The single cut rule is* admissible *in* BCK°.

Exercise 4.2.7. Outline the modifications (if there are any) in the proof of the admissibility of cut for LA in order to obtain a proof here.

Let the structuralized version of $(C\vdash)$ be denoted by $(C'\vdash)$.

$$\frac{\Gamma[\Delta, \Theta] \vdash \mathcal{A}}{\Gamma[\Theta, \Delta] \vdash \mathcal{A}}\; C'\vdash$$

Exercise 4.2.8. Compare BCK° to LA with $(C'\vdash)$ and $(K'\vdash)$ added and $\leftarrow$ disregarded. [Hint: You might consider whether $(C'\vdash)$ is derivable when $(C\vdash)$ is available. Another question is whether $(C'\vdash)$ can preclude the admissibility of the cut rule.]

None of the calculi in this section so far has theorems defined as provable formulas. One way to remedy the problem is to introduce a *zero-ary constant* that, by appearing in a certain place, will indicate that a wff is a theorem. Sometimes, this constant is denoted by $\mathbf{1}$, because it is a sort of (left) *identity element*. We will normally call this constant the *intensional truth* constant and we use the notation $\boldsymbol{t}$ for it. This constant can be also thought of as the *least theorem* among all the theorems, and this concept can be made more tangible when there is a conjunction in a logic.

$\boldsymbol{t}$ plays an important role in relevance logics—both in sequent and consecution calculus formulations of various fragments of relevance logics, and $\boldsymbol{t}$ is a very useful addition in the algebraization of these logics.

In LA, the order of the wff's in the antecedent of a sequent matters. We want to make $\boldsymbol{t}$ into a *left identity* for $\circ$, and to achieve this, we introduce a rule, which we label as $(\boldsymbol{t}\vdash)$.

$$\frac{\Gamma[\Delta] \vdash \mathcal{B}}{\Gamma[\boldsymbol{t}; \Delta] \vdash \mathcal{B}}\; \boldsymbol{t}\vdash$$

$LA + (\boldsymbol{t}\vdash)$ is the calculus that results from adding $(\boldsymbol{t}\vdash)$ to LA. The notion of a *proof* is unchanged, but we say that $\mathcal{A}$ is a *theorem* of $LA + (\boldsymbol{t}\vdash)$, when $\boldsymbol{t} \vdash \mathcal{A}$ has a proof. First, let us establish that $\boldsymbol{t}$ is a left identity for $\circ$.

$$\vdash\to\;\dfrac{\vdash\circ\;\dfrac{\boldsymbol{t}\vdash\boldsymbol{t}\qquad\mathcal{A}\vdash\mathcal{A}}{\boldsymbol{t};\mathcal{A}\vdash\boldsymbol{t}\circ\mathcal{A}}}{\boldsymbol{t}\vdash\mathcal{A}\to(\boldsymbol{t}\circ\mathcal{A})}
\qquad\qquad
\dfrac{\dfrac{\dfrac{\dfrac{\mathcal{A}\vdash\mathcal{A}}{\boldsymbol{t};\mathcal{A}\vdash\mathcal{A}}\;\boldsymbol{t}\vdash}{\boldsymbol{t}\circ\mathcal{A}\vdash\mathcal{A}}\;\circ\vdash}{\boldsymbol{t};\boldsymbol{t}\circ\mathcal{A}\vdash\mathcal{A}}\;\boldsymbol{t}\vdash}{\boldsymbol{t}\vdash(\boldsymbol{t}\circ\mathcal{A})\to\mathcal{A}}\;\vdash\to$$

Notice that in the proof of $\mathcal{A} \to (\boldsymbol{t}\circ\mathcal{A})$, $(\boldsymbol{t}\vdash)$ is not used at all. Algebraically speaking, the first proof shows that $\boldsymbol{t}$ is *upper* left identity for $\circ$, whereas the second proof establishes that it is also *lower* left identity. The proof on the left uses an instance of the (id) axiom, in which $\mathcal{A}$ is instantiated by $\boldsymbol{t}$.

Exercise* 4.2.9. We have already considered several calculi in this section. In each, there is an axiom schema, (id). Can this axiom schema be replaced in any of the calculi in this and in the previous section by $p \vdash p$, where p is a propositional variable? [Hint: If the answer is "yes," then the proof is more elaborate than if the answer is "no." The modification should not affect the set of provable sequents.]

It might appear that the replacement theorem for $LA + (\boldsymbol{t}\vdash)$ is inherited from LA. However, now we want to define the equivalence relation—that we hope to show to be a congruence relation—from the notion of theoremhood. Namely, $\mathcal{A}$ and $\mathcal{B}$ are *equivalent* iff $\boldsymbol{t}\vdash\mathcal{A}\to\mathcal{B}$ and $\boldsymbol{t}\vdash\mathcal{B}\to\mathcal{A}$ are both provable. If we try to prove $\boldsymbol{t}\vdash(\mathcal{C}\to\mathcal{A})\to\mathcal{C}\to\mathcal{B}$, for instance, then we will need to apply the cut rule, because none of the rules in $LA + (\boldsymbol{t}\vdash)$ allows taking apart a wff like $\mathcal{A}\to\mathcal{B}$, but the latter is not a subformula of our target formula, and neither is $\mathcal{B}\to\mathcal{A}$.

Let us assume for a moment that we have the single cut rule available to us. Then we can obtain from $\boldsymbol{t}\vdash\mathcal{A}\to\mathcal{B}$ the sequent $\boldsymbol{t};\mathcal{A}\vdash\mathcal{B}$ by a cut with $\mathcal{A}\to\mathcal{B};\mathcal{A}\vdash\mathcal{B}$. We get the following two proof chunks.

$$\vdash\to\text{'s}\;\dfrac{\to\vdash\;\dfrac{\mathcal{C}\vdash\mathcal{C}\qquad\begin{matrix}\vdots\\ \boldsymbol{t};\mathcal{A}\vdash\mathcal{B}\end{matrix}}{\boldsymbol{t};\mathcal{C}\to\mathcal{A};\mathcal{C}\vdash\mathcal{B}}}{\boldsymbol{t}\vdash(\mathcal{C}\to\mathcal{A})\to\mathcal{C}\to\mathcal{B}}
\qquad\qquad
\dfrac{\dfrac{\begin{matrix}\vdots\\ \boldsymbol{t};\mathcal{A}\vdash\mathcal{B}\end{matrix}\qquad\mathcal{C}\vdash\mathcal{C}}{\mathcal{B}\to\mathcal{C};\boldsymbol{t};\mathcal{A}\vdash\mathcal{C}}\;\to\vdash}{\mathcal{B}\to\mathcal{C};\boldsymbol{t}\vdash\mathcal{A}\to\mathcal{C}}\;\vdash\to$$

We succeeded in proving $(\mathcal{C}\to\mathcal{A})\to\mathcal{C}\to\mathcal{B}$, but we cannot obtain $(\mathcal{B}\to\mathcal{C})\to\mathcal{A}\to\mathcal{C}$ as a theorem. We could get instead $(\mathcal{A}\to\mathcal{C})\leftarrow\mathcal{B}\to\mathcal{C}$ or $(\mathcal{B}\to\mathcal{C})\to\boldsymbol{t}\to\mathcal{A}\to\mathcal{C}$ as theorems.

Another idea that may appear to be plausible is to show that if $\boldsymbol{t};\mathcal{A}\vdash\mathcal{B}$ is provable, then $\mathcal{A}\vdash\mathcal{B}$ is provable too.

Exercise 4.2.10. Explore whether it can be demonstrated that the provability of the sequent $\boldsymbol{t};\mathcal{A}\vdash\mathcal{B}$ implies the provability of the sequent $\mathcal{A}\vdash\mathcal{B}$.

If we try to prove the rest of the replacement theorem—still assuming that we have the cut rule available—then we run into problems again.

$$\dfrac{\dfrac{\dfrac{\begin{matrix}\vdots\\ t;\mathcal{A}\vdash\mathcal{B}\end{matrix}\quad \mathcal{C}\vdash\mathcal{C}}{t;\mathcal{A};\mathcal{C}\vdash\mathcal{B}\circ\mathcal{C}}\,{\scriptstyle\vdash\circ}}{t;\mathcal{A}\circ\mathcal{C}\vdash\mathcal{B}\circ\mathcal{C}}\,{\scriptstyle\circ\vdash}}{t\vdash(\mathcal{A}\circ\mathcal{C})\to(\mathcal{B}\circ\mathcal{C})}\,{\scriptstyle\vdash\to}\qquad\qquad \dfrac{\mathcal{C}\vdash\mathcal{C}\quad \begin{matrix}\vdots\\ t;\mathcal{A}\vdash\mathcal{B}\end{matrix}}{\mathcal{C};t;\mathcal{A}\vdash\mathcal{C}\circ\mathcal{B}}\,{\scriptstyle\vdash\circ}$$

We could continue the proof on the right to obtain either $t\vdash((\mathcal{C}\circ t)\circ\mathcal{A})\to(\mathcal{C}\circ\mathcal{B})$ or $t\vdash(\mathcal{C}\circ(t\circ\mathcal{A}))\to(\mathcal{C}\circ\mathcal{B})$, but we would like to have $t\vdash(\mathcal{C}\circ\mathcal{A})\to(\mathcal{C}\circ\mathcal{B})$.

$$\dfrac{\begin{matrix}\vdots\\ t;\mathcal{A}\vdash\mathcal{B}\end{matrix}\quad \mathcal{C}\vdash\mathcal{C}}{t;\mathcal{A};\mathcal{C}\leftarrow\mathcal{B}\vdash\mathcal{C}}\,{\scriptstyle\leftarrow\vdash}\qquad\qquad \dfrac{\mathcal{C}\vdash\mathcal{C}\quad \begin{matrix}\vdots\\ t;\mathcal{A}\vdash\mathcal{B}\end{matrix}}{t;\mathcal{C};\mathcal{A}\leftarrow\mathcal{C}\vdash\mathcal{B}}\,{\scriptstyle\leftarrow\vdash}$$

There are various formulas that can be proved from the two bottom sequents. Some of them are: $t\vdash\mathcal{A}\to(\mathcal{C}\leftarrow\mathcal{B})\to\mathcal{C}$, $t\vdash(\mathcal{C}\leftarrow\mathcal{B})\to(\mathcal{C}\leftarrow t)\leftarrow\mathcal{A}$, and $t\vdash\mathcal{C}\to(\mathcal{A}\leftarrow\mathcal{C})\to\mathcal{B}$, $t\vdash(\mathcal{A}\leftarrow\mathcal{C})\to(\mathcal{B}\leftarrow t)\leftarrow\mathcal{C}$.

Exercise 4.2.11. Convince yourself that the two sequents that we would like to prove, namely, $t\vdash(\mathcal{C}\leftarrow\mathcal{B})\to\mathcal{C}\leftarrow\mathcal{A}$ and $t\vdash(\mathcal{A}\leftarrow\mathcal{C})\to\mathcal{B}\leftarrow\mathcal{C}$ are not provable. [Hint: Try to continue the above proof segments and convince yourself that there are no other proofs either.]

The above proof-attempts illustrate very well the problems that emerge in $LA+(t\vdash)$, when we try to algebraize the logic. By this we mean that there is no way to circumvent (e.g., by inserting additional applications of the cut rule) the unprovability of the target sequents. Of course, one might observe that all the problems arise, simply because of the introduction of the t; perhaps, we can do without a truth constant.

An alternative approach, indeed, is to permit the antecedent to become empty, and treat that as the sign of a formula being a theorem. In certain cases, this simplification is possible, whereas in others such a simplification would yield unacceptable results. Then the empty sequence plainly disappears when inserted into the sequence of wff's, thereby, causing no problems. In effect, the empty sequence acts as a *left-right identity* (really like ε, the empty string, does for concatenation). We will return to such calculi later on, however, now we introduce LA_1^t.

The associative Lambek calculus with t, denoted by LA_1^t comprises the axiom and rules of LA, together with $(t\vdash)$ and three special structural rules, which are as follows.

$$\dfrac{\Gamma[\Delta;t]\vdash\mathcal{A}}{\Gamma[t;\Delta]\vdash\mathcal{A}}\,{\scriptstyle\overleftarrow{t}\vdash}\qquad\qquad \dfrac{\Gamma[t;\Delta]\vdash\mathcal{A}}{\Gamma[\Delta;t]\vdash\mathcal{A}}\,{\scriptstyle\overrightarrow{t}\vdash}\qquad\qquad \dfrac{\Gamma[t;t]\vdash\mathcal{A}}{\Gamma[t]\vdash\mathcal{A}}\,{\scriptstyle\overset{tt}{t}\vdash}$$

The notion of a *proof* is as before. $\mathcal{A}$ is a *theorem* when $t\vdash\mathcal{A}$ is provable.

Exercise 4.2.12. Assume that the single cut rule is admissible. Prove the replacement theorem for LA^t_1. [Hint: Finish the proof attempts for $LA + (\boldsymbol{t}\vdash)$, by using some of the new structural rules.]

To motivate the $(^{tt}_{\ t}\vdash)$ rule, we consider how we could emulate modus ponens. (Even if the rule of detachment is not sufficient to axiomatize LA or LA^t_1, it is a reasonable rule to include.)

$$\dfrac{\dfrac{\vdots \atop \boldsymbol{t} \vdash \mathcal{A} \qquad \dfrac{\vdots \atop \boldsymbol{t} \vdash \mathcal{A} \to \mathcal{B} \qquad \mathcal{A} \to \mathcal{B}; \mathcal{A} \vdash \mathcal{B}}{\boldsymbol{t}; \mathcal{A} \vdash \mathcal{B}}\,\text{cut}}{\boldsymbol{t}; \boldsymbol{t} \vdash \mathcal{B}}\,\text{cut}}{\boldsymbol{t} \vdash \mathcal{B}}\,{}^{tt}_{\ t}\vdash$$

For $LA + (\boldsymbol{t}\vdash)$ and LA^t_1, we assumed that the cut rule is available. This is justified by the next theorem.

Theorem 4.11. *The single cut rule is* admissible *in* LA^t_1.

Proof: The proof for LA is straightforward. However, looking at that proof and trying to add steps to it, we might think that we cannot prove the theorem; hence, it is flawed. Of course, the lack of a proof is not sufficient for the falsification of a theorem. We saw how to show that the cut rule is not admissible. The following example shows where we run into difficulties, when we try to re-use the previous proof with some new cases added.

$$\dfrac{\vdots \atop \Gamma \vdash \boldsymbol{t} \qquad \dfrac{\vdots \atop \Delta[\boldsymbol{t}; \Theta; \Lambda] \vdash \mathcal{A}}{\Delta[\Theta; \boldsymbol{t}; \Lambda] \vdash \mathcal{A}}\,\overset{\to}{\boldsymbol{t}}\vdash}{\Delta[\Theta; \Gamma; \Lambda] \vdash \mathcal{A}}\,\text{cut}$$

The usual move in a case like this is to swap the two rules—as, for instance, in Cases **2.1.1 (11)–(13)**, in Section 2.3. In other words, we would want to move the cut upward in the proof tree, and then apply the same rule (possibly, several times), and perhaps, make other necessary but permitted adjustments to the resulting sequent. Here this would lead to the following proof segment, where **?** shows that there is no rule (or series of rules) to obtain the previous bottom sequent.

$$\dfrac{\dfrac{\vdots \atop \Gamma \vdash \boldsymbol{t} \qquad \vdots \atop \Delta[\boldsymbol{t}; \Theta; \Lambda] \vdash \mathcal{A}}{\Delta[\Gamma; \Theta; \Lambda] \vdash \mathcal{A}}\,\text{cut}}{\Delta[\Theta; \Gamma; \Lambda] \vdash \mathcal{A}}\,?$$

We can permute $\boldsymbol{t}$ with any sequence of wff's, however, Γ and Θ cannot be guaranteed to comprise just some $\boldsymbol{t}$'s.

However, $\boldsymbol{t}$ is a *special formula* not only because there are special rules that are applicable only in the presence of $\boldsymbol{t}$, but also because it is neither a propositional variable, nor a compound formula. $\boldsymbol{t}$ alone may be introduced in exactly one way on the right-hand side of the turnstile, namely, by an instance of (id). A similar observation concerning the fixed point combinator led to the idea to augment the proof of the cut theorem (for multiple cut) in structurally free logics by an induction in [31]. An extra induction can prove that no matter which rules were used to obtain $\Gamma \vdash \boldsymbol{t}$, the steps can be replicated in the context of the antecedent of the right premise. Incidentally, if we would order the cases in the proof of the cut theorem requiring that the cases where $\varrho_l > 1$ are to be dealt with before the cases in which $\varrho_r > 1$, then we could show that the above situation does not need to be considered among the cases where $\varrho_r > 1$. qed

Exercise 4.2.13. Finish the proof of the cut theorem. [Hint: Section 7.3 contains a detailed example for the additional induction.]

Exercise* 4.2.14. Present the proof of the cut theorem so that the additional lemma is incorporated into the whole proof via an ordering of the cases.

We started off with the idea of adding a *left identity* for $\circ$, because the implication operation is the right residual of fusion. However, the desire to have the replacement theorem forced us to turn $\boldsymbol{t}$ into *identity*.

$$\vdash\to\ \dfrac{\overleftarrow{\boldsymbol{t}}\vdash\ \dfrac{\vdash\circ\ \dfrac{\mathcal{A}\vdash\mathcal{A} \qquad \boldsymbol{t}\vdash\boldsymbol{t}}{\mathcal{A};\boldsymbol{t}\vdash\mathcal{A}\circ\boldsymbol{t}}}{\boldsymbol{t};\mathcal{A}\vdash\mathcal{A}\circ\boldsymbol{t}}}{\boldsymbol{t}\vdash\mathcal{A}\to(\mathcal{A}\circ\boldsymbol{t})} \qquad \dfrac{\dfrac{\dfrac{\dfrac{\mathcal{A}\vdash\mathcal{A}}{\boldsymbol{t};\boldsymbol{t};\mathcal{A}\vdash\mathcal{A}}\ \boldsymbol{t}\vdash\text{'s}}{\boldsymbol{t};\mathcal{A};\boldsymbol{t}\vdash\mathcal{A}}\ \overrightarrow{\boldsymbol{t}}\vdash}{\boldsymbol{t};\mathcal{A}\circ\boldsymbol{t}\vdash\mathcal{A}}\ \circ\vdash}{\boldsymbol{t}\vdash(\mathcal{A}\circ\boldsymbol{t})\to\mathcal{A}}\ \vdash\to$$

The three latest structural rules specific to $\boldsymbol{t}$ can be added similarly to the other calculi that we have considered. Furthermore, in some cases, obvious simplifications are possible, because $(\overleftarrow{\boldsymbol{t}}\vdash)$ and $(\overrightarrow{\boldsymbol{t}}\vdash)$ are instances of $(C'\vdash)$, whereas $(\overset{tt}{\boldsymbol{t}}\vdash)$is an instance of $(W\vdash)$. The $(\boldsymbol{t}\vdash)$ rule is almost an instance of $(K\vdash)$; $\boldsymbol{t}$ is introduced on the left of Δ, whereas $\mathcal{B}$ in $(K\vdash)$ and Θ in $(K'\vdash)$ are added on the right.

As long as we do not add grouping to the sequences of wff's, that is, we keep them "flat," and we do not have more than one kind of sequence, that is, the only structural connective is ;, it is possible to introduce $\boldsymbol{t}$ without special structural rules. This approach is less general and it requires changing the whole calculus, but sometimes it is sufficient for simpler sequent calculi.

LA can be extended this way with $\boldsymbol{t}$, but we have to change some of our earlier definitions. The main modification is that we allow the antecedent of a sequent to be empty. Accordingly, $\Gamma, \Delta, \ldots$ range over sequences of wff's including the *empty sequence*. The BNF description of the sequences of formulas includes the space now (where wff rewrites to a formula in the language, which has been extended with $\boldsymbol{t}$).

$$\Gamma ::= \ \mid \text{wff} \mid \Gamma;\Gamma$$

The *sequent calculus* LA_2^t contains the axiom and the rules of LA together with the rule $(t\vdash)$ and the following axiom.

$$\vdash \boldsymbol{t} \quad \vdash t$$

The notion of a *proof* is the usual one, but now $\mathcal{A}$ is a *theorem* of LA_2^t, when $\vdash \mathcal{A}$ is provable.

We can similarly formulate LA_{W2}^t, LA_{K2}^t, $LL_{\to 2}^{\circ t}$, $BCK_2^{\circ t}$ and LA_{KW2}^t (where the last logic contains $(K\vdash)$ and $(W'\vdash)$ over LA_2^t). The addition of $\boldsymbol{t}$ to the logic of relevant implication will be dealt with in Section 4.3.

Theorem 4.12. *The single cut rule is* admissible *in* LA_2^t, LA_W^t, LA_K^t, $LL_{\to}^{\circ t}$, $BCK^{\circ t}$ *and* LA_{KW}^t.

Exercise 4.2.15. Prove the above theorem. [Hint: Try to discern the general pattern of how the proofs for the $\boldsymbol{t}$-less calculi have to be augmented to work for the new calculi.]

Exercise* 4.2.16. Prove that LA_1^t and LA_2^t, and the five extensions of each, which we have considered so far, are pairwise equivalent. [Hint: Recall that not only the sets of axioms and rules differ, but also the notions of theoremhood.]

Exercise 4.2.17.** Prove that all the logics, that is, LA_2^t, LA_{W2}^t, LA_{K2}^t, $LL_{\to 2}^{\circ t}$, $BCK_2^{\circ t}$ and LA_{KW2}^t are distinct. [Hint: The names for the logics give away what you can prove in them, but it may require a more complicated argument to show that a particular wff is not provable in one or another logic.]

Incidentally, all these calculi are also distinct from $LR_{\to}^t$, the logic of relevant implication with truth in Section 4.3. Finding a proof for this claim is Exercise 4.3.4.

The use of a *space* to indicate theoremhood is in line with the original approach in LK, however, it is not suitable for all logics, as we will see in Chapter 5. The double-entry bookkeeping approach to having $\boldsymbol{t}$, but "sort of not always," leads to the following lemma.

Lemma 4.13. *For any formula* $\mathcal{A}$, *in any of the logics* LA_2^t, LA_{W2}^t, LA_{K2}^t, $LL_{\to 2}^{\circ t}$, $BCK_2^{\circ t}$ *and* LA_{KW2}^t, $\boldsymbol{t} \vdash \mathcal{A}$ *is* provable *if and only if* $\vdash \mathcal{A}$ *is* provable *(in the same logic).*

Exercise 4.2.18. Prove the lemma. [Hint: One direction is obvious; for the other, use induction on the height of the proof tree.]

The rules in LA allow *different contexts*, and in the absence of the structural rules, $(\circ\vdash)$ and $(\vdash\circ)$ cannot be shown to be equivalent to their versions that stipulate the same context. Belnap, in [21], calls *structure-free rules* the versions of the rules for conjunction and disjunction with the same context and with one subformula in the left conjunction and the right disjunction introduction rules. The insight behind the label "structure-free" is that if the contexts in the premises of the rule are the same, and the introduction of the principal formula does not require the presence and does not force the introduction of a structural connective, then the rule is *independent* from the properties of the context. That is, such rules do not link in any way the structural connective and the connective introduced by the rules.

The conjunction and disjunction rules in LK and LJ are almost structure-free. The reason why they are not quite structure-free is because of Gentzen's fixation on the edges of sequences of wff's. However, that is easy to repair, given that we already have a suitably general notation to indicate an occurrence of a formula in a sequence of wff's—without specifying its position with respect to the other wff's (if there are any) in the sequence.

We add two new binary connectives, *conjunction* ($\wedge$) and *disjunction* ($\vee$) to the language. The set of wff's is defined as in Definition 4.1, where now $\div$ can also stand for these two connectives.

The *positive Lambek calculus* LA^t_{+2} is defined by (id) and ($\vdash t$) as axioms, and by the rules ($\rightarrow\vdash$), ($\vdash\rightarrow$), ($\circ\vdash$), ($\vdash\circ$), ($\leftarrow\vdash$), ($\vdash\leftarrow$) and ($t\vdash$), together with the following six rules for $\wedge$ and $\vee$. That is, LA^t_{+2} is LA^t_2 with ($\wedge\vdash_1$), ($\wedge\vdash_2$), ($\vdash\wedge$), ($\vee\vdash$), ($\vdash\vee_1$) and ($\vdash\vee_2$) added.

$$\frac{\Gamma[\mathcal{A}]\vdash\mathcal{C}}{\Gamma[\mathcal{A}\wedge\mathcal{B}]\vdash\mathcal{C}}\;\wedge\vdash_1 \qquad \frac{\Gamma[\mathcal{B}]\vdash\mathcal{C}}{\Gamma[\mathcal{A}\wedge\mathcal{B}]\vdash\mathcal{C}}\;\wedge\vdash_2 \qquad \frac{\Gamma\vdash\mathcal{A}\quad\Gamma\vdash\mathcal{B}}{\Gamma\vdash\mathcal{A}\wedge\mathcal{B}}\;\vdash\wedge$$

$$\frac{\Gamma[\mathcal{A}]\vdash\mathcal{C}\quad\Gamma[\mathcal{B}]\vdash\mathcal{C}}{\Gamma[\mathcal{A}\vee\mathcal{B}]\vdash\mathcal{C}}\;\vee\vdash \qquad \frac{\Gamma\vdash\mathcal{A}}{\Gamma\vdash\mathcal{A}\vee\mathcal{B}}\;\vdash\vee_1 \qquad \frac{\Gamma\vdash\mathcal{B}}{\Gamma\vdash\mathcal{A}\vee\mathcal{B}}\;\vdash\vee_2$$

The notion of a *proof* is unchanged and $\mathcal{A}$ is a *theorem* iff $\vdash\mathcal{A}$ is a provable sequent.

Theorem 4.14. *The single cut rule is* admissible *in* LA^t_{+2}.

Exercise 4.2.19. Prove the theorem. [Hint: Note the cases that have to be added to the proof of the cut theorem for LA^t_2.]

Exercise 4.2.20. Can LA^t_1 be similarly extended with $\wedge$ and $\vee$? Is it a well-behaved sequent calculus (i.e., is the cut rule admissible)?

Exercise* 4.2.21. Extend with $\vee$ and $\wedge$ the five other logics—LA^t_{Wi}, LA^t_{Ki}, $LL^{\circ t}_{\rightarrow i}$, $BCK^{\circ t}_i$ and LA^t_{KWi}, where i is 1 or 2. Prove that the cut rule is admissible. [Hint: Try to systematize the cases that have to be added to the proof.]

The two versions of the six logics that we considered in this section are equivalent with respect to the set of their theorems. This remains true of their extensions with the conjunction and disjunction rules given above.

Exercise* 4.2.22. Prove the last claim. [Hint: Recall that the $\wedge$ and $\vee$ rules are structure-free.]

We expanded the language with two new connectives. We would like to be able to prove that the equivalence relation that was sufficient for replacement before, is sufficient now.

Let us assume that we consider LA^t_{K+1}, which has no empty sequences of wff's. The equivalence relation between $\mathcal{A}$ and $\mathcal{B}$ obtains when $\boldsymbol{t} \vdash \mathcal{A} \to \mathcal{B}$ and $\boldsymbol{t} \vdash \mathcal{B} \to \mathcal{A}$ are provable. Here is a proof of $(\mathcal{A} \wedge \mathcal{C}) \to (\mathcal{B} \wedge \mathcal{C})$, which is one of the wff's that has to be proved as part of the replacement theorem.

$$\dfrac{\dfrac{\dfrac{\vdots}{t; \mathcal{A} \vdash \mathcal{B}}}{t; \mathcal{A} \wedge \mathcal{C} \vdash \mathcal{B}}\, {\scriptstyle \wedge\vdash_1} \qquad \dfrac{\dfrac{\mathcal{C} \vdash \mathcal{C}}{t; \mathcal{C} \vdash \mathcal{C}}\, {\scriptstyle t\vdash}}{t; \mathcal{A} \wedge \mathcal{C} \vdash \mathcal{C}}\, {\scriptstyle \wedge\vdash_2}}{\dfrac{t; \mathcal{A} \wedge \mathcal{C} \vdash \mathcal{B} \wedge \mathcal{C}}{\boldsymbol{t} \vdash (\mathcal{A} \wedge \mathcal{C}) \to (\mathcal{B} \wedge \mathcal{C})}\, {\scriptstyle \vdash\to}}\, {\scriptstyle \vdash\wedge}$$

The cut rule is admissible, hence, $t; \mathcal{A} \vdash \mathcal{B}$ has a proof. It is easy to see, that if we would start the left branch with $t; \mathcal{B} \vdash \mathcal{A}$, then we would get—by applications of the same rules—$\boldsymbol{t} \vdash (\mathcal{B} \wedge \mathcal{C}) \to (\mathcal{A} \wedge \mathcal{C})$.

The other sequents that have to be proved, given the assumption that $\mathcal{A}$ and $\mathcal{B}$ are equivalent, are $\boldsymbol{t} \vdash (\mathcal{C} \wedge \mathcal{A}) \to (\mathcal{C} \wedge \mathcal{B})$, $\boldsymbol{t} \vdash (\mathcal{C} \wedge \mathcal{B}) \to (\mathcal{C} \wedge \mathcal{A})$, $\boldsymbol{t} \vdash (\mathcal{C} \vee \mathcal{B}) \to (\mathcal{C} \vee \mathcal{A})$, $\boldsymbol{t} \vdash (\mathcal{C} \vee \mathcal{A}) \to (\mathcal{C} \vee \mathcal{B})$, $\boldsymbol{t} \vdash (\mathcal{A} \vee \mathcal{C}) \to (\mathcal{B} \vee \mathcal{C})$ and $\boldsymbol{t} \vdash (\mathcal{B} \vee \mathcal{C}) \to (\mathcal{A} \vee \mathcal{C})$.

Exercise 4.2.23. Finish the proof of the replacement theorem. [Hint: The proofs do not rely on any of the structural rules $(C\vdash)$, $(W'\vdash)$ or $(K\vdash)$.]

Exercise 4.2.24. Prove the replacement theorem for LA^t_{+2} (and its five extensions). [Hint: Try to find a way to systematically transform the proofs for LA^t_{+1}.]

Conjunction and disjunction are often called *lattice operations*. This label involves a slight (but harmless) confusion between the language of a logic and the Lindenbaum algebra that results from a logic. But once the difference has been made clear, the term should be unproblematic and it may be even useful. A lattice is a well-understood kind of algebra, and it can be defined by finitely many equations, that is, it is a *variety*. For example, we can prove that $\mathcal{A} \wedge \mathcal{A}$ is equivalent to $\mathcal{A}$, which leads to the algebraic property called *idempotence* for the type-lifted $\wedge$. Let us assume that we are working with one of the calculi in which the empty sequence is permitted. We easily prove the following.

$$\dfrac{\dfrac{\mathcal{A} \vdash \mathcal{A} \qquad \mathcal{A} \vdash \mathcal{A}}{\mathcal{A} \vdash \mathcal{A} \wedge \mathcal{A}}\, {\scriptstyle \vdash\wedge}}{\vdash \mathcal{A} \to (\mathcal{A} \wedge \mathcal{A})}\, {\scriptstyle \vdash\to} \qquad\qquad \dfrac{\dfrac{\mathcal{A} \vdash \mathcal{A}}{\mathcal{A} \wedge \mathcal{A} \vdash \mathcal{A}}\, {\scriptstyle \wedge\vdash}}{\vdash (\mathcal{A} \wedge \mathcal{A}) \to \mathcal{A}}\, {\scriptstyle \vdash\to}$$

Exercise 4.2.25. Prove that the following pairs of wff's are equivalent. (1) $(\mathcal{A} \wedge \mathcal{B}) \wedge \mathcal{C}$ and $\mathcal{A} \wedge (\mathcal{B} \wedge \mathcal{C})$, (2) $\mathcal{A} \wedge \mathcal{B}$ and $\mathcal{B} \wedge \mathcal{A}$, (3) $\mathcal{A} \vee \mathcal{A}$ and $\mathcal{A}$, (4) $\mathcal{A} \vee \mathcal{B}$ and $\mathcal{B} \vee \mathcal{A}$, (5) $(\mathcal{A} \vee \mathcal{B}) \vee \mathcal{C}$ and $\mathcal{A} \vee (\mathcal{B} \vee \mathcal{C})$.

The above formulas do not say anything about interactions between conjunction and disjunction. We might ask whether *absorption* is provable.

$$\wedge\vdash_1\ \dfrac{\mathcal{A}\vdash\mathcal{A}}{\mathcal{A}\wedge(\mathcal{B}\vee\mathcal{A})\vdash\mathcal{A}} \qquad\qquad \dfrac{\mathcal{A}\vdash\mathcal{A}\qquad \dfrac{\mathcal{A}\vdash\mathcal{A}}{\mathcal{A}\vdash\mathcal{B}\vee\mathcal{A}}\ {\vdash\vee_2}}{\mathcal{A}\vdash\mathcal{A}\wedge(\mathcal{B}\vee\mathcal{A})}\ {\vdash\wedge}$$

These proofs involve only some of the rules for $\wedge$ and $\vee$, hence, they are proofs in each calculus that we have considered so far in this section. The proofs establishing that $\mathcal{A}$ and $\mathcal{A}\vee(\mathcal{A}\wedge\mathcal{B})$ are equivalent do not require applications of the structural rules either. The proofs in the calculi with empty sequences can be expanded by a $(\vdash\rightarrow)$ step, and in the other calculi by a $(\boldsymbol{t}\vdash)$ and a $(\vdash\rightarrow)$ step.

In LK, the interaction of $\wedge$ and $\vee$ does not end with absorption. For example, $(\mathcal{A}\vee\mathcal{B})\wedge(\mathcal{A}\vee\mathcal{C})\vdash\mathcal{A}\vee(\mathcal{B}\wedge(\mathcal{A}\vee\mathcal{C}))$ and $\mathcal{A}\wedge(\mathcal{B}\vee\mathcal{C})\vdash(\mathcal{A}\wedge\mathcal{B})\vee(\mathcal{A}\wedge\mathcal{C})$ are provable sequents.

Exercise* 4.2.26. Is either sequent provable in the six calculi? [Hint: Prove the sequents in LK, and then try to re-construct the proofs in LL^t_{+i}, $BCK^{\circ t}_{+i}$ and LA^t_{WK+i}, where i is 1 or 2.]

LA introduced the three connectives $\leftarrow$, $\circ$ and $\rightarrow$ all at once, and the rules were formulated so that these connectives were linked by residuation. $\wedge$ and $\vee$ are not similarly connected to each other or to the three original connectives. However, certain relationships between $\wedge$ and $\vee$, on one hand, and $\leftarrow$, $\circ$ and $\rightarrow$, on the other, fall out from the rules.

$$\dfrac{\dfrac{\dfrac{\vdash\vee_1\ \dfrac{\vdash\circ\ \dfrac{\mathcal{A}\vdash\mathcal{A}\qquad\mathcal{B}\vdash\mathcal{B}}{\mathcal{A};\mathcal{B}\vdash\mathcal{A}\circ\mathcal{B}}}{\mathcal{A};\mathcal{B}\vdash(\mathcal{A}\circ\mathcal{B})\vee(\mathcal{A}\circ\mathcal{C})}\qquad \dfrac{\dfrac{\mathcal{A}\vdash\mathcal{A}\qquad\mathcal{C}\vdash\mathcal{C}}{\mathcal{A};\mathcal{C}\vdash\mathcal{A}\circ\mathcal{C}}\ {\vdash\circ}}{\mathcal{A};\mathcal{C}\vdash(\mathcal{A}\circ\mathcal{B})\vee(\mathcal{A}\circ\mathcal{C})}\ {\vdash\vee_2}}{\mathcal{A};\mathcal{B}\vee\mathcal{C}\vdash(\mathcal{A}\circ\mathcal{B})\vee(\mathcal{A}\circ\mathcal{C})}\ {\vee\vdash}}{\mathcal{A}\circ(\mathcal{B}\vee\mathcal{C})\vdash(\mathcal{A}\circ\mathcal{B})\vee(\mathcal{A}\circ\mathcal{C})}\ {\circ\vdash}}{\vdash(\mathcal{A}\circ(\mathcal{B}\vee\mathcal{C}))\rightarrow((\mathcal{A}\circ\mathcal{B})\vee(\mathcal{A}\circ\mathcal{C}))}\ {\vdash\rightarrow}$$

It is a remarkable feature of sequent calculi—already in these relatively simple extensions of LA—that $\circ$ and the implication-like connectives have *distribution properties* with respect to conjunction and disjunction.[6]

Exercise 4.2.27. Prove (1)–(5) in LA^t_+. (1) $\boldsymbol{t}\vdash((\mathcal{A}\circ\mathcal{B})\vee(\mathcal{A}\circ\mathcal{C}))\rightarrow(\mathcal{A}\circ(\mathcal{B}\vee\mathcal{C}))$, (2) $\boldsymbol{t}\vdash((\mathcal{B}\leftarrow\mathcal{A})\wedge(\mathcal{C}\leftarrow\mathcal{A}))\rightarrow((\mathcal{B}\wedge\mathcal{C})\leftarrow\mathcal{A})$, (3) $\boldsymbol{t}\vdash((\mathcal{B}\wedge\mathcal{C})\leftarrow\mathcal{A})\rightarrow((\mathcal{B}\leftarrow\mathcal{A})\wedge(\mathcal{C}\leftarrow\mathcal{A}))$, (4) $\boldsymbol{t}\vdash((\mathcal{B}\leftarrow\mathcal{A})\wedge(\mathcal{B}\leftarrow\mathcal{C}))\rightarrow(\mathcal{B}\leftarrow(\mathcal{A}\vee\mathcal{C}))$ and (5) $\boldsymbol{t}\vdash(\mathcal{B}\leftarrow(\mathcal{A}\vee\mathcal{C}))\rightarrow((\mathcal{B}\leftarrow\mathcal{A})\wedge(\mathcal{B}\leftarrow\mathcal{C}))$.

$\boldsymbol{t}$ is clearly the most important zero-ary logical constant. It is inherently connected to *theoremhood* in the way $\rightarrow$ is connected to *logical entailment*.

[6]Residuation and distribution types are at the heart of gaggle theory, which was introduced by Dunn in [78], and is presented in a comprehensive form in [38].

However, sometimes it is useful to enrich the vocabulary of a logic further. For example, for topological representation theorems, we might want to have at least a bounded lattice as a reduct of the algebra of the logic. If we permit the set of propositional variables to be infinite, then there is no guarantee that there are wff's $\mathcal{A}$ and $\mathcal{B}$ with the property that for any propositional variable p, $\mathcal{A} \to p$ and $p \to \mathcal{B}$.

Intensional truth implies *all theorems*, whereas extensional truth is implied by *all formulas*. *Extensional truth* is often denoted by $\boldsymbol{T}$ or $\top$. The latter shows that a wff that is implied by any wff is the top element among all the formulas. It is easy to see that if "all theorems are equal," so to speak, that is, a theorem implies any other theorem, then we cannot distinguish $\boldsymbol{t}$ and $\boldsymbol{T}$. If $(K\vdash)$ is a rule, then the formula $\mathcal{B}$ (or the sequence of wff's Θ in $(K'\vdash)$) may be another theorem. This phenomenon is well-known to relevance logicians, because one of the consequences of the lack of thinning is that there are theorems that do not imply each other.

The dual of extensional truth is *extensional falsity*, which is often denoted by $\boldsymbol{F}$ or $\bot$. $\boldsymbol{F}$ implies all formulas, and therefore, can be viewed as the bottom element among formulas. The inclusion of $\boldsymbol{T}$ and $\boldsymbol{F}$, with the above mentioned properties bounds a structure. If $\wedge$ and $\vee$ are in the language, with their usual properties, then $\boldsymbol{T}$ and $\boldsymbol{F}$ bound a lattice. In general, a lattice—even a distributive lattice—does not need to have either a least or a greatest element. $\boldsymbol{T}$ and $\boldsymbol{F}$ are not definable in LA^{t}_{+1} or in its extensions that we have considered so far.

If $\boldsymbol{t}$ is thought of as the theorem that implies all theorems, then its dual should be a logical falsity that is implied by all the logical falsities. The reader may wonder what a "logical falsity" is, and it should be acknowledged that it is an awkward term. If there were a negation in the language of these logics, then we could at least talk about the negations of theorems. The expression "logical falsity" seems to be encumbered with semantical connotations, which look inadequate in connection to an entirely syntactic calculus. We could, perhaps, use the term "contradiction," but that would also presuppose either negation or a semantic interpretation. Proof systems are, typically, geared toward producing theorems (and a logical consequence relation)—exactly the opposite of logical falsities. Hopefully, this informal explanation clarifies what is meant. The axioms and rules will fix the properties of the new zero-ary logical constant, which is usually denoted by $\boldsymbol{f}$. This constant often has a somewhat diminished role to play, because of its connection to wff's that are negations of theorems. (It may be interesting to note that in classical logic all these zero-ary connectives are definable, moreover, given the informal meaning described, they turn out to be equivalent to $\mathcal{A} \wedge \neg\mathcal{A}$ ($\boldsymbol{F}$ and $\boldsymbol{f}$) and to $\mathcal{A} \vee \neg\mathcal{A}$ ($\boldsymbol{T}$ and $\boldsymbol{t}$).)

In order to be able to add $\boldsymbol{f}$, we have to change once more the notion of a sequent. The dual of the $\boldsymbol{t}$ axiom would look like $\boldsymbol{f} \vdash$, however, the latter is not a sequent in LA or in its extensions considered so far. Just as in LJ in Chapter 3, a *sequent* is a finite (possibly empty) sequence of wff's followed by $\vdash$, and then by at most one formula.

The *sequent calculus* LA^{tTFf}_{+2} includes $\boldsymbol{T}$, $\boldsymbol{F}$ and f as wff's, and the set of axioms and rules for LA^{t}_{+} is extended by four axioms and a rule.

$$\Gamma[\boldsymbol{F}] \vdash \mathcal{A} \ \ {\scriptstyle F\vdash_1} \qquad\qquad \Gamma[\boldsymbol{F}] \vdash \ \ {\scriptstyle F\vdash_2} \qquad\qquad \Gamma \vdash \boldsymbol{T} \ \ {\scriptstyle \vdash T}$$

$$f \vdash \ \ {\scriptstyle f\vdash} \qquad\qquad \frac{\Gamma \vdash}{\Gamma \vdash f}\ {\scriptstyle \vdash f}$$

Note that two of the axioms and the upper sequent of the $\vdash f$ rule contain empty right-hand sides.[7]

Theorem 4.15. *The single cut rule is* admissible *in* LA^{tTFf}_{+2}.

Exercise 4.2.28. Prove the theorem. [Hint: Start from the proof of the cut theorem for LA^{t}_{+2}.]

Exercise*4.2.29. Develop similar extensions of the other logics in this section, and scrutinize the admissibility of the cut rule.

Exercise 4.2.30. We mentioned that in classical logic $\boldsymbol{T}$ and $\boldsymbol{t}$, as well as $\boldsymbol{F}$ and f turn out to be the same, and each is a definable constant. Are any of the constants provably equivalent in any of the LA logics?

The fusion operation is the intensional generalization of conjunction. We have already seen that $\circ$ is monotone, exactly as $\wedge$ is. $\boldsymbol{F}$ is the null element for $\wedge$ in classical logic, that is, $(\boldsymbol{F} \wedge \mathcal{A}) \supset \boldsymbol{F}$. The axiom $(\boldsymbol{F}\vdash_1)$ ensures that a similar wff is provable in LA^{tTFf}_{+2}.

$$\frac{\dfrac{\boldsymbol{F};\mathcal{A} \vdash \boldsymbol{F}}{\boldsymbol{F}\circ\mathcal{A} \vdash \boldsymbol{F}}\ {\scriptstyle \circ\vdash}}{\vdash (\boldsymbol{F}\circ\mathcal{A}) \to \boldsymbol{F}}\ {\scriptstyle \vdash\to} \qquad\qquad \frac{\dfrac{\mathcal{A};\boldsymbol{F} \vdash \mathcal{A}}{\mathcal{A}\circ\boldsymbol{F} \vdash \mathcal{A}}\ {\scriptstyle \circ\vdash}}{\vdash (\mathcal{A}\circ\boldsymbol{F}) \to \boldsymbol{F}}\ {\scriptstyle \vdash\to}$$

The proofs can be repeated in the context of Γ, where (possibly several) $(\circ\vdash)$ steps would be required to create the antecedents of the implicational wff's. If $\boldsymbol{F}$ is a component of a wff that is a fusion, then that wff is provably equivalent to $\boldsymbol{F}$. The converses of the implications are simply instances of $(\boldsymbol{F}\vdash_1)$. Algebraically, an operation that distributes over $\vee$ (hence, monotone) and preserves the bottom element is called a *normal operator*. The fusion of the LA logics is an example.

To conclude this section, Figure 4.1 overviews the relationships between the logics described here, with the exception of LA_{WC}, which has not been included in this section. LA_{WC} is almost like $LR_{\to}$, which we introduce in the next section. $LR_{\to}$ permits empty left-hand sides, but originally, does not include $\boldsymbol{t}$ or $\circ$.

[7] A closely related calculus—without $\boldsymbol{T}$ and $\boldsymbol{F}$, but with the cut rule included as one of the rules—has been called the "full Lambek calculus" in [86].

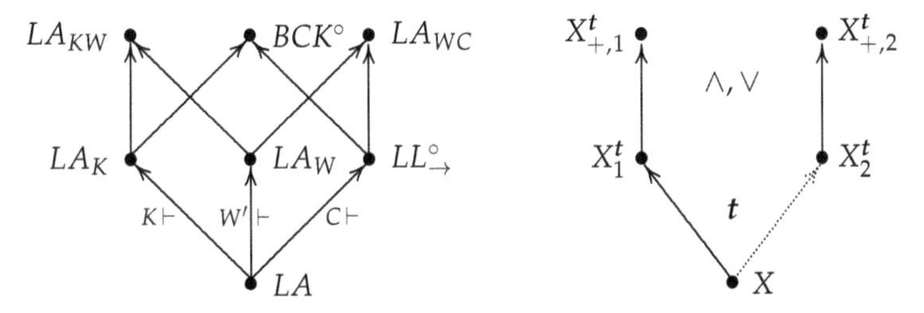

Figure 4.1. Relationships between extensions of LA

The right-hand side diagram in the above figure shows extensions of the LA calculi with t and $\wedge$, $\vee$. X functions as a variable that can stand for either of the six logics (LA, LA_K, LA_W, $LL^\circ_\rightarrow$, LA_{KW} and BCK°). The *dotted arrow* indicates that the notion of a sequent is changed at the same time as new rules are added.

Exercise 4.2.31. Complete Figure 4.1 by situating the extensions that contain T, F and f too.

Exercise 4.2.32. We did not consider LA, LA_K, LA_W, $LL^\circ_\rightarrow$, LA_{KW} and BCK° with empty sequences of wff's but without t. Would the similarly modified logics be reasonable and well behaved?

Exercise* 4.2.33. $LL^\circ_\rightarrow$ (i.e., BCI°) and BCK° often appear as fragments of various logics. Prove that for any t-free wff $\mathcal{A} \rightarrow \mathcal{B}$, the sequent $\mathcal{A} \vdash \mathcal{B}$ is provable in $LL^\circ_\rightarrow$ (or BCK°) iff $\mathcal{A} \rightarrow \mathcal{B}$ is a theorem of $LL^{\circ t}_{\rightarrow 2}$ (or $BCK^{\circ t}_2$).

4.3 Relevant implication and pure entailment

The extension of LA with both permutation and contraction (i.e., LA_{CW} in Figure 4.1) deserves a special treatment, which is why we devote this section to $LR_\rightarrow$ (which omits the connectives $\leftarrow$ and $\circ$), and $LR_{\overset{\sim}{\rightarrow}}$ (which extends $LR_\rightarrow$ with negation). Both $LR_\rightarrow$ and $LR_{\overset{\sim}{\rightarrow}}$ allow an empty sequence of formulas as an antecedent of a sequent.

The logic of relevant implication, $R_\rightarrow$ was introduced (though not named) by Church in [63]. Church formulated a pure implicational logic that he obtained from an axiomatization of the implicational fragment of intuitionistic logic by replacing $\mathcal{A} \rightarrow \mathcal{B} \rightarrow \mathcal{A}$ by $\mathcal{A} \rightarrow \mathcal{A}$. His modification amounts to excluding the possibility to insert irrelevant premises into a valid inference. It has been noted many times that Church's preferred version of his λ-calculi was (what is now called) the λI-calculus, not the λK-calculus.[8] The former

[8]See, for example, [8, §70 and §71] and Church's [62, pp. 6–7 and Ch. V].

does not permit vacuous λ's that do not bind free variables. Church's calculus may be labeled as $WBCI$ logic—with reference to the combinators, the principal type schemas of which are the axioms given by Church.[9] From results in combinatory logic, the non-definability of a cancellator from the base $\{\mathsf{W}, \mathsf{B}, \mathsf{C}, \mathsf{I}\}$ is obvious. Cancellators, which are constant functions with respect to at least one of their arguments, introduce irrelevance via the arguments on which they do not depend.

The *language of* $R_{\rightarrow}$ contains one binary connective $\rightarrow$, and a denumerable set of propositional variables. The set of wff's is as in Definition 4.1 with $\div$ ranging over $\{\rightarrow\}$. A *sequent* is a finite sequence of wff's followed by $\vdash$ and a wff. A finite sequence may be empty; hence, the left-hand side of a sequent may be empty too.[10]

The *axiom* and *rules* of $LR_{\rightarrow}$ are as follows.

$$\mathcal{A} \vdash \mathcal{A} \ \text{id}$$

$$\frac{\Gamma \vdash \mathcal{A} \qquad \Delta[\mathcal{B}] \vdash \mathcal{C}}{\Delta[\mathcal{A} \rightarrow \mathcal{B}; \Gamma] \vdash \mathcal{C}} \rightarrow\vdash \qquad \frac{\mathcal{A}; \Gamma \vdash \mathcal{B}}{\Gamma \vdash \mathcal{A} \rightarrow \mathcal{B}} \vdash\rightarrow$$

$$\frac{\Gamma[\mathcal{A}; \mathcal{B}] \vdash \mathcal{C}}{\Gamma[\mathcal{B}; \mathcal{A}] \vdash \mathcal{C}} \mathsf{C}\vdash \qquad \frac{\Gamma[\mathcal{A}; \mathcal{A}] \vdash \mathcal{B}}{\Gamma[\mathcal{A}] \vdash \mathcal{B}} \mathsf{W}\vdash$$

The rules are like the rules in the identically labeled rules in the previous section. We also retain the notion of a *proof*; $\mathcal{A}$ is a *theorem* when $\vdash \mathcal{A}$ is a provable sequent.

Exercise 4.3.1. Prove that (1)–(4) are theorems of $LR_{\rightarrow}$. (1) $(\mathcal{A} \rightarrow \mathcal{A} \rightarrow \mathcal{B}) \rightarrow \mathcal{A} \rightarrow \mathcal{B}$ (W, contraction), (2) $(\mathcal{A} \rightarrow \mathcal{B}) \rightarrow (\mathcal{C} \rightarrow \mathcal{A}) \rightarrow \mathcal{C} \rightarrow \mathcal{B}$ (B, prefixing), (3) $(\mathcal{A} \rightarrow \mathcal{B} \rightarrow \mathcal{C}) \rightarrow \mathcal{B} \rightarrow \mathcal{A} \rightarrow \mathcal{C}$ (C, permutation) and (4) $\mathcal{A} \rightarrow \mathcal{A}$ (I, self-identity).

Theorem 4.16. *The single cut rule is* admissible *in* $LR_{\rightarrow}$.

Exercise 4.3.2. Prove the theorem. [Hint: Proofs of similar theorems in the previous section, in Chapter 2 and in Section 7.2 may be helpful.]

$LR_{\rightarrow}$ may be formulated—similarly to classical and intuitionistic logics in Section 3.2—based on multisets (as in [41]). The necessary modifications are small, but they may be confusing. First, we re-interpret $\Gamma, \Delta, \ldots$ to range over finite multisets of wff's, including the empty multiset. Second, we omit $(\mathsf{C}\vdash)$ from among the structural rules.

---

[9]See Sections 9.3 and 5.2 on connections between combinators (or λ-terms) and implicational formulas. Incidentally, the term "relevant implication" seems to have originated in [18].

[10]The first sequent calculus for $R_{\rightarrow}$ was described by Kripke in [115]. We more or less follow the presentation by Dunn in [77].

The addition of t is straightforward. We expand the language, and add $(\vdash t)$ as an axiom and $(t\vdash)$ as a rule to obtain $LR^t_{\rightarrow}$.

$$\vdash t \;\; \vdash t \qquad\qquad \frac{\Gamma[\Delta] \vdash \mathcal{A}}{\Gamma[t;\Delta] \vdash \mathcal{A}}\; t\vdash$$

Depending on whether we extended $LR_{\rightarrow}$ based on sequences of wff's or multisets of wff's, we obtain one of two distinct but equivalent calculi.

Exercise 4.3.3. Prove that the single cut rule is admissible in $LR^t_{\rightarrow}$.

Exercise 4.3.4. Prove that the set of theorems of $LR^t_{\rightarrow}$ is distinct from that of LA^t_2, LA^t_{W2}, LA^t_{K2}, $LL^{\circ t}_{\rightarrow 2}$, BCK^t_2 and LA^t_{KW2}.

Once we have gotten rid of permutation, we might want to have formulations of $LR_{\rightarrow}$ and $LR^t_{\rightarrow}$, which do not contain a contraction rule. Sequent calculi without an explicit contraction rule are especially important for decidability. In Section 9.1, we present some decidability results, which can be proved using sequent calculi.

$LR_{\rightarrow}$ has exactly one way to introduce *multiplicity* on the left-hand side of $\vdash$, namely, by an application of $(\rightarrow\vdash)$. This means that the number of wff's that can be contracted depends on the number of implications that have been introduced by the $(\rightarrow\vdash)$ rule in a proof. (Of course, there might be other implications in a sequent, because (id) can be instantiated with wff's of arbitrary complexity.) Then, we can compensate for the omission of $(W\vdash)$ by incorporating a limited amount of contraction into the $(\rightarrow\vdash)$ rule.

The *calculus* $[LR_{\rightarrow}]$ is a modification of $LR_{\rightarrow}$ that omits $(W\vdash)$ and replaces $(\rightarrow\vdash)$ by $([\rightarrow\vdash])$.[11]

$$\frac{\Gamma \vdash \mathcal{A} \qquad \Delta[\mathcal{B}] \vdash \mathcal{C}}{[\Delta[\mathcal{A} \rightarrow \mathcal{B};\Gamma]] \vdash \mathcal{C}}\; [\rightarrow\vdash]$$

$[\Delta[\mathcal{A} \rightarrow \mathcal{B};\Gamma]]$ requires explanation. The outer brackets and the inner brackets are different notations, which accidentally look similar. The inner [ ] indicates an occurrence within a sequence. The role of the outer [ ] is to show that contraction may have been applied at the time when the implicational formula $\mathcal{A} \rightarrow \mathcal{B}$ was introduced into the lower sequent. The amount of permitted contraction is specified by (1)–(3), each of which must be satisfied.

(1) If $\mathcal{D}$ occurs in Γ or Δ, then it occurs in $[\Delta[\mathcal{A} \rightarrow \mathcal{B};\Gamma]]$, and $\mathcal{A} \rightarrow \mathcal{B}$ occurs in $[\Delta[\mathcal{A} \rightarrow \mathcal{B};\Gamma]]$;

(2) if $\mathcal{D}$ is $\mathcal{A} \rightarrow \mathcal{B}$, then $\mathcal{D}$ has 0, 1 or 2 fewer occurrences in $[\Delta[\mathcal{A} \rightarrow \mathcal{B};\Gamma]]$ than in $\Delta[\mathcal{A} \rightarrow \mathcal{B};\Gamma]$;

(3) if $\mathcal{D}$ is not $\mathcal{A} \rightarrow \mathcal{B}$, then $\mathcal{D}$ has 0 or 1 fewer occurrence in $[\Delta[\mathcal{A} \rightarrow \mathcal{B};\Gamma]]$ than in $\Delta[\mathcal{A} \rightarrow \mathcal{B};\Gamma]$.

[11] The brackets acquire a new role here, which is quite different from picking out an occurrence. There is a limited stock of delimiters to choose from; hence, we re-use the brackets. We follow [77] in the presentation of $[LR_{\rightarrow}]$, which is hinted at in [115].

The first clause is to clarify and to ensure that the number of occurrences cannot be negative or even zero for a wff that occurred in the premises, or for the principal formula itself. The 0's in (2) and (3) mean that the rule does not force contraction. (There are applications of $(\rightarrow\vdash)$ that cannot be followed by $(W\vdash)$, thus it would be unreasonable to exclude the possibility of an application of the $([\rightarrow\vdash])$ rule without contraction.)

The difference between (2) and (3) is rooted in the obvious fact that the $(\rightarrow\vdash)$ rule creates a new formula. If $\mathcal{A} \rightarrow \mathcal{B}$ were to occur in Γ and Δ too, then $\Delta[\mathcal{A} \rightarrow \mathcal{B}; \Gamma]$ would contain *three* occurrences of the wff, and there is no way to reduce the number of occurrences of this particular formula *before* the application of the rule. The latter brings together the principal wff, and the two old occurrences into one sequent.

At first glance, this rule may look fishy, because of the indeterminacy included. However, it may be thought of as a concise description of *denumerably many rules.* (The latter explains why we need to have some succinct formulation like the new bracket notation.) But there is nothing dubious about this rule.

One could contemplate whether it is problematic that the number of contracted formulas is not determined by the shape of the rule. This is clearly a difference from the $(W\vdash)$ rule, but the $(W'\vdash)$ rule is similar, because depending on the length of the contracted sequence Δ, the number may vary. Perhaps what is questionable is that the formulas that are contracted are not explicit in the rule. However, $\Delta[\mathcal{A} \rightarrow \mathcal{B}; \Gamma]$ is not less explicit than Δ is in $(W'\vdash)$. Notice also that (by the other use of $[\ ]$ as singling out an occurrence) we have long ago discarded the idea that the formulas affected by a rule (e.g., the subalterns in an operational rule) must be at a fixed, specifiable place in the sequence—such as the outer edges of sequents.

It is important though that the $[\Delta[\mathcal{A} \rightarrow \mathcal{B}; \Gamma]]$ notation always contains only *finitely many* contractions. In this respect, the brackets are a notational device that is not unlike a shift from multisets to sequences. Multisets of wff's assume a fixed enumeration of formulas. A multiset can be seen to stand for any sequence of wff's, which is composed of the same wff's. If we start with the sequence, in which the order of the wff's adheres to the enumeration, then the other sequences can be obtained by *finitely many permutations.*

A more concrete observation, which supports the view that there is nothing suspicious about the $([\rightarrow\vdash])$ rule, is that whenever the rule has been applied, it is *definite* whether the rule was applied to contract the principal formula or some parametric formulas (in addition to introducing $\mathcal{A} \rightarrow \mathcal{B}$).

The same method to do away with an explicit contraction rule can be applied to $LR^t_{\rightarrow}$ as it was shown in [41]. We denote the resulting calculus by $[LR^t_{\rightarrow}]$.

We want to prove the cut theorem for these calculi too. If the cut formula is $\mathcal{A} \rightarrow \mathcal{B}$, and it is the principal formula in both premises, then we would want to rearrange the proof into the following proof.

$$\dfrac{\begin{matrix}\vdots \\ \Theta \vdash \mathcal{A}\end{matrix} \qquad \dfrac{\begin{matrix}\vdots \\ \Gamma ; \mathcal{A} \vdash \mathcal{B}\end{matrix} \qquad \begin{matrix}\vdots \\ \Delta[\mathcal{B}] \vdash \mathcal{C}\end{matrix}}{\Delta[\Gamma ; \mathcal{A}] \vdash \mathcal{C}}\,\text{cut}}{\Delta[\Gamma ; \Theta] \vdash \mathcal{C}}\,\text{cut}$$

However, the original bottom sequent is $[\Delta[\Gamma ; \Theta]] \vdash \mathcal{C}$, where Δ and Θ together may have 0 or 1 fewer occurrences of any formula (without completely omitting a wff). This means that in order to prove the cut theorem by adapting the triple induction from Chapter 2, we have to reformulate the cut rule itself to contain some contraction.

$$\dfrac{\Gamma \vdash \mathcal{C} \qquad \Delta[\mathcal{C}] \vdash \mathcal{A}}{[\Delta[\Gamma]] \vdash \mathcal{A}}\,[\text{cut}]$$

Every formula that occurs in $\Delta[\Gamma]$ occurs in $[\Delta[\Gamma]]$, however, if $\mathcal{B}$ occurs both in Δ and Γ, then $\mathcal{B}$ occurs 0 or 1 fewer times in $[\Delta[\Gamma]]$ than in $\Delta[\Gamma]$. Any other wff occurs as many times in $[\Delta[\Gamma]]$ as in $\Delta[\Gamma]$. The single cut rule is a *special instance* of the [cut] rule.

Theorem 4.17. *The cut rule is* admissible *in* $[LR_{\rightarrow}]$ *and* $[LR^{t}_{\rightarrow}]$.

Proof: The proof can now be carried out by triple induction on δ, μ and ϱ. A difference from the proofs for $LR_{\rightarrow}$ and $LR^{t}_{\rightarrow}$ is that there is no separate case for contraction, since there is no separate $(W\vdash)$ rule. However, the cases that involve the $([\rightarrow\vdash])$ rule are more complicated. qed

Exercise 4.3.5. Check the details of the previous proof.

Exercise 4.3.6. Prove that the formulas—in effect, the axiom schemas for $R_{\rightarrow}$—are provable in $[LR_{\rightarrow}]$ (hence, also in $[LR^{t}_{\rightarrow}]$).

Exercise 4.3.7. Prove that if $\boldsymbol{t} \rightarrow \mathcal{A}$ is a theorem of $LR^{t}_{\rightarrow}$ (or of $[LR^{t}_{\rightarrow}]$), then $\mathcal{A}$ is also a theorem of that calculus. [Hint: The proof is straightforward using the cut rule.]

Theorem 4.18. $LR_{\rightarrow}$ *is* equivalent *to* $[LR_{\rightarrow}]$, *and* $LR^{t}_{\rightarrow}$ *is* equivalent *to* $[LR^{t}_{\rightarrow}]$.

Proof: 1. We prove the theorem by first noting that $([\rightarrow\vdash])$ is a derived rule in $LR_{\rightarrow}$ and $LR^{t}_{\rightarrow}$. Given a concrete application of the rule, we apply $(\rightarrow\vdash)$ first. If $([\rightarrow\vdash])$ included contractions, then having determined which wff's have to be contracted and how many times, we insert suitably many applications of the $(W\vdash)$ rule. (If we are considering calculi that are based on sequences of wff's, then finitely many permutation steps may have to be inserted too.)

2. For the converse, instead of proving that any combination of $(W\vdash)$ and $(\rightarrow\vdash)$, possibly intermingled with $(\vdash\rightarrow)$ and even $(C\vdash)$ steps, can be emulated by $([\rightarrow\vdash])$ steps, we show that if $\mathcal{A}$ is a theorem of $R^{t}_{\rightarrow}$, then it is a theorem of $[LR^{t}_{\rightarrow}]$. (We assume that the easily provable equivalence of $LR^{t}_{\rightarrow}$ and $R^{t}_{\rightarrow}$ is known.) The results of the last two exercises above, together with the admissibility of the cut rule and with the $(\boldsymbol{t}\vdash)$ rule, suffice. qed

$[LR_{\rightarrow}]$ and $[LR^t_{\rightarrow}]$ are crucial in showing the decidability of implicational relevance logics—including $E_{\rightarrow}$ and $T_{\rightarrow}$—as we will discuss in Section 9.1.

A sequent calculus for *pure entailment* was first formulated by Kripke in [115]. In the late 1940s, Curry pinpointed the exact place where multiplicity on the right should be prohibited in order to obtain a sequent calculus for positive intuitionistic logic. It is enough to require that Δ is empty in the $(\vdash \supset)$ rule in LK. Kripke uses a similar restriction in $LE_{\rightarrow}$.

Sequents in the *sequent calculus* $LE_{\rightarrow}$ have finite sequences of wff's on both sides of the $\vdash$. The axiom is (id); the structural rules are $(W\vdash)$, $(\vdash W)$, $(C\vdash)$ and $(\vdash C)$. The rules for implication are as follows.

$$\frac{\Gamma \vdash \Delta; \mathcal{A} \qquad \mathcal{B}; \Theta \vdash \Lambda}{\mathcal{A} \rightarrow \mathcal{B}; \Gamma; \Theta \vdash \Delta; \Lambda} \rightarrow\vdash \qquad\qquad \frac{\mathcal{A}; \overrightarrow{\Gamma} \vdash \mathcal{B}}{\overrightarrow{\Gamma} \vdash \mathcal{A} \rightarrow \mathcal{B}} \vdash\rightarrow^*$$

The decoration in the $(\vdash \rightarrow^*)$ rule on Γ indicates that the wff's in Γ (if there are any) must be implications, that is, they cannot be propositional variables. (We starred the label of the rule to distinguish it from the usual $(\vdash \rightarrow)$ rule.) The notions of a *proof* and of a *theorem* are as before.

Exercise 4.3.8. Show that (1) $(\mathcal{A} \rightarrow \mathcal{B}) \rightarrow (\mathcal{B} \rightarrow \mathcal{C}) \rightarrow \mathcal{A} \rightarrow \mathcal{C}$ and (2) $(\mathcal{A} \rightarrow (\mathcal{B} \rightarrow \mathcal{C}) \rightarrow \mathcal{D}) \rightarrow (\mathcal{B} \rightarrow \mathcal{C}) \rightarrow \mathcal{A} \rightarrow \mathcal{D}$ are theorems of $LE_{\rightarrow}$. [Hint: Notice how the restriction on $(\vdash \rightarrow^*)$ enters into the proof of (2).]

$LE_{\rightarrow}$ is decidable (as we will discuss it in Section 9.1). In particular, $(\mathcal{A} \rightarrow \mathcal{B} \rightarrow \mathcal{C}) \rightarrow \mathcal{B} \rightarrow \mathcal{A} \rightarrow \mathcal{C}$ is *not a theorem* of $LE_{\rightarrow}$. It should be clear—informally—from the previous exercise where an attempted proof fails.

The cut rule is as in LK (with ; replacing ,).

Theorem 4.19. *The cut rule is* admissible *in* $LE_{\rightarrow}$.

Exercise 4.3.9. Prove the theorem. [Hint: First, prove that if a sequent is provable, then it has exactly one wff on the right-hand side of the $\vdash$. Then, carefully order the cases when $\varrho > 2$.]

Other relevance logics (such as $T_{\rightarrow}$ and $B_{\rightarrow}$) do not seem to have sequent calculus formulations (at least in the usual sense), because they do not allow permutation.[12]

As we have already seen in Section 3.4, it is possible to have only *one side* in sequents, for example, in the case of classical logic. The original negation in the main systems of relevance logics is a *De Morgan negation*, which allows both double negation introduction and double negation elimination. We denote this negation by $\sim$. $LE_{\overset{\sim}{\rightarrow}}$ was formulated (and proved decidable) in [23]; $LR_{\overset{\sim}{\rightarrow}}$ may be found, for example, in [77]. First, we consider the latter.

The connectives in the language of $LR_{\overset{\sim}{\rightarrow}}$ are $\rightarrow$ and $\sim$, where the latter is unary. (I.e., Definition 4.1 would have a new clause: if $\mathcal{A}$ is a wff, so is $\sim\mathcal{A}$.)

[12] We briefly overview *merge-style* calculi in Section 4.8.

To accommodate the rules for negation, we modify the notion of a *sequent* from that in $LR_{\to}$ to one permitting multiple wff's on the right-hand side—like in $LE_{\to}$. A *sequent* is of the form $\Gamma \vdash \Delta$, where the metavariables range over finite multisets (including the empty multiset).

The *sequent calculus* $LR_{\overset{\sim}{\to}}$ comprises the following axiom and rules.

$$\mathcal{A} \vdash \mathcal{A} \ \text{id}$$

$$\frac{\Gamma \vdash \Delta; \mathcal{A} \qquad \mathcal{B}; \Theta \vdash \Lambda}{\mathcal{A} \to \mathcal{B}; \Gamma; \Theta \vdash \Delta; \Lambda} \to\vdash \qquad \frac{\mathcal{A}; \Gamma \vdash \Delta; \mathcal{B}}{\Gamma \vdash \Delta; \mathcal{A} \to \mathcal{B}} \vdash\to$$

$$\frac{\mathcal{A}; \mathcal{A}; \Gamma \vdash \Delta}{\mathcal{A}; \Gamma \vdash \Delta} \mathrm{W}\vdash \qquad \frac{\Gamma \vdash \Delta; \mathcal{A}; \mathcal{A}}{\Gamma \vdash \Delta; \mathcal{A}} \vdash\mathrm{W}$$

$$\frac{\Gamma \vdash \Delta; \mathcal{A}}{\sim\mathcal{A}; \Gamma \vdash \Delta} \sim\vdash \qquad \frac{\mathcal{A}; \Gamma \vdash \Delta}{\Gamma \vdash \Delta; \sim\mathcal{A}} \vdash\sim$$

The operational rules look very much like the rules for $\supset$ and $\neg$ in LK. However, we changed sequences to multisets, hence, the places of the subalterns and of the principal formulas do not matter—unlike in LK.

It would be swell if $LR_{\overset{\sim}{\to}}$ would be an extension of $LR_{\to}$ by a pair of operational rules for $\sim$. We define the notions of a *proof* and of a formula being a *theorem* in $LR_{\overset{\sim}{\to}}$ exactly as in $LR_{\to}$. Then any negation-free theorem of $LR_{\overset{\sim}{\to}}$ is a theorem of $LR_{\to}$.

Exercise 4.3.10. Prove that $LR_{\to}$ could have been formulated equivalently using sequents with multisets on the right-hand side of the $\vdash$.

The single cut rule takes the following form.

$$\frac{\Gamma \vdash \Delta; \mathcal{C} \qquad \mathcal{C}; \Theta \vdash \Lambda}{\Gamma; \Theta \vdash \Delta; \Lambda} \text{cut}$$

Theorem 4.20. *The cut rule is* admissible *in* $LR_{\overset{\sim}{\to}}$.

Exercise 4.3.11. Prove the theorem. [Hint: The proof for $LR_{\to}$ can be extended by taking into consideration the three new rules.]

Contraction may be built into the operational rules like in the case of $LR_{\to}$ and $LR^t_{\to}$. $[LR_{\overset{\sim}{\to}}]$ retains (id), omits the contraction rules, and replaces the implication and negation rules by the following ones. (We repeat the identity axiom for the sake of the completeness of the presentation.)

$$\mathcal{A} \vdash \mathcal{A} \ \text{id}$$

$$\frac{\Gamma \vdash \Delta; \mathcal{A} \qquad \mathcal{B}; \Theta \vdash \Lambda}{[\mathcal{A} \to \mathcal{B}; \Gamma; \Theta] \vdash [\Delta; \Lambda]} [\to\vdash] \qquad \frac{\mathcal{A}; \Gamma \vdash \Delta; \mathcal{B}}{\Gamma \vdash [\Delta; \mathcal{A} \to \mathcal{B}]} [\vdash\to]$$

$$\frac{\Gamma \vdash \Delta; \mathcal{A}}{[\sim\mathcal{A}; \Gamma] \vdash \Delta}\ [\sim\vdash] \qquad \frac{\mathcal{A}; \Gamma \vdash \Delta}{\Gamma \vdash [\Delta; \sim\mathcal{A}]}\ [\vdash\sim]$$

The brackets that surround some of the multisets in the rules indicate the possibility of contractions. The general idea is that a bracketed multiset in a lower sequent is just like its unbracketed version except that if the *possibility of the contraction* emerged because the rule has been applied, then that contraction is permitted. For instance, if Δ contains n occurrences of $\mathcal{A}$ and m occurrences of $\mathcal{B}$, whereas Λ has i-many $\mathcal{A}$'s, no $\mathcal{B}$'s, but k-many $\mathcal{C}$'s, then $[\Delta; \Lambda]$ has $n+i$-many or $n+i-1$-many $\mathcal{A}$'s, m-many $\mathcal{B}$'s and k-many $\mathcal{C}$'s. Notice that it is tacitly assumed that if contractions are to be applied somewhere in a proof, then they are applied as soon as possible.

Exercise 4.3.12. Scrutinize the rules of $LR_{\overset{\sim}{\to}}$ to verify that applications of $(W\vdash)$ and $(\vdash W)$ can be permuted upward, when they do not depend on the preceding rule.

The negation rules allow us to move all the formulas to one side. A left-handed sequent calculus for $R_{\overset{\sim}{\to}}$ is given in [8, §60]. Our focus on theoremhood favors a right-handed calculus. We could completely omit $\vdash$, because the left-hand side is always empty, but we retain the symbol to remind us which side is empty. $\vdash$ is followed by a multiset of wff's.

The *right-handed sequent calculus* $LR^r_{\overset{\sim}{\to}}$ is defined by the following axiom and rules.

$$\vdash \mathcal{A}; \sim\mathcal{A}\ \text{id}$$

$$\frac{\vdash \Gamma; \mathcal{A}; \mathcal{A}}{\vdash \Gamma; \mathcal{A}}\ W \qquad \frac{\vdash \Gamma; \mathcal{A}}{\vdash \Gamma; \sim\sim\mathcal{A}}\ \sim$$

$$\frac{\vdash \Gamma; \sim\mathcal{A}; \mathcal{B}}{\vdash \Gamma; \mathcal{A} \to \mathcal{B}}\ \to \qquad \frac{\vdash \Gamma; \mathcal{A} \qquad \vdash \Delta; \sim\mathcal{B}}{\vdash \Gamma; \Delta; \sim(\mathcal{A} \to \mathcal{B})}\ \overset{\sim}{\to}$$

At first glance it may seem that we have to change what we mean by a theorem, because any proof starts with a pair of wff's after $\vdash$. But it is not so; the notions of a *proof* and of a *theorem* are defined as before.

Example 4.21. Here is a simple proof in $LR^r_{\overset{\sim}{\to}}$.

$$\dfrac{\dfrac{\dfrac{\dfrac{\vdash \mathcal{A}; \sim\mathcal{A} \qquad \dfrac{\vdash \mathcal{A}; \sim\mathcal{A} \qquad \vdash \mathcal{B}; \sim\mathcal{B}}{\vdash \sim\mathcal{A}; \mathcal{B}; \sim(\mathcal{A} \to \mathcal{B})}\ \overset{\sim}{\to}}{\vdash \sim\mathcal{A}; \sim\mathcal{A}; \mathcal{B}; \sim(\mathcal{A} \to \mathcal{A} \to \mathcal{B})}\ \overset{\sim}{\to}}{\vdash \sim\mathcal{A}; \mathcal{B}; \sim(\mathcal{A} \to \mathcal{A} \to \mathcal{B})}\ W}{\vdash \mathcal{A} \to \mathcal{B}; \sim(\mathcal{A} \to \mathcal{A} \to \mathcal{B})}\ \to}{\vdash (\mathcal{A} \to \mathcal{A} \to \mathcal{B}) \to \mathcal{A} \to \mathcal{B}}\ \to$$

Exercise 4.3.13. Prove the following formulas. (1) $(\mathcal{A} \to {\sim}\mathcal{B}) \to \mathcal{B} \to {\sim}\mathcal{A}$, (2) ${\sim}{\sim}\mathcal{A} \to \mathcal{A}$ and (3) $(\mathcal{A} \to {\sim}\mathcal{A}) \to {\sim}\mathcal{A}$. [Hint: The addition of (1)–(3) to $R_{\to}$ yields $R_{\overset{\sim}{\to}}$.]

We have fewer rules than in $LR_{\overset{\sim}{\to}}$, and the labels are shorter. However, a price that we have to pay for having a more compact calculus is the obvious lack of the subformula property in cut-free proofs. The *cut rule* is as follows.

$$\frac{\vdash \Gamma ; \mathcal{A} \qquad \vdash \Delta ; {\sim}\mathcal{A}}{\vdash \Gamma ; \Delta}\ \text{cut}$$

Theorem 4.22. *The cut rule is* admissible *in* $LR^{r}_{\overset{\sim}{\to}}$.

Exercise 4.3.14. Prove the theorem. [Hint: Although the cut rule has a new shape and some of the subalterns are not subformulas of the principal formulas, the whole structure of some of the previous proofs can be retained.]

The logic of pure entailment may also be extended by negation—as in [23]. The *sequent calculus* $LE_{\overset{\sim}{\to}}$ extends $LE_{\to}$ by the rules $({\sim}\vdash)$ and $(\vdash{\sim})$.

There is also a right-handed calculus for $E_{\overset{\sim}{\to}}$ in [23]. We will use the notation $\Gamma^{\overset{\sim}{\to}}$ to indicate that every element in the sequence Γ is of the form ${\sim}(\mathcal{A} \to \mathcal{B})$, for some $\mathcal{A}$ and $\mathcal{B}$. $LE^{r}_{\overset{\sim}{\to}}$ comprises two axioms, two structural rules and two operational rules, as well as the cut rule. The latter makes $LE^{r}_{\overset{\sim}{\to}}$ somewhat unusual, because the cut rule is not eliminable from $LE^{r}_{\overset{\sim}{\to}}$.

$$\vdash {\sim}\mathcal{A} ; \mathcal{A}\ \text{id} \qquad\qquad \vdash {\sim}\mathcal{A} ; {\sim}(\mathcal{A} \to \mathcal{B}) ; \mathcal{B}\ \overset{\sim}{\to}$$

$$\frac{\vdash \Gamma ; \mathcal{A} ; \mathcal{B} ; \Delta}{\vdash \Gamma ; \mathcal{B} ; \mathcal{A} ; \Delta}\ \text{C} \qquad \frac{\vdash \Gamma ; \mathcal{A} ; \mathcal{A}}{\vdash \Gamma ; \mathcal{A}}\ \text{W} \qquad \frac{\vdash \Gamma ; \mathcal{A} \qquad \vdash \Delta ; {\sim}\mathcal{A}}{\vdash \Gamma ; \Delta}\ \text{cut}$$

$$\frac{\vdash \Gamma^{\overset{\sim}{\to}} ; {\sim}\mathcal{A} ; \mathcal{B}}{\vdash \Gamma^{\overset{\sim}{\to}} ; \mathcal{A} \to \mathcal{B}}\ \to^{*} \qquad \frac{\vdash \Gamma ; \mathcal{A}}{\vdash \Gamma ; {\sim}{\sim}\mathcal{A}}\ {\sim}$$

Exercise 4.3.15. Prove that the cut rule cannot be omitted from $LE^{r}_{\overset{\sim}{\to}}$ without the loss of theorems.

Exercise 4.3.16. Show that by using cut, the rule to introduce negated implications in $LR_{\overset{\sim}{\to}}$ can be recovered.

$[LE_{\overset{\sim}{\to}}]$ is like $[LR_{\overset{\sim}{\to}}]$ except that the right rule for $\to$ is $([\vdash\to^{*}])$ (and $(\text{C}\vdash)$ and $(\vdash\text{C})$ are explicit rules in $[LE_{\overset{\sim}{\to}}]$).

$$\frac{\mathcal{A} ; \Gamma^{\overset{\sim}{\to}} \vdash \mathcal{B}}{\Gamma^{\overset{\sim}{\to}} \vdash \mathcal{A} \to \mathcal{B}}\ [\vdash\to^{*}]$$

Theorem 4.23. *The cut rule is* admissible *in* $LE_{\overset{\sim}{\to}}$ *and* $[LE_{\overset{\sim}{\to}}]$.

Exercise 4.3.17. Prove the theorem. [Hint: The core of the proof is similar to the proof of the cut theorem for $LE_{\to}$, but the two proofs will use different formulations of the cut rule.]

4.4 Non-distributive logic of relevant implication

A version of the logic R was defined by R. K. Meyer as a sequent calculus in his [135], which he labeled as LR (adding various subscripts for fragments and extensions).[13] Combining rules for $\land$ and $\lor$ with rules for $\rightarrow$ in the absence of some of the structural rules (in particular, of thinning) does not allow a proof of the distributivity of conjunction and disjunction over each other. This is why LR is a *version* of R rather than a genuinely different relevance logic.

The *sequent calculus for non-distributive* R, which is denoted by LR, is based on sequents that comprise two finite (possibly, empty) multisets of wff's. The meta-variables for the multisets are $\Gamma, \Delta, \ldots$. The connectives are $\rightarrow$, $\circ$, $\land$, $\lor$ and $\sim$—all are binary except the last one, which is unary. The informal meaning and the names for the connectives are as before.[14]

The *axiom* and the *rules* of LR are the following.

$$\mathcal{A} \vdash \mathcal{A} \ \text{id}$$

$$\frac{\Gamma \vdash \mathcal{A}; \Delta \qquad \Theta; \mathcal{B} \vdash \Lambda}{\Gamma; \Theta; \mathcal{A} \rightarrow \mathcal{B} \vdash \Delta; \Lambda} \ {\rightarrow}{\vdash} \qquad \frac{\Gamma; \mathcal{A} \vdash \mathcal{B}; \Delta}{\Gamma \vdash \mathcal{A} \rightarrow \mathcal{B}; \Delta} \ {\vdash}{\rightarrow}$$

$$\frac{\Gamma; \mathcal{A}; \mathcal{B} \vdash \Delta}{\Gamma; \mathcal{A} \circ \mathcal{B} \vdash \Delta} \ {\circ}{\vdash} \qquad \frac{\Gamma \vdash \mathcal{A}; \Delta \qquad \Theta \vdash \mathcal{B}; \Lambda}{\Gamma; \Theta \vdash \mathcal{A} \circ \mathcal{B}; \Delta; \Lambda} \ {\vdash}{\circ}$$

$$\frac{\Gamma; \mathcal{A} \vdash \Delta}{\Gamma; \mathcal{A} \land \mathcal{B} \vdash \Delta} \ {\land}{\vdash_1} \qquad \frac{\Gamma; \mathcal{A} \vdash \Delta}{\Gamma; \mathcal{B} \land \mathcal{A} \vdash \Delta} \ {\land}{\vdash_2} \qquad \frac{\Gamma \vdash \mathcal{A}; \Delta \qquad \Gamma \vdash \mathcal{B}; \Delta}{\Gamma \vdash \mathcal{A} \land \mathcal{B}; \Delta} \ {\vdash}{\land}$$

$$\frac{\Gamma; \mathcal{A} \vdash \Delta \qquad \Gamma; \mathcal{B} \vdash \Delta}{\Gamma; \mathcal{A} \lor \mathcal{B} \vdash \Delta} \ {\lor}{\vdash} \qquad \frac{\Gamma \vdash \mathcal{A}; \Delta}{\Gamma \vdash \mathcal{A} \lor \mathcal{B}; \Delta} \ {\vdash}{\lor_1} \qquad \frac{\Gamma \vdash \mathcal{A}; \Delta}{\Gamma \vdash \mathcal{B} \lor \mathcal{A}; \Delta} \ {\vdash}{\lor_2}$$

$$\frac{\Gamma \vdash \mathcal{A}; \Delta}{\Gamma; {\sim}\mathcal{A} \vdash \Delta} \ {\sim}{\vdash} \qquad \frac{\Gamma; \mathcal{A} \vdash \Delta}{\Gamma \vdash {\sim}\mathcal{A}; \Delta} \ {\vdash}{\sim}$$

$$\frac{\Gamma; \mathcal{A}; \mathcal{A} \vdash \Delta}{\Gamma; \mathcal{A} \vdash \Delta} \ W{\vdash} \qquad \frac{\Gamma \vdash \mathcal{A}; \mathcal{A}; \Delta}{\Gamma \vdash \mathcal{A}; \Delta} \ {\vdash}W$$

The notion of a *proof* is as usual. $\mathcal{A}$ is a *theorem* of LR iff $\vdash \mathcal{A}$ is provable. The cut rule looks like the single cut rule in LK, but it can be formulated as

[13] Perhaps, somewhat confusingly, L stands for "lattice" not for "logistic" in the label LR. However, LR is the natural logistic sequent calculus that results by adding rules for $\land$ and $\lor$ (and for $\circ$) to $LR_{\overset{\sim}{\rightarrow}}$. It just so happens that LR is not equivalent to the axiomatic calculus R.

[14] [135] did not include $\circ$, but there is no reason not to include fusion. The sequent calculus formulation immediately ensures conservativeness, because of the cut theorem.

follows—if we keep in mind that a sequent comprises a pair of multisets.

$$\frac{\Gamma \vdash \mathcal{C}; \Delta \qquad \Theta; \mathcal{C} \vdash \Lambda}{\Gamma; \Theta \vdash \Delta; \Lambda} \text{ cut}$$

Theorem 4.24. *The cut rule is* admissible *in LR.*

Proof: The proof is by triple induction on δ, μ and ϱ—very much like for LK. The locations of the principal wff's or of their subalterns do not matter, because we use multisets. Furthermore, the calculus looks quite like the propositional part of LK (if we forget notational differences). The rules that are missing are the thinning rules, whereas the rules for fusion are new. We have seen that the $(\circ\vdash)$ and $(\vdash\circ)$ rules do not pose a problem in LA (or its extensions), thus, what remains to be verified is that they are unproblematic here too. qed

Exercise 4.4.1. Check the claims that we made in the above proof.

LR (without fusion) may be viewed as a logic that substantiates the claim, which is sometimes emphasized by people working with relevance logics, that an aim of relevance logic is to preserve as much from classical logic as possible (while abandoning what is not acceptable). Indeed, LR results exactly by omitting the objectionable $(K\vdash)$ and $(\vdash K)$ rules (and by introducing the connective $\circ$ of which $\to$ is the residual). It is an unfortunate side effect of the absence of the weakening rules that the distributivity of $\wedge$ and $\vee$ are no longer provable.

Meyer considered a special rule that would be added to the negation-free (and single right-handed) fragment of LR to recover distributivity. We denote the calculus by LR^d_+.

$$\frac{\Gamma \vdash \mathcal{A} \qquad \Gamma \vdash \mathcal{B} \vee \mathcal{C}}{\Gamma \vdash (\mathcal{A} \wedge \mathcal{B}) \vee \mathcal{C}} \text{ dis}$$

Exercise 4.4.2. Suppose that (dis) is added to the negation-free part of LR, in which the right-hand multisets are replaced by exactly one wff. Pinpoint the step where an attempted proof of the cut theorem fails to go through. Can you construct a proof of a sequent with cut, from which the cut is not eliminable? [Hint: Δ is omitted from the cut rule, and Λ is taken to be one formula.]

Meyer was interested in LR and LR^d_+, because he hoped that these sequent calculi might create a path to the decidability of R, and eventually, of E. At the same time he noted that "[g]iven Gödel's result, it would seem that a strongly constructive system is likely to be relatively trivial which prompts the hope that LR^+ [i.e., the positive fragment of R] and eventually R and E will be showed undecidable and hence of lasting interest." ([135, p. 136]) Many years later, Urquhart proved in [183] that R_+, R and E (among a whole range of logics in the neighborhood) are undecidable.

Exercise* 4.4.3. The rule (dis) could be varied by keeping the same premises, but changing $(\mathcal{A} \wedge \mathcal{B}) \vee \mathcal{C}$ to $(\mathcal{B} \wedge \mathcal{A}) \vee \mathcal{C}$, for instance. How many versions of (dis) are there? Is the cut rule admissible, if all those versions of (dis) are added to LR^d_+?

The logic of entailment, which was obtained by Anderson and Belnap from Ackermann's logic of strong implication, contains implicational axioms that are not principal types of *proper combinators*. Initially, the various *restricted* and *specialized* formulas seemed to give E a modal tint—not unlike Lewis's original systems of strict implication have. An attempt to uncover this modality was to extend R with $\Box$, where $\Box$ is an $S4$-like modality, except that the axioms are formulated with $\rightarrow$ rather than $\supset$. Necessity distributes over conjunction and implication.

The *sequent calculus* $LR^{\Box}$ is defined, in [135], by adding two rules to introduce $\Box$. The notation $\Gamma^{\Box}$ indicates that each element of the multiset Γ is of the form $\Box\mathcal{A}$, for some $\mathcal{A}$.

$$\frac{\Gamma; \mathcal{A} \vdash \Delta}{\Gamma; \Box\mathcal{A} \vdash \Delta}\ \Box\vdash \qquad\qquad \frac{\Gamma^{\Box} \vdash \mathcal{A}}{\Gamma^{\Box} \vdash \Box\mathcal{A}}\ \vdash\Box$$

The notions of a *proof* and of a *theorem* are unchanged from LR.

Exercise 4.4.4. Prove the wff's (K), (T) and (4). The labels are to indicate the similarity of these wff's to axioms in normal modal logics (which we will touch upon in Section 4.7). (K) $\Box(\mathcal{A} \rightarrow \mathcal{B}) \rightarrow \Box\mathcal{A} \rightarrow \Box\mathcal{B}$, (T) $\Box\mathcal{A} \rightarrow \mathcal{A}$ and (4) $\Box\mathcal{A} \rightarrow \Box\Box\mathcal{A}$.

Exercise 4.4.5. Show that $\Box(\mathcal{A} \wedge \mathcal{B}) \rightarrow (\Box\mathcal{A} \wedge \Box\mathcal{B})$ and $(\Box\mathcal{A} \wedge \Box\mathcal{B}) \rightarrow \Box(\mathcal{A} \wedge \mathcal{B})$ are theorems of $LR^{\Box}$. (The latter formula is labeled as $(\Box\wedge)$, which expresses that $\Box$ distributes over $\wedge$, like it does in normal modal logics. $(\Box\wedge)$ figures in the next three sections too.)

(K), (T), (4) and $(\Box\wedge)$ are the modal axioms in $R^{\Box}$, the extension of R with necessity, which includes the rule of necessitation ($\vdash \mathcal{A}$ implies $\vdash \Box\mathcal{A}$) too.

Exercise 4.4.6. Prove that the cut rule is admissible in $LR^{\Box}$.

Of course, an important question is whether the implications in $E_{\rightarrow}$ are indeed modal relevant implications. Let us consider $LR^{\Box}_{\underset{\rightarrow}{\sim}}$. $\sim$ requires non-singleton multisets on the right-hand side, but, at the same time, $\sim$ also allows the right-hand side multisets to be interpreted as wff's.

If $\mathcal{A}_1; \ldots; \mathcal{A}_n \vdash \mathcal{B}_1; \ldots; \mathcal{B}_{m-1}; \mathcal{B}_m$ is provable, then the corresponding wff is $\mathcal{A}_1 \rightarrow \ldots \rightarrow \mathcal{A}_n \rightarrow {\sim}\mathcal{B}_1 \rightarrow \ldots \rightarrow {\sim}\mathcal{B}_{m-1} \rightarrow \mathcal{B}_m$. A way to obtain the latter wff is to select a wff on the right, for example, $\mathcal{B}_m$ and move all the other wff's to the left by the $({\sim}\vdash)$ rule. Then $n + m - 1$ $(\vdash\rightarrow)$ steps yield the final wff.

Definition 4.25. A translation τ of wff's of $E_{\underset{\rightarrow}{\sim}}$ into wff's of $LR^{\Box}_{\underset{\rightarrow}{\sim}}$ is recursively defined by (1)–(3).

(1) If $\mathcal{A}$ is an atomic wff, then $\mathcal{A}^\tau$ is $\mathcal{A}$;

(2) if $\mathcal{A}$ is $\sim\mathcal{B}$, then $\mathcal{A}^\tau$ is $\sim\mathcal{B}^\tau$;

(3) if $\mathcal{A}$ is $\mathcal{B} \to \mathcal{C}$, then $\mathcal{A}^\tau$ is $\Box(\mathcal{B}^\tau \to \mathcal{C}^\tau)$.

The translation turns every $\to$ in $E_{\stackrel{\sim}{\to}}$ into a $\to$ with a $\Box$ prefixed. The next theorem is proved in [135].

Theorem 4.26. *If $\mathcal{A}^\tau$ is a theorem of $LR^{\Box}_{\stackrel{\sim}{\to}}$, then $\mathcal{A}$ is a* theorem *of $E_{\stackrel{\sim}{\to}}$.*

Exercise* 4.4.7. Prove the theorem. [Hint: Notice that $\vdash \Box(\mathcal{A} \to \mathcal{B})$ must have resulted by the $(\vdash \Box)$ rule.]

The naturalness of how LR combines the lattice operations with implication has proved to be very attractive. We return to the positive fragment of R in Section 4.6, but first, we consider a modal variation on the contraction-free version of LR, which gained a lot of interest in the past thirty years or so.

4.5 Linear logic

Linear logic, L, was invented by Girard in [96]. It is not difficult to place linear logic on the landscape we have charted so far. L extends LA with permutation, with $\wedge$ and $\vee$, negation, the four zero-ary constants $\boldsymbol{t}$, $\boldsymbol{T}$, $\boldsymbol{F}$ and $\boldsymbol{f}$, quantifiers as well as two modalities, $\Diamond$ and $\Box$.[15]

The popularity of linear logic is undoubtedly due to the fact that Girard's paper has more than a touch of philosophical obscurity, which seems to have captured the imagination of computer scientists (who are likely less accustomed to speculative imprecision than continental philosophers are). The first few sentences of the paper should give a hint of this flavor.[16]

> Linear logic is a logic behind logic, more precisely, it provides a continuation of the constructivization that began with intuitionistic logic. The [The/Linear] logic is as strong as the usual ones, i.e., intuitionistic logic can be translated into linear logic in a faithful way. Linear logic shows that the constructive features of intuitionistic operations are indeed due to the linear aspects of some intuitionistic connectives or quantifiers, and these linear features are put at the [at the/in a] prominent place. [96, p. 2]

Intuitionistic logic is a widely used name for a logic that was carved out from classical logic by A. Heyting, following some ideas of L. E. J. Brouwer, and which is formalized as a sequent calculus LJ. (See Chapter 3.) The history

---

[15] To be precise, Definition 1.15 in [96] does not have $\to$ or the modalities included, but in the same paper, in Definition 1.21 the modalities are added without a change to the name of the calculus.

[16] Corrections to the English in the quote are inserted to facilitate reading.

of intuitionism and constructivism *in mathematics* is lengthy and convoluted, and it is way outside of the scope of our book. However, we may note that Gödel in the early 1930s (e.g., in [100], [101] and [102]) investigated intuitionistic logic, some of its properties and the contribution of intuitionistic logic to mathematics. He remarked that the significance of intuitionism in mathematics is to be found beyond Heyting's arithmetic (hence, beyond intuitionistic logic), specifically, in the debarment of impredicative definitions and of uses of the axiom of choice.

Intuitionistic logic can be translated into other logics, including relevance logic and classical logic. (We overviewed some of the translations into classical logic in Section 3.7.) Unfortunately, "the strength of a logic" is not a standard piece of terminology, and the term can be made precise in various (and totally incompatible) ways. To start with, there is no agreement between logicians and mathematicians, whether R_1 or R_2 is stronger (or weaker) when $R_1 \subseteq R_2$ (or even $R_1 \subsetneq R_2$); context and personal taste seem to play a decisive role as to which adjective is preferred. A logic can be understood as a *set of theorems* (if it has one) or as a *consequence relation*. However, for the comparison of logics, the vocabularies have to be taken into account too. One might contend that classical first-order logic is weaker than intuitionistic first-order logic, because the vocabulary of classical logic can be reduced to *one operator* (such as NEXTAND or NALLOR), whereas a similar reduction is not possible for intuitionistic logic.[17]

The "linear aspects" of connectives have come simply to mean $BCI_{\to}$—possibly, against the intentions of Girard himself. Informally speaking, a premise may be and should be used once. In terms of the λ-calculus, each abstraction operator binds exactly one free occurrence of a variable in the term in its scope. This understanding and usage, which is now generally accepted, has nothing to do with negation, which was originally hailed as linear by Girard.

> The vault key of the system is surely *linear negation* $(\,.\,)^{\perp}$. This negation, although constructive, is involutive! All the desirable De Morgan formulas can be written with $(\,.\,)^{\perp}$, without losing the usual constructive features. [96, p. 5]

Speculative informal interpretations of various logics can be very entertaining, but again, this is way afar from what we are concerned with in this book.[18]

Without questioning or disputing that a translation of intuitionistic logic into linear logic is possible, let us give a quick rundown of how the connectives that are, in some informal sense, alike match up to each other. (Of course, the formal translation is different.) We use an algebraic mode of speech (without proofs of the observations, which are easy to reconstruct).

---

[17] Following Church's [64], by an operator, we mean a variable binding logical constant. NEXTAND is Schönfinkel's operator; for other, similarly powerful operators see [35].

[18] A "tour de force" description of the development of logic culminating in linear logic is Girard's *Blind Spot* [97]. This informal interpretation involves almost everything from autism to the French railways through Plato, Nietzsche, Madoff, Kubrick and Big Brother.

In intuitionistic logic, $\wedge$ and $\vee$ *distribute* over each other, and $\rightarrow$ is the residual of $\wedge$. Therefore, $\rightarrow$ enters many theorems, including the positive paradox, contraction, permutation, suffixing and prefixing—to name a few.

In linear logic, $\wedge$ and $\vee$ *do not distribute* over each other, and $\rightarrow$ is the residual of $\circ$, *not* of $\wedge$. Neither contraction nor the positive paradox are theorems of L.

In intuitionistic logic, $\neg$ is order inverting, which permits a form of contraposition (that in turn yields double-negation introduction). However, some other forms of contraposition, and double-negation elimination are not available. $\wedge$ and $\neg$ suffice to define $\boldsymbol{F}$, that is, a contradiction, $\mathcal{A} \wedge \neg\mathcal{A}$ implies anything. Some of the De Morgan laws do not hold, which also means that $\wedge$ and $\vee$ are not interdefinable with the help of $\neg$.

In linear logic, $(\,.\,)^{\perp}$ is a so-called *De Morgan negation*. As the name suggests, all the De Morgan laws hold. But $\mathcal{A} \wedge \mathcal{A}^{\perp}$ is *not* the bottom element of the lattice spanned by $\wedge$ and $\vee$, which is non-distributive. Also, all the contrapositions are available, and therefore, double negations can be introduced and eliminated.

The *language* of L contains five binary connectives, three unary connectives, four zero-ary constants, and two quantifiers.[19] The extensional ("additive") connectives are $\wedge$ and $\vee$ (conjunction and disjunction). The intensional ("multiplicative") connectives are $\circ$ (fusion), $\rightarrow$ (implication) and $+$ (fission). The unary connectives are the modalities ("exponentials") $\Diamond$ and $\Box$, plus negation $\sim$. The zero-ary constants are $\boldsymbol{t}$, $\boldsymbol{T}$, $\boldsymbol{F}$ and $\boldsymbol{f}$. The quantifiers are denoted by $\forall$ and $\exists$. The set of wff's is defined similarly as for LK—see Definition 2.4. *Sequents* comprise a pair of finite (possibly empty) multisets separated by $\vdash$.

The axioms and rules of the *sequent calculus* LL are as follows.

$$\mathcal{A} \vdash \mathcal{A} \ \text{id} \qquad \Gamma; \boldsymbol{F} \vdash \Delta \ {\scriptstyle F\vdash} \qquad \Gamma \vdash \boldsymbol{T}; \Delta \ {\scriptstyle \vdash T}$$

$$\vdash \boldsymbol{t} \ {\scriptstyle \vdash t} \qquad \boldsymbol{f} \vdash \ {\scriptstyle f\vdash}$$

$$\frac{\Gamma \vdash \Delta}{\Gamma; \boldsymbol{t} \vdash \Delta}\ {\scriptstyle t\vdash} \qquad \frac{\Gamma \vdash \Delta}{\Gamma \vdash \boldsymbol{f}; \Delta}\ {\scriptstyle \vdash f}$$

$$\frac{\Gamma \vdash \mathcal{A}; \Delta}{\Gamma; \sim\mathcal{A} \vdash \Delta}\ {\scriptstyle \sim\vdash} \qquad \frac{\Gamma; \mathcal{A} \vdash \Delta}{\Gamma \vdash \sim\mathcal{A}; \Delta}\ {\scriptstyle \vdash\sim}$$

$$\frac{\Gamma; \mathcal{A}; \mathcal{B} \vdash \Delta}{\Gamma; \mathcal{A} \circ \mathcal{B} \vdash \Delta}\ {\scriptstyle \circ\vdash} \qquad \frac{\Gamma \vdash \mathcal{A}; \Delta \qquad \Theta \vdash \mathcal{B}; \Lambda}{\Gamma; \Theta \vdash \mathcal{A} \circ \mathcal{B}; \Delta; \Lambda}\ {\scriptstyle \vdash\circ}$$

$$\frac{\Gamma; \mathcal{A} \vdash \Delta \qquad \Theta; \mathcal{B} \vdash \Lambda}{\Gamma; \Theta; \mathcal{A} + \mathcal{B} \vdash \Delta; \Lambda}\ {\scriptstyle +\vdash} \qquad \frac{\Gamma \vdash \mathcal{A}; \mathcal{B}; \Delta}{\Gamma \vdash \mathcal{A} + \mathcal{B}; \Delta}\ {\scriptstyle \vdash+}$$

[19] We do not use the notation of [96], which is out of line with standard notation. See instead [10], [180] and [4] for presentations of linear logic using conventional notation.

$$\frac{\Gamma\vdash\mathcal{A};\Delta \qquad \Theta;\mathcal{B}\vdash\Lambda}{\Gamma;\Theta;\mathcal{A}\to\mathcal{B}\vdash\Delta;\Lambda}\;{\to}{\vdash} \qquad\qquad \frac{\Gamma;\mathcal{A}\vdash\mathcal{B};\Delta}{\Gamma\vdash\mathcal{A}\to\mathcal{B};\Delta}\;{\vdash}{\to}$$

$$\frac{\Gamma;\mathcal{A}\vdash\Delta}{\Gamma;\mathcal{A}\wedge\mathcal{B}\vdash\Delta}\;{\wedge}{\vdash_1} \qquad \frac{\Gamma;\mathcal{B}\vdash\Delta}{\Gamma;\mathcal{A}\wedge\mathcal{B}\vdash\Delta}\;{\wedge}{\vdash_2} \qquad \frac{\Gamma\vdash\mathcal{A};\Delta \qquad \Gamma\vdash\mathcal{B};\Delta}{\Gamma\vdash\mathcal{A}\wedge\mathcal{B};\Delta}\;{\vdash}{\wedge}$$

$$\frac{\Gamma;\mathcal{A}\vdash\Delta \qquad \Gamma;\mathcal{B}\vdash\Delta}{\Gamma;\mathcal{A}\vee\mathcal{B}\vdash\Delta}\;{\vee}{\vdash} \qquad \frac{\Gamma\vdash\mathcal{A};\Delta}{\Gamma\vdash\mathcal{A}\vee\mathcal{B};\Delta}\;{\vdash}{\vee_1} \qquad \frac{\Gamma\vdash\mathcal{B};\Delta}{\Gamma\vdash\mathcal{A}\vee\mathcal{B};\Delta}\;{\vdash}{\vee_2}$$

$$\frac{\Gamma;\mathcal{A}(y)\vdash\Delta}{\Gamma;\forall x\,\mathcal{A}(x)\vdash\Delta}\;{\forall}{\vdash} \qquad\qquad \frac{\Gamma\vdash\mathcal{A}(y);\Delta}{\Gamma\vdash\forall x\,\mathcal{A}(x);\Delta}\;{\vdash}{\forall}^{\oslash}$$

$$\frac{\Gamma;\mathcal{A}(y)\vdash\Delta}{\Gamma;\exists x\,\mathcal{A}(x)\vdash\Delta}\;{\exists}{\vdash}^{\oslash} \qquad\qquad \frac{\Gamma\vdash\mathcal{A}(y);\Delta}{\Gamma\vdash\exists x\,\mathcal{A}(x);\Delta}\;{\vdash}{\exists}$$

$$\frac{\Gamma;\mathcal{A}\vdash\Delta}{\Gamma;\Diamond\mathcal{A}\vdash\Delta}\;{\Diamond}{\vdash} \qquad\qquad \frac{\Gamma^{\Diamond}\vdash\mathcal{A};\Delta^{\Box}}{\Gamma^{\Diamond}\vdash\Diamond\mathcal{A};\Delta^{\Box}}\;{\vdash}{\Diamond}$$

$$\frac{\Gamma^{\Diamond};\mathcal{A}\vdash\Delta^{\Box}}{\Gamma^{\Diamond};\Box\mathcal{A}\vdash\Delta^{\Box}}\;{\Box}{\vdash} \qquad\qquad \frac{\Gamma\vdash\mathcal{A};\Delta}{\Gamma\vdash\Box\mathcal{A};\Delta}\;{\vdash}{\Box}$$

$$\frac{\Gamma\vdash\Delta}{\Gamma;\Diamond\mathcal{A}\vdash\Delta}\;{\Diamond}K{\vdash} \qquad\qquad \frac{\Gamma\vdash\Delta}{\Gamma\vdash\Box\mathcal{A};\Delta}\;{\vdash}{\Box}K$$

$$\frac{\Gamma;\Diamond\mathcal{A};\Diamond\mathcal{A}\vdash\Delta}{\Gamma;\Diamond\mathcal{A}\vdash\Delta}\;{\Diamond}W{\vdash} \qquad\qquad \frac{\Gamma\vdash\Box\mathcal{A};\Box\mathcal{A};\Delta}{\Gamma\vdash\Box\mathcal{A};\Delta}\;{\vdash}{\Box}W$$

We use here some earlier notation. $^{\oslash}$ indicates a variable restriction exactly like in LK, namely, that y is not free in the lower sequent of the rule, hence, in any wff's in Γ or Δ. $\Gamma^{\Diamond}$ ($\Delta^{\Box}$) means that all the elements of Γ (of Δ) are formulas starting with $\Diamond$ ($\Box$).

The notions of a *proof* and of a *theorem* are unchanged from $LR^{\Box}_{\overset{\sim}{\to}}$.

As we mentioned earlier, $\sim$ is a De Morgan negation. For example, $\sim$ distributes over $\wedge$ into $\vee$.

$$\dfrac{\dfrac{\dfrac{\dfrac{\dfrac{\mathcal{A}\vdash\mathcal{A}}{\vdash\mathcal{A};{\sim}\mathcal{A}}}{\vdash\mathcal{A};{\sim}\mathcal{A}\vee{\sim}\mathcal{B}} \qquad \dfrac{\dfrac{\mathcal{B}\vdash\mathcal{B}}{\vdash\mathcal{B};{\sim}\mathcal{B}}}{\vdash\mathcal{B};{\sim}\mathcal{A}\vee{\sim}\mathcal{B}}}{\vdash\mathcal{A}\wedge\mathcal{B};{\sim}\mathcal{A}\vee{\sim}\mathcal{B}}}{{\sim}(\mathcal{A}\wedge\mathcal{B})\vdash{\sim}\mathcal{A}\vee{\sim}\mathcal{B}}}{\vdash{\sim}(\mathcal{A}\wedge\mathcal{B})\to({\sim}\mathcal{A}\vee{\sim}\mathcal{B})} \qquad \dfrac{\dfrac{\dfrac{\dfrac{\dfrac{\mathcal{A}\vdash\mathcal{A}}{\mathcal{A}\wedge\mathcal{B}\vdash\mathcal{A}}}{{\sim}\mathcal{A};\mathcal{A}\wedge\mathcal{B}\vdash} \qquad \dfrac{\dfrac{\mathcal{B}\vdash\mathcal{B}}{\mathcal{A}\wedge\mathcal{B}\vdash\mathcal{B}}}{{\sim}\mathcal{B};\mathcal{A}\wedge\mathcal{B}\vdash}}{{\sim}\mathcal{A}\vee{\sim}\mathcal{B};\mathcal{A}\wedge\mathcal{B}\vdash}}{{\sim}\mathcal{A}\vee{\sim}\mathcal{B}\vdash{\sim}(\mathcal{A}\wedge\mathcal{B})}}{\vdash({\sim}\mathcal{A}\vee{\sim}\mathcal{B})\to{\sim}(\mathcal{A}\wedge\mathcal{B})}$$

These proofs look exactly like the proofs of the distributivity of $\neg$ over $\wedge$ into $\vee$ in LK—if we replace $\neg$ with $\sim$, $\supset$ with $\to$ and , with $;$.

Exercise 4.5.1. Verify that with the indicated change of symbols, the above proofs are proofs in $LK^{[\,]}$. Then prove that $\sim$ distributes over $\vee$ into $\wedge$.

The intuitionistically unacceptable variant of contraposition is $(\sim\mathcal{A} \rightarrow \mathcal{B}) \rightarrow \sim\mathcal{B} \rightarrow \mathcal{A}$. Here is its proof in LL, which uses a two-element multiset on the right-hand side.

$$\dfrac{\dfrac{\dfrac{\dfrac{\mathcal{A} \vdash \mathcal{A}}{\vdash \sim\mathcal{A}; \mathcal{A}}\vdash\sim \qquad \mathcal{B} \vdash \mathcal{B}}{\sim\mathcal{A} \rightarrow \mathcal{B} \vdash \mathcal{A}; \mathcal{B}} \rightarrow\vdash}{\sim\mathcal{A} \rightarrow \mathcal{B}; \sim\mathcal{B} \vdash \mathcal{A}} \sim\vdash}{\vdash (\sim\mathcal{A} \rightarrow \mathcal{B}) \rightarrow \sim\mathcal{B} \rightarrow \mathcal{A}} \vdash\rightarrow\text{'s}$$

Exercise 4.5.2. Prove the other versions of contraposition. These are the wff's $(\mathcal{A} \rightarrow \sim\mathcal{B}) \rightarrow \mathcal{B} \rightarrow \sim\mathcal{A}$, $(\mathcal{A} \rightarrow \mathcal{B}) \rightarrow \sim\mathcal{B} \rightarrow \sim\mathcal{A}$ and $(\sim\mathcal{A} \rightarrow \sim\mathcal{B}) \rightarrow \mathcal{B} \rightarrow \mathcal{A}$.

The vocabulary of LL is almost as redundant as that of classical logic. $\circ$, $+$ and $\rightarrow$ are interdefinable using $\sim$. So are $\boldsymbol{T}$ and $\boldsymbol{F}$, $\boldsymbol{t}$ and $\boldsymbol{f}$, $\Diamond$ and $\Box$, as well as $\forall$ and $\exists$.

Exercise 4.5.3. Prove that $\mathcal{A} + \mathcal{B}$ is definable as $\sim(\sim\mathcal{A} \circ \sim\mathcal{B})$ and also as $\sim\mathcal{A} \rightarrow \mathcal{B}$. Then prove that the four pairs of the other logical constants are linked via abstract De Morgan-like laws. [Hint: For example, $\boldsymbol{T}$ is definable as $\sim\boldsymbol{F}$ and $\Diamond\mathcal{A}$ as $\sim\Box\sim\mathcal{A}$.]

Another important connection between some of the connectives is the interaction of $\circ$ and $+$. The structural connective ; can be interpreted by $\circ$ on the left, and by $+$ on the right.

Exercise 4.5.4. Prove the sequents $\vdash (\mathcal{A} \circ (\mathcal{B} + \mathcal{C})) \rightarrow ((\mathcal{A} \circ \mathcal{B}) + \mathcal{C})$ and $\vdash (\mathcal{A} \circ (\mathcal{B} + \mathcal{C})) \rightarrow ((\mathcal{A} \circ \mathcal{C}) + \mathcal{B})$. [Hint: The connection between these kinds of wff's and the admissibility of cut is discussed in detail in Section 5.5.]

The cut rule is of the same shape as in LR in the previous section.

Theorem 4.27. *The single cut rule is* admissible *in LL.*

Exercise* 4.5.5. Prove the theorem. [Hint: Recall what went into the proof of the cut theorem for LK beyond the triple induction.]

The vocabulary of LL is missing a connective that would be straightforward to think of and to add. Let us denote a binary connective by $\succ$. We would like to think of $\mathcal{A} \succ \mathcal{B}$ as being equivalent to $\mathcal{A} \circ \sim\mathcal{B}$. The rules for $\succ$ then should be the following ones.

$$\dfrac{\Gamma; \mathcal{A} \vdash \mathcal{B}; \Delta}{\Gamma; \mathcal{A} \succ \mathcal{B} \vdash \Delta} \succ\vdash \qquad \dfrac{\Gamma \vdash \mathcal{A}; \Delta \qquad \Theta; \mathcal{B} \vdash \Lambda}{\Gamma; \Theta \vdash \mathcal{A} \succ \mathcal{B}; \Delta; \Lambda} \vdash\succ$$

These rules suffice for us to prove that the intended relationship between $\circ$, $\sim$ and $\succ$ obtains.

$$\dfrac{\dfrac{\dfrac{\mathcal{A}\vdash\mathcal{A}\qquad \dfrac{\mathcal{B}\vdash\mathcal{B}}{\vdash\mathcal{B};{\sim}\mathcal{B}}\,{\scriptstyle\vdash\sim}}{\mathcal{A}\vdash\mathcal{B};\mathcal{A}\circ{\sim}\mathcal{B}}\,{\scriptstyle\vdash\circ}}{\mathcal{A}\succ\mathcal{B}\vdash\mathcal{A}\circ{\sim}\mathcal{B}}\,{\scriptstyle\succ\vdash}}{\vdash(\mathcal{A}\succ\mathcal{B})\rightarrow(\mathcal{A}\circ{\sim}\mathcal{B})}\,{\scriptstyle\vdash\rightarrow}
\qquad
\dfrac{\dfrac{\dfrac{\mathcal{A}\vdash\mathcal{A}\qquad \dfrac{\mathcal{B}\vdash\mathcal{B}}{{\sim}\mathcal{B};\mathcal{B}\vdash}\,{\scriptstyle\sim\vdash}}{\mathcal{A};{\sim}\mathcal{B}\vdash\mathcal{A}\succ\mathcal{B}}\,{\scriptstyle\vdash\succ}}{\mathcal{A}\circ{\sim}\mathcal{B}\vdash\mathcal{A}\succ\mathcal{B}}\,{\scriptstyle\circ\vdash}}{\vdash(\mathcal{A}\circ{\sim}\mathcal{B})\rightarrow(\mathcal{A}\succ\mathcal{B})}\,{\scriptstyle\vdash\rightarrow}$$

Exercise 4.5.6. Prove that the cut theorem remains true if the rules $(\succ\vdash)$ and $(\vdash\succ)$ are added to LL. [Hint: $\succ$ (and its pair $\prec$) will be introduced into another logic in Section 5.5—with a different outcome.]

Exercise 4.5.7. Prove that $\succ$ is the residual of $+$, exactly like $\rightarrow$ is the residual of $\circ$.

The logic LL is sometimes called *classical linear logic*.[20] *Intuitionistic linear logic*, LI is defined from LL by introducing a similar structural restriction as was used to get LJ from LK.

As expected, the symmetry of LL, which is reminiscent of the symmetry found in LK, is lost and together with that some logical components and some proofs will be lost too.

Exercise 4.5.8. Scrutinize LL to determine which connectives should be omitted, and how the remaining rules look.

Exercise 4.5.9. What kind of negation is $\sim$ in LI? [Hint: You may consider (beyond the De Morgan and contraposition laws) $(\mathcal{A}\natural{\sim}\mathcal{A})\rightarrow\mathcal{B}$ and $\mathcal{B}\rightarrow(\mathcal{A}\natural{\sim}\mathcal{A})$ with some binary connectives for $\natural$.]

The unusual feature of classical linear logic LL—compared to other logics we have considered—is not the combination of a two-sided sequent calculus with a commutative $\circ$ connective, and not even the addition of a modal operator. Rather the surprising rules are the *modalized structural rules*, which combine the unusual modal operators with thinning and contraction.

The modalities—whether we take $\Diamond$ or $\Box$ as the primitive—are hard to interpret. Indeed, the original ! and ? notation has been occasionally mistranslated into the $\Diamond$ and $\Box$ notation. First, we establish that $\Diamond$ does not distribute over $\vee$. The monotonicity of $\Diamond$ leads to the next proof.

$$\dfrac{\dfrac{{\scriptstyle\vdash\Diamond}\,\dfrac{{\scriptstyle\Diamond\vdash}\,\dfrac{{\scriptstyle\vdash\vee}\,\dfrac{\mathcal{A}\vdash\mathcal{A}}{\mathcal{A}\vdash\mathcal{A}\vee\mathcal{B}}}{\Diamond\mathcal{A}\vdash\mathcal{A}\vee\mathcal{B}}}{\Diamond\mathcal{A}\vdash\Diamond(\mathcal{A}\vee\mathcal{B})}\qquad \dfrac{\dfrac{\dfrac{\mathcal{B}\vdash\mathcal{B}}{\mathcal{B}\vdash\mathcal{A}\vee\mathcal{B}}\,{\scriptstyle\vdash\vee}}{\Diamond\mathcal{B}\vdash\mathcal{A}\vee\mathcal{B}}\,{\scriptstyle\Diamond\vdash}}{\Diamond\mathcal{B}\vdash\Diamond(\mathcal{A}\vee\mathcal{B})}\,{\scriptstyle\vdash\Diamond}}{\Diamond\mathcal{A}\vee\Diamond\mathcal{B}\vdash\Diamond(\mathcal{A}\vee\mathcal{B})}\,{\scriptstyle\vee\vdash}}{\vdash(\Diamond\mathcal{A}\vee\Diamond\mathcal{B})\rightarrow\Diamond(\mathcal{A}\vee\mathcal{B})}\,{\scriptstyle\vdash\rightarrow}$$

[20]See, for example, Troelstra's [180].

An attempt to prove the converse implication, which expresses the distributivity of $\Diamond$ over $\vee$, runs into a difficulty, because unless the left-hand side of the $\vdash$ is empty (or all the wff's are modalized), there is no way to modalize a wff with $\Diamond$ on the right-hand side.

Exercise 4.5.10. Verify that $\Box(\mathcal{A} \vee \mathcal{B})$ and $\Box\mathcal{A} \vee \Box\mathcal{B}$ are not provably equivalent. [Hint: You should show that the implication between the two formulas does not hold in (at least) one direction.]

In normal modal logics, which are extensions of classical logic, the $\Diamond$ operator has another property beyond its additive character (i.e., its distributivity over $\vee$). Namely, $\Diamond$ preserves the bottom element. At first, it may appear counterintuitive that this preservation property is somehow linked to the distributivity over $\vee$. (It is an accident that in classical logic $\boldsymbol{F}$ is definable as $\mathcal{A} \wedge \neg\mathcal{A}$.) A better way to think about the least element of a Boolean algebra or of a lattice is as the least upper bound of $\emptyset$. (Algebraically speaking, $\Diamond$ in the modal logics K, T, B, $S4$ and $S5$, for example, is a *normal operator* on a Boolean algebra. The meaning of "normality" is exactly the preservation of the least element of the lattice.)

It is easy to see that we have proofs of $\boldsymbol{F} \vdash \Diamond\boldsymbol{F}$ and $\Box\boldsymbol{T} \vdash \boldsymbol{T}$ from the axioms for $\boldsymbol{T}$ and $\boldsymbol{F}$. We also have the following.

$$\Diamond\vdash\ \frac{\boldsymbol{F} \vdash \boldsymbol{F}}{\Diamond\boldsymbol{F} \vdash \boldsymbol{F}} \qquad\qquad \frac{\boldsymbol{T} \vdash \boldsymbol{T}}{\boldsymbol{T} \vdash \Box\boldsymbol{T}}\ \vdash\Box$$

The further comparison with normal modal logics is hampered by the lack of distributivity of $\wedge$ and $\vee$. Provable sequents like $\vdash \mathcal{A} \to \Box\mathcal{A}$, $\vdash \Diamond\mathcal{A} \to \mathcal{A}$, and by transitivity, $\vdash \Diamond\mathcal{A} \to \Box\mathcal{A}$ do not match well the usual alethic, temporal, deontic, etc. interpretations.[21] A sentence saying that "what is possible is necessary" gives the impression of strong determinism. (There were thinkers in the not too long past who argued for such a view, but those views seem not to be in vogue nowadays.) A deontic reading is hardly better: "If one is permitted to do $\mathcal{A}$, then one should do $\mathcal{A}$." Even without considering fine points in the ontology of actions, it is clear that normally people are permitted to do a variety of things, but they are not obliged to do all of them (perhaps, not even any of them).

The combination of the modalities with the usual structural rules yields interesting provable wff's.

Example 4.28. Here are two proofs that show how some wff's, which have a certain resemblance to classical theorems with respect to the series of steps in their proofs, and with respect to their shape, emerge.

[21] The strongly implausible readings of the ! and ? symbols as modalities is likely the reason behind their mistranslations.

$$\dfrac{\dfrac{\dfrac{\mathcal{A} \vdash \mathcal{A}}{\mathcal{A}; \Diamond\mathcal{B} \vdash \mathcal{A}}\ \Diamond K\vdash}{\mathcal{A} \vdash \Diamond\mathcal{B} \to \mathcal{A}}\ \vdash\to}{\vdash \mathcal{A} \to \Diamond\mathcal{B} \to \mathcal{A}}\ \vdash\to$$

$$\dfrac{\dfrac{\dfrac{\dfrac{\dfrac{\dfrac{\mathcal{A} \vdash \mathcal{A}}{\mathcal{A} \vdash \Box\mathcal{B}; \mathcal{A}}\ \vdash\Box K}{\vdash \mathcal{A} \to \Box\mathcal{B}; \mathcal{A}}\ \vdash\to \qquad \mathcal{A} \vdash \mathcal{A}}{(\mathcal{A} \to \Box\mathcal{B}) \to \mathcal{A} \vdash \mathcal{A}; \mathcal{A}}\ \to\vdash}{(\mathcal{A} \to \Box\mathcal{B}) \to \mathcal{A} \vdash \Box\mathcal{A}; \Box\mathcal{A}}\ \vdash\Box\text{'s}}{(\mathcal{A} \to \Box\mathcal{B}) \to \mathcal{A} \vdash \Box\mathcal{A}}\ \vdash\Box W}{\vdash ((\mathcal{A} \to \Box\mathcal{B}) \to \mathcal{A}) \to \Box\mathcal{A}}\ \vdash\to$$

The wff $\mathcal{A} \to \Diamond\mathcal{B} \to \mathcal{A}$ shows how a trace of the thinning rule appears in the formula itself. On the other hand, the wff that has the same proof structure as Peirce's law has in *LK*—except that the structural rules are modalized, hence, so are some of the wff's—demonstrates the interaction between the introduction rules for $\Diamond$ and $\Box$, and the structural rules.

Exercise 4.5.11. Some other so-called paradoxes of material implication are $\mathcal{A} \supset \mathcal{B} \supset \mathcal{B}$, $\mathcal{A} \supset \neg\mathcal{A} \supset \mathcal{B}$ and $(\mathcal{A} \supset \mathcal{B}) \vee (\mathcal{B} \supset \mathcal{A})$. (The latter formula is often called *linearity*.) Sketch their proofs in LK and re-create those proofs in LL—using modal rules, when necessary. [Hint: Try to modalize the formulas as little as possible.]

Exercise 4.5.12. $(\mathcal{A} \to \mathcal{B}) \to (\mathcal{A} \to \mathcal{B} \to \mathcal{C}) \to \mathcal{A} \to \mathcal{C}$ and $(\mathcal{A} \to \mathcal{B} \to \mathcal{C}) \to (\mathcal{A} \to \mathcal{B}) \to \mathcal{A} \to \mathcal{C}$ are not theorems of LL, but with some modalities dispersed within these wff's they are provable. Find provable versions of these wff's—with the least amount of modality added.

Exercise 4.5.13. A major drawback of LL is the lack of distributivity between $\wedge$ and $\vee$. Explore which modalized versions of distributivity are provable—again limiting occurrences of $\Diamond$ and $\Box$ to the absolutely inevitable ones.

Exercise* 4.5.14. The De Morgan negation allows LL to be formulated as a one-sided sequent calculus—either with nothing on the left, or with nothing on the right. Work out both alternatives. [Hint: Recall that some of the connectives are definable in LL.]

4.6 Positive logic of relevant implication

Relevance logics emerged from the dissatisfaction with certain deficiencies of classical logic such as the unrestricted monotonicity of the consequence relation with respect to sets of premises. There has been a lot written about the paradoxes of material implication, "the fifth indemonstrable syllogism" of the Stoics and the rules "ex falso quodlibet" and "ex contradictione quodlibet."[22] However, the distributivity of conjunction and disjunction has not been questioned or blamed for irrelevant inferences. All the major relevance logics (i.e., E, R and T) that were introduced and defined in the late 1950s–early 1960s, underwrite the distributivity of $\wedge$ and $\vee$.

We have seen in Section 4.4 how LR^d_+ attempted to deal with the distributivity of $\wedge$ and $\vee$. On the other hand, LL recaptures classical logic with the modalities marking the subformulas, which were subjected to contraction and thinning. Neither a sequent calculus that requires the cut rule to be included, nor a sequent calculus that can only prove modalized versions of desirable wff's is satisfactory.

A suitable sequent calculus formalization of negation-free R was invented by Dunn in 1969 (published as [75]). It is, probably, obvious at this point in this book that the structure-free rules for conjunction and disjunction are not sufficient to prove their distributivity. However, a relevance logic does not have thinning, which precludes us from using structural manipulations to remedy this non-provability.

The most important insight to achieve a suitable formalization for R_+, the positive fragment of R, is to have *two structural connectives*, namely, ; and , at the same time. On the left-hand side of the turnstile, the former is linked to $\circ$, whereas the latter to $\wedge$. Of course, then it must be explained and defined how sequences of wff's formed by one or the other structural connective relate to each other. Formulas are finite, but they can be arbitrarily long. Similarly, sequences of wff's should be embeddable into each other to an arbitrary (but finite) depth.[23]

Definition 4.29. (Sequences) We define *extensional sequences* (e-sequences) and *intensional sequences* (i-sequences) by clauses (1)–(4).

(1) If $\gamma_1, \ldots, \gamma_n$ (where $n > 1$) are wff's or e-sequences, then $\gamma_1, \ldots, \gamma_n$ is an e-sequence;

(2) if $\gamma_1, \ldots, \gamma_n$ (where $n > 1$) are wff's or i-sequences, then $(\gamma_1), \ldots, (\gamma_n)$ is an e-sequence;

(3) if $\gamma_1, \ldots, \gamma_n$ (where $n > 1$) are wff's or i-sequences, then $\gamma_1; \ldots; \gamma_n$ is an i-sequence;

(4) if $\gamma_1, \ldots, \gamma_n$ (where $n > 1$) are wff's or e-sequences, then $(\gamma_1); \ldots; (\gamma_n)$ is an i-sequence.

As a notational convention, we assume that if a pair of parentheses would be inserted around a single wff by (2) or (4), then those parentheses are omitted.

[22] See, for example, [7] and [8], as well as some references therein.

[23] An excellent exposition of LR_+ by Dunn may be found in his [77].

Clauses (2) and (4) are formulated so as to ensure that complex antecedents of a sequent are not ambiguous, but by convention, parentheses that do not aid the parsing of a sequence are not inserted. Since both the extensional and intensional sequences are *sequences*, the lack of parentheses inside them does not lead to ambiguity.

We will call a single formula or the empty sequence or an e-sequence or an i-sequence a *structure*. Structures that comprise more that a single wff are complex. The simplest complex structures are "flat" e- or i-sequences, in which all the elements are wff's separated by , or ;, respectively. Other complex structures are stacks of alternating e- and i-sequences, with the depth of the stack being finite at each point.

A *sequent* is a structure followed by $\vdash$ and a wff. The meta-variables, $\Gamma, \Delta, \ldots$ range over structures. The connectives are $\circ$, $\rightarrow$, $\wedge$, $\vee$ and $\boldsymbol{t}$ (with the same arities as in previous languages).

The *sequent calculus* LR_+ comprises two axioms, ten operational rules and five structural rules.

$$\mathcal{A} \vdash \mathcal{A} \ \text{id} \qquad\qquad \vdash \boldsymbol{t} \ \vdash\! t$$

$$\frac{\Gamma[\mathcal{A}; \mathcal{B}] \vdash \mathcal{C}}{\Gamma[\mathcal{A} \circ \mathcal{B}] \vdash \mathcal{C}} \circ\vdash \qquad \frac{\Gamma \vdash \mathcal{A} \qquad \Delta \vdash \mathcal{B}}{\Gamma; \Delta \vdash \mathcal{A} \circ \mathcal{B}} \vdash\circ$$

$$\frac{\Gamma \vdash \mathcal{A} \qquad \Delta[\mathcal{B}] \vdash \mathcal{C}}{\Delta[\mathcal{A} \rightarrow \mathcal{B}; \Gamma] \vdash \mathcal{C}} \rightarrow\vdash \qquad \frac{\Gamma; \mathcal{A} \vdash \mathcal{B}}{\Gamma \vdash \mathcal{A} \rightarrow \mathcal{B}} \vdash\rightarrow$$

$$\frac{\Gamma[\mathcal{A}, \mathcal{B}] \vdash \mathcal{C}}{\Gamma[\mathcal{A} \wedge \mathcal{B}] \vdash \mathcal{C}} \wedge\vdash \qquad \frac{\Gamma \vdash \mathcal{A} \qquad \Gamma \vdash \mathcal{B}}{\Gamma \vdash \mathcal{A} \wedge \mathcal{B}} \vdash\wedge$$

$$\frac{\Gamma[\mathcal{A}] \vdash \mathcal{C} \qquad \Gamma[\mathcal{B}] \vdash \mathcal{C}}{\Gamma[\mathcal{A} \vee \mathcal{B}] \vdash \mathcal{C}} \vee\vdash \qquad \frac{\Gamma \vdash \mathcal{A}}{\Gamma \vdash \mathcal{A} \vee \mathcal{B}} \vdash\vee_1 \qquad \frac{\Gamma \vdash \mathcal{B}}{\Gamma \vdash \mathcal{A} \vee \mathcal{B}} \vdash\vee_2$$

$$\frac{\Gamma[\Delta] \vdash \mathcal{C}}{\Gamma[\boldsymbol{t}; \Delta] \vdash \mathcal{C}} t\vdash \qquad \frac{\Gamma[\Delta] \vdash \mathcal{C}}{\Gamma[\Theta, \Delta] \vdash \mathcal{C}} K\vdash$$

$$\frac{\Gamma[\Delta; \Theta] \vdash \mathcal{C}}{\Gamma[\Theta; \Delta] \vdash \mathcal{C}} C\vdash \qquad \frac{\Gamma[\Delta, \Theta] \vdash \mathcal{C}}{\Gamma[\Theta, \Delta] \vdash \mathcal{C}} C\vdash$$

$$\frac{\Gamma[\Delta; \Delta] \vdash \mathcal{C}}{\Gamma[\Delta] \vdash \mathcal{C}} W\vdash \qquad \frac{\Gamma[\Delta, \Delta] \vdash \mathcal{C}}{\Gamma[\Delta] \vdash \mathcal{C}} W\vdash$$

The notion of a *proof* is as before. $\mathcal{A}$ is a *theorem* when $\vdash \mathcal{A}$ has a proof.

The e- and i-sequences and their embedding into each other may seem confusing. Here is a sample proof to illustrate how the sequences intertwine. We will imitate a "backward proof search" way of building the proof. We want to prove the wff

$$((\mathcal{D} \rightarrow (\mathcal{A} \vee \mathcal{B})) \wedge (\mathcal{D} \rightarrow (\mathcal{A} \vee \mathcal{C}))) \rightarrow \mathcal{D} \rightarrow (\mathcal{A} \vee (\mathcal{B} \wedge \mathcal{C})).$$

The wff is a contextualized version of a formula the provability of which implies that $\wedge$ and $\vee$ distribute over each other.[24] (We omit mentioning permutation steps, which hints toward the fact that the two types of sequences could have been replaced by two types of multisets.)

$$
\dfrac{\dfrac{\dfrac{\dfrac{\dfrac{\begin{array}{c}\vdots\\ (\mathcal{D}\to(\mathcal{A}\vee\mathcal{B});\mathcal{D}),(\mathcal{D}\to(\mathcal{A}\vee\mathcal{C});\mathcal{D})\vdash\mathcal{A}\vee(\mathcal{B}\wedge\mathcal{C})\end{array}}{((\mathcal{D}\to(\mathcal{A}\vee\mathcal{B}),\mathcal{D}\to(\mathcal{A}\vee\mathcal{C}));\mathcal{D}),((\mathcal{D}\to(\mathcal{A}\vee\mathcal{B}),\mathcal{D}\to(\mathcal{A}\vee\mathcal{C}));\mathcal{D})\vdash\mathcal{A}\vee(\mathcal{B}\wedge\mathcal{C})}\,K\vdash\text{'s}}{(\mathcal{D}\to(\mathcal{A}\vee\mathcal{B}),\mathcal{D}\to(\mathcal{A}\vee\mathcal{C}));\mathcal{D}\vdash\mathcal{A}\vee(\mathcal{B}\wedge\mathcal{C})}\,W\vdash}{(\mathcal{D}\to(\mathcal{A}\vee\mathcal{B}))\wedge(\mathcal{D}\to(\mathcal{A}\vee\mathcal{C}));\mathcal{D}\vdash\mathcal{A}\vee(\mathcal{B}\wedge\mathcal{C})}\,\wedge\vdash}{(\mathcal{D}\to(\mathcal{A}\vee\mathcal{B}))\wedge(\mathcal{D}\to(\mathcal{A}\vee\mathcal{C}))\vdash\mathcal{D}\to(\mathcal{A}\vee(\mathcal{B}\wedge\mathcal{C}))}\,\vdash\to}{\vdash((\mathcal{D}\to(\mathcal{A}\vee\mathcal{B}))\wedge(\mathcal{D}\to(\mathcal{A}\vee\mathcal{C})))\to\mathcal{D}\to(\mathcal{A}\vee(\mathcal{B}\wedge\mathcal{C}))}\,\vdash\to
$$

The three steps upward from the bottom are straightforward, and as far as connective rules go, the only other possibility would have been to eliminate $\vee$ on the right, which would clearly be an unwise step at this point. The line containing formulas in "fine print" is an e-sequence containing two copies of an i-sequence.

By two applications of the $(\to\vdash)$ rule together with copies of $\mathcal{D}\vdash\mathcal{D}$, we get the sequent $\mathcal{A}\vee\mathcal{B},\mathcal{A}\vee\mathcal{C}\vdash\mathcal{A}\vee(\mathcal{B}\wedge\mathcal{C})$. This is, essentially, distributivity without a context.

$$
\dfrac{\dfrac{\dfrac{\mathcal{A}\vdash\mathcal{A}}{\mathcal{A},\mathcal{A}\vee\mathcal{C}\vdash\mathcal{A}}}{\mathcal{A},\mathcal{A}\vee\mathcal{C}\vdash\mathcal{A}\vee(\mathcal{B}\wedge\mathcal{C})}\qquad\dfrac{\dfrac{\dfrac{\mathcal{A}\vdash\mathcal{A}}{\mathcal{B},\mathcal{A}\vdash\mathcal{A}}}{\mathcal{B},\mathcal{A}\vdash\mathcal{A}\vee(\mathcal{B}\wedge\mathcal{C})}\qquad\dfrac{\dfrac{\dfrac{\mathcal{B}\vdash\mathcal{B}}{\mathcal{B},\mathcal{C}\vdash\mathcal{B}}\quad\dfrac{\mathcal{C}\vdash\mathcal{C}}{\mathcal{B},\mathcal{C}\vdash\mathcal{C}}}{\mathcal{B},\mathcal{C}\vdash\mathcal{B}\wedge\mathcal{C}}}{\mathcal{B},\mathcal{C}\vdash\mathcal{A}\vee(\mathcal{B}\wedge\mathcal{C})}}{\mathcal{B},\mathcal{A}\vee\mathcal{C}\vdash\mathcal{A}\vee(\mathcal{B}\wedge\mathcal{C})}}{\mathcal{A}\vee\mathcal{B},\mathcal{A}\vee\mathcal{C}\vdash\mathcal{A}\vee(\mathcal{B}\wedge\mathcal{C})}
$$

This part of the proof makes clear that we have to apply $(W\vdash)$, rather than $(\mathrm{W}\vdash)$ in the proof. Also, all the instances of (id), save $\mathcal{D}\vdash\mathcal{D}$, are followed by applications of $(K\vdash)$. Incidentally, in the last chunk of the proof, none of the applications of extensional thinning are connected to the choice of Ketonen's version of the rule $(\wedge\vdash)$.

Exercise 4.6.1. Show that the sequent $\vdash(\mathcal{A}\vee(\mathcal{B}\wedge(\mathcal{A}\vee\mathcal{C})))\to((\mathcal{A}\vee\mathcal{B})\wedge(\mathcal{A}\vee\mathcal{C}))$ is provable in LR_+.

It can happen that a sequent contains one formula on the left, or nothing at all. We allowed this possibility, but we did not classify formulas or the empty sequence of wff's as an e-sequence or an i-sequence. The status of the empty sequence is of particular concern, because, typically, whenever the empty sequence is permitted in a sequent calculus, it can disappear together

[24] A proof of a different formula expressing distributivity may be found in [77, p. 177].

with the structural connective that joins it to a non-empty sequence. On the other hand, a single formula may be shifted to the right-hand side (as the antecedent of a wff) by the $(\vdash\to)$ rule, hence, we want to think of that wff in the context of a sequent as if it were an i-sequence. The empty sequence of wff's cannot be made part of a complex structure and it can appear only in the context of $\vdash$. Both in the $(\vdash \boldsymbol{t})$ axiom and when it is a result of an application of the $(\vdash\to)$ rule, the empty sequence is to be considered as an i-sequence. Really, the empty sequence can be thought of as an *invisible* $\boldsymbol{t}$.

Let $\Gamma^{\boldsymbol{t}}$ denote Γ, if Γ is not the empty sequence of wff's, and $\boldsymbol{t}$, otherwise. A chief insight in [75] is that the empty sequence must be handled carefully, especially, in the cut rule, which now looks like the following.

$$\dfrac{\Gamma\vdash C \qquad \Delta[C]\vdash \mathcal{A}}{\Delta[\Gamma^{\boldsymbol{t}}]\vdash \mathcal{A}}\ \text{cut}$$

As it is pointed out in [77, p. 177], irrelevance can emerge easily if an empty Γ is not replaced by $\boldsymbol{t}$. Here is a concrete example.

$$\text{cut?}\ \dfrac{\dfrac{\dfrac{p\vdash p}{\vdash p\to p} \qquad \dfrac{p\to p\vdash p\to p}{q,p\to p\vdash p\to p}}{q\vdash p\to p}}{\vdash q\to p\to p} \qquad\qquad \dfrac{\dfrac{\dfrac{\dfrac{p\vdash p}{\vdash p\to p} \qquad \dfrac{p\to p\vdash p\to p}{q,p\to p\vdash p\to p}}{q,\boldsymbol{t}\vdash p\to p}\ \text{cut}}{q\wedge\boldsymbol{t}\vdash p\to p}}{\vdash (q\wedge\boldsymbol{t})\to p\to p}$$

The questionable cut allows a concrete instance of the *negative paradox* to be proved. The properly defined cut is unproblematic from the point of view of relevance logic, because $(q\wedge\boldsymbol{t})\to p\to p$ says that the least theorem (i.e., $\boldsymbol{t}$) conjoined with another formula implies $p\to p$.

The next question to consider is whether the more refined cut rule, which replaces an empty Γ with $\boldsymbol{t}$, is admissible.

Theorem 4.30. *The single cut rule is* admissible *in* LR_+.

Proof: The claim can be proved by multiple inductions like in the case of $LR^{\boldsymbol{t}}_{\to}$. Notice that the insertion of an extra $\boldsymbol{t}$ in an application of the cut rule is unproblematic for the proof of the theorem. If the proof is transformed into a proof where a cut is performed higher in the proof tree with its left premise having a non-empty antecedent, then a $\boldsymbol{t}$ still may be inserted into the sequent by an application of the $(\boldsymbol{t}\vdash)$ rule. qed

Exercise 4.6.2. Work out the details of the proof.

It may be thought that the insertion of an extra $\boldsymbol{t}$ into some proofs may interfere with establishing the equivalence of LR_+ to the axiomatic formulation. However, the following holds too.

Theorem 4.31. *If* $\boldsymbol{t}\vdash\mathcal{A}$ *is provable in* LR_+, *then* $\vdash\mathcal{A}$ *is* provable *too.*

Exercise 4.6.3. Prove the theorem. [Hint: In a cut-free proof, $\boldsymbol{t}$ can be introduced by either of the axioms or by $(\boldsymbol{t}\vdash)$.]

In the previous two sections, we mentioned modalities in the context of logics in which $\wedge$ and $\vee$ do not distribute over each other. We may wonder whether LR_+ can be extended to a sequent calculus that captures the negation-free fragment of $R^{\Box}$.

Having two structural connectives makes a difference with respect to the modal extension of R_+ too. Axiomatically, (K), (T), (4) and $(\Box\wedge)$ are added together with (Nec), the rule of necessitation, which allows a theorem to be prefixed with $\Box$.[25]

We used the notation $\Gamma^{\Box}$ before, however, we want to give it a slightly different meaning now. Γ may be an i-sequence of e-sequences of i-sequences, etc. The decomposition of a non-empty structure ends when a component is a wff, hence, it makes sense to consider the *atomic components* of a structure, which is a set of wff's. $\boldsymbol{t}$ may or may not be an element of the set of atomic components for a concrete structure; if it is, we disregard it. $\Gamma^{\Box}$ is a structure in which all atomic components, except possibly $\boldsymbol{t}$, are of the form $\Box\mathcal{A}$, for some $\mathcal{A}$.

The *sequent calculus* $LR_+^{\Box}$ is LR_+ extended with the two following rules, $(\Box\vdash)$ and $(\vdash\Box)$.

$$\frac{\Gamma[\mathcal{A}] \vdash \mathcal{B}}{\Gamma[\Box\mathcal{A}] \vdash \mathcal{B}}\ \Box\vdash \qquad\qquad \frac{\Gamma^{\Box} \vdash \mathcal{A}}{\Gamma^{\Box} \vdash \Box\mathcal{A}}\ \vdash\Box$$

The notions of a *proof* and of a *theorem* are as before.

We have seen in Section 3.1 that Ketonen's versions of the rules $(\wedge\vdash)$ and $(\vdash\vee)$ are equivalent to each other within LK. The provability of the four modal axioms and the necessitation rule do not depend on the choice of the Ketonen rule for $\wedge$ in $LR_+^{\Box}$. $\Box$ fully distributes over $\wedge$ in $LR_+^{\Box}$, and the proof can be carried out without applications of the intensional contraction rule. The next proofs show how to prove $(\Box\wedge)$ in $LR_+^{\Box}$, or with the structure-free $(\wedge\vdash)$ rules where only one of the conjuncts occurs in the upper sequent. Both proofs show that e-sequences are essential. (We suppress mentioning permutation steps.)

$$\vdash\to\ \dfrac{\wedge\vdash\ \dfrac{\vdash\Box\ \dfrac{\vdash\wedge\ \dfrac{\Box\vdash\ \dfrac{K\vdash\ \dfrac{\mathcal{A}\vdash\mathcal{A}}{\mathcal{A},\Box\mathcal{B}\vdash\mathcal{A}}}{\Box\mathcal{A},\Box\mathcal{B}\vdash\mathcal{A}} \qquad \dfrac{\dfrac{\mathcal{B}\vdash\mathcal{B}}{\Box\mathcal{A},\mathcal{B}\vdash\mathcal{B}}}{\Box\mathcal{A},\Box\mathcal{B}\vdash\mathcal{B}}}{\Box\mathcal{A},\Box\mathcal{B}\vdash\mathcal{A}\wedge\mathcal{B}}}{\Box\mathcal{A},\Box\mathcal{B}\vdash\Box(\mathcal{A}\wedge\mathcal{B})}}{\Box\mathcal{A}\wedge\Box\mathcal{B}\vdash\Box(\mathcal{A}\wedge\mathcal{B})}}{\vdash(\Box\mathcal{A}\wedge\Box\mathcal{B})\to\Box(\mathcal{A}\wedge\mathcal{B})}$$

$$\dfrac{\dfrac{\dfrac{\dfrac{\dfrac{\dfrac{\dfrac{\mathcal{A}\vdash\mathcal{A}}{\mathcal{A},\Box\mathcal{B}\vdash\mathcal{A}}}{\Box\mathcal{A},\Box\mathcal{B}\vdash\mathcal{A}} \qquad \dfrac{\dfrac{\mathcal{B}\vdash\mathcal{B}}{\Box\mathcal{A},\mathcal{B}\vdash\mathcal{B}}\ K\vdash}{\Box\mathcal{A},\Box\mathcal{B}\vdash\mathcal{B}}\ \Box\vdash}{\Box\mathcal{A},\Box\mathcal{B}\vdash\mathcal{A}\wedge\mathcal{B}}\ \vdash\wedge}{\Box\mathcal{A},\Box\mathcal{B}\vdash\Box(\mathcal{A}\wedge\mathcal{B})}\ \vdash\Box}{\Box\mathcal{A}\wedge\Box\mathcal{B},\Box\mathcal{A}\wedge\Box\mathcal{B}\vdash\Box(\mathcal{A}\wedge\mathcal{B})}\ \wedge\vdash\text{'s}}{\Box\mathcal{A}\wedge\Box\mathcal{B}\vdash\Box(\mathcal{A}\wedge\mathcal{B})}\ W\vdash}{\vdash(\Box\mathcal{A}\wedge\Box\mathcal{B})\to\Box(\mathcal{A}\wedge\mathcal{B})}\ \vdash\to$$

[25] These formulas were introduced on page 114. $LR_+^{\Box}$ may be found, for example, in [7, §29].

Exercise 4.6.4. Verify that the two kinds of structural connectives do not interfere with the provability of the other axioms, (K), (T) and (4).

Exercise 4.6.5. State and prove the cut theorem for $LR^{\Box}_{+}$.

The above proofs could be viewed—except the last step—as if they were carried out in LK with rules for $\Box$ added. It is accidental that there is no need for multiple right-hand sides in the above proofs. The ingenuity of the introduction of *multiple structural connectives* led to further developments in the proof theory of non-classical logics. First of all, other relevance logics that contain a $\wedge$ and a $\vee$, which distribute over each other can be formalized similarly—provided that there is a suitable sequent calculus for their implicational fragment, possibly enriched with $\boldsymbol{t}$ and $\circ$. Some of those calculi turn out to be consecution calculi and we present them in Section 5.4. Second, a core idea behind display logics is the availability of multiple structural connectives. We devote several sections in Chapter 6 to those sorts of calculi.

We conclude this section with a sequent calculus for *first-degree entailments*, which is usually abbreviated as **fde**. This is a common fragment of R, E, T and B, for example. The preference given by Anderson and Belnap to E (rather than to R or T) explains the name of this logic as well as the other acronym used for it, $E_{\mathbf{fde}}$. This logic includes negation, and we include it here to illustrate why LR_{+} (or $LR^{\Box}_{+}$) cannot be straightforwardly extended to formalize all of R (or $R^{\Box}$). (We more or less follow the presentation in [7, §17].)

A sequent is a pair of finite multisets (including the empty multiset) separated by $\vdash$. The connectives are $\wedge$, $\vee$ and $\sim$, with their arities and informal interpretations as in $LR_{\tilde{\to}}$ and LR_{+}. Notice that there is no $\boldsymbol{t}$ in the language, but more importantly, there is no $\to$ either. "First-degree" means that the formulas contain exactly one implication, which is the main connective. Here we simply trade $\vdash$ for the implication. (This is not unlike how we interpreted provable sequents in calculi without empty left-hand sides.) Of course, there is exactly one $\vdash$ in every sequent.

We will specify multisets of formulas by listing the elements separated by $,$, which is the notation we used in $LK^{[\,]}$, the classical sequent calculus formulated with multisets of wff's. Another structural connective is *, which is the alter ego of $\sim$ like $,$ is of $\wedge$ or $\vee$ (depending on the side). The interaction of the two structural connectives is different than the interaction between $;$ and $,$. E- and i-sequences were embedded as whole blocks into each other, whereas, * distributes over $,$. In other words, Γ^*, Δ^* is $(\Gamma, \Delta)^*$, and accordingly, if Θ is $\mathcal{A}_1, \ldots, \mathcal{A}_n$, then Θ^* is $\mathcal{A}_1^*, \ldots, \mathcal{A}_n^*$. $\Gamma, \Delta, \ldots$ range over multisets, and just as Γ may or may not contain $,$, it may or may not include *. We use the bracket notation to single out occurrences; in particular, if Γ is $\mathcal{A}, \mathcal{B}^*, \mathcal{C}^*, \mathcal{C}^{**}$, then $\Gamma[\mathcal{C}^{**}]$ indicates the second occurrence of $\mathcal{C}$, which is in the scope of two *'s, whereas $\Gamma[\mathcal{C}^*]$ may fix either occurrence of $\mathcal{C}^*$.

The axiom and rules of the *sequent calculus* $L\mathbf{fde}$ are as follows.

$$\mathcal{A} \vdash \mathcal{A} \ \text{id}$$

$$\frac{\Gamma, \mathcal{A}, \mathcal{B} \vdash \Delta}{\Gamma, \mathcal{A} \wedge \mathcal{B} \vdash \Delta} \wedge\vdash \qquad \frac{\Gamma \vdash \Delta, \mathcal{A} \qquad \Gamma \vdash \Delta, \mathcal{B}}{\Gamma \vdash \Delta, \mathcal{A} \wedge \mathcal{B}} \vdash\wedge$$

$$\frac{\Gamma, \mathcal{A} \vdash \Delta \qquad \Gamma, \mathcal{B} \vdash \Delta}{\Gamma, \mathcal{A} \vee \mathcal{B} \vdash \Delta} \vee\vdash \qquad \frac{\Gamma \vdash \Delta, \mathcal{A}, \mathcal{B}}{\Gamma \vdash \Delta, \mathcal{A} \vee \mathcal{B}} \vdash\vee$$

$$\frac{\Gamma\,[\mathcal{A}^*] \vdash \Delta}{\Gamma\,[\sim\mathcal{A}] \vdash \Delta} \sim\vdash \qquad \frac{\Gamma \vdash \Delta[\mathcal{A}^*]}{\Gamma \vdash \Delta[\sim\mathcal{A}]} \vdash\sim$$

$$\frac{\Gamma \vdash \Delta}{\Gamma, \mathcal{A} \vdash \Delta} K\vdash \qquad \frac{\Gamma \vdash \Delta}{\Gamma \vdash \Delta, \mathcal{A}} \vdash K$$

$$\frac{\Gamma, \mathcal{A}, \mathcal{A} \vdash \Delta}{\Gamma, \mathcal{A} \vdash \Delta} W\vdash \qquad \frac{\Gamma \vdash \Delta, \mathcal{A}, \mathcal{A}}{\Gamma \vdash \Delta, \mathcal{A}} \vdash W$$

$$\frac{\Gamma \vdash \Delta}{\Delta^* \vdash \Gamma^*} {*\vdash*} \qquad \frac{\Gamma\,[\Delta^{**}] \vdash \Theta}{\Gamma\,[\Delta] \vdash \Theta} {**\vdash} \qquad \frac{\Gamma \vdash \Delta[\Theta^{**}]}{\Gamma \vdash \Delta[\Theta]} {\vdash**}$$

Proofs are defined as before. If a sequent $\mathcal{A} \vdash \mathcal{B}$ is provable, then the first-degree formula $\mathcal{A} \to \mathcal{B}$ is a *theorem* of $L\mathbf{fde}$.

It is useful to scrutinize how * works. The $({\sim}\vdash)$ and $(\vdash{\sim})$ rules clearly link * to negation. But $({*}\vdash{*})$ reveals the real problem. It swaps the two sides and adds *'s throughout. In LR_+, the left-hand side is, essentially, an i-sequence, because of the $(\vdash\to)$ rule. Further, i- and e-sequences may be embedded into each other as many times as desired. It is not clear at all how one would specify the application of the * in LR_+.

The last three * rules, in effect, make possible the introduction and elimination of **'s without affecting the component formulas. These structural rules are "more structural," so to speak, than $(K\vdash)$ or $(W\vdash)$. They are similar to certain structural rules in consecution calculi (which we introduce in the next chapter) that *rearrange* parentheses, but do not change the number or the shape of any of the wff's. A difference between those and these rules is that given a finite tree with binary branching, there are finitely many ways to move atomic components around, whereas there is no limit on how many *'s can be accumulated. Star-like structural connectives are part of certain display logics too (which we present in Section 6.1).

Exercise 4.6.6. Show that a triple inductive proof for the usual formulation of the cut rule (as, e.g., in $LR_{\overset{\sim}{\to}}$) runs into a problem.

Exercise* 4.6.7. Find a suitable formulation of the cut rule and prove that that rule is admissible in $L\mathbf{fde}$.

4.7 Sequent calculi for modal logics

Modal logics emerged as systems of strict implication in the early 20th century, and developed into a vast field since then. From the uncountably many normal modal logics, we consider only a handful in this section; each of them is a conservative extension of classical logic.[26]

In order to further narrow down the scope of what we are to deal with here, we add only one unary connective to the language, namely, $\Box$. In many standard systems of modal logic, $\Diamond$ is definable from $\Box$ and negation, using, what we may call, the *modal De Morgan laws*. In the case of $LR^{\Box}$ and $LR^{\Box}_{+}$, we did not even mention $\Diamond$, partly, because the original motivation for a modal extension of R called for a necessity type modality. Also, R contains a De Morgan negation; hence, defining $\Diamond$ would be easy.

The *language* of the modal calculi contains $\neg$, $\wedge$, $\vee$, $\supset$ and $\Box$. Many well-known modal logics are *normal*, which means that if $\mathcal{A}$ is a theorem, then so is $\Box\mathcal{A}$. (Given the underlying classical logic, this is the same as $\Diamond(\mathcal{A} \wedge \neg\mathcal{A}) \supset (\mathcal{A} \wedge \neg\mathcal{A})$.) Another reasonable rule to entertain concerns implicational theorems: if $\mathcal{A} \supset \mathcal{B}$ is a theorem, then $\Box\mathcal{A} \supset \Box\mathcal{B}$ is a theorem too. Some of the formulas from the following list should look reminiscent of some earlier formulas (which had $\rightarrow$'s, rather than $\supset$'s). They are well-known axioms from the modal logic literature.[27]

(K) $\Box(\mathcal{A} \supset \mathcal{B}) \supset \Box\mathcal{A} \supset \Box\mathcal{B}$

(T) $\Box\mathcal{A} \supset \mathcal{A}$

(4) $\Box\mathcal{A} \supset \Box\Box\mathcal{A}$

(B) $\mathcal{A} \supset \Box\neg\Box\neg\mathcal{A}$

(5) $\neg\Box\mathcal{A} \supset \Box\neg\Box\mathcal{A}$

(Nec) $\vdash \mathcal{A}$ implies $\vdash \Box\mathcal{A}$

(IN) $\vdash \mathcal{A} \supset \mathcal{B}$ implies $\vdash \Box\mathcal{A} \supset \Box\mathcal{B}$

These axioms and rules are selectively added to some axiomatization of classical propositional logic. (K) is an axiom that is common to all the systems we consider. The axiom is labeled with (K) in honor of Kripke, and it expresses that $\Box$ distributes over $\wedge$, because of the interdefinability between conjunction and implication. Adding (T) and (IN), gives the logic $E2$ of Lemmon, which coincidentally has a somewhat similar label as the logic of entailment. If only (Nec) is added, then the system K results, which is the minimal normal modal logic in the sense that nothing else is provable in it

[26] Some other sequent calculi for modal logics are mentioned in [190], though most of them go beyond the type of logistic calculi that we consider in this chapter.

[27] Some useful sources on modal logics, in general, include [59], [46] and [47].

except what makes $\Box$ a dual normal operator. (That is, $\Box$ distributes over $\wedge$, and it preserves $\boldsymbol{T}$ or $\mathcal{A} \vee \neg\mathcal{A}$.)

The distinct modal logics that result from combinations of the above axioms extending K are KB, $K4$, $K5$, $K45$, $KB4$ ($= KB5$), T (i.e., $K + $(T)), B (i.e., $T + $(B)), $S4$ (i.e., $T + (4)$) and $S5$ (i.e., $B + (4)$). The first and the last four are the most widely known systems of modal logics. Undoubtedly, some of their fame comes from Gödel's elegant approach to formalize $S4$ as an extension of classical logic in [101], and from Kripke's semantics in [116] and [117], which not only provided sound and complete interpretations, but also spurred the imagination of many philosophers. It is easy to formalize $S4$, as we will see, and some other of the logics have been given sequent calculus formalizations, for example, by Goble in [99]. $S5$ turned out to be notoriously difficult to formalize as a sequent calculus; several approaches to this problem have been developed. In some of them, parametric formulas of some of the rules have to satisfy side conditions; or the whole calculus is based on a translation into monadic predicate logic, or the calculus involves a nomenclature of levels and indices.[28] We consider here only calculi, which fit straightforwardly among the calculi in this chapter.

Sometimes it has been lamented that modal logics are not easy to formalize by sequent calculi and there is little modularity in the sequent calculus constructions. This should not come as a surprise. Although there is a certain amount of interaction between $\Box$ and the classical connectives (hence, the structural connective $\,,$), there is no correspondence between $\,,$ and $\Box$ that would parallel the correspondence between $\,;$ and $\circ$.

The *sequent calculus* $LE2$ in [99] comprises the propositional part of LK and the following two rules.

$$\frac{\mathcal{A}, \Gamma \vdash \Delta}{\Box\mathcal{A}, \Gamma \vdash \Delta}\;\Box\vdash \qquad\qquad \frac{\mathcal{A}, \Gamma \vdash \mathcal{B}}{\Box\mathcal{A}, \Gamma^{\Box} \vdash \Box\mathcal{B}}\;\Box\vdash\Box'$$

We used the notation $\Gamma^{\Box}$ to indicate that all the formulas in $\Gamma^{\Box}$ are of the form $\Box\mathcal{A}$, for some $\mathcal{A}$. The relationship of $\Gamma^{\Box}$ and Γ in the above rule is as, probably, expected; that is, $\Gamma^{\Box}$ is the sequence of formulas, in which every element of Γ is prefixed with $\Box$. The notion of a *proof* is defined as in LK, and $\mathcal{A}$ is a *theorem* of $LE2$ iff $\vdash \mathcal{A}$ is provable.

A *sequent calculus* corresponding to K is obtained by extending LK with a rule, which is similar to ($\Box\vdash\Box'$) except that there may be no formula on the left in the premise. (We do not introduce a label for this calculus, because prefixing K with L would yield LK.)

$$\frac{\Gamma \vdash \mathcal{A}}{\Gamma^{\Box} \vdash \Box\mathcal{A}}\;\Box\vdash\Box$$

It is easy to see that this rule gives both (Nec) and (K). If Γ is empty, then from the provability of $\mathcal{A}$, the provability of $\Box\mathcal{A}$ follows. From two in-

[28] See [164], [124] and [190].

stances of the axiom (id), we can get $\mathcal{A} \supset \mathcal{B}, \mathcal{A} \vdash \mathcal{B}$, and then by $(\Box\vdash\Box)$ and by $(\vdash\supset)$'s, $\vdash \Box(\mathcal{A} \supset \mathcal{B}) \supset \Box\mathcal{A} \supset \Box\mathcal{B}$.

Similarly, $(\Box\vdash)$ yields in one step $\Box\mathcal{A} \vdash \mathcal{A}$, that is, the rule corresponds to (T). The sequent calculus formalizing T contains this rule together with $(\Box\vdash\Box)$.

An apparent problem with the axioms (4) and (5) is that they are implications where both the antecedent and the consequent of the formula is modalized. The solution in the case of (4) seems to have been accepted without objections, but (5) seem to be a persistent obstacle.[29]

A *sequent calculus* for $S4$ is defined by the addition of two rules, $(\Box\vdash)$ and $(\vdash\Box)$, to classical logic.

$$\frac{\Gamma^{\Box} \vdash \mathcal{A}}{\Gamma^{\Box} \vdash \Box\mathcal{A}} \vdash\Box$$

The rule $(\Box\vdash\Box)$ is a derived rule in this system, because Γ is finite, and so finitely many applications of $(\Box\vdash)$ can be followed by an application of $(\vdash\Box)$ to obtain the lower sequent of the $(\Box\vdash\Box)$ rule.

Exercise 4.7.1. Prove that the modal axioms and rules of $E2$, K, T and $S4$ are provable in the respective sequent calculi.

Exercise 4.7.2. State and prove the cut theorem for the sequent calculi for $E2$, K, T and $S4$. [Hint: The proof for LK is detailed in Chapter 2.]

The $(\vdash\Box)$ rule cannot be easily seen to match up with (4). However, the general tendency in the literature is to try to find *a pair of rules*, one to introduce $\Box$ on the left, and another to introduce it on the right. Strictly speaking, $(\vdash\Box)$ is not as pure an introduction rule as, let us say, $(\Box\vdash)$ is. To make a comparison between implicational relevance logics and normal modal logics, in the former, the axioms that add more and more features to $\to$ are reflected by more and more powerful structural rules, which affect sequences (or other data types) each formed by ;. The modal axioms we listed (as well as other modal axioms) play a similar role—they add features to $\Box$. The obstacle to obtaining an elegant sequent calculus formulation for a wide-range of modal logics is that , is simply $\wedge$ (or $\vee$) and does not have any "inherent modality" or some connection to the modalities. Another questionable feature of some proposed sequent calculi is the multiplicity of rules for the modal connectives.

Of course, there are philosophical debates about the meaning of connectives and whether that is determined by the introduction and the elimination rules. The question seems to be ill-posed even about natural deduction systems. There are natural deduction systems which have the same set of theorems, but some of which make explicit structural assumptions, which are left tacit in some others. The question appears even more vague, given that—as we will see in the next chapter—the seemingly bare-bone Lambek

[29] For instance, [83] does not mention any (logistic) sequent calculus for $S5$.

calculus LA has implicit assumptions. (We will disregard the latter sort of objections toward certain modal sequent calculi.)

The *sequent calculus* for B is defined in [176] by extending LK with two rules, $(\Box\vdash)$ and $(\Box\vdash\Box_B)$. (The $(\Box\vdash)$ rule is introduced above for $LE2$.)

$$\frac{\Gamma \vdash \Delta^{\Box}, \mathcal{A}}{\Gamma^{\Box} \vdash \Delta, \Box\mathcal{A}} \; \Box\vdash\Box_B$$

The relationship between $\Delta^{\Box}$ and Δ is as before, that is $\Delta^{\Box}$ is like the sequence Δ, but with each formula in Δ prefixed with $\Box$. In the context of the rule, $\Delta^{\Box}$ in the upper sequent gives a condition on the applicability of the rule. Δ in the lower sequent results by deleting the initial $\Box$ from each formula in $\Delta^{\Box}$.

The *sequent calculus* for $S5$ is defined in [145] by $(\Box\vdash)$ and a version of $(\Box\vdash\Box)$, in which there can be several formulas on the right-hand side of the turnstile.

$$\frac{\Gamma^{\Box} \vdash \Delta^{\Box}, \mathcal{A}}{\Gamma^{\Box} \vdash \Delta^{\Box}, \Box\mathcal{A}} \; \Box\vdash\Box_5$$

The cut rule (as we defined for LK) is not admissible in this sequent calculus (as was noted by the same authors in [146])—if no restrictions are imposed on proofs.

Exercise 4.7.3. Construct a proof of the sequent $\vdash \Box\neg\Box p, p$ using the cut rule. [Hint: $\Box p \vdash p$ is obviously provable.]

A roundabout way to deal with the problem is suggested in [146]. It is well-known that $S5$ has only six (distinct) modalities; hence, every theorem of $S5$ is equivalent to a formula with modal degree at most 1. (Of course, $S5$ first has to be defined as a logic semantically or syntactically for this claim to be provable.)

A *modified sequent calculus* $LS5^*$ for $S5$ is defined by extending LK with $(\Box\vdash)$ and $(\vdash\Box^*)$. The latter rule is applicable only when all the formulas in the sequents have a modal degree strictly less than 2.

$$\frac{\Gamma \vdash \Delta, \mathcal{A}}{\Gamma \vdash \Delta, \Box\mathcal{A}} \; \vdash\Box^*$$

Theorem 4.32. *The cut rule is* admissible *in* $LS5^*$.

Theorem 4.33. *If* $\mathcal{A}$ *is a theorem of* $S5$ *and* $\mathcal{A}^*$ *is equivalent to* $\mathcal{A}$ *of modal degree no greater than* 1*, then* $\mathcal{A}^*$ *is a* theorem *of* $LS5^*$.

Exercise* 4.7.4. Try to construct proofs for the previous two theorems (or scrutinize the proofs in [146]).

The problem caused by the $(\Box\vdash\Box)$ rule can be avoided by allowing a more complicated $(\vdash\Box_5)$ rule, the applicability of which cannot be determined by looking at the consequent of the sequent, or even by looking at all the formulas in the sequent. Rather, the whole proof has to be inspected to establish that formulas within a sequent are independent (in a precise sense defined in [55]). If $\mathcal{A}$ is independent of the elements of Γ and Δ, then those formulas do not need to be modalized for the rule to be applicable.

Exercise 4.7.5. Show that (B) is provable in the sequent calculus for B, and (5) is provable in the two calculi for $S5$.

A different sequent calculus is defined in [164], where two auxiliary unary connectives are introduced, namely, $^+$ and $^-$. For implicational formulas they behave as positive and negative occurrences, and distribute to subformulas accordingly. (E.g., $(\mathcal{A} \supset \mathcal{B})^-$ yields $\mathcal{A}^+ \supset \mathcal{B}^-$.) However, if a formula $\mathcal{A}$ is prefixed with $\Box$, then the plus and minus disappears, that is, $(\Box\mathcal{A})^-$ is $\Box\mathcal{A}$, and so is $(\Box\mathcal{A})^+$. On the other hand, p^- is $\boldsymbol{F}$, whereas p^+ is $\boldsymbol{T}$.

The modal rules are $(\Box\vdash)$ and $(\Box\vdash\Box_5)$ from above, and the next rule, which is a version of the $(\vdash \Box^*)$ rule.

$$\frac{\Gamma \vdash \Delta, \mathcal{A}^-}{\Gamma \vdash \Delta, \Box\mathcal{A}} \vdash\Box^-$$

It seems that there is some insight behind this rule which might be related to what is captured by independence. Unfortunately, [164] does not explain how the $^+$'s and $^-$'s are introduced into a sequent, and there are no rules introducing them as connectives either.

Exercise* 4.7.6. Amend or expand the calculus in [164] to provide a formalization for $S5$ so that the cut rule is admissible.

The cut rule is claimed to be admissible in the calculi for $S5$ with the restrictions taken into account. Another approach is to restrict the cut rule itself. The *analytic cut*, introduced by Smullyan in [172], allows the application of the cut rule only in cases where the cut formula is a subformula of the lower sequent of the cut. (We consider the analytic cut in Chapter 7 in more detail.) [176] shows that in the above sequent calculus for B and in the first sequent calculus for $S5$, every proof can be transformed into a proof of the same end sequent with the property that all applications of the cut rule and of the $(\Box\vdash\Box)$ obey the subformula property. For the cut rule, this means that the cut formula is a subformula of a formula in the lower sequent, whereas for the $(\Box\vdash\Box)$ rule it means that all elements of $\Delta^\Box$ are subformulas of $\Gamma^\Box, \Delta$ or $\Box\mathcal{A}$. Although the analytic cut is not eliminable, the subformula property turns out to be sufficient for proofs of some meta-theorems, such as the Craig interpolation lemma, for these logics.

4.8 Merge calculi

In this section, we briefly outline an approach to sequent calculi that does not change the data type in the sequents, rather it introduces a new operation that combines sequences of formulas. This operation is called the *merge* of sequences, which gives the name of the calculi. Metaphorically speaking, a merge of two sequences is like closing a zipper with unevenly spaced juts: the juts on each strip retain their place with respect to each other, but a group of juts on one strip can jostle between a pair of juts on the other strip. In applications, the merge operation will be more complicated, because formally, it turns out to be an associative operation, hence, applicable to more than two sequences at once.

The merge calculi were introduced in the late 1950s.[30] The logic of entailment, E does not prove $(\mathcal{A} \to \mathcal{B} \to \mathcal{C}) \to \mathcal{B} \to \mathcal{A} \to \mathcal{C}$, as we already saw in Section 4.3. However, a substitution instance of this formula in which $\mathcal{B}$ itself is an entailment, is a theorem of $E_{\to}$ (and E). The problem is how to capture this restriction in a sequent calculus. $LE_{\to}$ added a side condition to the right introduction rule, that we denoted by $(\vdash \to^*)$. There are side conditions in other sequent calculi, from LK to $LR^{\Box}$ or in the sequent calculus for $S4$ above.[31] It is of certain interest that pure entailment and pure ticket entailment can be formalized by merging the antecedents of the premises of the left introduction rule for $\to$, if certain side conditions are added. We consider the two mentioned implicational relevance logics, $E_{\to}$ and $T_{\to}$. $R_{\to}$ can be formulated as a merge system too, however, the merge operation plays a negligible role in $L_{\mu}R_{\to}$.

The metavariables $\Gamma, \Delta, \ldots$ range over finite (possibly empty) sequences of formulas. If Γ and Δ are two sequences, then their merge is a sequence too. However, merge is an *underdetermined function* yielding a sequence of wff's from a pair of sequences of wff's. By this strange expression we mean that although we may be sure that the result is a sequence of wff's, we can only say that it is one of finitely many possible sequences. In other words, if we start with a pair of concrete sequences of formulas and a sequence of formulas, then we can always decide whether the latter is a merge of the former two. The merge of two sequences is defined via a characterization of this decidable relation, that is, by describing when a sequence is a merge of two sequences.

The notation $\mu(\Gamma, \Delta)$ stands for the set of sequences each of which is a *merge* of Γ and Δ. The notation $\eta(\Gamma, \Delta)$ indicates a sequence that is a merge of Γ and Δ, that is, $\eta(\Gamma, \Delta) \in \mu(\Gamma, \Delta)$.

[30] The originator of these calculi is Belnap, and we (mostly) follow [7] in this section.

[31] A consecution calculus for $E^t_{\to}$ that does not require any side conditions is presented in Section 5.3, and $T^t_{\to}$ also has consecution formalizations without side conditions.

Definition 4.34. (Merge) Let Γ be $\mathcal{A}_1, \ldots, \mathcal{A}_n$ and let Δ be $\mathcal{B}_1, \ldots, \mathcal{B}_m$, where $m, n \in \mathbb{N}$. Let $\mathcal{A}_i \prec_\Gamma \mathcal{A}_j$ $(1 \leq i, j \leq n)$ indicate that $\mathcal{A}_i$ precedes $\mathcal{A}_j$ in Γ. Similarly, $\mathcal{B}_k \prec_\Delta \mathcal{B}_l$ $(1 \leq k, l \leq m)$ means that $\mathcal{B}_k$ comes before $\mathcal{B}_l$ in the sequence Δ. $\eta(\Gamma, \Delta) \in \mu(\Gamma, \Delta)$ iff $\eta(\Gamma, \Delta)$ is $\mathcal{C}_1, \ldots, \mathcal{C}_{m+n}$ and (1)–(3) hold. ($\prec_{\eta(\Gamma,\Delta)}$ is the linear order of the $\mathcal{C}$'s.)

(1) Each $\mathcal{C}_i$ $(1 \leq i \leq m+n)$ is exactly one of the $\mathcal{A}_l$'s $(1 \leq l \leq n)$ or of the $\mathcal{B}_k$'s $(1 \leq k \leq m)$;[32]

(2) if $\mathcal{A}_i$ is $\mathcal{C}_j$ and $\mathcal{A}_k$ is $\mathcal{C}_l$, then $\mathcal{A}_i \prec_\Gamma \mathcal{A}_k$ iff $\mathcal{C}_j \prec_{\eta(\Gamma,\Delta)} \mathcal{C}_l$;

(3) if $\mathcal{B}_i$ is $\mathcal{C}_j$ and $\mathcal{B}_k$ is $\mathcal{C}_l$, then $\mathcal{B}_i \prec_\Delta \mathcal{B}_k$ iff $\mathcal{C}_j \prec_{\eta(\Gamma,\Delta)} \mathcal{C}_l$.

μ has some properties that may seem surprising, but we should remember that μ is a *set of sequences*, which is not a singleton—unless Γ or Δ is empty. The list of interesting properties of μ includes (1)–(5).

(1) $\mu(\Gamma, \Delta) = \mu(\Delta, \Gamma)$, that is, μ is commutative;

(2) if $\mu(\{\Gamma_{i \in I}\}, \{\Delta_{k \in K}\}) = \{\eta(\Gamma_i, \Delta_k) \colon i \in I \wedge k \in K\}$, then $\mu(\{\Gamma\}, \mu(\{\Delta\}, \{\Theta\})) = \mu(\mu(\{\Gamma\}, \{\Delta\}), \{\Theta\})$, that is, μ, type-lifted to sets of sequences, is associative;

(3) $\mu(\Gamma, \emptyset) = \{\Gamma\}$, where $\emptyset$ denotes the empty sequence of wff's;

(4) if $\eta_1 \in \mu(\Gamma_1, \Delta_1)$ and $\eta_2 \in \mu(\Gamma_2, \Delta_2)$, then $\eta_1, \mathcal{A}, \eta_2 \in \mu(\langle \Gamma_1, \mathcal{A}, \Gamma_2 \rangle, \langle \Delta_1, \Delta_2 \rangle)$, where $\langle$ and $\rangle$ delineate the two arguments of μ;

(5) if $\Gamma_1, \mathcal{A}, \Gamma_2 \in \mu(\langle \Delta_1, \mathcal{A}, \Delta_2 \rangle, \Theta)$, then there are Θ_1 and Θ_2 such that Θ is Θ_1, Θ_2 and $\Gamma_1 \in \mu(\Delta_1, \Theta_1)$ and $\Gamma_2 \in \mu(\Delta_2, \Theta_2)$.

Exercise 4.8.1. Prove that μ has properties (1)–(2).

Exercise* 4.8.2. Prove that (4) and (5) hold for μ.

We expand the η notation. $\eta(\Gamma, \Delta, \mathcal{A})$ will denote an element of $\mu(\mu(\{\Gamma\}, \{\Delta\}), \{\mathcal{A}\})$. By (2) above, we could drop the second μ and the inner pair of parentheses. (Of course, $\mathcal{A}$ is identified with the one-element sequence that has only $\mathcal{A}$ as its element.) Now we are ready to define merge calculi for $E_\to$ and $T_\to$.

The *axiom and the rules* for $L_\mu E_\to$ and $L_\mu T_\to$ are the following, with the side conditions specified afterward.

$$\mathcal{A} \vdash \mathcal{A} \ \text{id} \qquad \frac{\Gamma, \mathcal{A}, \mathcal{A}, \Delta \vdash \mathcal{A}}{\Gamma, \mathcal{A}, \Delta \vdash \mathcal{A}} \ \text{W}\vdash$$

$$\frac{\Gamma, \Delta \vdash \mathcal{A} \qquad \Theta, \mathcal{B}, \Lambda \vdash \mathcal{C}}{\eta(\Gamma, \Theta, \mathcal{A} \to \mathcal{B}), \Delta, \Lambda \vdash \mathcal{C}} \ {\to}{\vdash_\eta} \qquad \frac{\Gamma, \mathcal{A} \vdash \mathcal{B}}{\Gamma \vdash \mathcal{A} \to \mathcal{B}} \ {\vdash}{\to}$$

[32] Some of the $\mathcal{A}$'s and $\mathcal{B}$'s may be occurrences of the same formula, but in the sequences they have unique indices.

$(\rightarrow\vdash_\eta)$ is restricted in both $L_\mu E_\rightarrow$ and $L_\mu T_\rightarrow$. In $L_\mu T_\rightarrow$, Δ must be *non-empty*. In $L_\mu E_\rightarrow$, if Δ is empty, then $\mathcal{A}$ *must contain* $\rightarrow$, that is, it must be of the form $\mathcal{D} \rightarrow \mathcal{E}$, for some $\mathcal{D}, \mathcal{E}$.

The notion of a *proof* is as before, and $\mathcal{A}$ is a *theorem* iff $\vdash \mathcal{A}$ has a proof.

Exercise 4.8.3. Verify that (1) from Exercise 4.3.8 on p. 108 is provable in $L_\mu T_\rightarrow$ (hence, also in $L_\mu E_\rightarrow$), whereas (2) is provable only in $L_\mu E_\rightarrow$.

Exercise 4.8.4. Recall the wff's from Exercise 4.3.1 on p. 104. Show that (1) and (2) are provable in both merge calculi, but (3) is provable in neither.

Of course, proving some axioms does not go even halfway in establishing the equivalence of a sequent calculus to an axiomatic calculus. The next crucial step is to prove a suitable cut theorem. The cut theorem for $L_\mu T_\rightarrow$ and $L_\mu E_\rightarrow$ turns out to be quite complicated. We will not return to merge calculi, and refer the interested reader to [7, §7] for further details.

Chapter 5

Consecution calculi for non-classical logics

All the calculi we have looked at in the previous chapters were built on data structures such as (finite) *sequences*, (finite) *multisets* or (finite) *sets*. Now, we look at calculi that add *grouping* to sequences, that is, they have certain trees as their basic data structure. We call these systems *consecution calculi*, when we want to emphasize this specific feature. First, it may appear that grouping is a complication that creates various obstacles, for instance, for a straightforward proof of the cut theorem. But soon it becomes clear that keeping the data structure free from any implicit assumptions brings within reach logics that could not be formalized earlier. Consecution calculi are not only exciting and rewarding to work with, but they provide a more *refined perspective* on sequent calculi than the calculi we have dealt with in the preceding chapters.

5.1 Non-associative Lambek calculus

Lambek introduced a new sequent calculus in his [121], which is nowadays called the *non-associative Lambek calculus*. We denote this calculus by LQ.[1] The motivations for the calculus came from linguistic observations that suggested that natural language sentences are not merely sequences of words, but rather *structured sequences of phrases*, which give rise to trees. Similarly, *function application* is not associative. Lambek seems to have been the first one to introduce this idea in a sequent calculus.

The classical sequent calculus LK can be presented by assuming a *polyadic* structural connective (often denoted by $,$). Indeed, this is how we defined LK in Chapter 2, following Gentzen's [89]. If the structural connective is taken to be binary but *associative*, then using a polyadic connective instead does not cause a loss of information as to how the parts of a sequence of formulas are related to each other. This is no longer true, if the structural connective is not associative. Now we use the term *structure* to refer to the

[1] LQ is not literally the calculus in [121]. We do not follow Lambek's notation and we also include the $(\circ\vdash)$ rule.

objects that are built up from formulas using a binary operation, the *structural connective*, which we will denote by ; instead of , .

Definition 5.1. (Structures) The set of *structures*, denoted by str, is defined by (1)–(2).

(1) If $\mathcal{A}$ is a wff, then $\mathcal{A} \in \text{str}$;

(2) if $\mathfrak{A}$ and $\mathfrak{B}$ are elements of str, then $(\mathfrak{A}; \mathfrak{B}) \in \text{str}$.

The *script* letters such as $\mathcal{A}$ are meta-variables for wff's (as earlier), and the *fraktur* letters such as $\mathfrak{A}$ are meta-variables for structures.[2]

Notice that we do not have an *empty structure* among the structures that belong to the set of structures. In certain calculi, we will introduce a special wff that will play the role of a placeholder (beyond certain other roles).

We intentionally started our presentation with this definition—despite the fact that we have not yet defined the set of well-formed formulas. This definition is crucial to all the calculi that we consider in this chapter. The label *consecution calculus* will emphasize that the antecedent of a *consecution* is a structure, which is not simply a sequence of formulas.

The set of wff's can vary from calculus to calculus, and each time, we will assume that Definition 5.1 is based on the appropriate set of wff's.

Structures can be viewed as *trees*. Since ; is a binary operation, the trees that correspond to structures are like the formation trees of wff's in a fragment of a propositional logic with a single binary connective. A good example here is implicational formulas (or simple types). Implicational fragments are of interest on their own, because of the connection between implication and logical consequence. Structures though, are more properly thought to correspond to formulas built by the *fusion* connective (or intensional conjunction).

LQ is a propositional calculus and it has the same set of connectives as Lambek's earlier calculus *LA* has, which we introduced in Section 4.1. There is a denumerable sequence of propositional variables, the elements of which will be denoted by $p, q, r, p_0, p_1, \ldots$. The connectives—each of them is binary —are denoted by $\leftarrow$, $\circ$ and $\rightarrow$. (Again, we do not follow Lambek's notation, which uses slashes instead of arrows. That notation is widely adopted by linguists.)

A *consecution* is a structure followed by $\vdash$ and a wff. Consecutions are very much like sequents, however, now we use the term "sequent" only if the antecedent is a sequence of wff's. Neither the left-hand side, nor the right-hand side of the $\vdash$ may be empty in a consecution. We call a calculus in which the right-hand side always contains exactly one wff a *single right-handed* calculus. Almost all the calculi in this chapter are of this kind.

[2]The use of pieces of notation, such as ; instead of , and $\mathfrak{A}$ instead Γ, is simply a choice. However, we hope that the differences in notation provide visual clues about some of the differences between logistic sequent calculi and consecution calculi.

The notation $\mathfrak{A}[\mathfrak{B}]$ indicates *one particular occurrence* of the structure $\mathfrak{B}$ in the structure $\mathfrak{A}$. Unless we consider concrete proofs, we are typically not interested in the whole shape of $\mathfrak{A}$ or where exactly $\mathfrak{B}$ is located within $\mathfrak{A}$. However, sometimes we want to make sure that $\mathfrak{B}$ *does occur* in $\mathfrak{A}$, and we have pinpointed one particular occurrence of $\mathfrak{B}$. Then, we use this notation in a limited context (such as a rule) also for the *result of replacing* the singled-out occurrence of the structure $\mathfrak{B}$ in the upper consecution by the structure shown in the lower consecution. In other words, $\mathfrak{A}[\mathfrak{B}]$ may stand for $\mathfrak{A}[\mathfrak{B}/\mathfrak{C}]$ too.

The *consecution calculus* LQ comprises an axiom and six rules.

$$\mathcal{A} \vdash \mathcal{A}$$

$$\frac{\mathfrak{A}[\mathcal{A};\mathcal{B}] \vdash \mathcal{C}}{\mathfrak{A}[\mathcal{A}\circ\mathcal{B}] \vdash \mathcal{C}}\ \circ\vdash \qquad \frac{\mathfrak{A} \vdash \mathcal{A} \quad \mathfrak{B} \vdash \mathcal{B}}{\mathfrak{A};\mathfrak{B} \vdash \mathcal{A}\circ\mathcal{B}}\ \vdash\circ$$

$$\frac{\mathfrak{A} \vdash \mathcal{A} \quad \mathfrak{B}[\mathcal{B}] \vdash \mathcal{C}}{\mathfrak{B}[\mathfrak{A};\mathcal{B}\leftarrow\mathcal{A}] \vdash \mathcal{C}}\ \leftarrow\vdash \qquad \frac{\mathcal{A};\mathfrak{A} \vdash \mathcal{B}}{\mathfrak{A} \vdash \mathcal{B}\leftarrow\mathcal{A}}\ \vdash\leftarrow$$

$$\frac{\mathfrak{A} \vdash \mathcal{A} \quad \mathfrak{B}[\mathcal{B}] \vdash \mathcal{C}}{\mathfrak{B}[\mathcal{A}\to\mathcal{B};\mathfrak{A}] \vdash \mathcal{C}}\ \to\vdash \qquad \frac{\mathfrak{A};\mathcal{A} \vdash \mathcal{B}}{\mathfrak{A} \vdash \mathcal{A}\to\mathcal{B}}\ \vdash\to$$

The notion of a *proof* is as in Definition 2.7, but, of course, with the axiom and the rules of LQ.

Exercise 5.1.1. Give proofs of the following consecutions in the LQ calculus. (a) $(\mathcal{A}\to(\mathcal{A}\circ\mathcal{B}))\circ\mathcal{A} \vdash \mathcal{A}\circ\mathcal{B}$, (b) $\mathcal{B}\to\mathcal{A} \vdash \mathcal{B}\to(\mathcal{B}\leftarrow(\mathcal{A}\to\mathcal{B}))$, (c) $\mathcal{B} \vdash (\mathcal{A}\leftarrow\mathcal{B})\to(\mathcal{B}\leftarrow(\mathcal{A}\to\mathcal{B}))$, (d) $(\mathcal{C}\to\mathcal{C})\leftarrow\mathcal{C} \vdash (\mathcal{C}\to(((\mathcal{C}\to\mathcal{C})\leftarrow\mathcal{C})\to(\mathcal{C}\to\mathcal{C})))\leftarrow\mathcal{C}$.

The main difference of LQ from LK is the *absence* of all structural rules—including any hidden associativity built into the notion of a sequence of wff's. LQ has full control over the structures and wff's in the consecutions. The *single cut rule* looks like the following.

$$\frac{\mathfrak{A} \vdash \mathcal{A} \qquad \mathfrak{B}[\mathcal{A}] \vdash \mathcal{C}}{\mathfrak{B}[\mathfrak{A}] \vdash \mathcal{C}}\ \text{cut}$$

The cut rule is *admissible* in LQ, that is, any provable consecution has a proof without any application of the cut rule.

Exercise 5.1.2. Prove the admissibility of the cut rule. [Hint: There is no contraction-like rule in LQ. Hence, the proof of Theorem 2.27 can be modified so as to omit μ altogether. See Section 7.2.]

Exercise* 5.1.3. Scrutinize the proof you gave in the previous exercise. Is there a way to simplify it? [Hint: Notice that every rule of LQ introduces a wff, hence, the sum of the degrees of wff's in a consecution is monotone increasing in a proof (from the leaves to the root of the proof tree).]

Cut-free proofs have the *subformula property*, moreover, LQ has the subformula property with respect to *occurrences* of wff's, because it has no contraction-like rules. If an occurrence of a wff entered a consecution, then in all consecutions below that one, there is an occurrence of the same wff (possibly, as a subformula of another formula) that matches the occurrence. The *decidability* of LQ is immediate.

Lambek's calculi were likely motivated—directly or indirectly—by Ajdukiewicz's calculus of semantic types, as well as by Church's type theory and by Curry's functional characters.[3] In particular, Ajdukiewicz's influence is apparent, because Lambek uses the types n (for name) and s (for sentences) in his English examples. He also enriches the set of types by adding a separate type for infinitives.

Example 5.2. If the lexicon assigns the type n to *Jim*, and $s \leftarrow n$ to *works*, then it should be provable that *Jim works* is a sentence. This is easy to prove.

$$\dfrac{\dfrac{n \vdash n \qquad s \vdash s}{n; s \leftarrow n \vdash s}}{n \circ (s \leftarrow n) \vdash s}$$

Here the types n and s function as propositional variables (and in linguistics, it is assumed that n is distinct from s). They indicate types or *categories* of well-formed expressions; therefore, there is no need to have s_1 to stand for one sentence, but s_2 for another sentence. Words or phrases are combined by $\circ$, and the type on the right-hand side of the turnstile is the derived type of the whole phrase.

This sample sentence cannot show the difference between LA and LQ. If we add an adverb, then LQ excludes a potential derivation, which would be viewed as an ad hoc re-grouping of words.

Example 5.3. Let us add *diligently* with type $(s \leftarrow n) \leftarrow (s \leftarrow n)$ to our earlier example. Giving this type to the adverb means that it modifies the verb, not the sentence. The derivation of the type of *Jim works diligently* as a sentence now includes grouping, which indicates—beyond the types themselves—the order of combining of the phrases into a sentence. The first proof shows that *Jim* (*works diligently*) is a sentence. The second proof would require the adverb to attach to the sentence as in (*Jim works*) *diligently*, and to have the type $s \leftarrow s$.

$$\dfrac{s \leftarrow n \vdash s \leftarrow n \qquad \dfrac{n \vdash n \quad s \vdash s}{n; s \leftarrow n \vdash s}}{n; (s \leftarrow n; (s \leftarrow n) \leftarrow (s \leftarrow n)) \vdash s} \qquad \dfrac{\dfrac{n \vdash n \quad s \vdash s}{n; s \leftarrow n \vdash s} \qquad s \vdash s}{(n; s \leftarrow n); s \leftarrow s \vdash s}$$

[3] Lambek did not provide these references, but we may presume that he could have meant Ajdukiewicz [3], Church [61] and Church [67].

At the time, when LA and LQ were formulated, formal syntactic theories for natural languages were only a little developed, but hopes for syntax-based translations were high. As Lambek stated in [121, p. 169], "[s]ince mechanical parsing of sentences is not only of conceptual but also of practical importance (in connection with mechanical translation of languages), it becomes of interest to know for which languages it is possible to replace all grammatical rules by type assignments in the dictionary." However, it was noted by Lambek himself that his calculi overgenerate, in the sense that sequences of words that are marginal at best (if not outright agrammatical) will be considered well-formed, because their type can be proved to be s. For instance, if there are words such as the adverb *diligently* in our example above that have a type $\tau \rightarrow \tau$ or $\tau \leftarrow \tau$ (for some τ), then such a word can be inserted into a sentence arbitrarily many times. The syntactic types would not preclude semantic incoherence either, such as in the case of the grammatical sentence *Three works brightly*, where the subject of the sentence is a positive integer, and it is dubious that the sentence has a meaning when the other component words are also taken to have their usual meaning.

Whether all natural languages can be described using Lambek's calculi might be an ill-posed question, because natural languages, and the set of their grammatical sentences is not well defined. However, it is known that *categorial grammars* originating from these calculi cannot describe *context-sensitive* languages or *computable languages* (i.e., "0-type languages"), in general. The latter are formally definable classes of languages, hence, the comparison makes sense. (We say a bit more about categorial grammars in connection to typed calculi in Section 9.3.2.)

LQ can be *extended* by reinstating some of the structural rules, or by adding new logical constants such as $\mathbf{t}$, $\wedge$, $\vee$, $\Box$, etc. For example, the addition of the following associativity rules to LQ would yield a calculus that is equivalent to LA. (Parentheses are omitted from left-associated structures.)

$$\frac{\mathfrak{A}[\mathfrak{B};(\mathfrak{C};\mathfrak{D})] \vdash A}{\mathfrak{A}[\mathfrak{B};\mathfrak{C};\mathfrak{D}] \vdash A} \qquad \frac{\mathfrak{A}[\mathfrak{B};\mathfrak{C};\mathfrak{D}] \vdash A}{\mathfrak{A}[\mathfrak{B};(\mathfrak{C};\mathfrak{D})] \vdash A}$$

Exercise 5.1.4. Prove that the cut rule remains admissible, if these rules are added to LQ. [Hint: Is it possible to extend the proof of Theorem 7.8 with a few cases?]

Exercise 5.1.5. Prove that the set of provable consecutions in the modified LQ calculus coincides with the set of consecutions provable in LA. [Hint: One direction is obvious.]

The following sections in this chapter deal with logics that can be viewed as extensions of LQ, sometimes, by adding both structural rules and new connectives at once. Many of those logics, however, emerged from independent motivations, and they ascribe different meanings to wff's.

5.2 Structurally free logics

The non-associative Lambek calculus may be viewed as implementing the idea that certain expressions are functions. That is, the $\circ$ and its structural pair, ; are analogs of function application. The preeminent *theory of functionality* is the theory of λ-calculi. In certain ways, combinatory logic is equivalent to λ-calculi, and in other ways, it is superior to λ-calculi, because it gives rise to a variety or a pre-ordered algebra (depending on whether weak equality or weak reduction are taken to be the main relation of interest on the set of combinatory terms).

The discovery of a connection between some combinators and some implicational formulas goes back at least to [69], and it has become known as the *formulas-as-types paradigm* or the *Curry–Howard isomorphism*. We will look at typed systems in more detail in Section 9.3, but we note for now that in all the typed systems, terms (in combinatory logic or in a λ-calculus of some kind) are linked to formulas—typically, to formulas of a non-classical logic.

Motivations for *structurally free logics*, which were introduced by Dunn and Meyer in [81], combine the potential for $\circ$ to be the *function application operation* on combinatory terms and for a residual of $\circ$ (namely, $\rightarrow$) to be the *type constructor operation*.

Curry in [68, 5C] used the same letters to label permutation, contraction and weakening rules in sequent systems that he used for three regular combinators, the ternary permutator, and the binary duplicator and the binary cancellator. We must point out that—as it might be expected from somebody like Curry, who paid a lot of attention to precision and detail—the letters are the same, but the typeface is not. Curry used sans serif letters for combinators, for example, in Section 3D in the same book, and that is the standard notation for combinators in [69], in [70] and elsewhere too. Clearly, there is a sensible reason for the labels for the rules, however, Curry made no claim about the rules implementing or corresponding to the combinators. *Combinatory logic* is a well-developed topic, and we cannot do justice to it as a whole here.[4] However, for the sake of the completeness of our exposition of structurally free logics, we introduce some core concepts here.

Definition 5.4. (Combinatory terms) Let $\mathbb{B}$ be a set of *primitive combinators*, and let $\langle x_i \rangle_{i \in \mathbb{N}}$ be a sequence of (distinct) *variables*. The set of *combinatory terms* (or CL-terms, for short) is inductively defined by (1)–(3).

(1) If $\mathsf{Z} \in \mathbb{B}$, then Z is a CL-term;

(2) if $x \in \langle x_i \rangle_{i \in \mathbb{N}}$, then x is a CL-term;

[4] Beyond the two volumes of the book *Combinatory Logic* already mentioned, see also Hindley and Seldin's [109] as well as [36] for more detailed and comprehensive presentations.

(3) if M and N are CL-terms, then so is (MN).

The set $\mathbb{B}$ is usually called a *combinatory base* and it is often taken to be finite. Indeed, one of the most important contributions of combinatory logic to our understanding of the problem of substitution and the problem of the definability of computable functions is that a finite set of combinators is sufficient for those purposes. The finite set can be "very finite" such as $\{\mathsf{S}, \mathsf{K}\}$ with merely two elements. $\{\mathsf{S}, \mathsf{K}\}$ is a *combinatorially complete* set of combinators, and these two combinators were found or invented by M. Schönfinkel, who was the first to consider combinators. Although he did not prove combinatorial completeness for any base, he established that all the other combinators, that appeared to be useful in solving the problem of the elimination of bound variables, can be defined from S and K.

CL-terms that are built purely from primitive combinators (without variables) are called *combinators*. It is not difficult to prove—given the above definition—that a CL-term is either a variable or a primitive combinator, or of the form (MN), for some M and N. The former are called *atomic terms*, whereas the latter are *compound* or *complex terms*. The *subterms* of a CL-term are the CL-terms that appear while the CL-term is constructed step by step, according to Definition 5.4, and a CL-term is a subterm of itself.

Variables are not always included in combinatory logic. First of all, combinatory logic completely avoids variable binding, which is the source of certain difficulties concerning substitution in λ-calculi (or in a logic with quantifiers). Second, when terms are typed, variables and combinators behave very differently, as we explain in Section 9.3. If types are *assigned* to terms, then the self-application of certain combinators is typable by simple types; moreover, a very elegant relationship emerges between theorems of some implicational logics and typable combinators over certain sets of combinators. The original Curry–Howard correspondence is an example of such a relationship between typable combinators over $\{\mathsf{S}, \mathsf{K}\}$ and theorems of *implicational intuitionistic logic*.

Curry formulated as sequent calculi the negation-free fragments of classical logic (what he labeled as LC_m) and of intuitionistic logic (what he denoted by LA_1). He called the latter the *absolute system* (hence, the A in the label). Curry's rules differ from Gentzen's, but we recall here only his structural rules.[5]

$$\frac{\Gamma \vdash \Delta}{\Gamma' \vdash \Delta}\ \mathsf{C}\vdash \qquad \frac{\Gamma \vdash \Delta}{\Gamma \vdash \Delta'}\ \vdash\mathsf{C}$$

$$\frac{\Gamma, \mathcal{A}, \mathcal{A} \vdash \Delta}{\Gamma, \mathcal{A} \vdash \Delta}\ \mathsf{W}\vdash \qquad \frac{\Gamma \vdash \mathcal{A}, \mathcal{A}, \Delta}{\Gamma \vdash \mathcal{A}, \Delta}\ \vdash\mathsf{W}$$

[5]As in Chapter 2, we make inessential modifications to Curry's notation in order to harmonize his notation with ours. $_m$ stands for multiple, $_1$ for single.

$$\frac{\Gamma \vdash \Delta}{\Gamma, \mathcal{A} \vdash \Delta}\ K\vdash \qquad\qquad \frac{\Gamma \vdash \Delta}{\Gamma \vdash \mathcal{A}, \Delta}\ \vdash K$$

Γ and Δ are sequences of formulas, and either may be empty. Γ' and Δ' are permutations of Γ and Δ, respectively. $(C\vdash)$ and $(\vdash C)$ here are clearly more general than the similarly labeled rules in LK or LJ, in the sense that one step here may require more than one step in Gentzen's systems. In the negation-free fragment of intuitionistic logic, the right structural rules are omitted, whereas in the left structural rules, Δ must be exactly one formula, that is, Curry's LA_1 is a right singular system. (Notice that in the absence of negation, there is no way to move a formula from the right-hand side of the $\vdash$ to its left-hand side.)

Curry writes in [68] on page 186: "It will be convenient to call these three rules C, W, K, after the combinators C, W and K, respectively." He refers to the rules that we denoted by $(C\vdash)$, $(W\vdash)$ and $(K\vdash)$. (Actually, later on, he uses $*$C for the left permutation rule and C$*$ for the right permutation rule, and similarly, he adds $*$'s to K and W.)

The combinators C, W and K—like any combinatory term—are thought to be *functions*. The primitive combinators come with a *fixed arity* and an *axiom* that characterizes what happens if the combinator is applied to as many terms as its arity. We will use variables in the axioms and assume that there is a rule of uniform substitution of terms for variables. Substitution is unproblematic here, because no variable clashes can occur without bound variables.

The axioms for C, W and K are as follows.

(1) $\mathsf{C}xyz \rhd xzy$ (2) $\mathsf{W}xy \rhd xyy$ (3) $\mathsf{K}xy \rhd x$

where $\rhd$ separates the two terms and points toward the resulting term. The careful reader surely noticed that neither $\mathsf{C}xyz$ nor xzy is a term according to Definition 5.4. We omitted all the parentheses as per the usual convention that parentheses are to be restored via association toward the left. When fully parenthesized, $\mathsf{C}xyz$ would look like $(((\mathsf{C}x)y)z)$. The label $*$C is clearly an allusion at best, because C permutes only its second and third arguments. Indeed, by itself, C is not sufficient to define all the permutations of a sequence, when we think of the sequence as successive arguments of a function. (For instance, the appearance of x, y and z in $\mathsf{C}xyz$ would amount to treating the string xyz this way. Of course, xyz is not a subterm of $\mathsf{C}xyz$, but the parentheses-free notation gives the suitable impression.)

Exercise* 5.2.1. Try to construct an argument toward showing that C is not sufficient to generate all permutations. [Hint: No matter how many C's are combined and in what way, no such combinator can take the place of Z, for example, in $\mathsf{Z}xy \rhd_{\mathsf{w}} yx$, where $\rhd_{\mathsf{w}}$ is weak reduction, a concept that we are about to introduce precisely. See Definition 5.5 below.]

$(W\vdash)$ and $(K\vdash)$ have more similarity to the axioms of the combinators W and K, respectively. If we assume either that Γ is one formula or that Γ

is somehow of the same sort of an entity as $\mathcal{A}$, which can be replaced by a variable, then an application of the rules should be looked at as if we were reading the combinatory axioms from right to left. The need for Γ to be exactly one formula suggests that the letters in the labels $(W\vdash)$ and $(K\vdash)$ are rather loose tropes, and even more so in $(\vdash W)$ and $(\vdash K)$.

Although Curry did not view the $\,,\,$ as an operation that parallels function application, his labels are *justifiable* in another way. Given the simple types (i.e., the implicational formulas) that link up with these three combinators, the identically labeled left rules are *necessary* to prove those formulas. (The letters in the right rules could be maintained on the basis of symmetry between $(W\vdash)$ and $(\vdash W)$, etc.)

Structurally free logics introduce the combinators into the language of the calculus. The combinators (or their letters) are no longer merely labels on rules, which may enter the annotation of proofs (if a proof is decorated). The calculus does not contain any structural rules, because permutation, thinning and contraction are replaced by suitable combinatory rules. The complete lack of structural rules should explain why they are called structurally free.

A combinator is a function with *fixed arity* and with a *preset effect*. Plus, the action of a combinator is thought to be independent from its *context* and from the concrete *shape* of its arguments.

Definition 5.5. (Reductions and equality) Let Z be a combinator with axiom $\mathsf{Z}x_1\ldots x_n \rhd M$. If $\mathsf{Z}x_1\ldots x_n[x_1,\ldots,x_n/N_1,\ldots,N_n]$ is a subterm of P, then P *one-step reduces* to P', that is, $P \rhd_1 P'$, where P' is $P[(\mathsf{Z}x_1\ldots x_n[x_1,\ldots,x_n/N_1,\ldots,N_n])/(M[x_1,\ldots,x_n/N_1,\ldots,N_n])]$.

Weak reduction, denoted by $\rhd_{\mathrm{w}}$, is the reflexive, transitive closure of one-step reduction. *Weak equality*, denoted by $=_{\mathrm{w}}$, is the transitive closure of the symmetric, reflexive closure of one-step reduction.

In words, P' is a term, in which one occurrence of the term $\mathsf{Z}N_1\ldots N_n$ is replaced by the right-hand side term in the axiom of Z, but with N's substituted for all the occurrences of the x's. Of course, any of the N's may itself be a variable, including the possibility that N_i is x_i.

<u>*Example*</u> 5.6. The term $\mathsf{C}(\mathsf{C}\mathsf{K}xz)xy$ one-step reduces to $\mathsf{C}\mathsf{K}xzyx$ or $\mathsf{C}(\mathsf{K}zx)xy$. Furthermore, $\mathsf{C}\mathsf{K}xzyx \rhd_1 \mathsf{K}zxyx \rhd_1 zyx$ and $\mathsf{C}(\mathsf{K}zx)xy \rhd_1 \mathsf{C}zxy \rhd_1 zyx$. $\mathsf{C}(\mathsf{C}\mathsf{K}xz)xy \rhd_{\mathrm{w}} \mathsf{K}zxyx$ and $\mathsf{C}(\mathsf{C}\mathsf{K}xz)xy \rhd_{\mathrm{w}} \mathsf{C}zxy$. All these terms are (pairwise) in the $=_{\mathrm{w}}$ relation.

For our purposes here, one-step reduction is the most important notion. However, in combinatory logic the core concept is weak reduction, which is often shortened to "reduction." Weak reduction does not reproduce all β-reduction steps that would be possible if a combinator would be swapped with the corresponding closed λ-term, hence, it is "weak." We will not use the notions of strong reduction here, or the notion that imitates β-reduction in combinatory logic, and so we will occasionally omit the adjective "weak."

We gave three examples of combinatory axioms, but we have not made precise the general shape of potential combinatory axioms. It is reasonable to exclude a combinator that introduces a new variable. Roughly speaking, a new variable popping up on the right-hand side of $\triangleright$ in an axiom would be like calculating with a function that is thought to depend on $x_1, \ldots, x_n$, but then suddenly realizing that some x_{n+1} is to be taken into consideration too. We would, probably, think that it was a mistake to start with to use only $x_1, \ldots, x_n$. This observation would suggest that M in $\mathsf{Z}x_1 \ldots x_n \triangleright M$ should be composed of some of $x_1, \ldots, x_n$. However, this would be an overly limited view. It may happen that in the process of the calculation with a function we rely on the same or on other known functions. Permitting combinators to occur in M is theoretically important in combinatory logic, and it is also necessary in establishing connections between certain implicational logics and combinatory type-assignment systems.

A combinator that has an axiom in which M, the right-hand side term, contains no other atomic components than some of $x_1, \ldots, x_n$ is called a *proper combinator*. Although the *fixed point combinator*, which is not a proper combinator, is a very important combinator for many reasons, we focus on proper combinators in the context of structurally free logics. First of all, a fixed point combinator is definable from a combinatorially complete base (even from a less expressive base), in which all the combinators are proper. Second, the structural rules that are usually included into sequent calculi do not resemble improper combinators (for the somewhat trivial reason that structural rules usually do not introduce any markers or special formulas into a sequent). Each of C, W and K, for instance, is proper.

In order to be able to formulate the sequent calculus rule that matches the axiom of an arbitrary proper combinator, we introduce a piece of notation.

Definition 5.7. Z is a *proper combinator* iff given its axiom $\mathsf{Z}x_1 \ldots x_n \triangleright M$, M is a CL-term in which all the atomic subterms are elements of $\{ x_1, \ldots, x_n \}$. To emphasize that Z is a proper combinator, we use the notation $\mathsf{Z}x_1 \ldots x_n \triangleright (\!| x_1, \ldots, x_n |\!)$.

The notation should not be taken to mean that each of $x_1, \ldots, x_n$ does occur in $(\!| x_1, \ldots, x_n |\!)$ or that no variable can have more than one occurrence in the term. We will extend the use of this notation so that the x's are replaced by structures, and instead of juxtaposition (indicating function application), the structural connective ; forms the structures.

Dunn and Meyer in [81] introduced a handful of combinatory rules. We give rules for combinators from the following list, all of which are proper combinators except Y, the fixed point combinator.

$\mathsf{I}x \triangleright x$	$\mathsf{B}xyz \triangleright x(yz)$	$\mathsf{S}xyz \triangleright xz(yz)$
$\mathsf{T}xy \triangleright yx$	$\mathsf{B}'xyz \triangleright y(xz)$	$\mathsf{S}'xyz \triangleright yz(xz)$
$\mathsf{M}x \triangleright xx$	$\mathsf{V}xyz \triangleright zxy$	$\mathsf{Y}x \triangleright x(\mathsf{Y}x)$

Typically, we assume that a structurally free calculus includes finitely many combinatory rules.[6] Following [81], we denote such a calculus by LC, and we make the set of combinators explicit in the notation, if that is not clear from the context.

Definition 5.8. Let $\mathbb{B}$, a combinatory base, be fixed, and let $\to$ and $\circ$ be two binary connectives. Let $\{ p_i \}_{i \in \mathbb{N}}$ be a set of propositional variables. The set of *well-formed formulas* (wff's) is inductively defined by (1)–(3).

(1) If p is a propositional variable, then p is a wff;

(2) if $\mathsf{Z} \in \mathbb{B}$, then Z is a wff;

(3) if $\mathcal{A}, \mathcal{B}$ are wff's, then $(\mathcal{A} \to \mathcal{B})$ and $(\mathcal{A} \circ \mathcal{B})$ are wff's.

The notion of *structures* is as in Definition 5.1 (with the current set of wff's), and a *consecution* is a structure and a wff with a $\vdash$ between them.

We define the *consecution calculus* LC to include the following axiom and the operational rules. There is one combinatory rule for each element of $\mathbb{B} = \{ \mathsf{I}, \mathsf{K}, \mathsf{C}, \mathsf{W}, \mathsf{B}, \mathsf{S}, \mathsf{T}, \mathsf{M} \}$.

$$\mathcal{A} \vdash \mathcal{A} \ \text{id}$$

$$\frac{\mathfrak{A} \vdash \mathcal{A} \quad \mathfrak{B}[\mathcal{B}] \vdash \mathcal{C}}{\mathfrak{B}[\mathcal{A} \to \mathcal{B}; \mathfrak{A}] \vdash \mathcal{C}} \to\vdash \qquad \frac{\mathfrak{A}; \mathcal{A} \vdash \mathcal{B}}{\mathfrak{A} \vdash \mathcal{A} \to \mathcal{B}} \vdash\to$$

$$\frac{\mathfrak{A}[\mathcal{A}; \mathcal{B}] \vdash \mathcal{C}}{\mathfrak{A}[\mathcal{A} \circ \mathcal{B}] \vdash \mathcal{C}} \circ\vdash \qquad \frac{\mathfrak{A} \vdash \mathcal{A} \quad \mathfrak{B} \vdash \mathcal{B}}{\mathfrak{A}; \mathfrak{B} \vdash \mathcal{A} \circ \mathcal{B}} \vdash\circ$$

$$\frac{\mathfrak{A}[\mathfrak{B}] \vdash \mathcal{A}}{\mathfrak{A}[\mathsf{I}; \mathfrak{B}] \vdash \mathcal{A}} \mathsf{I}\vdash \qquad \frac{\mathfrak{A}[\mathfrak{B}; \mathfrak{D}; (\mathfrak{C}; \mathfrak{D})] \vdash \mathcal{A}}{\mathfrak{A}[\mathsf{S}; \mathfrak{B}; \mathfrak{C}; \mathfrak{D}] \vdash \mathcal{A}} \mathsf{S}\vdash$$

$$\frac{\mathfrak{A}[\mathfrak{B}] \vdash \mathcal{A}}{\mathfrak{A}[\mathsf{K}; \mathfrak{B}; \mathfrak{C}] \vdash \mathcal{A}} \mathsf{K}\vdash \qquad \frac{\mathfrak{A}[\mathfrak{B}; (\mathfrak{C}; \mathfrak{D})] \vdash \mathcal{A}}{\mathfrak{A}[\mathsf{B}; \mathfrak{B}; \mathfrak{C}; \mathfrak{D}] \vdash \mathcal{A}} \mathsf{B}\vdash$$

$$\frac{\mathfrak{A}[\mathfrak{B}; \mathfrak{B}] \vdash \mathcal{A}}{\mathfrak{A}[\mathsf{M}; \mathfrak{B}] \vdash \mathcal{A}} \mathsf{M}\vdash \qquad \frac{\mathfrak{A}[\mathfrak{B}; \mathfrak{C}; \mathfrak{C}] \vdash \mathcal{A}}{\mathfrak{A}[\mathsf{W}; \mathfrak{B}; \mathfrak{C}] \vdash \mathcal{A}} \mathsf{W}\vdash$$

$$\frac{\mathfrak{A}[\mathfrak{C}; \mathfrak{B}] \vdash \mathcal{A}}{\mathfrak{A}[\mathsf{T}; \mathfrak{B}; \mathfrak{C}] \vdash \mathcal{A}} \mathsf{T}\vdash \qquad \frac{\mathfrak{A}[\mathfrak{B}; \mathfrak{D}; \mathfrak{C}] \vdash \mathcal{A}}{\mathfrak{A}[\mathsf{C}; \mathfrak{B}; \mathfrak{C}; \mathfrak{D}] \vdash \mathcal{A}} \mathsf{C}\vdash$$

The notion of a *proof* is as usual. A formula $\mathcal{A}$ is provable or a *theorem* of LC iff there is a proof of the sequent $\mathsf{I} \vdash \mathcal{A}$. Incidentally, we use the same *convention to omit parentheses* from structures as we use in combinatory terms; for instance, $\mathfrak{A}[\mathsf{S}; \mathfrak{B}; \mathfrak{C}; \mathfrak{D}]$ is short for $\mathfrak{A}[(((\mathsf{S}; \mathfrak{B}); \mathfrak{C}); \mathfrak{D})]$.

[6] Some choices for $\mathbb{B}$ may be more suitable for certain purposes, but in general, there are infinitely many, more precisely, $2^{\aleph_0}$-many structurally free logics.

The base $\{\mathsf{I},\mathsf{K},\mathsf{C},\mathsf{W},\mathsf{B},\mathsf{S},\mathsf{T},\mathsf{M}\}$ is obviously combinatorially complete, because K and S are its elements. This may compel us to consider the question if LC is sensible, or for all formulas $\mathcal{A}$, $\mathcal{A}$ is a theorem of LC. It is well known that a combinatory base does not need to be complete to lead to absolute inconsistency when combined with an implicational logic. Curry constructed a paradox (in two slightly different versions), which shows that a simple combination of λ-calculus (or of combinatory logic) with usual propositional logic is unfeasible. (See Section A.3 for a brief exposition of Curry's paradox.)

To reiterate what we already mentioned, a combinator like Y is definable from $\mathbb{B}$. Without going into the details of its construction, we note that the CL-term $\mathsf{BW(B(CB)M)(BW(B(CB)M))}$ matches the one-step reduction emerging from Y's axiom with $\triangleright_{\mathsf{w}}$, that is, $\mathsf{BW(B(CB)M)(BW(B(CB)M))}x \triangleright_{\mathsf{w}} x(\mathsf{BW(B(CB)M)(BW(B(CB)M))}x)$.

Exercise 5.2.2. Verify that the weak reduction just mentioned obtains.

The logic LC surely does not replicate Curry's paradox, because *expansion* steps (i.e., converses of reduction steps) are not provable, in general. It may appear that this contradicts how we informally motivated the combinatory rules as "reading them bottom-up." However, the left rules in a sequent calculus are analogs of elimination rules in natural deduction systems, which supports the informal view that the combinatory redexes are reduced.

The general form of a combinatory rule is as follows—assuming that Z is a proper combinator with axiom $\mathsf{Z}x_1 \ldots x_n \triangleright (\!| x_1, \ldots, x_n |\!)$.

$$\frac{\mathfrak{A}[(\!| \mathfrak{B}_1, \ldots, \mathfrak{B}_n |\!)] \vdash \mathcal{A}}{\mathfrak{A}[\mathsf{Z}; \mathfrak{B}_1; \ldots; \mathfrak{B}_n] \vdash \mathcal{A}}\ \mathsf{Z}\vdash$$

The $\vdash$ relation in provable consecutions may be thought of as a pre-order relation in (well-behaved) consecution calculi. The $\triangleright_{\mathsf{w}}$ relation is also a pre-order relation.

<u>*Example*</u> 5.9. We consider the following proof, where $\mathcal{A}, \mathcal{B}$ and $\mathcal{C}$ are arbitrary wff's.

$$\dfrac{\dfrac{\dfrac{\dfrac{\mathcal{A} \vdash \mathcal{A} \quad \mathcal{C} \vdash \mathcal{C}}{\mathcal{A}; \mathcal{C} \vdash \mathcal{A} \circ \mathcal{C}}\ \vdash\circ \qquad \mathcal{B} \vdash \mathcal{B}}{\mathcal{A}; \mathcal{C}; \mathcal{B} \vdash \mathcal{A} \circ \mathcal{C} \circ \mathcal{B}}\ \vdash\circ}{\mathsf{C}; \mathcal{A}; \mathcal{B}; \mathcal{C} \vdash \mathcal{A} \circ \mathcal{C} \circ \mathcal{B}}\ \mathsf{C}\vdash}{\mathsf{C} \circ \mathcal{A} \circ \mathcal{B} \circ \mathcal{C} \vdash \mathcal{A} \circ \mathcal{C} \circ \mathcal{B}}\ \circ\vdash\text{'s}$$

If we look at the formulas in the bottom sequent as CL-terms, that is, $\circ$ standing for the function application operation, and $\vdash$ replacing $\triangleright_1$, then we have a proof of the axiom of C.

The next derivation shows that LC proves a sequent in which C is on the left-hand side of the turnstile and its principal type schema is on the right-hand side.[7]

$$\dfrac{\dfrac{\dfrac{\mathcal{A}\vdash\mathcal{A}\qquad\dfrac{\mathcal{B}\vdash\mathcal{B}\qquad\mathcal{C}\vdash\mathcal{C}}{\mathcal{B}\to\mathcal{C};\mathcal{B}\vdash\mathcal{C}}\to\vdash}{\mathcal{A}\to\mathcal{B}\to\mathcal{C};\mathcal{A};\mathcal{B}\vdash\mathcal{C}}\to\vdash}{\mathsf{C};\mathcal{A}\to\mathcal{B}\to\mathcal{C};\mathcal{B};\mathcal{A}\vdash\mathcal{C}}\mathsf{C}\vdash}{\mathsf{C}\vdash(\mathcal{A}\to\mathcal{B}\to\mathcal{C})\to\mathcal{B}\to\mathcal{A}\to\mathcal{C}}\vdash\to\text{'s}$$

Exercise 5.2.3. Let p, q and r be distinct propositional variables. Prove that $p\circ q\circ r\vdash \mathsf{C}\circ p\circ r\circ q$ is not provable in LC.

Exercise* 5.2.4. Generalize the argument from the previous exercise to all the combinators in $\mathbb{B}$. [Hint: Each combinator in $\mathbb{B}$ is proper, and each rule conforms to the pattern given by the meta-rule $(\mathsf{Z}\vdash)$.]

The *single cut rule* in the following form is admissible in LC.

$$\dfrac{\mathfrak{A}\vdash\mathcal{C}\qquad\mathfrak{B}[\mathcal{C}]\vdash\mathcal{A}}{\mathfrak{B}[\mathfrak{A}]\vdash\mathcal{A}}\ \text{cut}$$

The admissibility of cut is proved in [81] via the admissibility of *multiple cut*. The angle brackets in $\mathfrak{B}\langle\mathfrak{A}\rangle$ function like the brackets do in $\mathfrak{B}[\mathfrak{A}]$ except that they indicate that at least one but *possibly several* occurrences of the structure $\mathfrak{A}$ have been selected and fixed in the structure $\mathfrak{B}$.

$$\dfrac{\mathfrak{A}\vdash\mathcal{C}\qquad\mathfrak{B}\langle\mathcal{C}\rangle\vdash\mathcal{A}}{\mathfrak{B}\langle\mathfrak{A}\rangle\vdash\mathcal{A}}\ \text{multi-cut}$$

Exercise 5.2.5. Prove that the two cut rules are equivalent, that is, the same sequents are provable with one as with the other. [Hint: One direction is obvious.]

The availability of the cut rule permits us to prove the replacement theorem for the calculus, thereby, it leads to an algebraization of LC.

Theorem 5.10. *Let* $\mathcal{A}\equiv\mathcal{B}$ *iff* $\mathcal{A}\vdash\mathcal{B}$ *and* $\mathcal{B}\vdash\mathcal{A}$ *are both provable in* LC. $\equiv$ *is an* equivalence relation *that is a* congruence *with respect to* $\to$ *and* $\circ$.

Proof: **1.** The reflexivity of $\equiv$ is immediate from the (id) axiom. Symmetry follows from the symmetry of the meta-language "and" in the definition of $\equiv$. For transitivity, we have to prove that $\mathcal{A}\vdash\mathcal{B}$ and $\mathcal{B}\vdash\mathcal{C}$ imply $\mathcal{A}\vdash\mathcal{C}$. This is immediate by an application of the cut rule.
2. There are four subcases to prove, two concerning each of $\to$ and $\circ$. We leave these for the next exercise. qed

[7] We explain in Section 9.3 in greater detail both the simple types and some typed calculi.

Exercise 5.2.6. Write out the details of step **2** in the proof above.

We formulate a lemma that provides useful insights about the behavior of the I combinator and about how complex structures emerge in the antecedent of a consecution.

Lemma 5.11. *If* $\mathsf{I}; \mathcal{A} \vdash \mathcal{B}$ *is provable in* LC *and there are no applications of the* $(\vdash \circ)$ *rule in the proof of this consecution, then* $\mathcal{A} \vdash \mathcal{B}$ *is* provable in LC.

Proof: The proof is by induction on the height of the proof of $\mathsf{I}; \mathcal{A} \vdash \mathcal{B}$, which we denote by χ.
1. If $\chi = 0$, then the claim is vacuously true, because there is no axiom in LC that contains a complex structure.
2. The inductive case is subdivided according to the last rule applied in the proof. We detail two possibilities.
2.1 Let us assume that the last step was by $(\mathsf{I}\vdash)$. If the I that is introduced by the rule is the leftmost structure, then the upper sequent in the rule ensures the truth of the claim. Otherwise, we have

$$\frac{\mathsf{I}; \mathfrak{A}[\mathfrak{B}] \vdash \mathcal{B}}{\mathsf{I}; \mathfrak{A}[\mathsf{I}; \mathfrak{B}] \vdash \mathcal{B}}\ \mathsf{I}\vdash$$

and by the hypothesis of the induction, $\mathfrak{A}[\mathfrak{B}] \vdash \mathcal{B}$ is provable. But then by $(\mathsf{I}\vdash)$ applied in the same place, we obtain $\mathfrak{A}[\mathsf{I}; \mathfrak{B}] \vdash \mathcal{B}$.
2.2 If the last step is by $(\to\vdash)$, then the lower sequent is $\mathsf{I}; \mathfrak{A}[\mathcal{D} \to \mathcal{C}; \mathfrak{B}] \vdash \mathcal{B}$, whereas the two premises are $\mathfrak{B} \vdash \mathcal{D}$ and $\mathsf{I}; \mathfrak{A}[\mathcal{C}] \vdash \mathcal{B}$. The latter sequent has a shorter proof that the end sequent, hence, there is a proof of $\mathfrak{A}[\mathcal{C}] \vdash \mathcal{B}$, thus, $\mathfrak{A}[\mathcal{D} \to \mathcal{C}; \mathfrak{B}] \vdash \mathcal{B}$ is provable. qed

Exercise 5.2.7. Complete the proof of the previous lemma. [Hint: All the combinatory rules (except $(\mathsf{I}\vdash)$, which we already dealt with) are alike for the purposes of this claim. The $(\circ\vdash)$ and $(\vdash\to)$ cases are easy.]

A simple example, which shows why the $(\vdash \circ)$ rule is excluded by a condition in the above lemma, is the following.

$$\frac{\mathsf{I}\vdash \dfrac{\vdash\to \dfrac{\mathcal{A} \vdash \mathcal{A}}{\mathsf{I}; \mathcal{A} \vdash \mathcal{A}}}{\mathsf{I} \vdash \mathcal{A} \to \mathcal{A}} \qquad \mathcal{B} \vdash \mathcal{B}}{\mathsf{I}; \mathcal{B} \vdash (\mathcal{A} \to \mathcal{A}) \circ \mathcal{B}}\ \vdash\circ$$

$(\vdash \circ)$ is excluded, because this rule allows a structure to be added to the antecedent as a result of enlarging the formula in the succedent.

Definition 5.12. The *Lindenbaum algebra of* LC is $\mathfrak{A}^{LC}_{\mathfrak{L}}$ that is obtained from the word algebra of the language of LC by taking its quotient with $\equiv$ and lifting the logical constants to the equivalence classes. The relation $\leq$ on equivalence classes is defined as $[\mathcal{A}] \leq [\mathcal{B}]$ iff $\mathcal{A} \vdash \mathcal{B}$ is provable in LC.

Exercise 5.2.8. Determine whether each of the equivalence classes with respect to $\equiv$ are singletons. [Hint: Either find two formulas $\mathcal{A}$ and $\mathcal{B}$ such that $\mathcal{A} \vdash \mathcal{B}$ and $\mathcal{B} \vdash \mathcal{A}$ are both provable, or show that this cannot happen, perhaps, because there are only left rules for the combinators.]

Lemma 5.13. $\mathfrak{A}_{\mathfrak{L}}^{LC} = \langle A; \leq, \circ, \rightarrow, \mathsf{I}, \mathsf{K}, \mathsf{C}, \mathsf{W}, \mathsf{B}, \mathsf{S}, \mathsf{T}, \mathsf{M} \rangle$ *is an ordered algebra of similarity type* $\langle 2, 2, 0, 0, 0, 0, 0, 0, 0, 0 \rangle$ *such that* (1)–(4) hold.

(1) $\leq$ *is a partial order;*

(2) $a \circ b \leq c$ *iff* $a \leq b \rightarrow c$;

(3) $a \leq d$ *implies* $b \circ a \leq b \circ d$ *and* $d \rightarrow b \leq a \rightarrow b$;

(4) $\mathsf{I} \circ a \leq a$; $\mathsf{K} \circ a \circ b \leq a$; $\mathsf{C} \circ a \circ b \circ c \leq a \circ c \circ b$; $\mathsf{W} \circ a \circ b \leq a \circ b \circ b$;

$\mathsf{B} \circ a \circ b \circ c \leq a \circ (b \circ c)$; $\mathsf{S} \circ a \circ b \circ c \leq a \circ c \circ (b \circ c)$ $\mathsf{T} \circ a \circ b \leq b \circ a$;

$\mathsf{M} \circ a \leq a \circ a$,

where a, b, c *and* d *range over the elements of* A.

Exercise 5.2.9. Prove the lemma. [Hint: Clause (4) is specific to LC, but it is easy to prove that those inequations hold.]

We have seen in Chapter 3 that judiciously chosen axioms can simplify a calculus and some proofs of meta-theorems. We may wonder whether adding axioms for the combinators would have been a better choice in LC. Our main worry is whether the cut rule would remain admissible.

In general, an inequation (with suitable terms) may be *turned into a rule* as the combinatory rules demonstrate. We assume that there is an axiom like (id). If we have the inequation $a \leq b$, then we know that for any c, $b \leq c$ implies $a \leq c$. In the first instance, the inequation is turned into

$$\frac{\mathcal{B} \vdash \mathcal{C}}{\mathcal{A} \vdash \mathcal{C}}.$$

The original inequation can be recovered by starting with the identity axiom instantiated with $\mathcal{B}$ (in place of $\mathcal{C}$). In one step, we get $\mathcal{A} \vdash \mathcal{B}$.

Of course, the combinatory rules show that this may be a good way to proceed in spirit, but it requires some further thought. The combinatory rule, for instance, for S, *does not have* the form

$$\frac{\mathcal{A}; \mathcal{C}; (\mathcal{B}; \mathcal{C}) \vdash \mathcal{D}}{\mathsf{S}; \mathcal{A}; \mathcal{B}; \mathcal{C} \vdash \mathcal{D}}.$$

We want to ensure that having a rule (instead of an axiom) does not enlarge the set of provable sequents. But at the same time, we want to have a proof of the admissibility of the cut rule.

If the rule derived from the inequation $a \leq b$ would look like the one above, then we could show immediately that there are no new provable consecutions—provided that we have the cut rule.

$$\frac{\mathcal{A} \vdash \mathcal{B} \qquad \mathcal{B} \vdash \mathcal{C}}{\mathcal{A} \vdash \mathcal{C}}$$

However, to ensure the admissibility of the cut rule, the simple form of the rule derived from the inequation is not sufficient. The rule must be *structuralized* and *contextualized*. These moves may appear to make the rule more powerful, because a structure may be complex, that is, not just a formula. The schematic shape of the contextualized (by $\mathfrak{D}[\ ]$) and structuralized (by $\mathfrak{B}$ and $\mathfrak{A}$) rule then is as follows.

$$\frac{\mathfrak{D}[\mathfrak{B}] \vdash \mathcal{C}}{\mathfrak{D}[\mathfrak{A}] \vdash \mathcal{C}}\ (*)$$

We can no longer make comparisons with the inequality, because so far we have not stipulated anything in the algebra to correspond to $;$. In LC, and in $\mathfrak{A}^{LC}_{\mathfrak{L}}$ there is a corresponding operation $\circ$, fusion, which is monotone in both argument places. Thus, the hole in the structure $\mathfrak{D}[\ ]$ itself is a place with respect to which the whole structure is monotone. Thus, to justify the new form of the rule, we have to expand the context of the inequation by including $\circ$ with the rules expressing its monotonicity as $a \leq b$ implies $a \circ c \leq b \circ c$ and $c \circ a \leq c \circ b$.

The inequation, given the $(*)$ rule, is derivable as before. The explanation as to why to use structures is that in the proof of the admissibility of the cut rule, switching an application of the cut rule and of the $(*)$ rule, may necessitate the $(*)$ rule to be applied to a structure, which took the place of the cut formula.

So far we were arguing informally for the admissibility of the cut being preserved via turning an axiom into an inequation. However, in the case of LC, we know that the cut rule is admissible and the combinatory inequations are emulated by rules, not axioms. Thus, the question is whether it is possible to formulate LC with axioms instead of rules.

There are *two or three ways* in which we can formulate axioms for the combinators in $\mathbb{B}$.

First of all, we could simply take the inequations for the combinators from $\mathfrak{A}^{LC}_{\mathfrak{L}}$ and capitalize the lowercase letters, so to speak. This would give us axioms such as $\mathsf{I} \circ \mathcal{A} \vdash \mathcal{A}$, $\mathsf{K} \circ \mathcal{A} \circ \mathcal{B} \vdash \mathcal{A}$, etc. An objection is that I could no longer play a role in the definition of theoremhood, because it is firmly attached to $\mathcal{A}$ in the axiom. Another problem is that the cut rule is obviously not admissible as the following proof shows.

$$\frac{\begin{matrix}\vdots \\ \mathsf{I}; \mathcal{A} \vdash \mathsf{I} \circ \mathcal{A}\end{matrix} \qquad \begin{matrix}\vdots \\ \mathsf{I} \circ \mathcal{A} \vdash \mathcal{A}\end{matrix}}{\mathsf{I}; \mathcal{A} \vdash \mathcal{A}}$$

Since, in this hypothetical situation, I can be introduced only by its axiom or as an instance of (id), the lower sequent has no cut-free proof. Furthermore, we could not prove sequents that would correspond to complex combinators. For example, the following segment shows some steps from a proof in LC.

$$\dfrac{\dfrac{\dfrac{\mathcal{C} \vdash \mathcal{C} \qquad \mathsf{B}'; \mathcal{A} \to \mathcal{B}; \mathcal{B} \to \mathcal{D}; \mathcal{A} \vdash \mathcal{D}}{\mathsf{B}'; (\mathcal{C} \to \mathcal{A} \to \mathcal{B}; \mathcal{C}); \mathcal{B} \to \mathcal{D}; \mathcal{A} \vdash \mathcal{D}} \to\vdash}{\mathsf{B}; \mathsf{B}'; \mathcal{C} \to \mathcal{A} \to \mathcal{B}; \mathcal{C}; \mathcal{B} \to \mathcal{D}; \mathcal{A} \vdash \mathcal{D}} \mathsf{B}\vdash}{\mathsf{B} \circ \mathsf{B}'; \mathcal{C} \to \mathcal{A} \to \mathcal{B}; \mathcal{C}; \mathcal{B} \to \mathcal{D}; \mathcal{A} \vdash \mathcal{D}} \circ\vdash$$

Lastly, we would not be able to prove the simple types of the combinators, because the combinators cannot be unfused from the rest of the formulas in their axioms. All this is more than enough for us not to consider such axioms.

We could allow ; to occur on the left-hand side of the $\vdash$ instead of $\circ$. Then the axioms would have the form $\mathsf{C}; \mathcal{A}; \mathcal{B}; \mathcal{C} \vdash \mathcal{A} \circ \mathcal{C} \circ \mathcal{B}$, $\mathsf{W}; \mathcal{A}; \mathcal{B} \vdash \mathcal{A} \circ \mathcal{B} \circ \mathcal{B}$, etc. Clearly, we can fuse the formulas on the left, thereby, recouping the axioms we considered earlier.

We could add I on the left to provable sequences by taking an appropriate instance of $(\mathsf{I}\vdash)$ and applying a cut. Using cut, we could prove the lower sequent in the above proof.

Exercise 5.2.10. Substantiate the claims about I and about the provability of $\mathsf{B} \circ \mathsf{B}'; \mathcal{C} \to \mathcal{A} \to \mathcal{B}; \mathcal{C}; \mathcal{B} \to \mathcal{D}; \mathcal{A} \vdash \mathcal{D}$.

We can also prove the simple types of the combinators—again assuming that we have the cut rule available.

Exercise 5.2.11. Prove the consecutions $\mathsf{S} \vdash (\mathcal{A} \to \mathcal{B} \to \mathcal{C}) \to (\mathcal{A} \to \mathcal{B}) \to \mathcal{A} \to \mathcal{C}$ and $\mathsf{T} \vdash \mathcal{A} \to (\mathcal{A} \to \mathcal{B}) \to \mathcal{B}$.

Exercise 5.2.12. Generalize the results from the previous two exercises concerning how to prove simple types for typable combinators. [Hint: Describe the general structure of such proofs with special attention to the steps where the cut rule is applied.]

It seems that the problems we run into with the completely fused combinatory axioms do not reappear with their unfused equivalents, if we assume that we can use the cut rule. But, unfortunately, we soon discover that the cut rule is not admissible, that is, we would have to stipulate it as a rule of the calculus. Although $\mathsf{I} \vdash \mathcal{A} \to \mathcal{A}$ is provable without an application of the cut rule, it is accidental. (An alternative view is that it is the result of I being a combinator which takes one argument and returns its argument without any change.) The use of the cut in creating complex combinators and in proving principal types is essential in practically all the cases.

Exercise 5.2.13. Provide examples for the non-eliminability of the cut rule based on the following three motives. (1) $\mathsf{I} \vdash \mathcal{A}$ defines $\mathcal{A}$ being a theorem,

(2) the effect of several combinators combined is simulated and (3) the simple types of typable combinators are provable.

The last variant of combinatory axioms that we might contemplate is to have structures in place of formulas. An obvious problem with this idea is that we have a right singular consecution calculus which can only accommodate formulas on the right-hand side of the $\vdash$.

Exercise 5.2.14. Design a workaround for this problem. [Hint: A structure can be thought to correspond to a unique formula.]

The inclusion of potentially complex structures on the left of the $\vdash$, however, does not solve the problem with the admissibility of the cut rule. The cut is always on a *formula*, which does not necessitate the combinatory axioms to include complex structures. The cut rule is *not admissible* in a calculus that implements this last idea concerning the shape of combinatory axioms.

Exercise 5.2.15. Show that this is the case. [Hint: You might want to rely on the solutions to the previous two exercises.]

To sum up, we can conclude that the *only reasonable way* to include the combinators is as in [81], that is, by rules, not by axioms. The rules express one-step reductions, which can be stitched together into weak reduction steps because of the admissibility of the cut rule.

Turning back to the question of consistency, it should be clear now that Curry's paradox cannot be reconstructed in LC, because the paradox includes *weak equality* in an essential way, whereas LC models only *weak reduction*.

Combinatory logic has been extended in various ways, and structurally free logics can be extended too. In Section 5.1, LQ contained $\leftarrow$ in addition to $\rightarrow$ and $\circ$. Function application is not a commutative operation, that is, fg, in general, is not the same function as gf. Algebraically speaking, this means that $\circ$ can have two distinct residuals. In LQ, the two arrows cannot be proved to be the same connective, because as it is easy to show, the sequents $\mathcal{A} \rightarrow \mathcal{B} \vdash \mathcal{B} \leftarrow \mathcal{A}$ and $\mathcal{A} \leftarrow \mathcal{B} \vdash \mathcal{B} \rightarrow \mathcal{A}$ are not provable. In LC, we may include some combinators, which permute some of their arguments; indeed, $\mathbb{B}$ above contains C and T, which are the prototypical permutators, and S—together with K—can create all the possible permutators. This combinatorial definability transfers straightforwardly into LC, where the permutators themselves are formulas in the consecution that results from applications of the corresponding rules.

As a first extension step, we add the same rules for $\leftarrow$ that we had in LQ. $LC^{\leftarrow}$ is LC with $(\leftarrow\vdash)$ and $(\vdash\leftarrow)$ added.

$$\frac{\mathfrak{A} \vdash \mathcal{A} \qquad \mathfrak{B}[\mathcal{B}] \vdash \mathcal{C}}{\mathfrak{B}[\mathfrak{A}; \mathcal{B} \leftarrow \mathcal{A}] \vdash \mathcal{C}} \leftarrow\vdash \qquad\qquad \frac{\mathcal{A}; \mathfrak{A} \vdash \mathcal{B}}{\mathfrak{A} \vdash \mathcal{B} \leftarrow \mathcal{A}} \vdash\leftarrow$$

Having both arrows and assuming the same set of combinatory rules as before, it makes sense to ask whether $\rightarrow$ and $\leftarrow$ are still distinct.

Exercise 5.2.16. Is either of the two sequents $\mathcal{A} \rightarrow \mathcal{B} \vdash \mathcal{B} \leftarrow \mathcal{A}$ and $\mathcal{A} \leftarrow \mathcal{B} \vdash \mathcal{B} \rightarrow \mathcal{A}$ (perhaps, with some combinators added) provable in $LC^{\leftarrow}$? [Hint: Notice that the combinatory rules insert combinators on the left-hand side of some structures.]

Combinators always apply *from left to right*, and in LC, they can be thought of as embodying the structural changes. Importantly, combinators take their arguments, as if the structure to which they are applied has been sliced up by association to the left. The inability of a combinator to "look inside" an argument should be obvious from the rules. For example, if the rule $(\mathsf{W}\vdash)$ is applied, then no matter whether $\mathfrak{B}$ and $\mathfrak{C}$ are atomic or complex structures, they are exactly the same in the upper and lower sequents. Similarly, $(\mathsf{B}\vdash)$ disassociates $\mathfrak{C}$ and $\mathfrak{D}$, but there is no rule that could reverse the effect of an application of this rule. Incidentally, you might have noticed that although LC is not defined in the context of sequences of formulas, we have not mentioned so far what happens to *associativity* or how to recover the associativity of ; in LC.

Usually, full associativity can be achieved not only via expressly postulating rules for associativity (as we suggested toward the end of Section 5.1 in the case of LQ). Associativity can result from having half of associativity and (full) permutation.

<u>*Example*</u> 5.14. Let $\mathfrak{A} = \langle A; \leq, \circ \rangle$ be an ordered algebra of similarity type $\langle 2 \rangle$, where (1) and (2) hold.

(1) $(a \circ b) \circ c \leq a \circ (b \circ c)$ (2) $a \circ b \leq b \circ a$

Then $a \circ (b \circ c) \leq (a \circ b) \circ c$ holds too. The following series of inequations (chained together for the sake of succinctness) demonstrates that it is so.

$$a \circ (b \circ c) \leq (b \circ c) \circ a \leq (c \circ b) \circ a \leq c \circ (b \circ a) \leq c \circ (a \circ b) \leq (a \circ b) \circ c$$

Could this mean that despite the original intention of ; representing function application within LC we have ended up with a fully associative ; ? The presence of the combinators in the lower consecutions in their rules saves us from such a disaster. The above example would have to be amended by the inclusion of certain combinators. In the algebra $\mathfrak{A}'$, T and B would be new distinguished elements of A entering the two inequations as $((\mathsf{B} \circ a) \circ b) \circ c \leq a \circ (b \circ c)$ and $(\mathsf{T} \circ a) \circ b \leq b \circ a$. The left-most term in the chain above should have all the triggers (i.e., justifications) for the later steps inserted.

$$(\mathsf{T} \circ a) \circ ((\mathsf{T} \circ (\mathsf{T} \circ b)) \circ (\mathsf{B} \circ (\mathsf{T} \circ c))) \leq ((\mathsf{T} \circ (\mathsf{T} \circ b)) \circ (\mathsf{B} \circ (\mathsf{T} \circ c))) \circ a \leq$$
$$((\mathsf{B} \circ (\mathsf{T} \circ c)) \circ (\mathsf{T} \circ b)) \circ a \leq (\mathsf{T} \circ c) \circ ((\mathsf{T} \circ b) \circ a) \leq (\mathsf{T} \circ c) \circ (a \circ b) \leq (a \circ b) \circ c$$

The very first term is quite a bit different from $a \circ (b \circ c)$, though it is an element of A.

The common notation for functions and their arguments such as $f(x_0, \ldots, x_{n-1})$ seems to suggest that functions are applied to their arguments from the left. But of course, there is *plenty* of *infix notation* used in mathematics, and some functions are written according to the conventions of *suffix notation*.

For example, the prime (i.e., $'$) is a favorite as a notation for a unary function. In Peano arithmetic, x' can be the successor of x, which is alternatively denoted by $s(x)$ or $x+1$. Leibniz used $'$ for the derivative, and with unary functions this notation for the derivative is concise and unambiguous.

Functions placed on the right-hand side of their arguments occur in computer science too. Introductory logic courses often include prefix, parentheses-free notation for propositional logic. (It is sometimes also called Polish or Łukasiewicz's notation.) Anybody who tried to determine whether a string of upper- and lowercase letters represents a well-formed formula in prefix notation knows that it is better to start at the end of the string chopping it into substrings (which are wff's). Alternatively, taking the *reverse* of the string changes prefix into suffix notation and facilitates the parsing to be started at the beginning of the string. With no claim to historical accuracy, we may speculate that thinking along these lines might have led to Hewlett-Packard's adoption of suffix notation in some of its languages.

Another example is the programming language APL, which was invented to facilitate calculations on matrices, and uses untyped operations. Some of the symbols are used for functions in infix or suffix positions. The execution of a program starts at the right edge, and proceeds as if the string would have been parenthesized by association to the right. APL is not widely used nowadays, and it was never intended to be a language for the masses. Beyond the fact that a program in APL has to be read from right to left, the preference for single symbols to denote functions and the language fully assimilating function composition into the syntax mean that, in APL, one can write extremely concise and cryptic looking code.

Perhaps, one could say that all the above examples are merely notational variations that may have some added convenience, but could be fully dispensed with in favor of an "all-and-only-prefix" style notation. However, there is a reason to think that it is not so—as long as we wish to have functions of arbitrary arity.

With the development of universal algebra, algebraic operations as well as the elements of the carrier set came to be understood more abstractly than numerical operations and numbers. Rings and fields emerged as generalizations of sets of numbers with generalizations of usual numerical operations. Groups were introduced in the early 19th century in connection with polynomial equations with integer coefficients, however, the operations of a group do not need to be numerical operations such as addition, multiplication or division. Groups are inherently connected to *permutations*—both in the work of Galois, and later on, in Cayley's representations. Groups are a variety, and in an economical axiomatization the identity element, let us say e, is postulated to be the *right identity* for the binary group operation (which is often termed multiplication, and denoted by $\cdot$). That is, $a \cdot e = a$, but $\cdot$ is not stipulated to be commutative (unless the group is Abelian), hence, the latter equation by itself does not mean that e is a left identity element. Of course, it is well known that together with the other axioms, it follows that e is an identity for $\cdot$. Those who are not interested in the fine points of an axiomatization of groups or in the axioms being independent may simply state or stipulate that there is an identity for $\cdot$.

All these examples support our claim that the introduction of a new kind of constant in structurally free logics is well motivated. Dunn and Meyer in [81, §8] introduced what they called *dual combinators*.

The addition of $\leftarrow$ and the connections of $\rightarrow$ to simple types forces the consideration of questions such as the following. If I is a left identity for $\circ$, is it also a right identity or is there a right identity at all? Is $(\mathcal{A} \leftarrow \mathcal{A} \leftarrow \mathcal{B}) \leftarrow \mathcal{B}$ a type of a (complex) combinator? In $LC^{\leftarrow}$, the answer to these questions is a resounding "no."

Exercise 5.2.17. Prove that, in $LC^{\leftarrow}$, (a) I is not a right identity for $\circ$ and (b) there is no wff Z, in which all the atomic wff's are (primitive) combinators, and $\mathsf{Z} \vdash (\mathcal{A} \leftarrow \mathcal{A} \leftarrow \mathcal{B}) \leftarrow \mathcal{B}$ is provable.

Dual combinators are denoted by lowercase sans serif letters and they are thought to act just as combinators do—except that they take their arguments from the left. For easy reference, we will use the small letter pair of a combinator for its dual. We list the duals of the combinators, with their axioms, that we introduced on page 147. (We keep the convention that we may omit parentheses from left-associated terms. This appears to introduce a visual asymmetry into some pairs of axioms, for example, for B and b. Complete symmetry reemerges when all the terms are fully parenthesized.)

$$
\begin{array}{rlrlrl}
x\mathsf{i} & \rhd x & x(y(z\mathsf{b})) & \rhd xyz & xyz\mathsf{s} & \rhd xy(xz) \\
x(y\mathsf{t}) & \rhd yx & x(y(z\mathsf{b}')) & \rhd xzy & xyz\mathsf{s}' & \rhd xz(xy) \\
x\mathsf{m} & \rhd xx & x(y(z\mathsf{v})) & \rhd y(zx) & x\mathsf{y} & \rhd (x\mathsf{y})x
\end{array}
$$

Exercise 5.2.18. What are the axioms for w, c and k? [Hint: The axioms for W, C and K are on page 145.]

Combinatory logic being extended with constants that go beyond the combinators over $\{\mathsf{S}, \mathsf{K}\}$ (or some other combinatorially complete base) is not unusual. In fact, Schönfinkel's U in [165] is a constant that is not a combinator. Other constants had been added from time to time to pure combinatory logic—from connectives and operators to constants that echo type statements. The addition of dual combinators preserves the purity of combinatory logic in the sense that dual combinators do not blend propositional or predicate logic into a theory of functions. Still, they lead to a rich theory and we will say a word or two about some of the results and whether they impact LC toward the end of this section.

The language of LC_{+} is that of $LC^{\leftarrow}$ extended by new zero-ary constants (by dual combinators), and by conjunction ($\wedge$) and disjunction ($\vee$). The logical constants are the binary connectives $\rightarrow, \circ, \leftarrow, \wedge, \vee$, and the zero-ary constants $\mathsf{I}, \mathsf{K}, \mathsf{C}, \mathsf{W}, \mathsf{B}, \mathsf{S}, \mathsf{T}, \mathsf{M}, \mathsf{i}, \mathsf{k}, \mathsf{c}, \mathsf{w}, \mathsf{b}', \mathsf{s}', \mathsf{v}$ and m.

The set of wff's is as in Definition 5.8, with the dual combinators included in $\mathbb{B}$, and with clause (3) incorporating $\leftarrow$, $\wedge$ and $\vee$, in the obvious way.

LC_+ denotes the consecution calculus for *positive structurally free logic with dual combinators*, and it is defined to extend LC with the rules $(\leftarrow\vdash)$, $(\vdash\leftarrow)$ (given above), and with the following rules for $\wedge$, $\vee$ and the dual combinators in $\mathbb{B}$.

$$\frac{\mathfrak{A}[\mathcal{A}] \vdash \mathcal{B}}{\mathfrak{A}[\mathcal{A} \wedge \mathcal{C}] \vdash \mathcal{B}} \wedge\vdash_1 \qquad \frac{\mathfrak{A}[\mathcal{A}] \vdash \mathcal{B}}{\mathfrak{A}[\mathcal{C} \wedge \mathcal{A}] \vdash \mathcal{B}} \wedge\vdash_2 \qquad \frac{\mathfrak{A} \vdash \mathcal{A} \quad \mathfrak{A} \vdash \mathcal{B}}{\mathfrak{A} \vdash \mathcal{A} \wedge \mathcal{B}} \vdash\wedge$$

$$\frac{\mathfrak{A}[\mathcal{A}] \vdash \mathcal{C} \quad \mathfrak{A}[\mathcal{B}] \vdash \mathcal{C}}{\mathfrak{A}[\mathcal{A} \vee \mathcal{B}] \vdash \mathcal{C}} \vee\vdash \qquad \frac{\mathfrak{A} \vdash \mathcal{A}}{\mathfrak{A} \vdash \mathcal{A} \vee \mathcal{B}} \vdash\vee_1 \qquad \frac{\mathfrak{A} \vdash \mathcal{A}}{\mathfrak{A} \vdash \mathcal{B} \vee \mathcal{A}} \vdash\vee_2$$

$$\frac{\mathfrak{A}[\mathfrak{B}] \vdash \mathcal{A}}{\mathfrak{A}[\mathfrak{B}; \mathsf{i}] \vdash \mathcal{A}} \mathsf{i}\vdash \qquad \frac{\mathfrak{A}[\mathfrak{B}] \vdash \mathcal{A}}{\mathfrak{A}[\mathfrak{C}; (\mathfrak{B}; \mathsf{k})] \vdash \mathcal{A}} \mathsf{k}\vdash$$

$$\frac{\mathfrak{A}[\mathfrak{B}; (\mathfrak{C}; \mathfrak{D})] \vdash \mathcal{A}}{\mathfrak{A}[\mathfrak{C}; (\mathfrak{B}; (\mathfrak{D}; \mathsf{c}))] \vdash \mathcal{A}} \mathsf{c}\vdash \qquad \frac{\mathfrak{A}[\mathfrak{B}; (\mathfrak{B}; \mathfrak{C})] \vdash \mathcal{A}}{\mathfrak{A}[\mathfrak{B}; (\mathfrak{C}; \mathsf{w})] \vdash \mathcal{A}} \mathsf{w}\vdash$$

$$\frac{\mathfrak{A}[\mathfrak{B}; \mathfrak{C}; \mathfrak{D}] \vdash \mathcal{A}}{\mathfrak{A}[\mathfrak{B}; (\mathfrak{D}; (\mathfrak{C}; \mathsf{b}'))] \vdash \mathcal{A}} \mathsf{b}'\vdash \qquad \frac{\mathfrak{A}[\mathfrak{B}; \mathfrak{C}; (\mathfrak{B}; \mathfrak{D})] \vdash \mathcal{A}}{\mathfrak{A}[\mathfrak{B}; (\mathfrak{D}; (\mathfrak{C}; \mathsf{s}'))] \vdash \mathcal{A}} \mathsf{s}'\vdash$$

$$\frac{\mathfrak{A}[\mathfrak{B}; (\mathfrak{C}; \mathfrak{D})] \vdash \mathcal{A}}{\mathfrak{A}[\mathfrak{D}; (\mathfrak{B}; (\mathfrak{C}; \mathsf{v}))] \vdash \mathcal{A}} \mathsf{v}\vdash \qquad \frac{\mathfrak{A}[\mathfrak{B}; \mathfrak{B}] \vdash \mathcal{A}}{\mathfrak{A}[\mathfrak{B}; \mathsf{m}] \vdash \mathcal{A}} \mathsf{m}\vdash$$

Exercise 5.2.19. Formulate the rules for the dual combinators s, b and t.

The notions of a *proof* and of a *theorem* are defined as before. The single *cut rule* is admissible in LC_+.

Example 5.15. We can prove the analogs of the axioms of dual combinators now. The last line is the "dual" of the consecution proved in Example 5.9.

$$\cfrac{\cfrac{\cfrac{\mathcal{C} \vdash \mathcal{C} \qquad \cfrac{\mathcal{B} \vdash \mathcal{B} \qquad \mathcal{A} \vdash \mathcal{A}}{\mathcal{B}; \mathcal{A} \vdash \mathcal{B} \circ \mathcal{A}} \vdash\circ}{\mathcal{C}; (\mathcal{B}; \mathcal{A}) \vdash \mathcal{C} \circ (\mathcal{B} \circ \mathcal{A})} \vdash\circ}{\mathcal{B}; (\mathcal{C}; (\mathcal{A}; \mathsf{c})) \vdash \mathcal{C} \circ (\mathcal{B} \circ \mathcal{A})} \mathsf{c}\vdash}{\mathcal{B} \circ (\mathcal{C} \circ (\mathcal{A} \circ \mathsf{c})) \vdash \mathcal{C} \circ (\mathcal{B} \circ \mathcal{A})} \circ\vdash\text{'s}$$

Exercise 5.2.20. Suppose that z is a proper dual combinator. Formulate the $(\mathsf{z}\vdash)$ rule, and prove the analog of the axiom for z in LC_+. [Hint: You may assume that proper dual combinators are defined as expected.]

The following proof is straightforward.

$$\dfrac{\dfrac{\mathcal{A} \vdash \mathcal{A} \qquad \dfrac{\mathcal{B} \vdash \mathcal{B} \qquad \mathcal{A} \vdash \mathcal{A}}{\mathcal{B}; \mathcal{A} \leftarrow \mathcal{B} \vdash \mathcal{A}} \leftarrow\vdash}{\dfrac{\mathcal{A}; \mathcal{B} \leftarrow \mathcal{A}; \mathcal{A} \leftarrow \mathcal{B} \vdash \mathcal{A}}{\mathcal{A}; (\mathcal{A} \leftarrow \mathcal{B}; (\mathcal{B} \leftarrow \mathcal{A}; \mathsf{b}')) \vdash \mathcal{A}} \mathsf{b}'\vdash} \leftarrow\vdash}{\mathsf{b}' \vdash ((\mathcal{A} \leftarrow \mathcal{A}) \leftarrow \mathcal{A} \leftarrow \mathcal{B}) \leftarrow \mathcal{B} \leftarrow \mathcal{A}} \vdash\leftarrow\text{'s}$$

The formula on the right-hand side of the $\vdash$ is an instance of the (dual) simple type of the dual combinator b'.

However, we can prove consecutions, which may be more surprising.

Example 5.16. In combinatory logic, $\mathsf{I}\,\mathsf{i} \rhd_1 \mathsf{I}$ and $\mathsf{I}\,\mathsf{i} \rhd_1 \mathsf{i}$. These reductions are represented by the bottom consecutions in the next two proofs.

$$\dfrac{\dfrac{\mathsf{I} \vdash \mathsf{I}}{\mathsf{I}; \mathsf{i} \vdash \mathsf{I}} \mathsf{i}\vdash}{\mathsf{I} \circ \mathsf{i} \vdash \mathsf{I}} \circ\vdash \qquad\qquad \dfrac{\dfrac{\mathsf{i} \vdash \mathsf{i}}{\mathsf{I}; \mathsf{i} \vdash \mathsf{i}} \mathsf{I}\vdash}{\mathsf{I} \circ \mathsf{i} \vdash \mathsf{i}} \circ\vdash$$

If $M \rhd_1 N$ then $M =_{\mathrm{w}} N$. The two reductions—taken together—are reminiscent of how the uniqueness of multiplicative identity is established in group theory. $e_1 \cdot e_2 = e_2$, because e_1 is a left identity, and $e_1 \cdot e_2 = e_1$, because e_2 is a right identity; therefore, $e_1 = e_2$. However, $=_{\mathrm{w}}$ includes expansion steps, and neither $\mathsf{I} \vdash \mathsf{i}$ nor $\mathsf{i} \vdash \mathsf{I}$ is a provable consecution in LC_+.

It is not difficult to show that the *equational* theory of combinatory logic with the base $\{\mathsf{S}, \mathsf{K}, \mathsf{k}, \mathsf{s}\}$ is inconsistent. That is, given two (distinct) variables x and y, $x =_{\mathrm{w}} y$.

Exercise 5.2.21. Prove the last claim. [Hint: There are many proofs for this claim. A proof may be found in [26]; see also [14] for a related proof.]

The analog of the above inconsistency of a set of terms with combinators and dual combinators with respect to $=_{\mathrm{w}}$ would be the provability of a sequent of the form $p \vdash q$, where p and q are distinct propositional variables.

Lemma 5.17. **(LC_+ is consistent)** *For distinct p and q, $p \vdash q$ is* not provable *in LC_+.*

Proof: By assumption, p and q are not the same propositional variables. If $p \vdash q$ would be provable, then it would have a cut-free proof. Then it must be an axiom, because no rule can result in this consecution. But $p \vdash q$ is not an instance of the axiom. qed

We know that LC_+ is a *well-behaved calculus.* Looking at the above proofs with the I's and i's, we can see that a new interesting phenomenon is modeled in LC_+. Combinatory logic (or more precisely, the most widely used versions of it) possesses the so-called *Church–Rosser property*. There are two equivalent formulations of the Church–Rosser theorem. First, if $M =_{\mathrm{w}} N$, then there is a term P such that both $M \rhd_{\mathrm{w}} P$ and $N \rhd_{\mathrm{w}} P$. For our purposes the other formulation is more suitable, because it does not mention weak equality. If $M \rhd_{\mathrm{w}} N$ and $M \rhd_{\mathrm{w}} P$, then there is a term Q such that $N \rhd_{\mathrm{w}} Q$ and $P \rhd_{\mathrm{w}} Q$. (The latter formulation is, perhaps, mathematically more elegant in the sense that it is a theorem establishing a certain property of one relation, namely, of $\rhd_{\mathrm{w}}$.) Given the definition of $=_{\mathrm{w}}$, the first formulation is seen to imply the second, because the assumption that $M \rhd_{\mathrm{w}} N$ and $M \rhd_{\mathrm{w}} P$ implies $N =_{\mathrm{w}} P$. However, the second also implies the first, because $=_{\mathrm{w}}$ is the transitive closure of the symmetric closure of $\rhd_{\mathrm{w}}$. That is, the two formulations of the Church–Rosser theorem are equivalent.

The Church–Rosser property, that is, the confluence of the $\rhd_{\mathrm{w}}$ relation, may be considered without any concern about consistency, because we do not need to include $=_{\mathrm{w}}$ into the theory. Unless we restrict the way that terms can be built from atomic components, there is no dual combinatory logic that has the Church–Rosser property. The proof of this claim is highly non-trivial and very different in its method from proofs of the Church–Rosser theorem for combinatory logics (or λ-calculi).

We gave various examples from mathematics and computer science, where functions are used in a way that is similar to the use of dual combinators. One might wonder whether the *order of the evaluation* of expressions makes a difference in those cases. In other words, can we get different numbers as the result of some calculations or computations in those sample cases? It seems to us that non-confluence does not emerge in those situations for two reasons. First, having all functions apply from the right causes very little change—it is simply looking at expressions in a mirror, so to speak. Second, when some functions apply from the left and others from the right, then the analog of overlapping redexes, which is a key component in every instance of non-confluence in combinatory systems, which include combinators and dual combinators, is excluded by conventions or by parenthesizing the expressions. In other words, no ambiguity is permitted with respect to the order of the evaluation.

The two proofs in Example 5.16 illustrate non-confluence in a simple, perhaps, in the simplest possible way. The next example includes two proofs, which further illustrate how combinators and dual combinators interact.

Example 5.18. The identity (or dually, the dual identity) combinator allows proofs like the next two.

$$
{\scriptstyle \circ\vdash\text{'s}}\,\dfrac{{\scriptstyle \mathsf{B}'\vdash}\,\dfrac{{\scriptstyle \mathsf{i}\vdash}\,\dfrac{{\scriptstyle \mathsf{i}\vdash}\,\dfrac{\mathsf{i}\vdash\mathsf{i}}{\mathsf{i};\mathsf{i}\vdash\mathsf{i}}}{\mathsf{i};(\mathsf{i};\mathsf{i})\vdash\mathsf{i}}}{\mathsf{B}';\mathsf{i};\mathsf{i};\mathsf{i}\vdash\mathsf{i}}}{\mathsf{B}'\circ\mathsf{i}\circ\mathsf{i}\circ\mathsf{i}\vdash\mathsf{i}}
\qquad\qquad
\dfrac{\dfrac{\dfrac{\dfrac{\mathsf{B}'\vdash\mathsf{B}'}{\mathsf{B}';\mathsf{i}\vdash\mathsf{B}'}\,{\scriptstyle \mathsf{i}\vdash}}{\mathsf{B}';\mathsf{i};\mathsf{i}\vdash\mathsf{B}'}\,{\scriptstyle \mathsf{i}\vdash}}{\mathsf{B}';\mathsf{i};\mathsf{i};\mathsf{i}\vdash\mathsf{B}'}\,{\scriptstyle \mathsf{i}\vdash}}{\mathsf{B}'\circ\mathsf{i}\circ\mathsf{i}\circ\mathsf{i}\vdash\mathsf{B}'}\,{\scriptstyle \circ\vdash\text{'s}}
$$

Notice that the proofs differ in their top consecution, which is an instance of the axiom and in effect, shows which combinator or dual combinator is treated as a "genuine" formula, and which combinators or dual combinators are "traces" of applications of structural changes in the antecedent.

Exercise 5.2.22. Construct other proofs that show interesting interactions between combinators and dual combinators. [Hint: It is open to interpretation what you deem to be interesting.]

The next proof illustrates that not only combinators and dual combinators interact, but the two connectives $\to$ and $\leftarrow$ do too.

$$\dfrac{\dfrac{\dfrac{\mathcal{C} \vdash \mathcal{C} \qquad \dfrac{\mathcal{A} \vdash \mathcal{A} \qquad \mathcal{B} \vdash \mathcal{B}}{\mathcal{A} \to \mathcal{B}; \mathcal{A} \vdash \mathcal{B}}\,{\to\vdash}}{\mathcal{A} \to \mathcal{B}; (\mathcal{C} \to \mathcal{A}; \mathcal{C}) \vdash \mathcal{B}}\,{\to\vdash}}{\mathcal{C} \to \mathcal{A}; (\mathcal{A} \to \mathcal{B}; (\mathcal{C}; \mathsf{c})) \vdash \mathcal{B}}\,{\mathsf{c}\vdash}}{\mathsf{c} \vdash ((\mathcal{B} \leftarrow \mathcal{C} \to \mathcal{A}) \leftarrow \mathcal{A} \to \mathcal{B}) \leftarrow \mathcal{C}}\,{\vdash\leftarrow\text{'s}}$$

We do not go into further investigations of types and typing here. In Section 9.3, we consider structurally free logics in the more general setting of typed calculi.

We mentioned that the *fixed point combinator* Y is definable from a combinatorially complete base. As we have already seen, it is definable from a more modest base too, and there is a defining term that exhibits the fixed point property with respect to weak reduction. Notwithstanding, one might wonder what the rule for Y would look like.[8] The rules for Y and y are as follows.

$$\dfrac{\mathfrak{A}[\mathfrak{B}; (\mathsf{Y}; \mathfrak{B})] \vdash \mathcal{A}}{\mathfrak{A}[\mathsf{Y}; \mathfrak{B}] \vdash \mathcal{A}}\,{\mathsf{Y}\vdash} \qquad\qquad \dfrac{\mathfrak{A}[\mathfrak{B}; \mathsf{y}; \mathfrak{B}] \vdash \mathcal{A}}{\mathfrak{A}[\mathfrak{B}; \mathsf{y}] \vdash \mathcal{A}}\,{\mathsf{y}\vdash}$$

The fixed point combinator has a certain similarity to combinators such as W and M, in other words, Y is a combinator with duplicative effect, though what is usually emphasized about Y is that it is not a proper combinator. The cut theorem was proved in [81] using multiple cut, in which several occurrences of the cut formula in the right premise may be replaced by the antecedent of the left premise. In the $(\mathsf{Y}\vdash)$ and $(\mathsf{y}\vdash)$ rules, Y and y, respectively, occur in the upper consecution and so the displayed occurrences of those combinators in the lower consecution are not really new. This makes a difference to the rank of the cut formula when the cut formula is the (dual) fixed point combinator. The *problem*, to put it quickly, is that Y is not a proper combinator, and even if the cuts on other (than the displayed) occurrences of Y (or y) are separated out, the cut on Y cannot be permuted with an application of $(\mathsf{Y}\vdash)$ (or $(\mathsf{y}\vdash)$). This problem led to crucial advances in proofs of the cut theorem.[9]

[8] The fixed point combinator and its dual were added to structurally free logics in [29].

[9] The way to deal with constants like Y was described in [31], and triple inductive proofs were introduced in [33].

Let us consider some more concrete proof segments with W and with Y.

$$\dfrac{\overset{\vdots}{\mathfrak{A}\vdash\mathsf{W}} \qquad \dfrac{\overset{\vdots}{\mathcal{C}\to\mathcal{B};\mathcal{C};\mathsf{W};\mathsf{W}\vdash(\mathcal{B}\circ\mathsf{W})\circ\mathsf{W}}}{\mathsf{W};(\mathcal{C}\to\mathcal{B};\mathcal{C});\mathsf{W}\vdash(\mathcal{B}\circ\mathsf{W})\circ\mathsf{W}}\;{\scriptstyle\mathsf{W}\vdash}}{\mathfrak{A};(\mathcal{C}\to\mathcal{B};\mathcal{C});\mathfrak{A}\vdash(\mathcal{B}\circ\mathsf{W})\circ\mathsf{W}}\;\text{multi-cut}$$

By an application of multi-cut, we replaced two occurrences of W—including the one that was introduced by the $(\mathsf{W}\vdash)$ rule in the second consecution on the right-hand side. We can separate that occurrence from the other one, and obtain the same end consecution with cuts that are available to us by the hypothesis of induction.

$$\dfrac{\overset{\vdots}{\mathfrak{A}\vdash\mathsf{W}} \qquad \dfrac{\dfrac{\overset{\vdots}{\mathfrak{A}\vdash\mathsf{W}} \quad \overset{\vdots}{\mathcal{C}\to\mathcal{B};\mathcal{C};\mathsf{W};\mathsf{W}\vdash(\mathcal{B}\circ\mathsf{W})\circ\mathsf{W}}}{\mathcal{C}\to\mathcal{B};\mathcal{C};\mathfrak{A};\mathfrak{A}\vdash(\mathcal{B}\circ\mathsf{W})\circ\mathsf{W}}\;\text{multi-cut}}{\mathsf{W};(\mathcal{C}\to\mathcal{B};\mathcal{C});\mathfrak{A}\vdash(\mathcal{B}\circ\mathsf{W})\circ\mathsf{W}}\;{\scriptstyle\mathsf{W}\vdash}}{\mathfrak{A};(\mathcal{C}\to\mathcal{B};\mathcal{C});\mathfrak{A}\vdash(\mathcal{B}\circ\mathsf{W})\circ\mathsf{W}}\;\text{multi-cut}$$

An analog proof segment with Y looks like the following.

$$\dfrac{\overset{\vdots}{\mathfrak{A}\vdash\mathsf{Y}} \qquad \dfrac{\overset{\vdots}{\mathcal{C}\to\mathcal{B};\mathcal{C};(\mathsf{Y};(\mathcal{C}\to\mathcal{B};\mathcal{C}));\mathsf{Y}\vdash(\mathcal{B}\circ(\mathsf{Y}\circ\mathcal{B}))\circ\mathsf{Y}}}{\mathsf{Y};(\mathcal{C}\to\mathcal{B};\mathcal{C});\mathsf{Y}\vdash(\mathcal{B}\circ(\mathsf{Y}\circ\mathcal{B}))\circ\mathsf{Y}}\;{\scriptstyle\mathsf{Y}\vdash}}{\mathfrak{A};(\mathcal{C}\to\mathcal{B};\mathcal{C});\mathfrak{A}\vdash(\mathcal{B}\circ(\mathsf{Y}\circ\mathcal{B}))\circ\mathsf{Y}}\;\text{multi-cut}$$

We can try to mimic the previous modification by separating the cut that is on the Y that is used in the $(\mathsf{Y}\vdash)$ step and the cut on the other occurrence of Y. What we get is a proof of the same end consecution, however, the second cut does not have a lower rank, indeed, it has a strictly higher rank than the earlier one.

$$\dfrac{\overset{\vdots}{\mathfrak{A}\vdash\mathsf{Y}} \qquad \dfrac{\dfrac{\overset{\vdots}{\mathfrak{A}\vdash\mathsf{Y}} \quad \overset{\vdots}{\mathcal{C}\to\mathcal{B};\mathcal{C};(\mathsf{Y};(\mathcal{C}\to\mathcal{B};\mathcal{C}));\mathsf{Y}\vdash(\mathcal{B}\circ(\mathsf{Y}\circ\mathcal{B}))\circ\mathsf{Y}}}{\mathcal{C}\to\mathcal{B};\mathcal{C};(\mathsf{Y};(\mathcal{C}\to\mathcal{B};\mathcal{C}));\mathfrak{A}\vdash(\mathcal{B}\circ(\mathsf{Y}\circ\mathcal{B}))\circ\mathsf{Y}}\;\text{cut}}{\mathsf{Y};(\mathcal{C}\to\mathcal{B};\mathcal{C});\mathfrak{A}\vdash(\mathcal{B}\circ(\mathsf{Y}\circ\mathcal{B}))\circ\mathsf{Y}}\;{\scriptstyle\mathsf{Y}\vdash}}{\mathfrak{A};(\mathcal{C}\to\mathcal{B};\mathcal{C});\mathfrak{A}\vdash(\mathcal{B}\circ(\mathsf{Y}\circ\mathcal{B}))\circ\mathsf{Y}}\;\text{cut}$$

We could entertain the idea of cutting all the occurrences of Y at once in the first application of the cut above. This would also result in a proof, but not in a proof of the same end consecution. Moreover, there is no way to delete a copy of $\mathcal{C}\to\mathcal{B};\mathcal{C}$ from the antecedent.

$$\dfrac{\overset{\vdots}{\mathfrak{A}\vdash\mathsf{Y}} \quad \overset{\vdots}{\mathcal{C}\to\mathcal{B};\mathcal{C};(\mathsf{Y};(\mathcal{C}\to\mathcal{B};\mathcal{C}));\mathsf{Y}\vdash(\mathcal{B}\circ(\mathsf{Y}\circ\mathcal{B}))\circ\mathsf{Y}}}{\mathcal{C}\to\mathcal{B};\mathcal{C};(\mathfrak{A};(\mathcal{C}\to\mathcal{B};\mathcal{C}));\mathfrak{A}\vdash(\mathcal{B}\circ(\mathsf{Y}\circ\mathcal{B}))\circ\mathsf{Y}}\;\text{multi-cut}$$

Since there are no other options for what to do with the right premise, it is reasonable to turn our attention to the left premise. We may notice that Y

can occur on the right-hand side of the turnstile only because it was introduced by $\mathsf{Y} \vdash \mathsf{Y}$. Also, all the rules applied in the process of arriving at the left premise of the cut must have been left rules. These observations can be shaped as a lemma stating the eliminability of a cut on Y. The exact formulation of this lemma and its proof are in Section 7.3. In Chapter 7, we will also discuss how the proof of the cut theorem may be structured differently.

5.3 More implicational relevance logics

We have seen the *pure logic of relevant implication*, that is, $R_{\rightarrow}$ in Section 4.3. R has an important place in the family of relevance logics: it is completely relevant in the sense of not allowing arbitrary premises to be fused into a valid inference.

It is well known that $R_{\rightarrow}$ can be axiomatized in various ways, including by wff's that are principal type schemas of the combinators B, W, C and I.[10] The base $\{\mathsf{B}, \mathsf{W}, \mathsf{C}, \mathsf{I}\}$ is not combinatorially complete, however, it is "almost complete." The replacement of I by K restores completeness. (The reverse move, that is, taking $\mathcal{A} \rightarrow \mathcal{A}$ as an axiom instead of $\mathcal{A} \rightarrow \mathcal{B} \rightarrow \mathcal{A}$, is how Church was led to his axiom system.) Given structurally free logics, which we described in the previous section, and the neat match between combinators and the axioms of $R_{\rightarrow}$, we may wonder if we could obtain a consecution calculus that is inspired by structurally free logics, but free of the burden of carrying around the combinators.

We introduce several consecution calculi in this section that were invented in the last decade or so together with a couple of new ones. Historically, $LT^t_{\rightarrow}$ and $LT_{\rightarrow}$ were introduced first, in [32]. Then $LB^t_{\rightarrow}$ appeared as part of $LB^{\circ t}_{+}$ in [38, Ch. 2]. $LE^t_{\rightarrow}$ was introduced in [33]. Lastly, $LR^t_{\rightarrow;}$ and $LT^{\textcircled{t}}_{\rightarrow}$ were defined in [41]. We start with the consecution calculus $LR^t_{\rightarrow;}$, which formalizes $R^t_{\rightarrow}$, which is also formalized by the sequent calculus $LR^t_{\rightarrow}$ from Section 4.3.

$LR^t_{\rightarrow;}$ is a *consecution calculus* and its structural connective is $;$, which appears in the label for the calculus. The notion of structures is as in Definition 5.1; a consecution is a structure followed by $\vdash$ and a wff. A consecution is provable if it has a *proof* in the same sense as before. $\mathcal{A}$ is a *theorem* iff the consecution $\boldsymbol{t} \vdash \mathcal{A}$ is provable. The axiom and rules are as follows.

$$\mathcal{A} \vdash \mathcal{A} \ \text{id}$$

$$\frac{\mathfrak{A} \vdash \mathcal{A} \quad \mathfrak{B}[\mathcal{B}] \vdash \mathcal{C}}{\mathfrak{B}[\mathcal{A} \rightarrow \mathcal{B}; \mathfrak{A}] \vdash \mathcal{C}} \rightarrow\vdash \qquad \frac{\mathfrak{A}; \mathcal{A} \vdash \mathcal{B}}{\mathfrak{A} \vdash \mathcal{A} \rightarrow \mathcal{B}} \vdash\rightarrow$$

[10] This is the original axiomatization by Church in [63], though he did not use the label $R_{\rightarrow}$.

$$\frac{\mathfrak{A}[\mathfrak{B};(\mathfrak{C};\mathfrak{D})] \vdash \mathcal{A}}{\mathfrak{A}[\mathfrak{B};\mathfrak{C};\mathfrak{D}] \vdash \mathcal{A}}\ \mathsf{B}\vdash \qquad \frac{\mathfrak{A}[\mathfrak{B};\mathfrak{C};\mathfrak{C}] \vdash \mathcal{A}}{\mathfrak{A}[\mathfrak{B};\mathfrak{C}] \vdash \mathcal{A}}\ \mathsf{W}\vdash \qquad \frac{\mathfrak{A}[\mathfrak{B};\mathfrak{C};\mathfrak{D}] \vdash \mathcal{A}}{\mathfrak{A}[\mathfrak{B};\mathfrak{D};\mathfrak{C}] \vdash \mathcal{A}}\ \mathsf{C}\vdash$$

$$\frac{\mathfrak{A}[\mathfrak{B}] \vdash \mathcal{A}}{\mathfrak{A}[\boldsymbol{t};\mathfrak{B}] \vdash \mathcal{A}}\ \mathsf{Kl}_t\vdash \qquad \frac{\mathfrak{A}[\boldsymbol{t};\boldsymbol{t}] \vdash \mathcal{A}}{\mathfrak{A}[\boldsymbol{t}] \vdash \mathcal{A}}\ \mathsf{M}_t\vdash$$

There are important differences between $LR^t_{\to;}$ and $LR^t_{\to}$, on the one hand, and $LR^t_{\to;}$ and LQ, on the other.

First of all, a wff $\mathcal{A}$ is a theorem of $LR^t_{\to}$, when the sequent $\vdash \mathcal{A}$ is provable. In $LR^t_{\to}$, there is $\boldsymbol{t}$, nonetheless, we did not require $\boldsymbol{t}$ to appear on the left-hand side of the turnstile for the formula on the right to be provable. This means that we need a *special axiom* for $\boldsymbol{t}$ such as $\vdash \boldsymbol{t}$ in $LR^t_{\to}$. Of course, in $LR^t_{\to;}$, we cannot have such an axiom, because in $LR^t_{\to;}$ there is no empty structure to fill the left-hand side. In LQ, we did not have $\boldsymbol{t}$ and we could not talk about a formula being a theorem.

The antecedents in $LR^t_{\to}$ and in $LR^t_{\to;}$ are *different data types*—multisets and structures, respectively. Thus $LR^t_{\to}$ has only one structural rule, whereas $LR^t_{\to;}$ has five structural rules. In contrast, LQ has none.

Furthermore, $LR^t_{\to;}$ contains the *special contraction rule* $(\mathsf{M}_t\vdash)$, which is not an instance of $(\mathsf{W}\vdash)$. This raises the question about the mismatch between the structural rules $(W\vdash)$ in $LR^t_{\to}$ and $(\mathsf{W}\vdash)$ in $LR^t_{\to;}$. Since there are no empty structures in $LR^t_{\to;}$, $(\mathsf{W}\vdash)$ is more restricted than $(W\vdash)$, in which Γ (not including the two occurrences of $\mathcal{A}$) may be empty. There is a similar discrepancy between the $(\mathsf{C}\vdash)$ rule in $LR_{\to}$ (as we first formulated it in Section 4.3) and $(\mathsf{C}\vdash)$ in $LR^t_{\to;}$. Again, the $(\mathsf{C}\vdash)$ rule appears to be more restricted, because the structure $\mathfrak{B}$ cannot be empty. This restriction does not prevent us from proving the principal type schema of C, which is a theorem of $R^t_{\to}$, because the $(\mathsf{C}\vdash)$ rule in $LR^t_{\to;}$ corresponds exactly to that theorem. We may conjecture that the difference can be explained by the persistent presence of $\boldsymbol{t}$ in proofs in $LR^t_{\to;}$.

Let us consider the following segments of proofs.

$$\cfrac{\cfrac{\vdots}{\mathfrak{A}[\mathfrak{B};\mathfrak{B}] \vdash \mathcal{A}}}{\cfrac{\mathfrak{A}[\boldsymbol{t};\mathfrak{B};\mathfrak{B}] \vdash \mathcal{A}}{\mathfrak{A}[\boldsymbol{t};\mathfrak{B}] \vdash \mathcal{A}}\ \mathsf{W}\vdash}\ \mathsf{Kl}_t\vdash \qquad \cfrac{\cfrac{\vdots}{\mathfrak{A}[\mathfrak{B};\mathfrak{C}] \vdash \mathcal{A}}}{\cfrac{\mathfrak{A}[\boldsymbol{t};\mathfrak{B};\mathfrak{C}] \vdash \mathcal{A}}{\mathfrak{A}[\boldsymbol{t};\mathfrak{C};\mathfrak{B}] \vdash \mathcal{A}}\ \mathsf{C}\vdash}\ \mathsf{Kl}_t\vdash$$

In both cases, we added a $\boldsymbol{t}$ into the proof. We may contemplate whether an extra $\boldsymbol{t}$ is harmful. The next lemma answers the question.

Lemma 5.19. *In a proof of a wff $\mathcal{A}$ in $LR^t_{\to;}$, $(\mathsf{M}\vdash)$ and $(\mathsf{T}\vdash)$ are* admissible rules.

Proof: The two rules are the analogs of the combinatory rules in structurally free logics—except that no combinators are inserted into the lower consecutions.

$$\frac{\mathfrak{A}[\mathfrak{B};\mathfrak{B}] \vdash \mathcal{A}}{\mathfrak{A}[\mathfrak{B}] \vdash \mathcal{A}}\ \mathsf{M}\vdash \qquad \frac{\mathfrak{A}[\mathfrak{B};\mathfrak{C}] \vdash \mathcal{A}}{\mathfrak{A}[\mathfrak{C};\mathfrak{B}] \vdash \mathcal{A}}\ \mathsf{T}\vdash$$

1. The above two proof chunks are sufficient to show that with a t added on the left the contraction of the $\mathfrak{B}$'s and the permutation of $\mathfrak{B}$ and $\mathfrak{C}$ can be carried out.
2. The stipulation that we are dealing with a proof of a formula guarantees that somewhere in the proof a t is introduced, and finally a t will appear on the left-hand side of the $\vdash$. We show that the extra t we introduced to simulate $(\mathsf{M}\vdash)$ and $(\mathsf{T}\vdash)$ withers away with the help of another t.

The following two derivation fragments suffice.

$$\cfrac{\cfrac{\cfrac{\cfrac{\vdots}{\mathfrak{A}[\mathfrak{B};t] \vdash \mathcal{A}}}{\mathfrak{A}[t;\mathfrak{B};t] \vdash \mathcal{A}}\ \mathsf{KI}_t\vdash}{\mathfrak{A}[t;t;\mathfrak{B}] \vdash \mathcal{A}}\ \mathsf{C}\vdash}{\mathfrak{A}[t;\mathfrak{B}] \vdash \mathcal{A}}\ \mathsf{M}_t\vdash \qquad \cfrac{\cfrac{\cfrac{\cfrac{\mathfrak{A}[\mathfrak{B};(t;\mathfrak{C})] \vdash \mathcal{A}}{\mathfrak{A}[\mathfrak{B};t;\mathfrak{C}] \vdash \mathcal{A}}\ \mathsf{B}\vdash}{\mathfrak{A}[t;\mathfrak{B};t;\mathfrak{C}] \vdash \mathcal{A}}\ \mathsf{KI}_t\vdash}{\mathfrak{A}[t;t;\mathfrak{B};\mathfrak{C}] \vdash \mathcal{A}}\ \mathsf{C}\vdash}{\mathfrak{A}[t;\mathfrak{B};\mathfrak{C}] \vdash \mathcal{A}}\ \mathsf{M}_t\vdash$$

What we have shown is that if t is to the right from a structure, then it can be moved to the left of that structure. Furthermore, if t is grouped with another structure as in $\mathfrak{B};(t;\mathfrak{C})$, then one application of the rule $(\mathsf{B}\vdash)$ gives $\mathfrak{B};t;\mathfrak{C}$. Then $\mathfrak{B}$ and t may be permuted as before.

One might think that we have to show also that $t;(\mathfrak{C};\mathfrak{B})$ may be obtained from $t;\mathfrak{C};\mathfrak{B}$. However, it is important to remember that we are considering only proofs with an end sequent of the form $t \vdash \mathcal{A}$, for some $\mathcal{A}$. In $LR^t_{\to;}$, there are few ways to introduce ; and to create new formulas. If $t;\mathfrak{B};\mathfrak{C}$ occurs inside a structure, let us say, as in $\mathfrak{A};(t;\mathfrak{B};\mathfrak{C})$, then we do not need to obtain $t;\mathfrak{A};(\mathfrak{B};\mathfrak{C})$, because in a proof of a theorem, $\mathfrak{B}$ and $\mathfrak{C}$ must be disassociated. The $(\mathsf{KI}_t\vdash)$ rule is applicable to the same structure, no matter if there is or there isn't a t already in the consecution. The structural rules remain applicable too, possibly, taking $t;\mathfrak{B}$ in place of $\mathfrak{B}$. There is no change in applications of the rule $(\vdash\to)$, when the antecedent structure ends in a wff. However, a complex structure like $\mathfrak{A};(\mathfrak{B};\mathfrak{C})$ must be turned into a left-associated one before $\mathfrak{C}$ or $\mathfrak{B}$ (if they are wff's) can be moved to the right as a subformula of an $\to$ formula. When that happens, t can be moved further to the left as in the above derivation. qed

$R_\to$ has been axiomatized by *four* sets of independent axioms in [7, §8.4]. t can be added by two rules to each of the axiom systems: if $\mathcal{A}$ is provable, then so is $t \to \mathcal{A}$, and the other way around.

$LR^t_{\to;}$ matches $R_{\to 1}$ in [7, §8.4]. We could define slightly modified versions of $LR^t_{\to;}$ in parallel to the other axiomatizations. This possibility arises, because structures—unlike sequences or multisets—can make *finer distinctions* between structural rules. We introduce one more rule, and then we will select subsets of rules.

The *prefixing rule*, $(\mathsf{B}'\vdash)$ is the following. (This rule is an instance of $(\mathsf{Z}\vdash)$ with B' in place of Z from the previous section, except that the combinator does not appear in the lower consecution.)

$$\frac{\mathfrak{A}[\mathfrak{B};(\mathfrak{C};\mathfrak{D})]\vdash\mathcal{A}}{\mathfrak{A}[\mathfrak{C};\mathfrak{B};\mathfrak{D}]\vdash\mathcal{A}}\ \mathsf{B}'\vdash$$

The following lemma is, in effect, in [41, §3].

Lemma 5.20. *In a proof of a wff $\mathcal{A}$ in $LR^{t}_{\to;}$, $(\mathsf{B}'\vdash)$ is an* admissible rule.

Proof: Consider the following proof segment.

$$\cfrac{\cfrac{\cfrac{\cfrac{\vdots}{\mathfrak{A}[\mathfrak{C};(\mathfrak{B};\mathfrak{D})]\vdash\mathcal{A}}}{\mathfrak{A}[\mathfrak{C};\mathfrak{B};\mathfrak{D}]\vdash\mathcal{A}}\ \mathsf{B}\vdash}{\mathfrak{A}[t;\mathfrak{C};\mathfrak{B};\mathfrak{D}]\vdash\mathcal{A}}\ \mathsf{KI}_t\vdash}{\mathfrak{A}[t;\mathfrak{B};\mathfrak{C};\mathfrak{D}]\vdash\mathcal{A}}\ \mathsf{C}\vdash$$

Once again, $\mathfrak{C};(\mathfrak{B};\mathfrak{D})$ may be embedded into $\mathfrak{A}$ in different ways, however, an argument that is exactly as in the previous admissibility proof shows that the additional t does not cause any problems in a proof of a wff. qed

There are *six axioms* in the four axiom systems. We list them—together with the combinator of which they are the principal types.

(A1) $\mathcal{A}\to\mathcal{A}$ [I]

(A2) $(\mathcal{A}\to\mathcal{B})\to(\mathcal{C}\to\mathcal{A})\to\mathcal{C}\to\mathcal{B}$ [B]

(A3) $(\mathcal{A}\to\mathcal{B})\to(\mathcal{B}\to\mathcal{C})\to\mathcal{A}\to\mathcal{C}$ [B′]

(A4) $(\mathcal{A}\to\mathcal{A}\to\mathcal{B})\to\mathcal{A}\to\mathcal{B}$ [W]

(A5) $(\mathcal{A}\to\mathcal{B}\to\mathcal{C})\to\mathcal{B}\to\mathcal{A}\to\mathcal{C}$ [C]

(A6) $\mathcal{A}\to(\mathcal{A}\to\mathcal{B})\to\mathcal{B}$ [T]

In each axiomatization, the rule is detachment, that is, $\mathcal{A}\to\mathcal{B}$ and $\mathcal{A}$ imply $\mathcal{B}$. $R_{\to 2}$ includes (A1), (A3), (A4) and (A6); $R_{\to 3}$ consists of (A1), (A3), (A4) and (A5) and $R_{\to 4}$ has (A1), (A2), (A4) and (A6). The four axiom systems show that prefixing and suffixing can be paired with specialized assertion and permutation in every possible way—in the context of self-implication and contraction.

We define the parallel consecution calculi in the expected way. Each of them contains (id), $(\to\vdash)$, $(\vdash\to)$, $(\mathsf{KI}_t\vdash)$ and $(\mathsf{M}_t\vdash)$. The other structural rules in $LR^{t}_{\to;2}$ are $(\mathsf{W}\vdash)$, $(\mathsf{B}'\vdash)$ and $(\mathsf{T}\vdash)$. $LR^{t}_{\to;3}$ contains $(\mathsf{W}\vdash)$, $(\mathsf{B}'\vdash)$ and $(\mathsf{C}\vdash)$. Lastly, $LR^{t}_{\to;4}$ is like $LR^{t}_{\to;}$ save that $(\mathsf{T}\vdash)$ replaces $(\mathsf{C}\vdash)$.

Exercise 5.3.1. Prove that each of the four consecution calculi have as their theorems (A1)–(A6). [Hint: Six small proofs do the job.]

Exercise* 5.3.2. Consider each of the four consecution calculi that we have defined. In each case, determine whether the rules that belong to the other calculi are or are not admissible. Prove your claims.

Exercise 5.3.3. Prove each of the following four formulas in each calculus. (1) $(\mathcal{A} \to \mathcal{B} \to \mathcal{C}) \to (\mathcal{A} \to \mathcal{B}) \to \mathcal{A} \to \mathcal{C}$, (2) $(\mathcal{A} \to \mathcal{B}) \to (\mathcal{A} \to \mathcal{B} \to \mathcal{C}) \to \mathcal{A} \to \mathcal{C}$, (3) $(\mathcal{A} \to \mathcal{B}) \to (\mathcal{C} \to (\mathcal{A} \to \mathcal{B}) \to \mathcal{D}) \to \mathcal{C} \to \mathcal{D}$ and (4) $((\mathcal{A} \to \mathcal{A}) \to \mathcal{B}) \to \mathcal{B}$. [Hint: These wff's are sometimes called the self-distribution of implication on the major and minor, restricted conditioned modus ponens and specialized assertion. Although consecution calculi confine formulas more tightly than sequent calculi do, the proofs are not unique. Somewhat curiously, 15 proofs suffice.]

The next theorem expands Theorem 3.2 in [41].

Theorem 5.21. *The single cut rule is* admissible *in $LR^t_{\to;}$, as well as in its three variants that we defined above.*

Proof: The single cut rule has the same form as in LQ or in the structurally free logics.

The theorem may be proved by induction, namely, by a triple induction on δ, μ and ϱ, and by an induction on χ, the height of a proof tree of $\mathfrak{A} \vdash t$. We do not go into the details here. The general structure of such proofs is given in Section 7.3. qed

Exercise 5.3.4. Consider the proof in Section 7.3, and explain what modifications are needed specifically for the four variants of the $LR^t_{\to;}$ calculus.

We may notice that each of the four $LR^t_{\to;}$ calculi contained the two structural rules, $(\mathsf{KI}_t \vdash)$ and $(\mathsf{M}_t \vdash)$. On the other hand, each axiom system contained $\mathcal{A} \to \mathcal{A}$, but we seem not to have added a corresponding rule to the calculi. The explanation, of course, is that these rules by themselves do not correspond to an axiom. However, self-implication looks very much like the axiom (id), and for $\mathcal{A} \to \mathcal{A}$ to be a theorem of $LR^t_{\to;}$, we have to be able to get a t into the consecution and introduce $\to$ on the right. $(\mathsf{KI}_t \vdash)$ is the rule that allows t to be inserted in front of $\mathcal{A}$, and the $(\vdash \to)$ rule is in consonance with t appearing on the left of a structure.

$(\mathsf{KI}_t \vdash)$ is used similarly for each implicational theorem. t itself turns out to be a theorem in virtue of an instance of the axiom (id), $t \vdash t$.

t's *primary role* is to be a visible and countable place holder. The wff $((((\mathcal{A} \to \mathcal{A}) \to \mathcal{B}) \to \mathcal{B}) \to (\mathcal{A} \to \mathcal{A} \to \mathcal{B})) \to \mathcal{A} \to \mathcal{B}$ is a theorem of $R^t_{\to}$. Let us assume that we are using $LR^t_{\to;4}$. We could build a proof bottom-up (as if we were "applying" the rules backward) as follows.

$$\dfrac{\dfrac{\dfrac{\dfrac{\vdots}{t; (\mathcal{A}\to\mathcal{A})\to\mathcal{B}\vdash\mathcal{B}}}{t\vdash((\mathcal{A}\to\mathcal{A})\to\mathcal{B})\to\mathcal{B}}\,{\scriptstyle\vdash\to} \qquad \dfrac{\dfrac{\vdots}{\mathcal{A}\to\mathcal{A}\to\mathcal{B};\mathcal{A};\mathcal{A}\vdash\mathcal{B}}}{\mathcal{A}\to\mathcal{A}\to\mathcal{B};\mathcal{A}\vdash\mathcal{B}}\,{\scriptstyle W\vdash}}{\dfrac{(((\mathcal{A}\to\mathcal{A})\to\mathcal{B})\to\mathcal{B})\to(\mathcal{A}\to\mathcal{A}\to\mathcal{B});t;\mathcal{A}\vdash\mathcal{B}}{t;(((\mathcal{A}\to\mathcal{A})\to\mathcal{B})\to\mathcal{B})\to(\mathcal{A}\to\mathcal{A}\to\mathcal{B});\mathcal{A}\vdash\mathcal{B}}\,{\scriptstyle T\vdash}}\,{\scriptstyle\to\vdash}}{t\vdash((((\mathcal{A}\to\mathcal{A})\to\mathcal{B})\to\mathcal{B})\to(\mathcal{A}\to\mathcal{A}\to\mathcal{B}))\to\mathcal{A}\to\mathcal{B}}\,{\scriptstyle\vdash\to\text{'s}}$$

The chunk of the proof demystifies the wff by showing how it is obtained from splicing together contraction and specialized assertion with the help of t. Incidentally, it may be useful to emphasize that the wff does not say that specialized assertion implies contraction, which is *not* a theorem of $R^t_{\to}$.

Exercise 5.3.5. Start to build a proof from the bottom up for the wff $(((\mathcal{A}\to\mathcal{A})\to\mathcal{B})\to\mathcal{B})\to(\mathcal{C}\to\mathcal{C}\to\mathcal{D})\to\mathcal{C}\to\mathcal{D}$ and pinpoint where the attempt fails. [Hint: t may be thinned into a consecution, but an arbitrary theorem cannot.]

The inclusion of t into the consecution calculi forces us to scrutinize t's behavior, including its interaction with the cut rule. But another aspect of t's behavior is its "omnipresence." In order to algebraize $LR^t_{\to;}$, we have to prove that provably equivalent formulas can stand in each other's place. In other words, we have to prove the *replacement theorem*—assuming that the equivalence relation we want to use obtains between $\mathcal{A}$ and $\mathcal{B}$ when $t\vdash\mathcal{A}\to\mathcal{B}$ and $t\vdash\mathcal{B}\to\mathcal{A}$ are both provable.

Theorem 5.22. *If* $t\vdash\mathcal{A}\to\mathcal{B}$ *and* $t\vdash\mathcal{B}\to\mathcal{A}$ *are both provable, then* $t\vdash(\mathcal{A}\to\mathcal{C})\to\mathcal{B}\to\mathcal{C}$ *and* $t\vdash(\mathcal{C}\to\mathcal{A})\to\mathcal{C}\to\mathcal{B}$ *are* provable.

Proof: Notice that $\mathcal{A}$ and $\mathcal{B}$ occur symmetrically in $t\vdash\mathcal{A}\to\mathcal{B}$ and $t\vdash\mathcal{B}\to\mathcal{A}$. Therefore, if we can show the provability of the consecutions in the statement of the theorem, then we know that $\mathcal{A}$ and $\mathcal{B}$ can be switched in those formulas without losing provability.

We use the cut rule in the following proof, since according to Theorem 5.21, cut is an admissible rule. One application of the cut rule allows us to *invert* the $(\vdash\to)$ rule, which must have been the last step in a proof of $t\vdash\mathcal{A}\to\mathcal{B}$.

$$\dfrac{\dfrac{\vdots}{t\vdash\mathcal{A}\to\mathcal{B}} \qquad \dfrac{\vdots}{\mathcal{A}\to\mathcal{B};\mathcal{A}\vdash\mathcal{B}}}{t;\mathcal{A}\vdash\mathcal{B}}$$

The application of the cut is allowable here, because we assumed that $\mathcal{B}\to\mathcal{A}$ and its converse are theorems. The proof of the other premise is obvious.

We prove $(\mathcal{C} \to \mathcal{A}) \to \mathcal{C} \to \mathcal{B}$.

$$\dfrac{\dfrac{\dfrac{\mathcal{C}\vdash\mathcal{C}\qquad t;\mathcal{A}\vdash\mathcal{B}}{t;(\mathcal{C}\to\mathcal{A};\mathcal{C})\vdash\mathcal{B}}\,{\to}{\vdash}}{t;\mathcal{C}\to\mathcal{A};\mathcal{C}\vdash\mathcal{B}}\,\mathsf{B}{\vdash}}{t\vdash(\mathcal{C}\to\mathcal{A})\to\mathcal{C}\to\mathcal{B}}\,{\vdash}{\to}\text{'s}$$

Note that we have to use the structural rule $(\mathsf{B}\vdash)$ to introduce $\to$, because of the t on the left in the antecedent. Of course, there is no problem, because this is a rule of $LR^t_{\to;}$.

In the sequent calculus $LR^t_{\to}$ in Section 4.3, the cut rule can be used to delete an occurrence of t. However, a similar argument does not work now, because $\vdash t$ is not an axiom of the consecution calculus $LR^t_{\to;}$. That is, even if $\mathcal{A} \to \mathcal{B}$ is provable, the consecution $\mathcal{A} \vdash \mathcal{B}$ may not be assumed to be provable by cut. $LR^t_{\to;}$ includes $(\mathsf{C}\vdash)$ (not $(\mathsf{T}\vdash)$), which requires us to use $(\mathsf{M}_t\vdash)$ and $(\mathsf{Kl}_t\vdash)$ too in the next proof.

$$\dfrac{\dfrac{\dfrac{\dfrac{t;\mathcal{B}\vdash\mathcal{A}\qquad \dfrac{\mathcal{C}\vdash\mathcal{C}}{t;\mathcal{C}\vdash\mathcal{C}}\,\mathsf{Kl}_t{\vdash}}{t;(\mathcal{A}\to\mathcal{C};(t;\mathcal{B}))\vdash\mathcal{C}}\,{\to}{\vdash}}{t;\mathcal{A}\to\mathcal{C};t;\mathcal{B}\vdash\mathcal{C}}\,\mathsf{B}{\vdash}\text{'s}}{t;t;\mathcal{A}\to\mathcal{C};\mathcal{B}\vdash\mathcal{C}}\,\mathsf{C}{\vdash}}{t;\mathcal{A}\to\mathcal{C};\mathcal{B}\vdash\mathcal{C}}\,\mathsf{M}_t{\vdash}$$

qed

Exercise 5.3.6. Prove $(\mathcal{C} \to \mathcal{A}) \to \mathcal{C} \to \mathcal{B}$ using the rule $(\mathsf{B}'\vdash)$—assuming, of course, that $\mathcal{A}$ and $\mathcal{B}$ are provably equivalent.

Exercise 5.3.7. Prove $(\mathcal{A} \to \mathcal{C}) \to \mathcal{B} \to \mathcal{C}$ in $LR^t_{\to;i}$, for $i \in \{2,3,4\}$.

The role of the rule $(\mathsf{M}_t\vdash)$ becomes even clearer when we try to prove that the consecution calculus can *emulate detachment*. If we have empty left-hand sides and theoremhood is simply a wff being provable on the right-hand side, then $\vdash \mathcal{A} \to \mathcal{B}$ and $\vdash \mathcal{A}$ has to yield $\vdash \mathcal{B}$. Now, we have to show that from proofs of $t \vdash \mathcal{A} \to \mathcal{B}$ and $t \vdash \mathcal{A}$, we can construct a proof of $t \vdash \mathcal{B}$. We can proceed to build the following proof.

$$\dfrac{\dfrac{t\vdash\mathcal{A}\qquad \dfrac{t\vdash\mathcal{A}\to\mathcal{B}\qquad \mathcal{A}\to\mathcal{B};\mathcal{A}\vdash\mathcal{B}}{t;\mathcal{A}\vdash\mathcal{B}}\,\text{cut}}{t;t\vdash\mathcal{B}}\,\text{cut}}{t\vdash\mathcal{B}}\,\mathsf{M}_t{\vdash}$$

Lastly, we note that although t is in the language of each consecution calculus for $R^t_{\to}$, these calculi capture $R_{\to}$ too in the sense that in a proof of a theorem of $R_{\to}$, all the occurrences of t are related to structural modifications in the antecedent and to t filling in the left-hand side of the $\vdash$ in the end consecution.

After looking at the logic of pure relevant implication, now we turn to the logic of *pure entailment*. The logic of entailment E gave the title to the books [7] and [8]. E, together with its fragments and extensions, has been investigated from many angles, and it seems to have been *the favored* or *most liked* logic by A. R. Anderson and N. D. Belnap. The implicational fragment itself will reveal why it is more difficult to understand and use E than R.

We have listed on page 167 some axiomatizations of $R_\to$, with $\boldsymbol{t}$ added by rules. We list six formulas that enter into axiomatizations of $E_\to$ and $E^{\boldsymbol{t}}_\to$. $E_{\to 1}$–$E_{\to 4}$ are independent axiom systems from [7, §8.4]. (For the sake of easy reference, we continue the numbering of the formulas from p. 167.)

$$\text{(A7)}\ \ (\mathcal{A} \to \mathcal{B} \to \mathcal{C}) \to (\mathcal{A} \to \mathcal{B}) \to \mathcal{A} \to \mathcal{C} \qquad [\mathsf{S}]$$

$$\text{(A8)}\ \ (\mathcal{A} \to \mathcal{B}) \to ((\mathcal{A} \to \mathcal{B}) \to \mathcal{C}) \to \mathcal{C} \qquad [1]$$

$$\text{(A9)}\ \ ((\mathcal{A} \to \mathcal{A}) \to \mathcal{B}) \to \mathcal{B} \qquad [4]$$

$$\text{(A10)}\ \ (\mathcal{A} \to (\mathcal{B} \to \mathcal{C}) \to \mathcal{D}) \to (\mathcal{B} \to \mathcal{C}) \to \mathcal{A} \to \mathcal{D} \qquad [2]$$

$$\text{(A11)}\ \ (\mathcal{A} \to \mathcal{B}) \to (\mathcal{B} \to \mathcal{C}) \to ((\mathcal{A} \to \mathcal{C}) \to \mathcal{D}) \to \mathcal{D} \qquad [5]$$

$$\text{(A12)}\ \ (\boldsymbol{t} \to \boldsymbol{t}) \to \boldsymbol{t} \qquad [\Box \boldsymbol{t}]$$

Some of the above formulas have traditional names. The numbers are recently introduced labels for combinators. None of these combinators is widely known by a label or name, not the least, because they are *not proper* combinators. This should not cause panic, since they do not introduce new variables; still, to define their reduction pattern, we have to assume some other combinators such as B' and I.

(A8) and (A9) are refinements of *assertion*, which is (A6), but in two different ways. (A8) is *restricted assertion*, whereas (A9) is *specialized assertion*. (A10)'s relation to (A5) is as (A8)'s to (A6), namely, (A10) is *restricted permutation*. (A11) does not seem to have a particular name, but it has a certain resemblance to *suffixing*, (A3). Finally, (A12) expresses the *necessity of* $\boldsymbol{t}$, when the necessity of $\mathcal{A}$ is defined as $(\mathcal{A} \to \mathcal{A}) \to \mathcal{A}$.

The four independent axiom systems, which are defined in [7, §8], are the following sets of axioms together with detachment. $E_{\to 1} = \{\,\text{(A1)}, \text{(A3)}, \text{(A7)}, \text{(A8)}\,\}$; $E_{\to 2} = \{\,\text{(A3)}, \text{(A4)}, \text{(A9)}\,\}$; $E_{\to 3} = \{\,\text{(A1)}, \text{(A4)}, \text{(A11)}\,\}$; $E_{\to 4} = \{\,\text{(A1)}, \text{(A3)}, \text{(A4)}, \text{(A10)}\,\}$.

Exercise* 5.3.8. Prove that the axiom systems are equivalent. [Hint: This book is not about axiomatic presentations of logics, but doing (part of) this exercise will surely convince you of the advantages of sequent calculi.]

Due to the peculiarities of E, $\boldsymbol{t}$'s addition requires the axiom (A12), which we labeled as $(\Box \boldsymbol{t})$, and two rules. Namely, if $E_\to \vdash \mathcal{A}$, then $E_\to \vdash \boldsymbol{t} \to \mathcal{A}$, and vice versa.

The following two formulas are also theorems of $E_\to$.

$$\text{(A13)}\ \ (\mathcal{A} \to \mathcal{B}) \to (\mathcal{A} \to \mathcal{B} \to \mathcal{C}) \to \mathcal{A} \to \mathcal{C} \qquad [\mathsf{S}']$$

(A14) $(\mathcal{A} \to \mathcal{B}) \to (\mathcal{C} \to (\mathcal{A} \to \mathcal{B}) \to \mathcal{D}) \to \mathcal{C} \to \mathcal{D}$ [3]

(A13) is the *self-distribution of implication on the minor*. (A14) is another restricted formula; it is called *restricted conditioned modus ponens*. Conditioned modus ponens is the wff $\mathcal{A} \to (\mathcal{B} \to \mathcal{A} \to \mathcal{C}) \to \mathcal{B} \to \mathcal{C}$. The condition is $\mathcal{A}$, and in (A14) it is restricted, because it cannot be an atomic formula, rather it must be an implicational formula. The term "restricted" is used coherently to mean that a particular subformula has to contain a $\to$.

"Specialized" also has a precise meaning. It refers to a formula, which may be obtained from the unspecialized formula, by a detachment of the principal type of I, that is, of $\mathcal{A} \to \mathcal{A}$. For instance, assertion is $\mathcal{A} \to (\mathcal{A} \to \mathcal{B}) \to \mathcal{B}$. An instance of this formula that allows us to detach from it self-implication is $(\mathcal{A} \to \mathcal{A}) \to ((\mathcal{A} \to \mathcal{A}) \to \mathcal{B}) \to \mathcal{B}$. Thus specialized assertion is $((\mathcal{A} \to \mathcal{A}) \to \mathcal{B}) \to \mathcal{B}$.

Assertion itself can be viewed as a specialized formula, though it is not usually considered as such. However, we want to continue to emphasize connections to combinators, and so it is useful to go beyond the traditional understanding. Assertion is *specialized permutation*. In order to detach self-implication from (A5), we have to unify $\mathcal{A}$ and $\mathcal{B} \to \mathcal{C}$. The consequent of $((\mathcal{B} \to \mathcal{C}) \to (\mathcal{B} \to \mathcal{C})) \to \mathcal{B} \to (\mathcal{B} \to \mathcal{C}) \to \mathcal{C}$ is assertion. Indeed, in combinatory logic, T is definable as CI.

So far the only interesting example of an improper combinator was Y, the fixed point combinator. There is no guarantee that when combining proper combinators into a combinatory term, the result is a proper combinator. The preceding claim is an obvious consequence of Y's definability as a term over the base $\{\mathsf{S}, \mathsf{K}\}$, for instance. Now we have some other improper combinators to consider. ($\mathsf{1}, \mathsf{2}$ and $\mathsf{3}$ were introduced in [32] and $\mathsf{4}$ in [34].)

Definition 5.23. (1, 2, 3, 4, 5) Recall that B' and I are the combinators with axioms $\mathsf{B}'xyz \rhd y(xz)$ and $\mathsf{I}x \rhd x$. The *axioms* for $\mathsf{1}, \mathsf{2}, \mathsf{3}, \mathsf{4}$ and $\mathsf{5}$ are as follows.

(1) $\mathsf{1}xy \rhd y(\mathsf{B}'\mathsf{I}x)$ (2) $\mathsf{2}xyz \rhd xz(\mathsf{B}'\mathsf{I}y)$ (3) $\mathsf{3}xyz \rhd yz(\mathsf{B}'\mathsf{I}x)$

(4) $\mathsf{4}x \rhd x\mathsf{I}$ (5) $\mathsf{5}xyz \rhd z(\mathsf{B}'xy)$

Given that all these combinators are improper, the corresponding combinatory rules for $\mathsf{1}$ and $\mathsf{5}$ look like the following in structurally free logic.

$$\frac{\mathfrak{A}[\mathfrak{B}; (\mathsf{B}'; \mathsf{I}; \mathfrak{C})] \vdash \mathcal{A}}{\mathfrak{A}[\mathsf{1}; \mathfrak{C}; \mathfrak{B}] \vdash \mathcal{A}}\ \mathsf{1}\vdash \qquad \frac{\mathfrak{A}[\mathfrak{D}; (\mathsf{B}'; \mathfrak{B}; \mathfrak{C})] \vdash \mathcal{A}}{\mathfrak{A}[\mathsf{5}; \mathfrak{B}; \mathfrak{C}; \mathfrak{D}] \vdash \mathcal{A}}\ \mathsf{5}\vdash$$

Omitting the combinators turns these rules into structural rules, but into structural rules for other combinators than $\mathsf{1}$ or $\mathsf{5}$.

$$\frac{\mathfrak{A}[\mathfrak{B}; \mathfrak{C}] \vdash \mathcal{A}}{\mathfrak{A}[\mathfrak{C}; \mathfrak{B}] \vdash \mathcal{A}}\ \mathsf{CI}\vdash \qquad \frac{\mathfrak{A}[\mathfrak{D}; (\mathfrak{B}; \mathfrak{C})] \vdash \mathcal{A}}{\mathfrak{A}[\mathfrak{B}; \mathfrak{C}; \mathfrak{D}] \vdash \mathcal{A}}\ \mathsf{B'B'C}\vdash$$

Running ahead a bit, let us assume that the connective that naturally corresponds to ; in $LR^t_{\to ;}$ is added to the language of the logic. (We will add fusion, $\circ$, to the implicational logics in the next section.) The desired interaction between $\to$ and $\circ$ is that the former is the *right residual* of the latter. In R, we can omit the adjective "right," but in E we cannot. If carried out properly, then the addition of $\circ$ is benign in the sense that the resulting extension is *conservative*. Let us suppose for a moment that two rules are added to $E^t_\to$ to introduce and to eliminate $\circ$. If $E^{\circ t}_\to \vdash \mathcal{A} \to \mathcal{B} \to \mathcal{C}$, then $E^{\circ t}_\to \vdash (\mathcal{A} \circ \mathcal{B}) \to \mathcal{C}$.

We can show that $\boldsymbol{t}$ has certain properties that are reflected by wff's involving $\circ$. First of all, $\mathcal{A} \to \mathcal{A}$ is a theorem of $E^t_\to$, hence, so is $\boldsymbol{t} \to \mathcal{A} \to \mathcal{A}$. Then $E^{\circ t}_\to$ proves $(\boldsymbol{t} \circ \mathcal{A}) \to \mathcal{A}$. Algebraically, this means that $\boldsymbol{t}$ is a *lower left identity* for $\circ$. $\boldsymbol{t}$ is also an *upper left identity*, because $\mathcal{A} \to (\boldsymbol{t} \circ \mathcal{A})$ is provable. $(\boldsymbol{t} \circ \mathcal{A}) \to (\boldsymbol{t} \circ \mathcal{A})$ is an instance of self-implication, from which we get $\boldsymbol{t} \to \mathcal{A} \to (\boldsymbol{t} \circ \mathcal{A})$, by a rule for fusion. But $\boldsymbol{t}$ can be detached, because it is a theorem; hence, $E^{\circ t}_\to \vdash \mathcal{A} \to (\boldsymbol{t} \circ \mathcal{A})$.

If we have permutation, that is, axiom (A5) in $R_\to$, then it is easy to see that $\boldsymbol{t}$ is a lower and upper right identity. From $\boldsymbol{t} \to \mathcal{A} \to \mathcal{A}$ we obtain $\mathcal{A} \to \boldsymbol{t} \to \mathcal{A}$, by modus ponens from a suitable instance of (A5). Then by a fusion rule we have, $(\mathcal{A} \circ \boldsymbol{t}) \to \mathcal{A}$. For $\mathcal{A} \to (\mathcal{A} \circ \boldsymbol{t})$, we start with $(\mathcal{A} \circ \boldsymbol{t}) \to (\mathcal{A} \circ \boldsymbol{t})$, and insert a permutation into the series of steps that led to $\mathcal{A} \to (\boldsymbol{t} \circ \mathcal{A})$.

E, on the other hand, does not have (A5) as an axiom or as a theorem. Still, $E^{\circ t}_\to \vdash \mathcal{A} \to \boldsymbol{t} \to (\mathcal{A} \circ \boldsymbol{t})$. $\Box \boldsymbol{t}$ may be detached from $((\boldsymbol{t} \to \boldsymbol{t}) \to \boldsymbol{t}) \to (\boldsymbol{t} \to (\mathcal{A} \circ \boldsymbol{t})) \to (\boldsymbol{t} \to \boldsymbol{t}) \to (\mathcal{A} \circ \boldsymbol{t})$, which is an instance of (A3), that is, of suffixing. (A3) is an axiom in three of the axiomatizations listed above (and, of course, a theorem of the fourth), and it is known to be a very powerful theorem. Another instance of (A3) is $(\mathcal{A} \to (\boldsymbol{t} \to (\mathcal{A} \circ \boldsymbol{t}))) \to ((\boldsymbol{t} \to (\mathcal{A} \circ \boldsymbol{t})) \to ((\boldsymbol{t} \to \boldsymbol{t}) \to (\mathcal{A} \circ \boldsymbol{t}))) \to \mathcal{A} \to (\boldsymbol{t} \to \boldsymbol{t}) \to (\mathcal{A} \circ \boldsymbol{t})$, from which by two detachments we get $\mathcal{A} \to (\boldsymbol{t} \to \boldsymbol{t}) \to (\mathcal{A} \circ \boldsymbol{t})$. Then we need an instance of (A10), specifically, $(\mathcal{A} \to (\boldsymbol{t} \to \boldsymbol{t}) \to (\mathcal{A} \circ \boldsymbol{t})) \to (\boldsymbol{t} \to \boldsymbol{t}) \to \mathcal{A} \to (\mathcal{A} \circ \boldsymbol{t})$. In two further steps we arrive at the formula $\mathcal{A} \to (\mathcal{A} \circ \boldsymbol{t})$. The path to this theorem is less than obvious in the axiomatic setting, and the complications may incline us to think that, perhaps, $E^{\circ t}_\to \vdash \mathcal{A} \to \boldsymbol{t} \to \mathcal{A}$ too.

We put forward a somewhat heavy-handed argument that $\boldsymbol{t}$ is not a lower right identity for $\circ$ in E. If it were, then E and R would be the same logics, but we know that they are not. (This latter part of the argument is where we skip including details here and we simply appeal to the abundance of results about E and R in the literature that show that they differ.[11])

Let us assume—hypothetically—that $\mathcal{A} \to \boldsymbol{t} \to \mathcal{A}$ is provable in $E^{\circ t}_\to$, in addition to $\mathcal{A} \to (\mathcal{A} \circ \boldsymbol{t})$. Restricted permutation is a theorem of $E_\to$, and this suggests that if $\mathcal{B}$ in (A5) can be equivalently viewed as an implicational formula, then we may obtain $\vec{\mathcal{B}} \to \mathcal{A} \to \mathcal{C}$. What remains is to revert $\vec{\mathcal{B}}$ to $\mathcal{B}$. This is basically the right idea.

[11]As a handful of references, see for instance, [7], [8], [163], [38] and [34].

We give a series of formulas, which simulates a proof of (A5); however, we leave proving some of the intermediate justifications as exercises.

(1) $(((\mathcal{A} \to \mathcal{B} \to \mathcal{C}) \circ \mathcal{A}) \circ \mathcal{B}) \to \mathcal{C}$

(2) $(((\mathcal{A} \to \mathcal{B} \to \mathcal{C}) \circ \mathcal{A}) \circ (\mathcal{B} \circ \boldsymbol{t})) \to \mathcal{C}$

(3) $((\mathcal{B} \circ ((\mathcal{A} \to \mathcal{B} \to \mathcal{C}) \circ \mathcal{A})) \circ \boldsymbol{t}) \to \mathcal{C}$

(4) $((((\mathcal{A} \to \mathcal{B} \to \mathcal{C}) \circ \mathcal{B}) \circ \mathcal{A}) \circ \boldsymbol{t}) \to \mathcal{C}$

(5) $(((\mathcal{A} \to \mathcal{B} \to \mathcal{C}) \circ \mathcal{B}) \circ \mathcal{A}) \to \mathcal{C}$

Exercise 5.3.9. Prove that if $\mathcal{A} \to \boldsymbol{t} \to \mathcal{A}$ were provable for any $\mathcal{A}$ in $E_{\to}^{\circ t}$, then $((\mathcal{D} \circ \mathcal{B}) \to \mathcal{C}) \to (\mathcal{D} \circ (\mathcal{B} \circ \boldsymbol{t})) \to \mathcal{C}$ would be provable too. [Hint: This wff could be used to justify the step from (1) to (2) above.]

Exercise 5.3.10. Prove that $E_{\to}^{\circ t} \vdash ((\mathcal{A} \circ \mathcal{B}) \circ \mathcal{C}) \to (\mathcal{B} \circ (\mathcal{A} \circ \mathcal{C}))$. [Hint: Notice how the formula resembles the combinatory B' axiom. The wff justifies the steps from (2) to (3) and then to (4). Instances of the wff's (A1) and (A2) on p. 167, that is, instances of self-identity and prefixing, may be useful.]

The step from (4) to (5) is justified by $\boldsymbol{t}$ being an upper right identity for $\circ$. From $\mathcal{A} \to (\mathcal{A} \circ \boldsymbol{t})$, by modus ponens from an instance of the principal type of B', we obtain $((\mathcal{A} \circ \boldsymbol{t}) \to \mathcal{C}) \to \mathcal{A} \to \mathcal{C}$.

To sum up, we have established that $\boldsymbol{t}$ is a *left identity* and an *upper right identity* (but not a lower right identity) for fusion in $E_{\to}^{\circ t}$.

We do not have fusion in $E_{\to}^{t}$, but it may be expected that the rules for $\circ$ will be like the rules for $\circ$ in LQ. Then in order to prove that $\boldsymbol{t}$ is an upper right identity, we need a rule (or a combination of rules) that licenses the next step.

$$\dfrac{\vdots \atop \boldsymbol{t}; (\mathcal{A}; \boldsymbol{t}) \vdash \mathcal{A} \circ \boldsymbol{t}}{\boldsymbol{t}; \mathcal{A} \vdash \mathcal{A} \circ \boldsymbol{t}}$$

There is an obvious analogy between $\boldsymbol{t}$, when it is a *left identity*, and the combinator I. However, I is not a right identity at all in combinatory terms; $M\mathsf{I}$ does not reduce to M and there is no way to slip an I into a term on its right-hand side. But $\boldsymbol{t}$ is *one wff* in $E_{\to}^{t}$, which in various places fulfills identity roles. If we would ignore that I is not $\boldsymbol{t}$, then we could consider a transformation from a combinatory rule for $\mathsf{4}$ to a structural rule for $LE_{\to}^{t}$ along the following lines.

$$\dfrac{\mathfrak{A}[\mathfrak{B}; \mathsf{I}] \vdash \mathcal{A}}{\mathfrak{A}[\mathsf{4}; \mathfrak{B}] \vdash \mathcal{A}}\ \mathsf{4}\vdash \qquad \rightsquigarrow \qquad \dfrac{\mathfrak{A}[\mathfrak{B}; \boldsymbol{t}] \vdash \mathcal{A}}{\mathfrak{A}[\mathfrak{B}] \vdash \mathcal{A}}\ \hat{\boldsymbol{t}}_r\vdash$$

Each of the other numbered combinators has an axiom in which the reduct contains B'. But we do not have a constant or formula that could assume the role of B'. The rule $(\hat{\boldsymbol{t}}_r\vdash)$ is the characteristic rule in the consecution calculus $LE^t_{\to}$—compared to $LT^t_{\to}$. The motivation to introduce this rule was presented differently (and more briefly) in [33], where $LE^t_{\to}$ was defined.

The consecution calculus $LE^t_{\to}$ for *pure entailment with truth* comprises the following axiom and rules.

$$\mathcal{A} \vdash \mathcal{A} \ \text{id}$$

$$\frac{\mathfrak{A} \vdash \mathcal{A} \quad \mathfrak{B}[\mathcal{B}] \vdash \mathcal{C}}{\mathfrak{B}[\mathcal{A} \to \mathcal{B}; \mathfrak{A}] \vdash \mathcal{C}} \to\vdash \qquad \frac{\mathfrak{A}; \mathcal{A} \vdash \mathcal{B}}{\mathfrak{A} \vdash \mathcal{A} \to \mathcal{B}} \vdash\to$$

$$\frac{\mathfrak{A}[\mathfrak{B}; (\mathfrak{C}; \mathfrak{D})] \vdash \mathcal{A}}{\mathfrak{A}[\mathfrak{C}; \mathfrak{B}; \mathfrak{D}] \vdash \mathcal{A}} \mathsf{B}'\vdash \qquad \frac{\mathfrak{A}[\mathfrak{B}; \mathfrak{C}; \mathfrak{C}] \vdash \mathcal{A}}{\mathfrak{A}[\mathfrak{B}; \mathfrak{C}] \vdash \mathcal{A}} \mathsf{W}\vdash$$

$$\frac{\mathfrak{A}[\mathfrak{B}] \vdash \mathcal{A}}{\mathfrak{A}[\boldsymbol{t}; \mathfrak{B}] \vdash \mathcal{A}} \check{\boldsymbol{t}}_l\vdash \qquad \frac{\mathfrak{A}[\mathfrak{B}; \boldsymbol{t}] \vdash \mathcal{A}}{\mathfrak{A}[\mathfrak{B}] \vdash \mathcal{A}} \hat{\boldsymbol{t}}_r\vdash$$

The notions of a *proof* and a *wff being a theorem* of $LE^t_{\to}$ are like those in $LR^t_{\to;}$—with the above set of rules.

The implicational rules are exactly as in $LR^t_{\to;,}$ and so is $(\mathsf{W}\vdash)$. We used here the label $(\check{\boldsymbol{t}}_l\vdash)$ from [33], which is pleasingly symmetrical to $(\hat{\boldsymbol{t}}_r\vdash)$; the rule itself has the same shape as the rule that was labeled in $LR^t_{\to;}$ as $(\mathsf{KI}_{\boldsymbol{t}}\vdash)$. The $(\mathsf{B}'\vdash)$ rule is taken to be primitive here; its form is precisely as before. Notice that there is no need to stipulate the $(\mathsf{M}_{\boldsymbol{t}}\vdash)$ rule in $LE^t_{\to}$, because the rule is merely a special instance of $(\hat{\boldsymbol{t}}_r\vdash)$ (with $\mathfrak{B}$ being $\boldsymbol{t}$).

$LE^t_{\to}$, as defined above, closely mirrors $E_{\to 2}$ with $\boldsymbol{t}$ added, of course. All the formulas (A1)–(A14) save (A5) and (A6) are provable in $LE^t_{\to}$. However, this consecution calculus is trickier to work with than $LR^t_{\to;,}$, because of the $(\hat{\boldsymbol{t}}_r\vdash)$ rule.

Example 5.24. Prefixing is easily provable in $LR^t_{\to;}$ using the rule $(\mathsf{B}\vdash)$. Here is a proof of the same formula in $LE^t_{\to}$.

$$\dfrac{\dfrac{\dfrac{\dfrac{\dfrac{\mathcal{C} \vdash \mathcal{C} \qquad \dfrac{\dfrac{\dfrac{\dfrac{\mathcal{A} \vdash \mathcal{A} \quad \mathcal{B} \vdash \mathcal{B}}{\mathcal{A} \to \mathcal{B}; \mathcal{A} \vdash \mathcal{B}} \to\vdash}{\boldsymbol{t}; (\mathcal{A} \to \mathcal{B}; \mathcal{A}) \vdash \mathcal{B}} \check{\boldsymbol{t}}_l\vdash}{\mathcal{A} \to \mathcal{B}; \boldsymbol{t}; \mathcal{A} \vdash \mathcal{B}} \mathsf{B}'\vdash}{}}{\mathcal{A} \to \mathcal{B}; \boldsymbol{t}; (\mathcal{C} \to \mathcal{A}; \mathcal{C}) \vdash \mathcal{B}} \to\vdash}{\mathcal{C} \to \mathcal{A}; (\mathcal{A} \to \mathcal{B}; \boldsymbol{t}); \mathcal{C} \vdash \mathcal{B}} \mathsf{B}'\vdash}{\mathcal{A} \to \mathcal{B}; \mathcal{C} \to \mathcal{A}; \boldsymbol{t}; \mathcal{C} \vdash \mathcal{B}} \mathsf{B}'\vdash}{\dfrac{\dfrac{\mathcal{A} \to \mathcal{B}; \mathcal{C} \to \mathcal{A}; \mathcal{C} \vdash \mathcal{B}}{\boldsymbol{t}; \mathcal{A} \to \mathcal{B}; \mathcal{C} \to \mathcal{A}; \mathcal{C} \vdash \mathcal{B}} \check{\boldsymbol{t}}_l\vdash}{\boldsymbol{t} \vdash (\mathcal{A} \to \mathcal{B}) \to (\mathcal{C} \to \mathcal{A}) \to \mathcal{C} \to \mathcal{B}} \vdash\to\text{'s}} \hat{\boldsymbol{t}}_r\vdash$$

We annotated the proof to make transparent the use of the structural rules, especially the use of $(\mathsf{B}'\vdash)$ and $(\hat{\boldsymbol{t}}_r\vdash)$. The proof of prefixing from suffixing and specialized assertion is no less arduous in $E_{\to 2}$.

Exercise 5.3.11. Is the rule $(\mathsf{B}\vdash)$ admissible in $LE^t_\to$ proofs of theorems? [Hint: Notice the interaction between the $(\check{t}_l\vdash)$ and the $(\mathsf{B}'\vdash)$ rules in the proof above.]

Exercise* 5.3.12. Prove in $LE^t_\to$ the wff's (A8), (A10), (A11) and (A14), that is, restricted assertion and restricted permutation, the principal type of 5, as well as restricted conditioned modus ponens. [Hint: Some of these proofs are a bit challenging, because some occurrences of $\boldsymbol{t}$ may disappear from a proof. However, $\boldsymbol{t}$ must appear in the end consecution.]

Theorem 5.25. *The single cut rule is* admissible *in* $LE^t_\to$.

Proof: The proof of this theorem was first presented in some detail in [33], where it is the proof of Theorem 1.6. The general idea is to use triple induction on the rank of the cut, on the degree of the cut formula and on the contraction measure of the cut, and to show by a separate induction that a cut, in which the cut formula is $\boldsymbol{t}$, is directly eliminable. (For a similar proof see Section 7.3.) qed

Typically, the admissibility of the cut rule provides the subformula property for a calculus. For $LE^t_\to$, we have to phrase the subformula property more carefully than usual, because of the $(\hat{\boldsymbol{t}}_r\vdash)$ rule. This rule does not interfere with the proof of the cut theorem, however, specific occurrences of $\boldsymbol{t}$ may disappear without a trace. An example of this is the top part of the proof of prefixing above. We could have ended the proof with the consecution that resulted by the $(\hat{\boldsymbol{t}}_r\vdash)$ rule, which has no $\boldsymbol{t}$ in it. The tree would be a *proof,* but it would not be a *proof of a formula,* but only a proof of the consecution $\mathcal{A}\to\mathcal{B};\mathcal{C}\to\mathcal{A};\mathcal{C}\vdash\mathcal{B}$.

Lemma 5.26. *In a proof of a theorem $\mathcal{A}$ in $LE^t_\to$, all the formulas appearing in the proof are* subformulas of a formula *in the bottom consecution.*

Proof: Given the preceding discussion, it is clear that only the $(\hat{\boldsymbol{t}}_r\vdash)$ rule could cause a problem. The disaster is averted by the proof being a proof of a theorem, which contains at least one occurrence of $\boldsymbol{t}$. qed

The lemma may seem a bit like cheating, because what cut-free proofs provide is a possibility to track occurrences of formulas and subformulas across a proof. However, another way to view this issue is that it reflects some of the extraordinary features of entailment. Incidentally, the impossibility to track all the $\boldsymbol{t}$'s is what makes the proofs in Exercise 5.3.12 more difficult than those in other consecution calculi. (They are still much less difficult than proofs of the same wff's in $E_{\to 2}$ though.)

Entailment is the logic that Anderson and Belnap created from Ackermann's strong implication, and it has its own beauty.[12] However, I, personally, like the *logic of ticket entailment* even better, which explains why I worked (and continue to work) with this logic extensively.

[12] See Ackermann's [1], and the two volumes [7] and [8].

Section 4.3 does not contain a sequent calculus for any fragment of ticket entailment. In many ways, T and its fragments form a dividing line within the relevance logic family, which is already apparent when we consider the implicational fragment of T. The positive fragment is near the lower edge where undecidability starts, whereas the decidability of $T_{\rightarrow}$ was—until recently—the last major open problem from amongst those conceived in the 1950s and 1960s. "Weaker" and "stronger" implicational relevance logics had been proved decidable long ago.

The *logic of ticket entailment* was first defined in Anderson [5]. Anderson obtained T from Ackermann's strong implication by going a step further than in the case of entailment—not only omitting a rule called γ, but also omitting the rule δ. Conceptually, it can be thought of as omitting the modal component of entailment, which explains the title of Anderson's abstract: "Entailment shorn of modality."

As it is explained in [7, §6], the word "ticket" in the name of the logic alludes to certain implicational formulas functioning as "inference-tickets" in the process of reasoning. Distinctions between the roles of various kinds of premises were considered by the 20th century philosopher G. Ryle, and he used the term "inference-ticket," in particular. One can argue that a propositional variable is not capable of expressing a law or a necessary truth in a logical system, hence, its role is always secondary with respect to an entailment or an implicational formula. However, implicational formulas may be primary or secondary, depending on the whole composition of an inference. A moral to be drawn from this is that the *permutation of the premises* in an inference cannot be allowed, in general. Without going into further discussions about motivations and philosophical justifications for $T_{\rightarrow}$, we wish to highlight some of the distinguishing properties $T_{\rightarrow}$ has.[13]

Pure ticket entailment, $T_{\rightarrow}$ comprises the rule modus ponens and the axioms (A1)–(A4). (A7) and (A13) are the only other formulas from our list that are theorems of $T_{\rightarrow}$. Intensional truth can be added to $T_{\rightarrow}$ by the two rules, which are summarized as $T^t_{\rightarrow} \vdash \mathcal{A}$ if and only if $T^t_{\rightarrow} \vdash \mathbf{t} \rightarrow \mathcal{A}$. In the light of the above informal explanation, ($\Box \mathbf{t}$) is clearly an undesirable formula for $T^t_{\rightarrow}$, and so it is not added as an axiom (and it is not a theorem of $T^t_{\rightarrow}$).

Moving freely back and forth between $R^t_{\rightarrow}$, $E^t_{\rightarrow}$ and $T^t_{\rightarrow}$, that is, using the obvious matching between $\rightarrow_R$ and $\rightarrow_E$, etc., the sets of theorems of these logics are related by the following inclusions. (We use the labels for the logics to indicate their sets of theorems too.)

$$T^t_{\rightarrow} \subsetneq E^t_{\rightarrow} \subsetneq R^t_{\rightarrow}$$

$R^t_{\rightarrow}$ is the *strongest* among these logics in the sense of having all the theorems the others have and more. $T^t_{\rightarrow}$ is the *strongest* among these logics in the

[13] [7, §6] contains enlightening quotes from Ryle's work. [32, §3] attempts to show how the axioms reflect the laws–vs–facts distinction.

usual set-theoretic sense, where the smaller the set of tuples is, the stronger is the relation it determines.

If we focus on *permutation*, then $R^t_\to$ has all the permutations possible. $E^t_\to$ has restricted permutation as well as twice specialized permutation. $T^t_\to$ has none of these formulas as a theorem, nor does it have twice restricted permutation $((\mathcal{A} \to \mathcal{B}) \to (\mathcal{C} \to \mathcal{D}) \to \mathcal{E}) \to (\mathcal{C} \to \mathcal{D}) \to (\mathcal{A} \to \mathcal{B}) \to \mathcal{E}$. It has though thrice specialized permutation, which is simply self-implication.

Despite the fact that permutation and some of its derivatives are not theorems of $T^t_\to$, there is some possibility for permutation in $T^t_\to$, which stems from *suffixing*. To put it roughly, a difference is that $E^t_\to$ allows permutations based on the *shape* of a formula, whereas $T^t_\to$ allows some permutations of formulas based on their *location* among other formulas in the inference.

We may contrast these three logics from the point of view of $\boldsymbol{t}$'s behavior in them. $\boldsymbol{t}$ is a left identity in $T^t_\to$, whereas in $E^t_\to$, it is also an upper right identity (as we discussed above). In $R^t_\to$, $\boldsymbol{t}$ is simply identity.

The axiomatization of $T^t_\to$ by (A1)–(A4) and the three rules mentioned above suggests using three structural rules in a consecution calculus that correspond to the combinators of which (A2)–(A4) are principal types. $LT^t_\to$ was defined—together with consecution calculi for other fragments of negation-free ticket entailment—in [32]. [8, §67] contains a consecution calculus for LTW_+ due to S. Giambrone. As it turns out, in his Ph.D. thesis [94], which is not widely available, there is a consecution calculus for T_+ too. His calculi differ from the ones in this chapter, especially, with respect to handling $\boldsymbol{t}$.

The *consecution calculus* $LT^t_\to$ comprises the following axiom and rules.

$$\mathcal{A} \vdash \mathcal{A} \ \text{id}$$

$$\frac{\mathfrak{A} \vdash \mathcal{A} \quad \mathfrak{B}[\mathcal{B}] \vdash \mathcal{C}}{\mathfrak{B}[\mathcal{A} \to \mathcal{B}; \mathfrak{A}] \vdash \mathcal{C}} \ {\to}{\vdash} \qquad \frac{\mathfrak{A}; \mathcal{A} \vdash \mathcal{B}}{\mathfrak{A} \vdash \mathcal{A} \to \mathcal{B}} \ {\vdash}{\to}$$

$$\frac{\mathfrak{A}[\mathfrak{B}] \vdash \mathcal{A}}{\mathfrak{A}[\boldsymbol{t}; \mathfrak{B}] \vdash \mathcal{A}} \ \mathsf{Kl}_t{\vdash} \qquad \frac{\mathfrak{A}[\boldsymbol{t}; \boldsymbol{t}] \vdash \mathcal{A}}{\mathfrak{A}[\boldsymbol{t}] \vdash \mathcal{A}} \ \mathsf{M}_t{\vdash}$$

$$\frac{\mathfrak{A}[\mathfrak{B}; (\mathfrak{C}; \mathfrak{D})] \vdash \mathcal{A}}{\mathfrak{A}[\mathfrak{B}; \mathfrak{C}; \mathfrak{D}] \vdash \mathcal{A}} \ \mathsf{B}{\vdash} \qquad \frac{\mathfrak{A}[\mathfrak{B}; (\mathfrak{C}; \mathfrak{D})] \vdash \mathcal{A}}{\mathfrak{A}[\mathfrak{C}; \mathfrak{B}; \mathfrak{D}] \vdash \mathcal{A}} \ \mathsf{B}'{\vdash} \qquad \frac{\mathfrak{A}[\mathfrak{B}; \mathfrak{C}; \mathfrak{C}] \vdash \mathcal{A}}{\mathfrak{A}[\mathfrak{B}; \mathfrak{C}] \vdash \mathcal{A}} \ \mathsf{W}{\vdash}$$

The notions of a *proof* and of a *theorem* in $LT^t_\to$ are defined as before. $T_\to$ may also be axiomatized by replacing (A4) with (A7) and (A13). (This system is denoted by $T_{\to 1}$ in [7, §8.3].)

Exercise 5.3.13. Prove that (A7) and (A13) are theorems of $LT^t_\to$.

It is easy to see that $T_{\to 1}$, really, (A1) and (A13) with modus ponens yield (A4), because $(\mathcal{A} \to \mathcal{A}) \to (\mathcal{A} \to \mathcal{A} \to \mathcal{B}) \to \mathcal{A} \to \mathcal{B}$ is an instance of (A13) from which $\mathcal{A} \to \mathcal{A}$ may be detached.

We could define an alternative consecution calculus formulation $LT^t_{\to 2}$ for $T^t_\to$ by replacing the $(\mathsf{W}{\vdash})$ rule with $(\mathsf{S}'{\vdash})$.

$$\frac{\mathfrak{A}[\mathfrak{B};\mathfrak{C};(\mathfrak{D};\mathfrak{C})] \vdash \mathcal{A}}{\mathfrak{A}[\mathfrak{D};\mathfrak{B};\mathfrak{C}] \vdash \mathcal{A}}\ \mathsf{S}'\vdash$$

Exercise 5.3.14. Prove that $(\mathsf{S}'\vdash)$ is a derived rule in $LT^t_{\rightarrow}$.

Exercise 5.3.15. Prove that in a proof of a theorem, $(\mathsf{W}\vdash)$ is an admissible rule in $LT^t_{\rightarrow 2}$. [Hint: Recall that we considered similar admissibility claims earlier.]

Theorem 5.27. *The single cut rule is* admissible *in* $LT^t_{\rightarrow}$.

The proof goes along the same general lines as the similar proofs for $LR^t_{\rightarrow}$; and $LE^t_{\rightarrow}$. (See [32].)

It is not very common to try to solve a decidability problem by extending the language of a logic. The decidability of $T_{\rightarrow}$ has been a famous open problem until recently; we say more about the problem and its solution in Section 9.1. Thus, it was natural to ask the question, once $LT^t_{\rightarrow}$ has been defined, whether somehow $\boldsymbol{t}$ could be excluded from the consecution calculus. [32] answered this question positively.

The *consecution calculus* $LT_{\rightarrow}$ is defined as $LT^t_{\rightarrow}$, but without the rules $(\mathsf{KI}_t\vdash)$ and $(\mathsf{M}_t\vdash)$, and with including a new rule $(\mathsf{M}\vdash)$.

$$\frac{\mathfrak{A}[\mathfrak{B};\mathfrak{B}] \vdash \mathcal{A}}{\mathfrak{A}[\mathfrak{B}] \vdash \mathcal{A}}\ \mathsf{M}\vdash$$

The notion of a *proof* is unaltered. However, theoremhood cannot be defined as before, since $\boldsymbol{t}$ is not in the language of the logic. $\mathcal{A} \rightarrow \mathcal{B}$ is a *theorem* of $LT_{\rightarrow}$ iff $\mathcal{A} \vdash \mathcal{B}$ is provable. This definition is unproblematic in the sense that all the theorems of $T_{\rightarrow}$ have at least one $\rightarrow$, because $\boldsymbol{t}$ is not in the language. On the other hand, to prove the equivalence of $LT_{\rightarrow}$ to $T_{\rightarrow}$ becomes complicated, because what needs to be proved is that if $\mathcal{A} \vdash \mathcal{B}$ and $\mathcal{A} \rightarrow \mathcal{B} \vdash \mathcal{C} \rightarrow \mathcal{D}$, then $\mathcal{C} \vdash \mathcal{D}$ is provable in $LT_{\rightarrow}$. It is likely that it would be useful if the cut rule were admissible.

Theorem 5.28. *The single cut rule is* admissible *in* $LT_{\rightarrow}$.

Proof: A proof using double induction can be applied here straightforwardly for *multiple cut*. The shape of the multi-cut rule is as earlier—on page 150. (Since $\boldsymbol{t}$ is not in the language, there is no need to deal with cuts, in which $\boldsymbol{t}$ is the cut formula.) qed

Exercise* 5.3.16. Verify that multiple cut may be proved admissible for $LT_{\rightarrow}$ by double induction on the rank of the cut and on the degree of the cut formula. [Hint: A triple induction is detailed in Chapter 2, and the use of mix and multiple cut is discussed in some detail in Section 7.2.]

In [32], there is no direct proof that $LT_{\to}$ can emulate modus ponens.[14]

Lemma 5.29. **($LT_{\to}$ is equivalent to $T_{\to}$)** *If $\mathcal{A} \to \mathcal{B}$ is a theorem of $T_{\to}$, then $LT_{\to}$ proves $\mathcal{A} \vdash \mathcal{B}$, and the other way around.*

Proof: We know that $LT^{t}_{\to}$ is equivalent to $T^{t}_{\to}$, and the latter is a conservative extension of $T_{\to}$. That is, $LT^{t}_{\to}$ proves exactly the same $\boldsymbol{t}$-free formulas as $T_{\to}$ does. If we can show that $LT_{\to}$ proves no more theorems than $LT^{t}_{\to}$ does, but it does prove each of them, then we may conclude that the equivalence of $T_{\to}$ and $LT_{\to}$ obtains.
1. We have already mentioned fusion. In the next section we introduce $LT^{\circ t}_{+}$, which is a conservative extension of $LT^{t}_{\to}$. It is easy to prove that the rule $(\mathsf{M}\vdash)$ is admissible in $LT^{\circ t}_{+}$ in proofs of theorems, because of the rule $(\mathsf{b}_t\vdash)$. This means that the extra rule in $LT_{\to}$ does not allow us to prove *more* implicational theorems than $LT^{t}_{\to}$ does.
2. Let us assume that $\mathcal{A} \to \mathcal{B}$ is a theorem of $T_{\to}$. Then $\boldsymbol{t} \vdash \mathcal{A} \to \mathcal{B}$ is provable in $LT^{t}_{\to}$ (without any application of the cut rule). The last rule in a proof must have been $(\vdash\to)$, and furthermore, $\boldsymbol{t}$ must have come from $(\mathsf{KI}_t\vdash)$, because $\boldsymbol{t}$ is not a subformula of $\mathcal{A} \to \mathcal{B}$. $LT_{\to}$ has all the rules that $LT^{t}_{\to}$ has—except those in which $\boldsymbol{t}$ occurs. We have to consider whether this $\boldsymbol{t}$ could mean that an $LT^{t}_{\to}$ proof cannot be simulated in $LT_{\to}$. $\boldsymbol{t}$ cannot be a subaltern in $(\to\vdash)$ or $(\vdash\to)$, nor can it be $\mathfrak{A}$ in the former rule. If $\boldsymbol{t}$ were one of the structures in a $(\mathsf{B}\vdash)$ or $(\mathsf{B}'\vdash)$ rule, then we simply omit that step in $LT_{\to}$. $\boldsymbol{t}$ cannot be the contracted structure in $(\mathsf{W}\vdash)$, because $\boldsymbol{t}$ is not a right identity. If $\boldsymbol{t}$ were $\mathfrak{B}$ in the contraction rule, then without $\boldsymbol{t}$ the rule is an instance of $(\mathsf{M}\vdash)$. In sum, a proof of $\mathcal{A} \to \mathcal{B}$ in $LT^{t}_{\to}$ can be simulated step by step in $LT_{\to}$. qed

Exercise 5.3.17. Prove that (A1)–(A4) are theorems of $LT_{\to}$.

Although $LT_{\to}$ captures exactly $T_{\to}$ by the above lemma, the proof is not very elegant, because it uses results about extensions of $LT_{\to}$. A brief look at where a direct proof runs into difficulty will highlight the interaction of $(\to\vdash)$, grouping and contraction (once more).

Let us assume that $\mathcal{A}_1 \to \mathcal{A}_2 \vdash \mathcal{B}_1 \to \mathcal{B}_2$ and $\mathcal{A}_1 \vdash \mathcal{A}_2$ are provable in $LT_{\to}$. We have to show that $\mathcal{B}_1 \vdash \mathcal{B}_2$ is provable too. The assumption that there is an implicational wff $\mathcal{B}_1 \to \mathcal{B}_2$ in the first consecution is justified, because we intend to emulate modus ponens applied to *theorems* of $T_{\to}$. Then we have $\mathcal{A}_1 \to \mathcal{A}_2; \mathcal{B}_1 \vdash \mathcal{B}_2$, by cut. It is tempting to assume at this point that this consecution has been obtained by $(\to\vdash)$. If that were the case,

[14] In 2005, I thought that $LT_{\to}$ would be a more suitable system to solve the decidability problem of $T_{\to}$ than $LT^{t}_{\to}$ would be; hence, I developed an indirect proof to show that $LT_{\to}$ is equivalent to $T_{\to}$. Five years later, I changed my mind about which of $LT^{t}_{\to}$ and $LT_{\to}$ is more useful for the decidability problem, and so the proof of this lemma has not been improved.

then we would indeed have a proof of $\mathcal{B}_1 \vdash \mathcal{B}_2$—using the premises in the purported application of the $(\rightarrow\vdash)$ rule.

$$\dfrac{\text{cut}\ \dfrac{\mathcal{B}_1 \vdash \mathcal{A}_1 \qquad \mathcal{A}_1 \vdash \mathcal{A}_2}{\mathcal{B}_1 \vdash \mathcal{A}_2} \qquad \mathcal{A}_2 \vdash \mathcal{B}_2}{\mathcal{B}_1 \vdash \mathcal{B}_2}\ \text{cut}$$

However, here is an example of why the last assumption is unwarranted. There is a proof of $(\mathcal{A} \rightarrow \mathcal{A}) \rightarrow \mathcal{A} \rightarrow \mathcal{A}$ by detachment of $\mathcal{A} \rightarrow \mathcal{A}$ from $(\mathcal{A} \rightarrow \mathcal{A}) \rightarrow (\mathcal{A} \rightarrow \mathcal{A}) \rightarrow \mathcal{A} \rightarrow \mathcal{A}$. This is an unimaginative and unexciting way to prove this formula, but it is a proof, nonetheless. This means that we start with $\mathcal{A} \rightarrow \mathcal{A} \vdash (\mathcal{A} \rightarrow \mathcal{A}) \rightarrow \mathcal{A} \rightarrow \mathcal{A}$ and $\mathcal{A} \vdash \mathcal{A}$ as provable consecutions. By cut, we get $\mathcal{A} \rightarrow \mathcal{A}; \mathcal{A} \rightarrow \mathcal{A} \vdash \mathcal{A} \rightarrow \mathcal{A}$, but unfortunately, if we hypothesize that this consecution was obtained by an application of the rule $(\rightarrow\vdash)$, then we end up with the tree below.

$$\dfrac{\mathcal{A} \rightarrow \mathcal{A} \overset{?}{\vdash} \mathcal{A} \qquad \mathcal{A} \overset{?}{\vdash} \mathcal{A} \rightarrow \mathcal{A}}{\mathcal{A} \rightarrow \mathcal{A}; \mathcal{A} \rightarrow \mathcal{A} \vdash \mathcal{A} \rightarrow \mathcal{A}}$$

Neither $\mathcal{A} \rightarrow \mathcal{A} \vdash \mathcal{A}$, nor $\mathcal{A} \vdash \mathcal{A} \rightarrow \mathcal{A}$ is provable with an arbitrary $\mathcal{A}$, in $LT_{\rightarrow}$.

Exercise* 5.3.18. Determine whether there are some $\mathcal{A}$'s for which one or the other consecution is provable in $LT_{\rightarrow}$. [Hint: For one of the consecutions the answer is very easy.]

Thus, we have to scrutinize what happens if some other rule has been applied. The $(\mathsf{B}\vdash)$ and $(\mathsf{B}'\vdash)$ rules are excluded, because there are not enough structures on the left-hand side of the turnstile. However, $(\vdash\rightarrow)$, $(\mathsf{W}\vdash)$ and $(\mathsf{M}\vdash)$ are possibilities. None of these rules seem to help with the provability of $\mathcal{B}_1 \vdash \mathcal{B}_2$. If we look further up into the proof, then $(\mathsf{B}\vdash)$ and $(\mathsf{B}'\vdash)$ have to be taken into account too, because $(\mathsf{W}\vdash)$ and $(\mathsf{M}\vdash)$ may be applied without limit. With sufficiently many structures created, $(\mathsf{B}\vdash)$ and $(\mathsf{B}'\vdash)$ can be applied, which may group together an implicational wff with a structure for a $(\rightarrow\vdash)$ rule, etc.

Exercise 5.3.19.** Try to finish the direct proof that we started. [Hint: This is an open-ended exercise. A satisfactory argument completing the proof may or may not be overly complicated or excessively lengthy.]

We leave $LT_{\rightarrow}$ for now, and turn back to the consecution calculi that have $\boldsymbol{t}$ in them. In Section 9.1, we will introduce $LT^{\textcircled{t}}_{\rightarrow}$, an extension of $LT^{t}_{\rightarrow}$, which formalizes $R^{t}_{\rightarrow}$. From among the "weaker" than R relevance logics, we consider $L^{t}_{\rightarrow}$ and $B^{t}_{\rightarrow}$ here.

$L_{\rightarrow}$ has been known and considered as a logic in its own right at least since the beginning of the 1960s, for instance, by Meredith and Prior in [133]. $L_{\rightarrow}$ is also known as *implicational* BCI logic, because it is axiomatized by (A2), (A5) and (A1). The notation L comes from the recent popularity of this logic under the name the *implicational fragment of linear logic*. (We considered $L_{\rightarrow}$ as a sequent calculus in Section 4.2, and we devoted Section 4.5 to linear logic and its variations.)

$T_{\rightarrow}$ and $L_{\rightarrow}$ can be thought of as omitting "orthogonal" axioms from an axiomatization of $R_{\rightarrow}$. It is not difficult to see that (A2) and (A3) are related via their first two antecedents. In other words, suffixing can be obtained by detaching prefixing from a suitable instance of permutation. In terms of combinatory logic, B' is definable as CB. If we take (A1)–(A5) with modus ponens as a (slightly redundant) axiomatization of $R_{\rightarrow}$, then $T_{\rightarrow}$ drops (A5), whereas $L_{\rightarrow}$ omits (A4). As a result, both $T_{\rightarrow}$ and $L_{\rightarrow}$ have theorems that are not theorems of the other logic.

The presence of permutation and association to the right in $L_{\rightarrow}$ is what makes it possible to formalize $L_{\rightarrow}$ with an associative structural connective, that is, by a sequent calculus. However, it is also possible to use the rules for B, C and I—without the combinators—to create a consecution calculus for $L^{t}_{\rightarrow}$.

The *consecution calculus* $LL^{t}_{\rightarrow}$ is defined by the axiom (id), and by the rules $(\rightarrow\vdash)$, $(\vdash\rightarrow)$, $(\mathsf{B}\vdash)$, $(\mathsf{C}\vdash)$, $(\mathsf{KI}_t\vdash)$ and $(\mathsf{M}_t\vdash)$.[15] The notions of a *proof* and of *theoremhood* are as in $LT^{t}_{\rightarrow}$.

Theorem 5.30. *The single cut rule is* admissible *in* $LL^{t}_{\rightarrow}$.

Exercise 5.3.20. Modify the proof of Theorem 5.27 for $LL^{t}_{\rightarrow}$. [Hint: From the point of view of the proof of the theorem, dropping $(\mathsf{W}\vdash)$ is a bonus.]

Exercise 5.3.21. Prove that (A3) is a theorem of $LL^{t}_{\rightarrow}$.

Exercise* 5.3.22. The axiom system $L^{t}_{\rightarrow}$ is $L_{\rightarrow}$ extended with the rules, which are concisely formulated as $L^{t}_{\rightarrow} \vdash \mathcal{A}$ iff $L^{t}_{\rightarrow} \vdash \boldsymbol{t} \rightarrow \mathcal{A}$. Prove that $L^{t}_{\rightarrow}$ and $LL^{t}_{\rightarrow}$ have the same theorems.

B plays the role of a *minimal relevance logic* in a similar sense as K is the *minimal normal modal logic*. K states nothing special about $\Box$—except the distributivity of $\Box$ across $\supset$, and that theorems may be $\Box$'d. In a Kripke-style semantics for K, the accessibility relation has no additional properties beyond being a binary relation. Similarly, B does not stipulate special properties for $\rightarrow$, and in its Routley–Meyer-style semantics, the ternary accessibility relation does not have extra properties beyond its interaction with the set representing $\boldsymbol{t}$.

$LB^{\circ t}_{+}$ was introduced in [38, Ch. 2], and we can extract the implicational fragment of B from that consecution calculus. One might think that we can simply omit the three $\boldsymbol{t}$-less structural rules from $LT^{t}_{\rightarrow}$. However, that would leave us with a calculus that has strange properties. The provability of $\boldsymbol{t} \vdash$

[15]This consecution calculus is not the same as the implicational fragment of the consecution calculus $LBCI$ defined in [38, Ch. 3], where instead of $(\mathsf{C}\vdash)$ there are two rules, $(\mathsf{T}\vdash)$ and $(\mathsf{b}\vdash)$. The latter rule reverses $(\mathsf{B}\vdash)$'s effect.

$\mathcal{A} \to \mathcal{B}$ and $t \vdash \mathcal{B} \to \mathcal{A}$ would not be enough for replacement, and it is not clear how to strengthen that relation to make the replacement theorem provable.

The *consecution calculus* $LB^t_\to$ comprises the axiom (id), the rules $(\to\vdash)$, $(\vdash\to)$, $(\mathsf{KI}_t\vdash)$ and $(\mathsf{M}_t\vdash)$ together with two new rules, $(\mathsf{B}_t\vdash)$ and $(\mathsf{B}'_t\vdash)$.

$$\frac{\mathfrak{A}[\mathfrak{B};(t;\mathfrak{C})] \vdash \mathcal{A}}{\mathfrak{A}[t;\mathfrak{B};\mathfrak{C}] \vdash \mathcal{A}}\ \mathsf{B}'_t\vdash \qquad \frac{\mathfrak{A}[t;(\mathfrak{B};\mathfrak{C})] \vdash \mathcal{A}}{\mathfrak{A}[t;\mathfrak{B};\mathfrak{C}] \vdash \mathcal{A}}\ \mathsf{B}_t\vdash$$

Exercise 5.3.23. Verify that the two rules are sufficient to prove the replacement theorem for $LB^t_\to$. [Hint: You may use cut, which is admissible (as stated below).]

$B_\to$ is axiomatized by (A1), that is, self-identity together with detachment and the rules (R1) and (R2). t may be added by the two rules as before—yielding $B^t_\to$.

(R1) $B^t_\to \vdash \mathcal{A} \to \mathcal{B}$ implies $B^t_\to \vdash (\mathcal{C} \to \mathcal{A}) \to \mathcal{C} \to \mathcal{B}$

(R2) $B^t_\to \vdash \mathcal{A} \to \mathcal{B}$ implies $B^t_\to \vdash (\mathcal{B} \to \mathcal{C}) \to \mathcal{A} \to \mathcal{C}$

Theorem 5.31. *The single cut rule is* admissible *in* $LB^t_\to$.

Exercise 5.3.24. What modifications should be made to the proofs of Theorems 5.25 and 5.27 to adapt them to $LB^t_\to$? [Hint: Aim at obtaining a proof as simple as possible.]

Lastly, we consider once again the *implicational* BCK *logic*, which does not belong to the relevant family.[16] It can be conceptualized as a logic that omits contraction (rather that weakening) from implicational intuitionistic logic or as one that extends associative Lambek calculus (as in Section 4.2). The presence of B and C in the label for this logic explains why sequences of formulas may be used in its formalization. On the other hand, we may take seriously the combinatory label for the logic and proceed to use the corresponding structural rules in a consecution calculus.

BCK is axiomatized by (A2), (A5) and (A15).

(A15) $\mathcal{A} \to \mathcal{B} \to \mathcal{A}$

Exercise 5.3.25. Prove that (A1) is a theorem of BCK. [Hint: The reduction $\mathsf{C}KMN \rhd_w \mathsf{K}NM$ may be exploited to obtain (infinitely) many distinct proofs of $\mathcal{A} \to \mathcal{A}$.]

The intensional truth constant t may be added by the rules already mentioned. A closely related logic has been termed *direct logic* in [126], and linear logic with weakening has been called *affine logic*, for example, in [114] and in

[16] We place BCK here, even though the title of this section does not apply to this logic, because BCK is a natural extension of $L_\to$.

[187]. We will use the latter name and the label *LW* for this logic, though the name (and the logic itself) is somewhat of a puzzle. Informally, affine logic excludes the possibility for a premise to be used *repeatedly*, but it does permit (possibly) *unrelated* premises to be inserted. For example, "If it is Tuesday, then it is Tuesday, provided that it is Wednesday" is perfectly OK, whereas "If it is Tuesday, then it is Tuesday, provided it is Tuesday" is OK only as a variant of the previous example.

The *consecution calculus* $LLW^t_{\to}$ is defined as $LL^t_{\to}$ (on p. 182) with thinning, that is, the rule $(\mathsf{K}\vdash)$ added.

$$\frac{\mathfrak{A}[\mathfrak{B}] \vdash \mathcal{A}}{\mathfrak{A}[\mathfrak{B};\mathfrak{C}] \vdash \mathcal{A}}\ \mathsf{K}\vdash$$

The notions of a *proof* and of a *theorem* are as before. This logic may be simplified by including the $(\mathsf{T}\vdash)$ rule instead of $(\mathsf{C}\vdash)$ and by omitting the $(\mathsf{KI}_t\vdash)$ rule. The latter is *not* an instance of $(\mathsf{K}\vdash)$, because $\boldsymbol{t}$ appears on the other side of $\mathfrak{B}$, nonetheless, $(\mathsf{KI}_t\vdash)$ turns out to be a derived rule.

Exercise 5.3.26. Find the proof steps that lead from $\mathfrak{A}[\mathfrak{B}] \vdash \mathcal{A}$ to $\mathfrak{A}[\boldsymbol{t};\mathfrak{B}] \vdash \mathcal{A}$ without $(\mathsf{T}\vdash)$ replacing $(\mathsf{C}\vdash)$.

Theorem 5.32. *The single cut rule is* admissible *in* $LLW^t_{\to}$.

Exercise 5.3.27. What modifications are necessary in the proof of the cut theorem for $LL^t_{\to}$ to obtain a proof for $LLW^t_{\to}$?

5.4 Positive entailment logics

The implication connective together with $\boldsymbol{t}$ and the structural rules creates room to investigate a variety of logics, as we have seen in the previous section. In relevance logics, $\to$ may be thought to express that a formula is implied or entailed by another formula. Then it is straightforward to consider another binary connective, $\circ$, which is often called *fusion* or *cotenability*; fusion joins the premises in an inference. (We already mentioned $\circ$ in connection to $LR^t_{\to}$, $LE^t_{\to}$ and $LLW^t_{\to}$ in the previous section.)

In an axiom system, we add the rules (R1) and (R2). ($\vdash$ stands for *provability* in the axiomatic system in question, and we might add the label of a logic in concrete cases.)

(R1) If $\vdash (\mathcal{A} \circ \mathcal{B}) \to \mathcal{C}$, then $\vdash \mathcal{A} \to \mathcal{B} \to \mathcal{C}$;

(R2) if $\vdash \mathcal{A} \to \mathcal{B} \to \mathcal{C}$, then $\vdash (\mathcal{A} \circ \mathcal{B}) \to \mathcal{C}$.

The permutation of the antecedents of an implicational formula, which is a theorem of $LR^t_{\to}$, now expresses the *commutativity* of $\circ$. Then, it seems that in $LE^t_{\to}$ and $LT^t_{\to}$, we should be able to have two implications—or at least, two arrows.

Example 5.33. The analogs of the above rules in classical logic are the so-called *exportation* and *importation* rules. (1) and (2) are theorems of LK.

(1) $(\mathcal{A} \supset \mathcal{B} \supset \mathcal{C}) \supset (\mathcal{A} \wedge \mathcal{B}) \supset \mathcal{C}$

(2) $((\mathcal{A} \wedge \mathcal{B}) \supset \mathcal{C}) \supset \mathcal{A} \supset \mathcal{B} \supset \mathcal{C}$

Of course, $\wedge$ is commutative, and (3) and (4) are also theorems of LK.

(3) $(\mathcal{A} \supset \mathcal{B} \supset \mathcal{C}) \supset (\mathcal{B} \wedge \mathcal{A}) \supset \mathcal{C}$

(4) $((\mathcal{A} \wedge \mathcal{B}) \supset \mathcal{C}) \supset \mathcal{B} \supset \mathcal{A} \supset \mathcal{C}$

Algebraically speaking, $\supset$ is the *left* and *right residual* of $\wedge$. Although $\wedge$ and $\circ$ have certain similarities, especially, in R, one of the features of $\circ$ that fades away in other relevance logics is its commutativity. This is a similar phenomenon that we have already seen with respect to associativity in LA and LQ.

Exercise 5.4.1. Prove (1)–(4) in LK. Which of the proofs require an application of $(\mathsf{C}\vdash)$? [Hint: Recall that $(\supset\vdash)$ and $(\vdash\supset)$ (in Section 2.1) are not quite like the $(\to\vdash)$ and $(\vdash\to)$ rules from the previous section.]

We expand the language with two new *binary connectives*, $\circ$ (*fusion*) and $\leftarrow$ (*co-implication*). The latter connective has several other names such as *pre-implication* or *backward implication*. The adjectives may suggest that $\leftarrow$ is not only different from $\to$, but somehow inferior to $\to$. It turns out that in the natural extension of $E^t_{\to}$, for instance, they cannot be proved to be the same connective (which is a good thing, given that we have two symbols). The question of inferiority is intriguing though. $\boldsymbol{t}$ is not only a placeholder in $LE^t_{\to}$, but $\boldsymbol{t}$ signals that a wff $\mathcal{A}$ is a theorem, when $\boldsymbol{t} \vdash \mathcal{A}$ is provable. By including the rule $(\mathsf{KI}_t\vdash)$, rather than $(\mathsf{K}_t\vdash)$, which would introduce $\boldsymbol{t}$ on the right-hand side of a structure, we made $\leftarrow$ into a lesser arrow than $\to$ is, even before introducing it. In Section 5.2, we mentioned *dual combinators*. We could dualize the rules in $LE^t_{\to}$ so that $\leftarrow$ would turn out to be "the true entailment." There is really no reason to prefer $\to$ over $\leftarrow$—except, perhaps, notational tradition, familiarity or aesthetic considerations.

An axiom system that contains $\to$ and $\circ$ may be further extended by the rules (R3) and (R4). However, we could go directly from $\to$ to $\leftarrow$ too, if we were to add (R5) and (R6).

(R3) If $\vdash (\mathcal{A} \circ \mathcal{B}) \to \mathcal{C}$, then $\vdash \mathcal{B} \to (\mathcal{C} \leftarrow \mathcal{A})$;

(R4) if $\vdash \mathcal{A} \to (\mathcal{B} \leftarrow \mathcal{C})$, then $\vdash (\mathcal{C} \circ \mathcal{A}) \to \mathcal{B}$;

(R5) if $\vdash \mathcal{A} \to (\mathcal{B} \to \mathcal{C})$, then $\vdash \mathcal{B} \to (\mathcal{C} \leftarrow \mathcal{A})$;

(R6) if $\vdash \mathcal{A} \to (\mathcal{B} \leftarrow \mathcal{C})$, then $\vdash \mathcal{C} \to (\mathcal{A} \to \mathcal{B})$.

We denote by $E^{\circ t}_{\leftrightarrows}$ the axiom system obtained from one of the axiomatizations of $E_{\to}$ with $\boldsymbol{t}$ added (by the two earlier rules and the $(\Box\boldsymbol{t})$ axiom) together with the rules (R1)–(R4).

Exercise 5.4.2. Prove that the following wff's are theorems of $E^{\circ t}_{\leftrightarrows}$. (1) $\mathcal{A} \to \mathcal{B} \to (\mathcal{A} \circ \mathcal{B})$, (2) $\mathcal{A} \to ((\mathcal{B} \leftarrow \mathcal{C}) \leftarrow (\mathcal{A} \to (\mathcal{B} \leftarrow \mathcal{C})))$ and (3) $\mathcal{A} \to ((\mathcal{A} \to (\mathcal{B} \circ \mathcal{A})) \leftarrow (\mathcal{A} \to \mathcal{B}))$.

The *consecution calculus* $LE^{\circ t}_{\leftrightarrows}$ is defined as an extension of $LE^{t}_{\to}$ (see p. 175) by a pair or rules for $\circ$ and for $\leftarrow$, plus two new structural rules. The rules, which are added, are as follows.

$$\frac{\mathfrak{A}[\mathcal{B};\mathcal{C}] \vdash \mathcal{A}}{\mathfrak{A}[\mathcal{B}\circ\mathcal{C}] \vdash \mathcal{A}}\ \circ\vdash \qquad \frac{\mathfrak{A} \vdash \mathcal{A} \qquad \mathfrak{B} \vdash \mathcal{B}}{\mathfrak{A};\mathfrak{B} \vdash \mathcal{A}\circ\mathcal{B}}\ \vdash\circ$$

$$\frac{\mathfrak{A} \vdash \mathcal{A} \qquad \mathfrak{B}[\mathcal{B}] \vdash \mathcal{C}}{\mathfrak{B}[\mathfrak{A};\mathcal{B}\leftarrow\mathcal{A}] \vdash \mathcal{C}}\ \leftarrow\vdash \qquad \frac{\mathcal{A};\mathfrak{A} \vdash \mathcal{B}}{\mathfrak{A} \vdash \mathcal{B}\leftarrow\mathcal{A}}\ \vdash\leftarrow$$

$$\frac{\mathfrak{A}[\boldsymbol{t};\mathfrak{B};\mathfrak{C}] \vdash \mathcal{A}}{\mathfrak{A}[\boldsymbol{t};(\mathfrak{B};\mathfrak{C})] \vdash \mathcal{A}}\ \mathsf{b}_t\vdash \qquad \frac{\mathfrak{A}[\boldsymbol{t};\mathfrak{B};\mathfrak{C}] \vdash \mathcal{A}}{\mathfrak{A}[\mathfrak{B};(\boldsymbol{t};\mathfrak{C})] \vdash \mathcal{A}}\ \mathsf{b(cb)}_t\vdash$$

The notions of a *proof* and of a wff $\mathcal{A}$ being a *theorem* are as in $LE^{t}_{\to}$.

The rules for $\circ$ and $\leftarrow$ look like the corresponding rules in LQ, because we formulated the non-associative Lambek calculus in a notation that is in line with the other consecution calculi in this chapter. The two structural rules, which are labeled with dual combinators, are added to ensure that the replacement theorem can be proved.

Theorem 5.34. *The single cut rule is* admissible *in* $LE^{\circ t}_{\leftrightarrows}$.

Exercise 5.4.3. Do the new rules, especially the $\boldsymbol{t}$-specific rules, cause a problem in the proof? If they do not, then explain why; if they do, then devise a strategy to overcome those difficulties. [Hint: There are no structural rules $(\mathsf{b}\vdash)$ or $(\mathsf{b(cb)}\vdash)$ in $LE^{\circ t}_{\leftrightarrows}$.]

Exercise 5.4.4. Let $\mathcal{A}$ and $\mathcal{B}$ be equivalent iff $\boldsymbol{t} \vdash \mathcal{A} \to \mathcal{B}$ and $\boldsymbol{t} \vdash \mathcal{B} \to \mathcal{A}$ are both provable. Prove the cases with $\circ$ and $\leftarrow$ in the replacement theorem, and note where the $(\mathsf{b}\vdash)$ and $(\mathsf{b(cb)}\vdash)$ rules are used.

Exercise 5.4.5. Prove that $E^{\circ t}_{\leftrightarrows}$ and $LE^{\circ t}_{\leftrightarrows}$ are equivalent.

Exercise 5.4.6. $(\mathsf{B}\vdash)$ is not a rule of $LE^{\circ t}_{\leftrightarrows}$. Is this rule admissible once $\circ$ and $\leftarrow$ added? [Hint: Either create a concrete counterexample or prove that the $(\mathsf{B}\vdash)$ rule is admissible.]

The introduction of $\circ$ has the advantage of allowing us to view a consecution as a formula. First of all, $(\circ\vdash)$ is an *invertible* rule, as the following shows.

$$\dfrac{\dfrac{\mathcal{A}\vdash\mathcal{A}\qquad \mathcal{B}\vdash\mathcal{B}}{\mathcal{A};\mathcal{B}\vdash\mathcal{A}\circ\mathcal{B}}\qquad \begin{array}{c}\vdots\\ \mathfrak{A}[\mathcal{A}\circ\mathcal{B}]\vdash\mathcal{C}\end{array}}{\mathfrak{A}[\mathcal{A};\mathcal{B}]\vdash\mathcal{C}}$$

If $\mathfrak{A}\vdash\mathcal{A}$ is a provable consecution, then we can associate with this consecution the formula that we obtain in three steps: by n applications of $(\circ\vdash)$ (where n is the number of ;'s in $\mathfrak{A}$), then, by an application of $(\mathsf{Kl}_t\vdash)$, and finally, by an application of the $(\vdash\rightarrow)$ rule.

Definition 5.35. Let $\mathfrak{A}$ be a structure. $\mathfrak{A}^\circ$, the *formula corresponding to* $\mathfrak{A}$, is defined by (1) and (2).

(1) If $\mathfrak{A}$ is $\mathcal{A}$, then $\mathfrak{A}^\circ$ is $\mathcal{A}$;

(2) if $\mathfrak{A}$ is $\mathfrak{B};\mathfrak{C}$, then $\mathfrak{A}^\circ$ is $\mathfrak{B}^\circ\circ\mathfrak{C}^\circ$.

Using this notation, we can sum up how to associate a wff with the consecution $\mathfrak{A}\vdash\mathcal{A}$. We first generate $\mathfrak{A}^\circ\vdash\mathcal{A}$, then in two steps, we obtain $t\vdash\mathfrak{A}^\circ\rightarrow\mathcal{A}$.

The addition of $\leftarrow$ (without $\circ$), however, does not ensure that each provable consecution is equivalent to a theorem. (For this equivalence, we want to require that a corresponding theorem retains all the occurrences of all formulas in the structure on the left of the turnstile.)

Exercise 5.4.7. Assume that $\circ$, as well as $(\circ\vdash)$ and $(\vdash\circ)$ are omitted from $LE^{\circ t}_{\leftrightarrows}$. Create a concrete proof of a consecution that cannot be turned into a theorem, which would correspond to the consecution in a reasonable sense. [Hint: The only rules available to create the theorem are $(\vdash\rightarrow)$, $(\vdash\leftarrow)$ and $(\mathsf{Kl}_t\vdash)$, because the structural rules alter a sequent without a trace.]

There is a sort of "modularity" in the definition of the consecution calculi for relevance logics with $\circ$ and $\leftarrow$ added, because the new connectives have the same relationship to $\rightarrow$ in each logic. Let $B^{\circ t}_{\leftrightarrows}$ be the axiom system $B^t_\rightarrow$ with (R1)–(R4) added. Similarly, let $T^{\circ t}_{\leftrightarrows}$ be $T^t_\rightarrow$ expanded by the same rules. The corresponding consecution calculi, $LT^{\circ t}_{\leftrightarrows}$ and $LB^{\circ t}_{\leftrightarrows}$, are obtained by adding $(\circ\vdash)$, $(\vdash\circ)$, $(\leftarrow\vdash)$ and $(\vdash\leftarrow)$ as operational rules, and $(\mathsf{b}_t\vdash)$ and $(\mathsf{b}(\mathsf{cb})_t\vdash)$ as structural rules.

Exercise 5.4.8. Verify that the single cut rule is admissible both in $LT^{\circ t}_{\leftrightarrows}$ and in $LB^{\circ t}_{\leftrightarrows}$. [Hint: Your solution to Exercise 5.4.3 may be helpful here.]

The positive fragment of R was "Gentzenized" by Dunn in 1969; we described the sequent calculus LR_+ in Section 4.6. Having consecution calculi for $E^t_{\rightarrow}$, $T^t_{\rightarrow}$ and $B^t_{\rightarrow}$, we might wish to add conjunction ($\land$) and disjunction ($\lor$) in order to obtain formalizations of the positive fragments of E, T and B.[17]

As we mentioned while describing LR_+, relevance logics tend to accept the distributivity of conjunction and disjunction (when those connectives are in the language of the logic). This is true in the case of E, T and B too. The overall formulation of a sequent calculus may depend on whether ; is associative (or commutative, etc.), however, the properties of ; do not alter how ; and another structural connective , interact with each other.

So far, we have been using one kind of structure in this chapter, namely, structures that we introduced in Definition 5.1. We will call them now *intensional structures* or *i-str*'s, for short. *Extensional structures* or *e-str*'s are defined similarly to extensional *sequences*, because we think of them as the counterparts of wff's that are conjunctions.

Definition 5.36. (Structures) The set of *structures* (*str*'s), of *intensional structures* (*i-str*'s) and of *extensional structures* (*e-str*'s) are defined by (1)–(4).

(1) If $\mathcal{A}$ is a wff, then $\mathcal{A}$ is an e-str and an i-str;

(2) if $\mathfrak{A}$ and $\mathfrak{B}$ are str's, then $(\mathfrak{A};\mathfrak{B})$ is an i-str, where $\mathfrak{A}$ and $\mathfrak{B}$ is $(\mathfrak{A})$ and $(\mathfrak{B})$, respectively, if either is of the from $\mathfrak{C}_1, \ldots, \mathfrak{C}_n$ (for some $\mathfrak{C}$'s);

(3) if $\mathfrak{A}_1, \ldots, \mathfrak{A}_n$ (where $n \geq 2$) are str's, then $\mathfrak{A}_1, \ldots, \mathfrak{A}_n$ is an e-str;

(4) if $\mathfrak{A}$ is an i-str or an e-str, then $\mathfrak{A}$ is a str.

If a structure comprises just one formula, then we cannot tell whether it is an e-str or an i-str; hence, according to (1), we consider $\mathcal{A}$ to be both. However, once ; or , occurs in a structure, then we can determine whether it is an e-str or an i-str, but not the other. Although structures are generated from formulas in a *bottom-up* fashion, in every concrete case, we are interested in parsing a structure in a *top-down* order.

We formulated clause (2) so that a pair of parentheses are added when a complex e-str becomes an element of an i-str. While a similar convention was necessary in Definition 4.29, here it is merely convenience, because i-str's are always parenthesized.

Exercise 5.4.9. Demonstrate that there is no ambiguity in str's if the proviso about $(\mathfrak{A})$ and $(\mathfrak{B})$ is omitted.

[17] Zero-ary constants such as $\boldsymbol{t}$ and $\boldsymbol{T}$ are not always included in the main systems of relevance logics. If they are not included, then it would be more precise to talk about the positive (or negation-free) fragments of E^t, T^t and B^t here.

The *consecution calculus* LE^t_+, which includes $\circ$ but not $\leftarrow$ as a connective, is defined from $LE^{\circ t}_{\leftrightarrows}$ by omitting the rules $(\leftarrow\vdash)$, $(\vdash\leftarrow)$ and $(\mathsf{b(cb)}_t\vdash)$, and at the same time, by adding the rules below.[18]

$$\frac{\mathfrak{A}[\mathcal{A}] \vdash \mathcal{C}}{\mathfrak{A}[\mathcal{A}\wedge\mathcal{B}] \vdash \mathcal{C}}\ \wedge\vdash_1 \qquad \frac{\mathfrak{A}[\mathcal{B}] \vdash \mathcal{C}}{\mathfrak{A}[\mathcal{A}\wedge\mathcal{B}] \vdash \mathcal{C}}\ \wedge\vdash_2 \qquad \frac{\mathfrak{A} \vdash \mathcal{A} \qquad \mathfrak{A} \vdash \mathcal{B}}{\mathfrak{A} \vdash \mathcal{A}\wedge\mathcal{B}}\ \vdash\wedge$$

$$\frac{\mathfrak{A}[\mathcal{A}] \vdash \mathcal{C} \qquad \mathfrak{A}[\mathcal{B}] \vdash \mathcal{C}}{\mathfrak{A}[\mathcal{A}\vee\mathcal{B}] \vdash \mathcal{C}}\ \vee\vdash \qquad \frac{\mathfrak{A} \vdash \mathcal{A}}{\mathfrak{A} \vdash \mathcal{A}\vee\mathcal{B}}\ \vdash\vee_1 \qquad \frac{\mathfrak{A} \vdash \mathcal{B}}{\mathfrak{A} \vdash \mathcal{A}\vee\mathcal{B}}\ \vdash\vee_2$$

$$\frac{\mathfrak{A}[\mathfrak{B},\mathfrak{C}] \vdash \mathcal{A}}{\mathfrak{A}[\mathfrak{C},\mathfrak{B}] \vdash \mathcal{A}}\ C\vdash \qquad \frac{\mathfrak{A}[\mathfrak{B},\mathfrak{B}] \vdash \mathcal{A}}{\mathfrak{A}[\mathfrak{B}] \vdash \mathcal{A}}\ W\vdash \qquad \frac{\mathfrak{A}[\mathfrak{B}] \vdash \mathcal{A}}{\mathfrak{A}[\mathfrak{B},\mathfrak{C}] \vdash \mathcal{A}}\ K\vdash$$

The notion of a *provable consecution* is as before. Also, $\mathcal{A}$ is a *theorem* iff $t \vdash \mathcal{A}$ is provable.

Theorem 5.37. *The single cut rule is* admissible *in* LE^t_+.

Exercise 5.4.10. Prove the theorem. [Hint: You may modify the proof for $LE^{\circ t}_{\leftrightarrows}$ by omitting and adding steps—in parallel to how LE^t_+ emerges from $LE^{\circ t}_{\leftrightarrows}$.]

The *axiomatic formulation* of E^t_+ may be obtained from $E^{\circ t}_{\leftrightarrows}$ by omitting the two rules that involve $\leftarrow$, and then, by adding the following eight axioms and the rule of adjunction. (We continue the numbering of the axioms and rules from the previous section.)

(A16) $(\mathcal{A}\wedge\mathcal{B}) \to \mathcal{A}$

(A17) $(\mathcal{A}\wedge\mathcal{B}) \to \mathcal{B}$

(A18) $((\mathcal{C}\to\mathcal{A})\wedge(\mathcal{C}\to\mathcal{B})) \to \mathcal{C} \to (\mathcal{A}\wedge\mathcal{B})$

(A19) $\mathcal{A} \to (\mathcal{A}\vee\mathcal{B})$

(A20) $\mathcal{B} \to (\mathcal{A}\vee\mathcal{B})$

(A21) $((\mathcal{A}\to\mathcal{C})\wedge(\mathcal{B}\to\mathcal{C})) \to (\mathcal{A}\vee\mathcal{B}) \to \mathcal{C}$

(A22) $(\mathcal{A}\wedge(\mathcal{B}\vee\mathcal{C})) \to ((\mathcal{A}\wedge\mathcal{B})\vee(\mathcal{A}\wedge\mathcal{C}))$

(A23) $(((\mathcal{A}\to\mathcal{A})\to\mathcal{A})\wedge((\mathcal{B}\to\mathcal{B})\to\mathcal{B})) \to$
$((\mathcal{A}\wedge\mathcal{B})\to(\mathcal{A}\wedge\mathcal{B}))\to(\mathcal{A}\wedge\mathcal{B})$

(R7) $\mathcal{A}$ and $\mathcal{B}$ imply $\mathcal{A}\wedge\mathcal{B}$

Exercise 5.4.11. Prove that LE^t_+ and E^t_+ are one and the same logic in the sense that their sets of theorems coincide.

[18] This calculus was originally defined in [34].

If we add (A16)–(A22) to $T^{\circ t}_{\leftrightarrows}$, then we obtain the calculus $T^{\circ t}_{\mathbf{+}}$, which was denoted by $T^{\circ t}_{\overset{+}{\leftarrow}}$ in [32, §3.6]. Here we omit $\leftarrow$ to avoid cramming too many symbols into the label, but we boldify $+$. The corresponding consecution calculus $LT^{\circ t}_{\mathbf{+}}$ is obtained by adding the conjunction rules $((\wedge\vdash_1), (\wedge\vdash_2)$ and $(\vdash\wedge))$, the disjunction rules $((\vee\vdash), (\vdash\vee_1)$ and $(\vdash\vee_2))$, as well as the three extensional structural rules $((C\vdash), (W\vdash)$ and $(K\vdash))$ to $T^{\circ t}_{\leftrightarrows}$.

Theorem 5.38. *The single cut rule is* admissible *in* $LT^{\circ t}_{\mathbf{+}}$.

Theorem 5.39. $T^{\circ t}_{\mathbf{+}}$ *and* $LT^{\circ t}_{\mathbf{+}}$ *are formalizations of the same logic.*

Exercise* 5.4.12. Prove the previous two theorems.

$B^t_{\rightarrow}$ and $B^{\circ t}_{\leftrightarrows}$ may be similarly extended to B^t_+ and $B^{\circ t}_{\mathbf{+}}$, respectively.

Exercise 5.4.13.** Formulate the consecution calculi LB^t_+ and $LB^{\circ t}_{\mathbf{+}}$, and state and prove appropriate meta-theorems about them.

To conclude this section we very briefly overview some other logics. The contraction-free fragment of positive ticket entailment, TW_+, attracted some attention as a logic on its own. Early on, it was viewed as a sort of minimal logic in the relevance family (though it is not the same logic as B_+), because of the so-called $P\!-\!W$ *conjecture*, which was proved in [129]. The $P\!-\!W$ theorem states that if $\mathcal{A}\rightarrow\mathcal{B}$ and $\mathcal{B}\rightarrow\mathcal{A}$ are both theorems of $TW_{\rightarrow}$, then $\mathcal{A}$ and $\mathcal{B}$ are identical to each other.[19] Another source of interest might have been the aim to get a grasp on the difficulty that was inherent in the decidability problem of $T_{\rightarrow}$.

The language of the *consecution calculus* LTW^t_+ contains the connectives $\rightarrow, \circ, \wedge, \vee$ and $\boldsymbol{t}$. It is formulated by omitting $(\mathsf{W}\vdash)$ as well as the $(\leftarrow\vdash)$ and $(\vdash\leftarrow)$ and $(\mathsf{b}(\mathsf{cb})_{\boldsymbol{t}}\vdash)$ rules from $LT^t_{\mathbf{+}}$. This calculus differs from Giambrone's calculus for the same logic in [8, §67] or in [95] not only notationally, but in the way $\boldsymbol{t}$ is dealt with. We have two special structural rules $(\mathsf{M}_{\boldsymbol{t}}\vdash)$ and $(\mathsf{b}_{\boldsymbol{t}}\vdash)$, whereas Giambrone's calculus has a rule, which is like $(\hat{\boldsymbol{t}}_r\vdash)$ except that it allows a $\boldsymbol{t}$ to disappear from the left-hand side of a structure. Using our notation, his $(\boldsymbol{t}^-\vdash)$ rule is $(\hat{\boldsymbol{t}}_l\vdash)$.

$$\frac{\mathfrak{A}[\,\boldsymbol{t},\mathfrak{B}]\vdash\mathcal{A}}{\mathfrak{A}[\mathfrak{B}]\vdash\mathcal{A}}\ \hat{\boldsymbol{t}}_l\vdash$$

Exercise 5.4.14. Show that if $(\hat{\boldsymbol{t}}_l\vdash)$ is added to LTW^t_+, then $(\mathsf{M}_{\boldsymbol{t}}\vdash)$ and $(\mathsf{b}_{\boldsymbol{t}}\vdash)$ are derived rules.

Exercise* 5.4.15. Investigate whether the single cut rule may be proved admissible using the technique in Section 7.3 for the calculus that is almost like LTW^t_+, but has $(\hat{\boldsymbol{t}}_l\vdash)$ instead of our two special structural rules.

[19] At some point in the past, T was denoted by P, and W with or without a hyphen was attached to indicate *the absence of contraction*. (This notational convention is not related to the W in the label LW in the previous section, where the W alludes to *weakening*.)

A disadvantage of the $(\hat{t}_l \vdash)$ rule is that even if the cut theorem holds, the subformula property does not hold.

Exercise* 5.4.16. Extend the other implicational calculi from the previous section ($LR^t_{\to;}$, $LL^t_{\to}$ and $LLW^t_{\to}$) with $\circ$, $\leftarrow$, $\wedge$ and $\vee$ and prove the cut theorem for them. [Hint: The methods used for E, T and B will be applicable.]

Exercise* 5.4.17. Formulate axiomatic versions of the calculi in the previous exercise and prove that the consecution and the axiomatic calculi are pairwise equivalent with respect to their sets of theorems. [Hint: Depending on how you structure your proofs and how much detail you include, this exercise may be somewhat lengthy.]

Figure 5.1 (below) summarizes the relationships between the logics in the last two sections. The omission of super- and subscripts are justified by the following theorem.

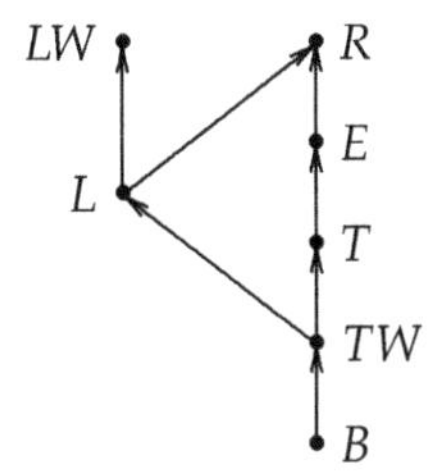

Figure 5.1. Relationships between logics.

Theorem 5.40. *If X is one of the logics in this section, then $LX^{\circ t}_{\leftrightarrows}$ and LX^t_{+} are* conservative extensions *of $LX^t_{\to}$. Furthermore, $LX^{\circ t}_{+}$ is a* conservative extension *of $LX^t_{\to}$, $LX^{\circ t}_{\leftrightarrows}$ and LX^t_{+}.*

The truth of this theorem follows from the cut theorems, and demonstrates that consecution calculi are well-suited to build up a logic gradually.

5.5 Calculi with multiple right-hand side

The sequent calculus for classical logic exhibits a *pleasant symmetry* in the sense that there can be finitely many formulas on the left-hand side of the turnstile, and also on its right-hand side. The interpretations of the multiplicities are conjunction and disjunction, respectively.

Conjunction and disjunction are usually thought to have certain properties that may be summarized by saying that they are *lattice operations*. A lattice may be distributive, and accordingly, conjunction and disjunction may distribute over each other. We have seen in Section 4.6 that given a usual formulation of the rules for $\wedge$ and $\vee$, the proof of the distributivity of $\wedge$ over $\vee$ requires each of the three structural rules (i.e., thinning, contraction and permutation).

Let us assume that the rules for $\wedge$ and $\vee$ are as in LK, and we have the three left structural rules too, though we hide the permutation steps. The next tree is a proof of the distributivity of $\vee$ over $\wedge$.

$$
\dfrac{\dfrac{\dfrac{\dfrac{\dfrac{\mathcal{A}\vdash\mathcal{A}}{\mathcal{A}\vdash\mathcal{A}\vee(\mathcal{B}\wedge\mathcal{C})}}{\mathcal{A},\mathcal{A}\vee\mathcal{C}\vdash\mathcal{A}\vee(\mathcal{B}\wedge\mathcal{C})} \qquad \dfrac{\dfrac{\dfrac{\mathcal{A}\vdash\mathcal{A}}{\mathcal{A}\vdash\mathcal{A}\vee(\mathcal{B}\wedge\mathcal{C})}}{\mathcal{B},\mathcal{A}\vdash\mathcal{A}\vee(\mathcal{B}\wedge\mathcal{C})} \quad \dfrac{\dfrac{\dfrac{\mathcal{B}\vdash\mathcal{B}}{\mathcal{B},\mathcal{C}\vdash\mathcal{B}} \quad \dfrac{\mathcal{C}\vdash\mathcal{C}}{\mathcal{B},\mathcal{C}\vdash\mathcal{C}}}{\mathcal{B},\mathcal{C}\vdash\mathcal{B}\wedge\mathcal{C}}}{\mathcal{B},\mathcal{C}\vdash\mathcal{A}\vee(\mathcal{B}\wedge\mathcal{C})}}{\mathcal{B},\mathcal{A}\vee\mathcal{C}\vdash\mathcal{A}\vee(\mathcal{B}\wedge\mathcal{C})}}{\mathcal{A}\vee\mathcal{B},\mathcal{A}\vee\mathcal{C}\vdash\mathcal{A}\vee(\mathcal{B}\wedge\mathcal{C})}}{\mathcal{A}\vee\mathcal{B},(\mathcal{A}\vee\mathcal{B})\wedge(\mathcal{A}\vee\mathcal{C})\vdash\mathcal{A}\vee(\mathcal{B}\wedge\mathcal{C})}}{(\mathcal{A}\vee\mathcal{B})\wedge(\mathcal{A}\vee\mathcal{C}),(\mathcal{A}\vee\mathcal{B})\wedge(\mathcal{A}\vee\mathcal{C})\vdash\mathcal{A}\vee(\mathcal{B}\wedge\mathcal{C})}}{(\mathcal{A}\vee\mathcal{B})\wedge(\mathcal{A}\vee\mathcal{C})\vdash\mathcal{A}\vee(\mathcal{B}\wedge\mathcal{C})}
$$

Exercise 5.5.1. Pinpoint in the above proof the applications of the structural rules—including where the $(\mathrm{C}\vdash)$ rule was used tacitly.

The other "half" of the distributivity of disjunction over conjunction is provable though without any structural rules.

$$
\dfrac{\dfrac{\dfrac{\mathcal{A}\vdash\mathcal{A}}{\mathcal{A}\vdash\mathcal{A}\vee\mathcal{B}} \quad \dfrac{\dfrac{\mathcal{B}\vdash\mathcal{B}}{\mathcal{B}\vdash\mathcal{A}\vee\mathcal{B}}}{\mathcal{B}\wedge\mathcal{C}\vdash\mathcal{A}\vee\mathcal{B}}}{\mathcal{A}\vee(\mathcal{B}\wedge\mathcal{C})\vdash\mathcal{A}\vee\mathcal{B}} \qquad \dfrac{\dfrac{\mathcal{A}\vdash\mathcal{A}}{\mathcal{A}\vdash\mathcal{A}\vee\mathcal{C}} \quad \dfrac{\dfrac{\mathcal{C}\vdash\mathcal{C}}{\mathcal{C}\vdash\mathcal{A}\vee\mathcal{C}}}{\mathcal{B}\wedge\mathcal{C}\vdash\mathcal{A}\vee\mathcal{C}}}{\mathcal{A}\vee(\mathcal{B}\wedge\mathcal{C})\vdash\mathcal{A}\vee\mathcal{C}}}{\mathcal{A}\vee(\mathcal{B}\wedge\mathcal{C})\vdash(\mathcal{A}\vee\mathcal{B})\wedge(\mathcal{A}\vee\mathcal{C})}
$$

Of course, this proof shows that $(\mathcal{A}\vee(\mathcal{B}\wedge\mathcal{C}))\supset((\mathcal{A}\vee\mathcal{B})\wedge(\mathcal{A}\vee\mathcal{C}))$ is not really distributivity, in the lattice-theoretic sense of the word.

Exercise* 5.5.2. Establish that if in a lattice $a\wedge(b\vee c)\leq(a\wedge b)\vee(a\wedge c)$, for any a,b and c, then also $(a\vee b)\wedge(a\vee c)\leq a\vee(b\wedge c)$, and vice versa. [Hint: The proof is not difficult, however, this book does not cover the presupposed algebraic background; see, for instance, [38, App. C] and references therein.]

From the above, it seems that each of the contraction, thinning and permutation rules is necessary to prove distributivity; also, there is exactly one formula on the right-hand side of each sequent, that is, multiplicity on the right was not utilized.

Let us recall from Chapter 3 that one of the roles that multiple formulas on the right-hand side can play was to obtain the specifically classical properties of $\supset$ and $\neg$. Although intuitionistic implication is not the same as the classical $\supset$, it retains its close relationship to $\wedge$. In particular, the structural alter ego of $\wedge$, that is, $,$ can still serve as the separator between formulas on the left of the $\vdash$.

In the logics in Section 5.4, the structural connective that is related to $\rightarrow$ is $;$. If we would like to add a connective that is the dual of $\circ$ in the way that $\vee$ is the dual of $\wedge$, then it is straightforward to contemplate a structural connective, namely, $;$ appearing on the right-hand side of the $\vdash$, which is to be interpreted as the dual of fusion. In the relevance logic literature, this dual connective has been called *fission* and denoted by $+$. In R, it is definable using the (De Morgan) negation of R and fusion.

The cut rule has been investigated in an algebraic setting by Dunn and Hardegree in [79], and they have described *hemi-distributivity* principles that must hold for cut. The algebraic principles, as inequations, are very clear.

Definition 5.41. (Hemi-distribution) Let $\mathfrak{A} = \langle A; \leq, \circ, + \rangle$ be an ordered algebra of similarity type $\langle 2, 2 \rangle$, where $\circ$ and $+$ are monotone operations. The *hemi-distributivity* of $\circ$ and $+$ is expressed by (1)–(4).

(1) $(a+b) \circ c \leq (a \circ c) + b$ (2) $a \circ (b+c) \leq b + (a \circ c)$

(3) $(a+b) \circ c \leq a + (b \circ c)$ (4) $a \circ (b+c) \leq (a \circ b) + c$

We already have rules for $\circ$, for instance, $(\circ\vdash)$ and $(\vdash\circ)$ from Section 5.4, and we can "dualize" those rules to obtain rules for $+$, the dual of $\circ$.

$$\frac{\mathcal{A} \vdash \mathfrak{A} \qquad \mathcal{B} \vdash \mathfrak{B}}{\mathcal{A} + \mathcal{B} \vdash \mathfrak{A}; \mathfrak{B}}\ {+\vdash} \qquad\qquad \frac{\mathfrak{A} \vdash \mathfrak{B}[\mathcal{A}; \mathcal{B}]}{\mathfrak{A} \vdash \mathfrak{B}[\mathcal{A} + \mathcal{B}]}\ {\vdash +}$$

Of course, simply combining the two pairs of rules (and the (id) axiom) leaves us with a very restricted system, because the rule for $\circ$ assumed exactly one formula on the right, and the rules above for $+$, assume a single formula on the left. Still, using these rules and a cut rule, we can see that the hemi-distributivity principles follow.

$$\dfrac{\dfrac{{+\vdash}\ \dfrac{\mathcal{A} \vdash \mathcal{A} \qquad \mathcal{B} \vdash \mathcal{B}}{\mathcal{A} + \mathcal{B} \vdash \mathcal{A}; \mathcal{B}} \qquad \dfrac{\mathcal{A} \vdash \mathcal{A} \qquad \mathcal{C} \vdash \mathcal{C}}{\mathcal{A}; \mathcal{C} \vdash \mathcal{A} \circ \mathcal{C}}\ {\vdash\circ}}{\mathcal{A} + \mathcal{B}; \mathcal{C} \vdash \mathcal{A} \circ \mathcal{C}; \mathcal{B}}\ \text{cut}}{(\mathcal{A} + \mathcal{B}) \circ \mathcal{C} \vdash (\mathcal{A} \circ \mathcal{C}) + \mathcal{B}}\ {\circ\vdash, \vdash+}$$

The next two hemi-distributivity principles are also provable.

$$\dfrac{\dfrac{\dfrac{\mathcal{B} \vdash \mathcal{B} \quad \mathcal{C} \vdash \mathcal{C}}{\mathcal{B} + \mathcal{C} \vdash \mathcal{B}; \mathcal{C}} \qquad \dfrac{\mathcal{A} \vdash \mathcal{A} \quad \mathcal{C} \vdash \mathcal{C}}{\mathcal{A}; \mathcal{C} \vdash \mathcal{A} \circ \mathcal{C}}}{\mathcal{A}; \mathcal{B} + \mathcal{C} \vdash \mathcal{B}; \mathcal{A} \circ \mathcal{C}}}{\mathcal{A} \circ (\mathcal{B} + \mathcal{C}) \vdash \mathcal{B} + (\mathcal{A} \circ \mathcal{C})} \qquad \dfrac{\dfrac{\dfrac{\mathcal{A} \vdash \mathcal{A} \quad \mathcal{B} \vdash \mathcal{B}}{\mathcal{A} + \mathcal{B} \vdash \mathcal{A}; \mathcal{B}} \qquad \dfrac{\mathcal{B} \vdash \mathcal{B} \quad \mathcal{C} \vdash \mathcal{C}}{\mathcal{B}; \mathcal{C} \vdash \mathcal{B} \circ \mathcal{C}}}{\mathcal{A} + \mathcal{B}; \mathcal{C} \vdash \mathcal{A}; \mathcal{B} \circ \mathcal{C}}}{(\mathcal{A} + \mathcal{B}) \circ \mathcal{C} \vdash \mathcal{A} + (\mathcal{B} \circ \mathcal{C})}$$

Exercise 5.5.3. We have listed one more hemi-distributivity principle above. Prove that too. Are there any other similar principles?

The careful reader surely noticed that we "cheated" in these three derivations. In the first one, we explicitly labeled a step as cut. However, the step does not conform precisely to the cut rule in the right singular systems, where the left premise of the cut cannot have some other formula next to the cut formula.

For example, we would need a cut rule in the following form to justify the first proof.

$$\frac{\mathcal{A} \vdash \mathcal{B};\mathcal{C} \quad \mathcal{B};\mathcal{D} \vdash \mathcal{E}}{\mathcal{A};\mathcal{D} \vdash \mathcal{E};\mathcal{C}}$$

Cut rules along these lines (though not exactly as given here) were considered by Lambek in [122]. The above rule—together with three others—would suffice for the proof of the hemi-distributivity principles above. But we do not want to restrict all consecutions to a pair of formulas, and most importantly, we would like to have a calculus permitting *multiple succedents* with a *cut rule admissible.* The cut rule will take the following shape, and it should be obvious that it is a rather straightforward generalization of the cut rule in LK.

$$\frac{\mathfrak{A} \vdash \mathfrak{B}[C] \quad \mathfrak{C}[C] \vdash \mathfrak{D}}{\mathfrak{C}[\mathfrak{A}] \vdash \mathfrak{B}[\mathfrak{D}]} \text{ algebraic cut}$$

We have used the name "multi-cut" to refer to a version of the cut rule, in which C, the cut formula, may have several occurrences in the right premise, and one or more of those occurrences are replaced by the antecedent of the left premise. The label "algebraic cut" comes from [79] and bespeaks the original context of the investigations into the cut rule in that book. To put it differently, the *algebraic cut* is the single cut rule for *consecutions with multiple right-hand sides.*

The generalization of the multi-cut is equally straightforward. Recall that $\mathfrak{A}\langle\mathfrak{B}\rangle$ indicates at least one but possibly more than one occurrence of $\mathfrak{B}$ in the structure $\mathfrak{A}$. The next version of the cut rule is a generalization of the mix rule too, but without the assumption that all the occurrences of C have been selected either in the left or in the right premise.

$$\frac{\mathfrak{A} \vdash \mathfrak{B}\langle C\rangle \quad \mathfrak{C}\langle C\rangle \vdash \mathfrak{D}}{\mathfrak{C}\langle\mathfrak{A}\rangle \vdash \mathfrak{B}\langle\mathfrak{D}\rangle} \text{ symmetric cut}$$

The name "symmetric cut" for this rule originates, again, in [79], though we have abbreviated the label slightly by omitting "algebraic."[20]

There are *several threads of research* that braid together at this point. We have already mentioned that the cut rule has been investigated in an abstract algebraic way, which led to the two latest formulations of the cut rule. Another motivation to consider principles, such as the hemi-distributivity

[20]To be precise, we shortened both "single algebraic cut" to "algebraic cut" and "symmetric algebraic cut" to "symmetric cut."

principles, stems from dualizing an ordered residuated groupoid. Lastly, linguists working with categorial grammars and extending LQ re-discovered these and similar principles; for an example, see Moortgat's [139]. The main concerns for linguists are phenomena, which show that a natural language allows certain permutations of words without generating all the possible sequences of words.

The calculus we are to present in this section (and another one in Section 6.3) is from [40]. The historical predecessor of that paper is [39], which is a semantical (rather than proof-theoretic) investigation of some of the same algebras, in particular, of their representations by sets.

Residuated semi-groups and *residuated groupoids* are ubiquitous in algebraic considerations of logic. Perhaps, the prototypical examples, at least from the point of view of this and the previous chapters, are the algebras of LA and LQ. The Lambek calculi generalize the implication-conjunction fragment of LJ. Algebraically, one might seek to generalize Heyting algebras.[21]

Definition 5.42. Let $\mathfrak{A}_\mathfrak{g} = \langle A; \leq, \leftarrow, \circ, \rightarrow, \succ, +, \prec \rangle$ be an ordered algebra of similarity type $\langle 2,2,2,2,2,2 \rangle$. $\mathfrak{A}_\mathfrak{g}$ satisfies *eight quasi-inequations*, which are summarized in (1) and (2).

$$(1)\ b \leq c \leftarrow a \quad \Leftrightarrow \quad a \circ b \leq c \quad \Leftrightarrow \quad a \leq b \rightarrow c$$

$$(2)\ a \succ b \leq c \quad \Leftrightarrow \quad a \leq b + c \quad \Leftrightarrow \quad c \prec a \leq b$$

This algebra comprises two copies of a partially ordered residuated groupoid, however, they are spliced together in a particular way. Namely, in one copy, the order relation is inverted, which puts $\circ$ and $+$ on different sides of $\leq$. When operations are connected to each other via residuation, certain *tonicity* properties follow. Given a partial order $\leq$, an operation o is *monotone* iff $a \leq b$ implies $o(a) \leq o(b)$, whereas o is *antitone* iff $o(b) \leq o(a)$ is implied. If an operation has either of these properties, then it has a tonicity.

Exercise 5.5.4. Determine whether the operations in $\mathfrak{A}_\mathfrak{g}$ have a tonicity, and if so, which one. [Hint: There is a pattern as to how to use residuation to find tonicities.]

Purely mathematically, there are twelve other ways to pair two of the six operations to resemble the hemi-distributivity principles. We list them continuing the numbering from Definition 5.41.

(5) $(a \rightarrow b) \circ c \leq a \rightarrow (b \circ c)$	(6) $a \circ (b \rightarrow c) \leq b \rightarrow (a \circ c)$
(7) $(b \leftarrow a) \circ c \leq (b \circ c) \leftarrow a$	(8) $a \circ (c \leftarrow b) \leq (a \circ c) \leftarrow b$
(9) $(a + b) \succ c \leq (a \succ c) + b$	(10) $(a + b) \succ c \leq a + (b \succ c)$
(11) $a \prec (b + c) \leq (a \prec b) + c$	(12) $a \prec (b + c) \leq b + (a \prec c)$

[21] Grishin proposed a particular generalization, which was somewhat modified in [39].

(13) $(a \to b) \succ c \leq a \to (b \succ c)$ (14) $(b \leftarrow a) \succ c \leq (b \succ c) \leftarrow a$

(15) $a \prec (b \to c) \leq b \to (a \prec c)$ (16) $a \prec (c \leftarrow b) \leq (a \prec c) \leftarrow b$

These inequations are *independent* from $\mathfrak{A}_\mathfrak{g}$ in the sense that they do not follow from $\leftarrow$ and $\to$ being the residuals of $\circ$, and dually, $\prec$ and $\succ$ being the residuals of $+$. Furthermore, for all i, j, where $1 \leq i, j \leq 16$, (i) iff (j) is *not derivable* in $\mathfrak{A}_\mathfrak{g}$ when i and j are distinct.

Now, having some algebras for reference, we turn back to formulating consecution calculi with multiple right-hand sides.

The notion of structures remains as in Definition 5.1. A *consecution* is a pair of structures with $\vdash$ between them. We define two calculi; the first is a *minimal hemi-distributive calculus*, denoted by L_{HD}, and the second is a *residuated hemi-distributive calculus*, denoted by $L_{HD?}$. The latter is an extension of the former. (The notation for this latter calculus is motivated by the fact that it is not as well behaved as we would like it to be.)

The axiom and the rules of L_{HD} are as follows.

$$\mathcal{A} \vdash \mathcal{A} \ \text{id}$$

$$\frac{\mathfrak{A}[\mathcal{A};\mathcal{B}] \vdash \mathfrak{B}}{\mathfrak{A}[\mathcal{A} \circ \mathcal{B}] \vdash \mathfrak{B}} \circ\vdash \qquad \frac{\mathfrak{A} \vdash \mathfrak{B}[\mathcal{A}] \quad \mathfrak{C} \vdash \mathfrak{D}[\mathcal{B}]}{\mathfrak{A};\mathfrak{C} \vdash \mathfrak{D}[\mathfrak{B}[\mathcal{A} \circ \mathcal{B}]]} \vdash\circ_1 \qquad \frac{\mathfrak{A} \vdash \mathfrak{B}[\mathcal{A}] \quad \mathfrak{C} \vdash \mathfrak{D}[\mathcal{B}]}{\mathfrak{A};\mathfrak{C} \vdash \mathfrak{B}[\mathfrak{D}[\mathcal{A} \circ \mathcal{B}]]} \vdash\circ_2$$

$$\frac{\mathfrak{A}[\mathcal{A}] \vdash \mathfrak{B} \quad \mathfrak{C}[\mathcal{B}] \vdash \mathfrak{D}}{\mathfrak{A}[\mathfrak{C}[\mathcal{A} + \mathcal{B}]] \vdash \mathfrak{B};\mathfrak{D}} +\vdash_1 \qquad \frac{\mathfrak{A}[\mathcal{A}] \vdash \mathfrak{B} \quad \mathfrak{C}[\mathcal{B}] \vdash \mathfrak{D}}{\mathfrak{C}[\mathfrak{A}[\mathcal{A} + \mathcal{B}]] \vdash \mathfrak{B};\mathfrak{D}} +\vdash_2 \qquad \frac{\mathfrak{A} \vdash \mathfrak{B}[\mathcal{A};\mathcal{B}]}{\mathfrak{A} \vdash \mathfrak{B}[\mathcal{A} + \mathcal{B}]} \vdash+$$

There are other connectives, for instance, $\wedge$ and $\vee$ in LK, which come with three operational rules. However, here the duplication of the right rule for $\circ$ and of the left rule for $+$ happens not because of a variation in the premises. Rather, the two structures containing the subalterns are embedded into each other by giving priority, once to $\mathcal{A}$'s, then to $\mathcal{B}$'s structure. There is no similar problem with $\wedge$ and $\vee$, because the contexts are the same in the premises.

The algebraic cut is admissible in L_{HD}, however, the proof of this theorem is quite a bit more complicated than the proof for LK, for instance. (See Section 7.5 for the details.)

$L_{HD?}$ extends L_{HD} with rules for the residuals of $\circ$ and $+$.

$$\frac{\mathfrak{A} \vdash \mathcal{A} \quad \mathfrak{B}[\mathcal{B}] \vdash \mathfrak{C}}{\mathfrak{B}[\mathcal{A} \to \mathcal{B};\mathfrak{A}] \vdash \mathfrak{C}} \to\vdash \qquad \frac{\mathfrak{A};\mathcal{A} \vdash \mathfrak{B}[\mathcal{B}]}{\mathfrak{A} \vdash \mathfrak{B}[\mathcal{A} \to \mathcal{B}]} \vdash\to$$

$$\frac{\mathfrak{A} \vdash \mathcal{A} \quad \mathfrak{B}[\mathcal{B}] \vdash \mathfrak{C}}{\mathfrak{B}[\mathfrak{A};\mathcal{B} \leftarrow \mathcal{A}] \vdash \mathfrak{C}} \leftarrow\vdash \qquad \frac{\mathcal{A};\mathfrak{A} \vdash \mathfrak{B}[\mathcal{B}]}{\mathfrak{A} \vdash \mathfrak{B}[\mathcal{B} \leftarrow \mathcal{A}]} \vdash\leftarrow$$

$$\frac{\mathfrak{A}[\mathcal{A}] \vdash \mathcal{B};\mathfrak{B}}{\mathfrak{A}[\mathcal{A} \succ \mathcal{B}] \vdash \mathfrak{B}} \succ\vdash \qquad \frac{\mathfrak{A} \vdash \mathfrak{B}[\mathcal{A}] \quad \mathcal{B} \vdash \mathfrak{C}}{\mathfrak{A} \vdash \mathfrak{B}[\mathfrak{C};\mathcal{A} \succ \mathcal{B}]} \vdash\succ$$

$$\frac{\mathfrak{A}[\mathcal{A}] \vdash \mathfrak{B}; \mathcal{B}}{\mathfrak{A}[\mathcal{B} \prec \mathcal{A}] \vdash \mathfrak{B}} \prec\vdash \qquad \frac{\mathfrak{A} \vdash \mathfrak{B}[\mathcal{A}] \quad \mathcal{B} \vdash \mathfrak{C}}{\mathfrak{A} \vdash \mathfrak{B}[\mathcal{B} \prec \mathcal{A}; \mathfrak{C}]} \vdash\prec$$

The cut rule is *not admissible* in $L_{HD?}$. It is easy to prove this claim, once we have an idea for a counterexample. For the sake of transparency, we choose three meta-variables, $\mathcal{A}$, $\mathcal{B}$ and $\mathcal{C}$. Here is a chunk of a proof that involves an application of the algebraic cut.

$$\dfrac{\dfrac{\mathcal{C} \vdash \mathcal{C} \quad \mathcal{A} \vdash \mathcal{A}}{\mathcal{C} \vdash \mathcal{A}; \mathcal{C} \succ \mathcal{A}} \vdash\succ \qquad \dfrac{\mathcal{A} \vdash \mathcal{A} \quad \mathcal{B} \vdash \mathcal{B}}{\mathcal{A} \to \mathcal{B}; \mathcal{A} \vdash \mathcal{B}} \to\vdash}{\mathcal{A} \to \mathcal{B}; \mathcal{C} \vdash \mathcal{B}; \mathcal{C} \succ \mathcal{A}} \text{ algebraic cut}$$

Exercise 5.5.5. Verify that the consecution $\mathcal{A} \to \mathcal{B}; \mathcal{C} \vdash \mathcal{B}; \mathcal{C} \succ \mathcal{A}$ has no proof without an application of the cut rule.

We always aim at defining a sequent calculus so that the cut rule is admissible, but the counterexample is hardly a surprise. Everybody knows that there are sixteen binary connectives in classical propositional logic, though usually less than half of them are named and denoted by a symbol. Occasionally, $\not\supset$ is introduced as a connective. For example, Church in [64] has this connective. As the notation may suggest, $\mathcal{A} \not\supset \mathcal{B}$ is $\neg(\mathcal{A} \supset \mathcal{B})$; hence, it is equivalent to $\mathcal{A} \wedge \neg\mathcal{B}$. However, let us treat $\not\supset$ as a symbol with no separate components, that is, as a primitive binary connective exactly like $\wedge$, $\vee$ or $\supset$ are. We add the following two rules to LK.

$$\frac{\mathcal{A}, \Gamma \vdash \Delta, \mathcal{B}}{\mathcal{A} \not\supset \mathcal{B}, \Gamma \vdash \Delta} \not\supset\vdash \qquad \frac{\Gamma \vdash \Delta, \mathcal{A} \qquad \mathcal{B}, \Theta \vdash \Lambda}{\Gamma, \Theta \vdash \Delta, \Lambda, \mathcal{A} \not\supset \mathcal{B}} \vdash\not\supset$$

The extension seems innocuous and natural. We even made the $(\vdash\not\supset)$ rule to resemble closely the $(\supset\vdash)$ rule in LK by stipulating different contexts in the premises. Beyond exclusively formal or aesthetic considerations, we may note that $\not\supset$ behaves as informally foretold. Here is a short proof of the sequent $\mathcal{A} \wedge \neg\mathcal{B} \vdash \mathcal{A} \not\supset \mathcal{B}$.

$$\dfrac{\dfrac{\dfrac{\dfrac{\dfrac{\dfrac{\dfrac{\mathcal{A} \vdash \mathcal{A} \qquad \mathcal{B} \vdash \mathcal{B}}{\mathcal{A} \vdash \mathcal{B}, \mathcal{A} \not\supset \mathcal{B}} \vdash\not\supset}{\mathcal{A} \vdash \mathcal{A} \not\supset \mathcal{B}, \mathcal{B}} \vdash C}{\neg\mathcal{B}, \mathcal{A} \vdash \mathcal{A} \not\supset \mathcal{B}} \neg\vdash}{\mathcal{A} \wedge \neg\mathcal{B}, \mathcal{A} \vdash \mathcal{A} \not\supset \mathcal{B}} \wedge\vdash_2}{\mathcal{A}, \mathcal{A} \wedge \neg\mathcal{B} \vdash \mathcal{A} \not\supset \mathcal{B}} C\vdash}{\mathcal{A} \wedge \neg\mathcal{B}, \mathcal{A} \wedge \neg\mathcal{B} \vdash \mathcal{A} \not\supset \mathcal{B}} \wedge\vdash_1}{\mathcal{A} \wedge \neg\mathcal{B} \vdash \mathcal{A} \not\supset \mathcal{B}} W\vdash$$

Exercise 5.5.6. Prove the following sequents (1) $\mathcal{A} \not\supset \mathcal{B} \vdash \mathcal{A} \wedge \neg\mathcal{B}$, (2) $\neg(\mathcal{A} \supset \mathcal{B}) \vdash \mathcal{A} \not\supset \mathcal{B}$ and (3) $\mathcal{A} \not\supset \mathcal{B} \vdash \neg(\mathcal{A} \supset \mathcal{B})$ in LK.

The analog of the $L_{HD?}$ consecution $\mathcal{A} \to \mathcal{B}; \mathcal{C} \vdash \mathcal{B}; \mathcal{C} \succ \mathcal{A}$ is the sequent $\mathcal{A} \supset \mathcal{B}, \mathcal{C} \vdash \mathcal{B}, \mathcal{C} \not\supset \mathcal{A}$. The previous proof using cut—with some obvious modifications—yields a proof of the latter sequent in LK. However, $\mathcal{A} \supset \mathcal{B}, \mathcal{C} \vdash \mathcal{B}, \mathcal{C} \not\supset \mathcal{A}$ has a cut-free proof, in fact, it has two straightforward cut-free proofs, because $\supset$ and $\not\supset$ are so much alike.

$$\dfrac{\dfrac{\dfrac{\dfrac{\mathcal{C} \vdash \mathcal{C} \quad \mathcal{A} \vdash \mathcal{A}}{\mathcal{C} \vdash \mathcal{A}, \mathcal{C} \not\supset \mathcal{A}}\vdash\not\supset}{\mathcal{C} \vdash \mathcal{C} \not\supset \mathcal{A}, \mathcal{A}}\vdash\mathrm{C} \quad \mathcal{B} \vdash \mathcal{B}}{\mathcal{A} \supset \mathcal{B}, \mathcal{C} \vdash \mathcal{C} \not\supset \mathcal{A}, \mathcal{B}}\supset\vdash}{\mathcal{A} \supset \mathcal{B}, \mathcal{C} \vdash \mathcal{B}, \mathcal{C} \not\supset \mathcal{A}}\vdash\mathrm{C}
\qquad
\dfrac{\dfrac{\mathcal{C} \vdash \mathcal{C} \quad \dfrac{\dfrac{\mathcal{A} \vdash \mathcal{A} \quad \mathcal{B} \vdash \mathcal{B}}{\mathcal{A} \supset \mathcal{B}, \mathcal{A} \vdash \mathcal{B}}\supset\vdash}{\mathcal{A}, \mathcal{A} \supset \mathcal{B} \vdash \mathcal{B}}\mathrm{C}\vdash}{\mathcal{C}, \mathcal{A} \supset \mathcal{B} \vdash \mathcal{B}, \mathcal{C} \not\supset \mathcal{A}}\vdash\not\supset}{\mathcal{A} \supset \mathcal{B}, \mathcal{C} \vdash \mathcal{B}, \mathcal{C} \not\supset \mathcal{A}}\mathrm{C}\vdash$$

Now it should be obvious why similar proofs do not exist in $L_{HD?}$. In the two-premise rules for $\to$ and $\succ$ (or those for $\leftarrow$ and $\prec$), one of the premises has a subaltern alone. Perhaps, those rules are not sufficiently general and can be further generalized.

Exercise 5.5.7.** Formulate the rules for $\to$, $\leftarrow$, $\succ$ and $\prec$ so that the cut rule can be proved to be admissible.

Exercise* 5.5.8. We listed sixteen inequations (including the four hemi-distributivity principles). Which of those inequations (or more precisely, which of the corresponding consecutions) are provable in $L_{HD?}$? [Hint: None of the cases is difficult, but they are not all the same, and checking them adds up.]

Exercise* 5.5.9. Delineate which parts of $L_{HD?}$ are necessary to be able to construct counterexamples to the admissibility of the algebraic cut rule.

In [40], $L_{HD?}$ has been reformulated as a *display calculus*, which we describe in Section 6.3.

Chapter 6

Display calculi and hypersequents

Sequent and consecution calculi provide excellent control over proofs, however, their components have to be assembled in a careful manner. Even with a very delicate approach, some logics seem not to be formalizable using the kind of calculi that we have seen in the last two chapters.

A great idea, which originated in the middle of the 20th century, is *residuation*. Sometimes, operations or connectives can be defined—explicitly or contextually—from each other. A simple example is the definability of any truth-functional connective of classical logic using $\neg$ and $\wedge$. In many nonclassical logics, the connectives are not definable from each other; for instance, $\leftarrow$, $\circ$ and $\rightarrow$ in LA cannot be defined from each other. An insight behind the general theory of *relational semantics* for a wide range of nonclassical logics is that residuation is sufficient to model a pair of operations from a shared relation.[1]

Display logics exploit the idea of residuation in the setting of sequent calculi—by slightly expanding the latter notion. We consider several logics formalized as display calculi in this chapter.

Another extension of sequent calculi permits *hyper sequents*, which comprise several sequents in parallel. We consider two versions of a hypersequent calculus for the normal modal logic $S5$ toward the end of this chapter.

6.1 Display logics with star

Display logics were invented by Belnap in [19].[2] The name for these calculi derives from their *characteristic property*, namely, that every formula in a consecution can be displayed. We will make this claim more precise, however, it should already be clear that the logistic sequent and the consecution calculi that we considered so far, do not allow, in general, a formula to be exhibited as the antecedent or the succedent. In LK, we can select a formula

[1] *Gaggle theory* was invented by Dunn in [78], then further developed by him in several papers. For a comprehensive exposition, see [38].

[2] See also [8, §62]. We more or less follow the cited works, but we adjust the notation.

in a sequent and move all the other formulas surrounding it to the other side of the turnstile, but each move will add a $\neg$ to a formula, which remains part of the formula from that point on in the proof (if the proof is cut-free). In display logics, rules mimicking residuation allow formulas to be moved around without leaving an indelible mark on them. We will focus on display calculi for relevance logics, starting with DL_R, which is a display logic corresponding to CR, the Boolean extension of the logic of relevant implication.

Definition 6.1. The *language* of DL_R contains as connectives $\neg$ (Boolean negation), $\wedge$ (conjunction), $\vee$ (disjunction), $\supset$ (material implication), $\boldsymbol{F}$ (absurdity), $\boldsymbol{T}$ (triviality), $\boldsymbol{t}$ (truth), $\boldsymbol{f}$ (falsity), $\rightarrow$ (implication), $+$ (fission), $\circ$ (fusion) and $\sim$ (De Morgan negation). The *set of formulas* is inductively generated from the set of propositional variables by these connectives.

Display logics typically contain several structural connectives, even going beyond the two structural connectives (i.e., $,$ and $;$) appearing together in sequent calculi for certain positive relevance logics.

Definition 6.2. The set of *structural connectives* for DL_R contains $,$, *, $\mathbf{0}$, $;$, $^\times$ and $\mathbf{I}$. $\mathbf{0}$ and $\mathbf{I}$ are zero-ary, * and $^\times$ are unary, whereas $,$ and $;$ are binary. The *set of structures* is inductively generated from the set of formulas plus $\mathbf{0}$ and $\mathbf{I}$, by *, $^\times$, $,$ and $;$.

We used $,$ in sequences of formulas. Now this structural connective is binary, and we do not assume that it is associative (or that it may be viewed as a polyadic connective). There is *no empty structure*; hence, a consecution always contains at least one formula or structural constant on each side of the $\vdash$.

The conventions concerning omitting parentheses are as before. Concretely, implicational formulas are associated to the right, whereas structures are associated to the left—in the absence of parentheses. $\mathfrak{A}, \mathfrak{B}, \mathfrak{C}, \ldots$ are *meta-variables* for structures. A *consecution* is a pair of structures separated by $\vdash$.

The *display calculus* DL_R is defined by a set of *axioms, connective rules, structural rules* and *display rules*. The axioms and connective rules are as follows.

$$\mathcal{A} \vdash \mathcal{A} \ \text{id}$$

$$\frac{\mathbf{0} \vdash \mathfrak{A}}{\boldsymbol{T} \vdash \mathfrak{A}}\ \boldsymbol{T}\vdash \qquad \mathbf{0} \vdash \boldsymbol{T}\ \vdash\boldsymbol{T} \qquad \boldsymbol{F} \vdash \mathbf{0}\ \boldsymbol{F}\vdash \qquad \frac{\mathfrak{A} \vdash \mathbf{0}}{\mathfrak{A} \vdash \boldsymbol{F}}\ \vdash\boldsymbol{F}$$

$$\frac{\mathcal{A}, \mathcal{B} \vdash \mathfrak{A}}{\mathcal{A} \wedge \mathcal{B} \vdash \mathfrak{A}}\ \wedge\vdash \qquad \frac{\mathfrak{A} \vdash \mathcal{A} \quad \mathfrak{B} \vdash \mathcal{B}}{\mathfrak{A}, \mathfrak{B} \vdash \mathcal{A} \wedge \mathcal{B}}\ \vdash\wedge$$

$$\frac{\mathcal{A} \vdash \mathfrak{A} \quad \mathcal{B} \vdash \mathfrak{B}}{\mathcal{A} \vee \mathcal{B} \vdash \mathfrak{A}, \mathfrak{B}}\ \vee\vdash \qquad \frac{\mathfrak{A} \vdash \mathcal{A}, \mathcal{B}}{\mathfrak{A} \vdash \mathcal{A} \vee \mathcal{B}}\ \vdash\vee$$

$$\frac{\mathfrak{A} \vdash \mathcal{A} \quad \mathcal{B} \vdash \mathfrak{B}}{\mathcal{A} \supset \mathcal{B} \vdash \mathfrak{A}^*, \mathfrak{B}}\ \supset\vdash \qquad \frac{\mathfrak{A}, \mathcal{A} \vdash \mathcal{B}}{\mathfrak{A} \vdash \mathcal{A} \supset \mathcal{B}}\ \vdash\supset$$

$$\frac{\mathcal{A}^* \vdash \mathfrak{A}}{\neg\mathcal{A} \vdash \mathfrak{A}}\ \neg\vdash \qquad \frac{\mathfrak{A} \vdash \mathcal{A}^*}{\mathfrak{A} \vdash \neg\mathcal{A}}\ \vdash\neg$$

$$\frac{\mathbf{I} \vdash \mathcal{A}}{t \vdash \mathcal{A}}\ t\vdash \qquad \mathbf{I} \vdash t\ \ \vdash t \qquad f \vdash \mathbf{I}\ \ f\vdash \qquad \frac{\mathfrak{A} \vdash \mathbf{I}}{\mathfrak{A} \vdash f}\ \vdash f$$

$$\frac{\mathcal{A}; \mathcal{B} \vdash \mathfrak{A}}{\mathcal{A} \circ \mathcal{B} \vdash \mathfrak{A}}\ \circ\vdash \qquad \frac{\mathfrak{A} \vdash \mathcal{A} \quad \mathfrak{B} \vdash \mathcal{B}}{\mathfrak{A}; \mathfrak{B} \vdash \mathcal{A} \circ \mathcal{B}}\ \vdash\circ$$

$$\frac{\mathcal{A} \vdash \mathfrak{A} \quad \mathcal{B} \vdash \mathfrak{B}}{\mathcal{A} + \mathcal{B} \vdash \mathfrak{A}; \mathfrak{B}}\ +\vdash \qquad \frac{\mathfrak{A} \vdash \mathcal{A}; \mathcal{B}}{\mathfrak{A} \vdash \mathcal{A} + \mathcal{B}}\ \vdash+$$

$$\frac{\mathfrak{A} \vdash \mathcal{A} \quad \mathcal{B} \vdash \mathfrak{B}}{\mathcal{A} \to \mathcal{B} \vdash \mathfrak{A}^\times; \mathfrak{B}}\ \to\vdash \qquad \frac{\mathfrak{A}; \mathcal{A} \vdash \mathcal{B}}{\mathfrak{A} \vdash \mathcal{A} \to \mathcal{B}}\ \vdash\to$$

$$\frac{\mathcal{A}^\times \vdash \mathfrak{A}}{\sim\mathcal{A} \vdash \mathfrak{A}}\ \sim\vdash \qquad \frac{\mathfrak{A} \vdash \mathcal{A}^\times}{\mathfrak{A} \vdash \sim\mathcal{A}}\ \vdash\sim$$

The axioms and rules (save (id)) look like as if we would have made two copies of them—with changing a symbol here and there. This is a *virtue of display logic*, because the rules may be viewed from a more abstract point of view. Another noticeable feature of the rules is that most of the subalterns and all the principal formulas take up the whole antecedent or the whole succedent of a consecution.

The *structural rules* are the following.

$$\frac{\mathfrak{A} \vdash \mathfrak{B}}{\mathbf{0}, \mathfrak{A} \vdash \mathfrak{B}}\ \check{0}\vdash \qquad \frac{\mathbf{0}, \mathfrak{A} \vdash \mathfrak{B}}{\mathfrak{A} \vdash \mathfrak{B}}\ \hat{0}\vdash \qquad \frac{\mathbf{0}^* \vdash \mathfrak{A}}{\mathbf{0} \vdash \mathfrak{A}}\ \hat{0^*}\vdash$$

$$\frac{\mathfrak{A} \vdash \mathfrak{B}}{\mathfrak{A}, \mathfrak{C} \vdash \mathfrak{B}}\ K\vdash \qquad \frac{\mathfrak{A}, \mathfrak{B}, \mathfrak{B} \vdash \mathfrak{C}}{\mathfrak{A}, \mathfrak{B} \vdash \mathfrak{C}}\ W\vdash$$

$$\frac{\mathfrak{A} \vdash \mathfrak{B}}{\mathbf{I}; \mathfrak{A} \vdash \mathfrak{B}}\ \check{\mathrm{I}}\vdash \qquad \frac{\mathbf{I}; \mathfrak{A} \vdash \mathfrak{B}}{\mathfrak{A} \vdash \mathfrak{B}}\ \hat{\mathrm{I}}\vdash \qquad \frac{\mathbf{I} \vdash \mathfrak{A}}{\mathbf{I}^\times \vdash \mathfrak{A}}\ \check{\mathrm{I}^\times}\vdash$$

$$\frac{\mathfrak{A}; (\mathfrak{B}; \mathfrak{C}) \vdash \mathfrak{D}}{\mathfrak{A}; \mathfrak{B}; \mathfrak{C} \vdash \mathfrak{D}}\ B\vdash \qquad \frac{\mathfrak{A}; \mathfrak{B}; \mathfrak{C} \vdash \mathfrak{D}}{\mathfrak{A}; \mathfrak{C}; \mathfrak{B} \vdash \mathfrak{D}}\ C\vdash \qquad \frac{\mathfrak{A}; \mathfrak{B}; \mathfrak{B} \vdash \mathfrak{C}}{\mathfrak{A}; \mathfrak{B} \vdash \mathfrak{C}}\ W\vdash$$

The last group of rules delivers the display property. The consecutions on a line are *display equivalent*, meaning that any two of them can be paired into a rule, and either can be placed as the upper consecution and the other as the lower consecution.

$$\mathfrak{A}, \mathfrak{B} \vdash \mathfrak{C} \qquad \mathfrak{A} \vdash \mathfrak{B}^*, \mathfrak{C}$$

$$\mathfrak{A} \vdash \mathfrak{B}, \mathfrak{C} \qquad \mathfrak{A}, \mathfrak{B}^* \vdash \mathfrak{C} \qquad \mathfrak{A} \vdash \mathfrak{C}, \mathfrak{B}$$

$$\mathfrak{A} \vdash \mathfrak{B} \qquad \mathfrak{B}^* \vdash \mathfrak{A}^* \qquad \mathfrak{A}^{**} \vdash \mathfrak{B}$$

$$\mathfrak{A}; \mathfrak{B} \vdash \mathfrak{C} \qquad \mathfrak{A} \vdash \mathfrak{B}^\times; \mathfrak{C}$$

$$\frac{\mathfrak{A} \vdash \mathfrak{B}; \mathfrak{C}}{\mathfrak{A} \vdash \mathfrak{B}} \qquad \frac{\mathfrak{A}; \mathfrak{B}^\times \vdash \mathfrak{C}}{\mathfrak{B}^\times \vdash \mathfrak{A}^\times} \qquad \frac{\mathfrak{A} \vdash \mathfrak{C}; \mathfrak{B}}{\mathfrak{A}^{\times\times} \vdash \mathfrak{B}}$$

The notion of a *proof* is the usual one, with the rules including the display equivalences. A formula $\mathcal{A}$ is a *theorem* of DL_R iff $\boldsymbol{t} \vdash \mathcal{A}$ is provable.

In order to make the display property precise we need to distinguish between two types of occurrences of formulas and structures.

Definition 6.3. (Antecedent and succedent occurrences) Let $\mathfrak{A} \vdash \mathfrak{B}$ be a given consecution, which we abbreviate by σ. Then $\mathfrak{A}$ has an *antecedent occurrence in* σ, or $\mathfrak{a}(\mathfrak{A})$ (in σ), whereas $\mathfrak{B}$ has a *succedent occurrence in* σ, or $\mathfrak{s}(\mathfrak{B})$ (in σ). $\mathfrak{a}$ and $\mathfrak{s}$ are iteratively extended to substructures by (1)–(4).

(1) If $\mathfrak{a}(\mathfrak{A}, \mathfrak{B})$ or $\mathfrak{a}(\mathfrak{A}; \mathfrak{B})$, then $\mathfrak{a}(\mathfrak{A})$ and $\mathfrak{a}(\mathfrak{B})$;

(2) if $\mathfrak{s}(\mathfrak{A}, \mathfrak{B})$ or $\mathfrak{s}(\mathfrak{A}; \mathfrak{B})$, then $\mathfrak{s}(\mathfrak{A})$ and $\mathfrak{s}(\mathfrak{B})$;

(3) if $\mathfrak{a}(\mathfrak{A}^*)$ or $\mathfrak{a}(\mathfrak{A}^\times)$, then $\mathfrak{s}(\mathfrak{A})$;

(4) if $\mathfrak{s}(\mathfrak{A}^*)$ or $\mathfrak{s}(\mathfrak{A}^\times)$, then $\mathfrak{a}(\mathfrak{A})$.

Given a consecution, each occurrence of an atomic structure (which is a wff or **0** or **I**) is either an $\mathfrak{a}$ or an $\mathfrak{s}$ occurrence, but not both.

Exercise 6.1.1. Prove that this claim is true. [Hint: A structure has a unique formation tree, because , is binary.]

The *display property* means that given a consecution and an antecedent occurrence (a succedent occurrence) of a structure, there is a consecution derivable from the given one, in which that antecedent occurrence (that succedent occurrence) is the entire antecedent (succedent) of the consecution.

Theorem 6.4. *If σ is a consecution and $\mathfrak{a}(\mathfrak{A})$, then* there is a consecution σ', *which is derivable from σ, and is of the form $\mathfrak{A} \vdash \mathfrak{B}$ (with some $\mathfrak{B}$). Dually, if $\mathfrak{s}(\mathfrak{A})$, then* there is a derivable consecution *of the form $\mathfrak{B} \vdash \mathfrak{A}$ (with some $\mathfrak{B}$).*

Proof: We prove the claim by structural induction paralleling the definition of antecedent and succedent occurrences of structures.

1. Let the given consecution be $\mathfrak{C} \vdash \mathfrak{D}$. If $\mathfrak{a}(\mathfrak{A})$, because $\mathfrak{A}$ is $\mathfrak{C}$, then the claim is obvious. Similarly, if $\mathfrak{s}(\mathfrak{A})$, and $\mathfrak{A}$ is $\mathfrak{D}$, then the given consecution suffices.

2. Let us assume that any structure of which $\mathfrak{A}$ is a proper substructure can be displayed. If $\mathfrak{a}(\mathfrak{A})$, because of $\mathfrak{a}(\mathfrak{A}; \mathfrak{C})$, then $\mathfrak{A}; \mathfrak{C} \vdash \mathfrak{D}$ (with some $\mathfrak{D}$), by the hypothesis, hence, by a display equivalence, $\mathfrak{A} \vdash \mathfrak{C}^\times; \mathfrak{D}$. If $\mathfrak{a}(\mathfrak{A})$, because of $\mathfrak{a}(\mathfrak{C}; \mathfrak{A})$, then $\mathfrak{C}; \mathfrak{A} \vdash \mathfrak{D}$, for some $\mathfrak{D}$. The display equivalences yield the following series of consecutions, each derivable from $\mathfrak{C}; \mathfrak{A} \vdash \mathfrak{D}$. $\mathfrak{C} \vdash \mathfrak{A}^\times; \mathfrak{D}$, $\mathfrak{C}; \mathfrak{D}^\times \vdash \mathfrak{A}^\times$, $\mathfrak{A} \vdash (\mathfrak{C}; \mathfrak{D}^\times)^\times$. The latter consecution has $\mathfrak{A}$ as its antecedent, which proves the claim.

If $\mathfrak{s}(\mathfrak{A})$, because of $\mathfrak{s}(\mathfrak{A};\mathfrak{C})$, then for some $\mathfrak{D}$, $\mathfrak{D} \vdash \mathfrak{A};\mathfrak{C}$. Then we can derive the following consecutions by display equivalences. $\mathfrak{D} \vdash \mathfrak{C};\mathfrak{A}$, $\mathfrak{D};\mathfrak{C}^\times \vdash \mathfrak{A}$. If we have $\mathfrak{s}(\mathfrak{C};\mathfrak{A})$, then one fewer step yields $\mathfrak{D};\mathfrak{C}^\times \vdash \mathfrak{A}$.

If $\mathfrak{a}(\mathfrak{A})$ from $\mathfrak{s}(\mathfrak{A}^\times)$, then for some $\mathfrak{C}$, $\mathfrak{C} \vdash \mathfrak{A}^\times$, by hypothesis. Obviously, $\mathfrak{A}^{\times\times} \vdash \mathfrak{C}^\times$; hence, $\mathfrak{A} \vdash \mathfrak{C}^\times$. Dually, if $\mathfrak{s}(\mathfrak{A})$, because $\mathfrak{A}$ is the substructure of $\mathfrak{A}^\times$, then by hypothesis, $\mathfrak{A}^\times \vdash \mathfrak{C}$, for some $\mathfrak{C}$. The following series of consecutions proves the claim. $\mathfrak{C}^\times \vdash \mathfrak{A}^{\times\times}$, $\mathfrak{A}^{\times\times\times} \vdash \mathfrak{C}^{\times\times}$, $\mathfrak{A}^\times \vdash \mathfrak{C}^{\times\times}$, $\mathfrak{C}^\times \vdash \mathfrak{A}$.

We leave the rest of the proof as an exercise. qed

Exercise 6.1.2. Finish the proof of the display theorem. [Hint: There are more possibilities for a structure $\mathfrak{A}$ to have an antecedent or a succedent occurrence than we went through in the proof.]

The display property—beyond its potential beauty—facilitates a formulation of rules with the principal formulas (and subalterns) displayed. This holds for the cut rule too. The *display cut rule* has the following shape.

$$\frac{\mathfrak{A} \vdash \mathcal{C} \qquad \mathcal{C} \vdash \mathfrak{B}}{\mathfrak{A} \vdash \mathfrak{B}} \text{ display cut}$$

If the antecedent and the consequent of every consecution can be turned into a formula, then the display cut rule may be seen to express the transitivity of an implication. However, the display cut rule, in general, is not equivalent to the single cut rule, as it will become clear from Section 6.3.

Theorem 6.5. *The display cut rule is* admissible *in DL_R.*

The proof may be obtained by verifying that the conditions of the general proof of the cut theorem from Section 7.4 are satisfied by the rules of the calculus.

The logic of entailment is often out of step with the other relevance logics, partly because of the special status of implicational formulas.

Although it is known, by a result of Edwin D. Mares, that the extension of E with a Boolean negation is not conservative, Belnap designed a display logic in the style of DL_R, which formalizes E. The trick is to restrict considerations to wff's in the *original vocabulary* of E together with a restriction of a structural rule.

The *language* of DL_E contains all the connectives in DL_R, however, a proper subset of formulas is carved out.

Definition 6.6. (E-formulas) The set of propositional variables has a subset, the elements of which are called *e-variables*. The set of *e-formulas* (e-wff's, for short) is inductively defined by (1)–(3).

(1) If $\mathcal{A}$ is an e-variable, then $\mathcal{A}$ is an e-wff;

(2) if $\mathcal{A}$ is an e-wff, then $\sim\mathcal{A}$ is an e-wff;

(3) if $\mathcal{A}$ and $\mathcal{B}$ are e-wff's, then $(\mathcal{A} \wedge \mathcal{B})$, $(\mathcal{A} \vee \mathcal{B})$ and $(\mathcal{A} \to \mathcal{B})$ are e-wff's.

The e-wff's are the wff's that are in the original language of E. In particular, no e-wff contains $\neg$ (which is Boolean negation).

The axioms and rules of the *display calculus* DL_E include all the axioms and the connective rules of DL_R, all the display equivalences of DL_R, the five structural rules involving $\mathbf{0}$ and $,$, as well as the following structural rules (where p_e is an e-variable).

$$\frac{p_e \vdash \mathfrak{B}}{\mathbf{I}; p_e \vdash \mathfrak{B}}\ \check{\mathbf{I}}_e\vdash \qquad \frac{\mathbf{I}; \mathfrak{A} \vdash \mathfrak{B}}{\mathfrak{A} \vdash \mathfrak{B}}\ \hat{\mathbf{I}}\vdash \qquad \frac{\mathbf{I} \vdash \mathfrak{A}}{\mathbf{I}^\times \vdash \mathfrak{A}}\ \mathbf{I}^{\check{\times}}\vdash \qquad \frac{\mathbf{I}^\times \vdash \mathfrak{A}}{\mathbf{I} \vdash \mathfrak{A}}\ \mathbf{I}^{\hat{\times}}\vdash$$

$$\frac{\mathfrak{A}; \mathbf{I} \vdash \mathfrak{B}}{\mathbf{I}; \mathfrak{A} \vdash \mathfrak{B}}\ T_{\mathbf{I}}\vdash \qquad \frac{\mathfrak{A}; (\mathfrak{B}; \mathfrak{C}) \vdash \mathfrak{D}}{\mathfrak{B}; \mathfrak{A}; \mathfrak{C} \vdash \mathfrak{D}}\ B'\vdash \qquad \frac{\mathfrak{A}; \mathfrak{B}; \mathfrak{B} \vdash \mathfrak{C}}{\mathfrak{A}; \mathfrak{B} \vdash \mathfrak{C}}\ W\vdash \qquad \frac{\mathfrak{A} \vdash \mathfrak{B}; \mathfrak{B}}{\mathfrak{A} \vdash \mathfrak{B}}\ \vdash M$$

E-wff's play a crucial role in DL_E formalizing E.

Theorem 6.7. *The display cut rule is* admissible *in* DL_E.

Exercise 6.1.3. Prove the claim. [Hint: Notice that DL_R and DL_E have a lot in common.]

Lemma 6.8. *If* $\mathcal{A}$ *is a theorem of* E, *then* $\mathbf{I} \vdash \mathcal{A}$ *is* a theorem of DL_E.

Here E is thought to be formulated as, for example, in [8, §R], that is, only the connectives of e-wff's occur in $\mathcal{A}$. In other words, from the point of view of DL_E, $\mathcal{A}$ is an e-wff.

Exercise 6.1.4. Prove the claim for $E_\rightarrow$, the axioms of which are in Section 5.3.

Exercise* 6.1.5. Prove the claim for E. [Hint: See the axioms in [8, §R].]

Of course, the previous lemma only guarantees that DL_E is sufficiently powerful. However, a more serious concern—especially, in the light of the Boolean extension of E producing new theorems in the original vocabulary of E—is that DL_E may prove e-wff's, which are not theorems of E. We only state the following lemma (fact 6, from [8, §62.5.4]), because its proof goes through semantics.[3] Once again, speaking of theorems of E, we mean wff's of the original E; that is, $\mathcal{A}$ must be an e-wff.

Lemma 6.9. *If* $\mathcal{A}$ *is not a theorem of* E, *then* $\mathbf{I} \vdash \mathcal{A}$ *is* not provable in DL_E.

The *display calculus* DL_T formalizes the Boolean extension of the logic of ticket entailment. By a result of Steve Giambrone, it is known to be conservative over T itself. The language of DL_T is like the language of DL_R. The set of axioms and rules differs from DL_R with respect to the *structural rules* that involve $\mathbf{I}$ or $;$. Each of these structural rules have already been mentioned, hence, we list them by their labels. $(\check{\mathbf{I}}\vdash)$, $(\hat{\mathbf{I}}\vdash)$, $(\mathbf{I}^{\check{\times}}\vdash)$, $(\mathbf{I}^{\hat{\times}}\vdash)$, $(B\vdash)$, $(B'\vdash)$, $(W\vdash)$ and $(\vdash M)$.

[3] The argument itself is not difficult, but its complete presentation would not only take us into ternary relational semantics, but it would also mean a lengthy detour.

Exercise 6.1.6. Prove that the display cut rule is admissible.

Exercise 6.1.7. Prove that all the theorems of $T_{\rightarrow}$ are provable in DL_T. [Hint: The axioms of $T_{\rightarrow}$ are (A1)–(A4) on p. 167.]

Exercise* 6.1.8. Prove that all the theorems of T are provable in DL_T. [Hint: The axioms of T may be found, for example, in [8, §R].]

6.2 Display logic for linear logic

The previous section should already be convincing as to how powerful display logics are. In this and the next section, we overview some other logics that can be formalized as display logics. We start with linear logic, which has been "displayed" in [20]. (We described a logistic sequent calculus for linear logic in Section 4.5.)

The language of linear logic contains the zero-ary constants $\boldsymbol{F}$, $\boldsymbol{T}$, $\boldsymbol{f}$ and $\boldsymbol{t}$, the unary connectives $\sim$, $\Diamond$ and $\Box$, the binary connectives $\wedge$, $\vee$, $\circ$, $+$ and $\rightarrow$, and the quantifiers $\forall$ and $\exists$. The set of well-formed formulas is generated as usual.

Unlike in relevance logics, conjunction and disjunction do not distribute over each other, which means that the structure-free rules for those connectives suffice in a sequent calculus. The *structural connectives* in DL_L are $\mathbf{I}$, which is zero-ary, *, which is unary, and $;$, which is binary. The set of structures is generated from the set of formulas and $\mathbf{I}$ in the usual way.

Definition 6.10. (Modal restriction) Given a consecution $\mathfrak{A} \vdash \mathfrak{B}$, if a structure $\mathfrak{C}$ is *modally restricted*, then for any atomic structure of $\mathfrak{C}$ that is a wff, let us say $\mathcal{A}$, (1) or (2) holds.

(1) If $\mathfrak{a}(\mathcal{A})$, then it is of the form $\Diamond\mathcal{B}$, for some $\mathcal{B}$;

(2) if $\mathfrak{s}(\mathcal{A})$, then it is of the form $\Box\mathcal{B}$, for some $\mathcal{B}$.

A modally restricted structure is superscripted with $\boxtimes$ (as in $\mathfrak{A}^{\boxtimes}$).

The axioms and rules of the *display calculus* DL_L are the following.

$$\mathcal{A} \vdash \mathcal{A} \ \text{id}$$

$$\frac{\mathbf{I} \vdash \mathfrak{A}}{\boldsymbol{t} \vdash \mathfrak{A}}\ {\scriptstyle t\vdash} \qquad \mathbf{I} \vdash \boldsymbol{t}\ {\scriptstyle \vdash t} \qquad \boldsymbol{f} \vdash \mathbf{I}^*\ {\scriptstyle f\vdash} \qquad \frac{\mathfrak{A} \vdash \mathbf{I}^*}{\mathfrak{A} \vdash \boldsymbol{f}}\ {\scriptstyle \vdash f}$$

$$\frac{\mathbf{I}^* \vdash \mathfrak{A}}{\boldsymbol{T} \vdash \mathfrak{A}}\ {\scriptstyle T\vdash} \qquad \mathbf{I}^* \vdash \boldsymbol{T}\ {\scriptstyle \vdash T} \qquad \boldsymbol{F} \vdash \mathbf{I}\ {\scriptstyle F\vdash} \qquad \frac{\mathfrak{A} \vdash \mathbf{I}}{\mathfrak{A} \vdash \boldsymbol{F}}\ {\scriptstyle \vdash F}$$

$$\frac{\mathcal{A}^* \vdash \mathfrak{A}}{\sim\mathcal{A} \vdash \mathfrak{A}}\ {\scriptstyle \sim\vdash} \qquad \frac{\mathfrak{A} \vdash \mathcal{A}^*}{\mathfrak{A} \vdash \sim\mathcal{A}}\ {\scriptstyle \vdash\sim}$$

$$\frac{\mathcal{A} \vdash \mathfrak{A}}{\mathcal{A} \wedge \mathcal{B} \vdash \mathfrak{A}} \wedge\vdash_1 \qquad \frac{\mathcal{B} \vdash \mathfrak{A}}{\mathcal{A} \wedge \mathcal{B} \vdash \mathfrak{A}} \wedge\vdash_2 \qquad \frac{\mathfrak{A} \vdash \mathcal{A} \quad \mathfrak{A} \vdash \mathcal{B}}{\mathfrak{A} \vdash \mathcal{A} \wedge \mathcal{B}} \vdash\wedge$$

$$\frac{\mathcal{A} \vdash \mathfrak{A} \quad \mathcal{B} \vdash \mathfrak{A}}{\mathcal{A} \vee \mathcal{B} \vdash \mathfrak{A}} \vee\vdash \qquad \frac{\mathfrak{A} \vdash \mathcal{A}}{\mathfrak{A} \vdash \mathcal{A} \vee \mathcal{B}} \vdash\vee_1 \qquad \frac{\mathfrak{A} \vdash \mathcal{B}}{\mathfrak{A} \vdash \mathcal{A} \vee \mathcal{B}} \vdash\vee_2$$

$$\frac{\mathcal{A}; \mathcal{B} \vdash \mathfrak{A}}{\mathcal{A} \circ \mathcal{B} \vdash \mathfrak{A}} \circ\vdash \qquad \frac{\mathfrak{A} \vdash \mathcal{A} \quad \mathfrak{B} \vdash \mathcal{B}}{\mathfrak{A}; \mathfrak{B} \vdash \mathcal{A} \circ \mathcal{B}} \vdash\circ$$

$$\frac{\mathcal{A} \vdash \mathfrak{A} \quad \mathcal{B} \vdash \mathfrak{B}}{\mathcal{A} + \mathcal{B} \vdash \mathfrak{A}; \mathfrak{B}} +\vdash \qquad \frac{\mathfrak{A} \vdash \mathcal{A}; \mathcal{B}}{\mathfrak{A} \vdash \mathcal{A} + \mathcal{B}} \vdash+$$

$$\frac{\mathfrak{A} \vdash \mathcal{A} \quad \mathcal{B} \vdash \mathfrak{B}}{\mathcal{A} \to \mathcal{B} \vdash \mathfrak{A}^*; \mathfrak{B}} \to\vdash \qquad \frac{\mathfrak{A}; \mathcal{A} \vdash \mathcal{B}}{\mathfrak{A} \vdash \mathcal{A} \to \mathcal{B}} \vdash\to$$

$$\frac{\mathcal{A}(y) \vdash \mathfrak{A}}{\forall x\, \mathcal{A}(x) \vdash \mathfrak{A}} \forall\vdash \qquad \frac{\mathfrak{A} \vdash \mathcal{A}(y)}{\mathfrak{A} \vdash \forall x\, \mathcal{A}(x)} \vdash\forall^{\oslash} \qquad \frac{\mathcal{A}(y) \vdash \mathfrak{A}}{\exists x\, \mathcal{A}(x) \vdash \mathfrak{A}} \exists\vdash^{\oslash} \qquad \frac{\mathfrak{A} \vdash \mathcal{A}(y)}{\mathfrak{A} \vdash \exists x\, \mathcal{A}(x)} \vdash\exists$$

$$\frac{\mathcal{A} \vdash \mathfrak{A}}{\Diamond\mathcal{A} \vdash \mathfrak{A}} \Diamond\vdash \qquad \frac{\mathfrak{A}^{\boxtimes} \vdash \mathcal{A}}{\mathfrak{A}^{\boxtimes} \vdash \Diamond\mathcal{A}} \vdash\Diamond \qquad \frac{\mathcal{A} \vdash \mathfrak{A}^{\boxtimes}}{\Box\mathcal{A} \vdash \mathfrak{A}^{\boxtimes}} \Box\vdash \qquad \frac{\mathfrak{A} \vdash \mathcal{A}}{\mathfrak{A} \vdash \Box\mathcal{A}} \vdash\Box$$

The $^{\oslash}$ indicates the usual restriction on the quantifier rules, namely, y may not occur free in the lower consecution. The double lines in the display equivalences (below) indicate that the rules can be applied to the upper consecution resulting in the lower one, as well as the other way around.

$$\frac{\mathfrak{A}; \mathfrak{B} \vdash \mathfrak{C}}{\overline{\mathfrak{A} \vdash \mathfrak{B}^*; \mathfrak{C}}} \qquad \frac{\mathfrak{A}; \mathfrak{B}^* \vdash \mathfrak{C}}{\overline{\mathfrak{A} \vdash \mathfrak{B}; \mathfrak{C}}} \qquad \frac{\mathfrak{A}; \mathfrak{B}^* \vdash \mathfrak{C}}{\overline{\mathfrak{A} \vdash \mathfrak{C}; \mathfrak{B}}} \qquad \frac{\mathfrak{B}^* \vdash \mathfrak{A}^*}{\overline{\mathfrak{A} \vdash \mathfrak{B}}} \qquad \frac{\mathfrak{B}^* \vdash \mathfrak{A}^*}{\overline{\mathfrak{A}^{**} \vdash \mathfrak{B}}}$$

$$\frac{\mathfrak{A} \vdash \mathfrak{B}}{\mathbf{I}; \mathfrak{A} \vdash \mathfrak{B}} \check{\mathbf{I}}\vdash \qquad \frac{\mathbf{I}; \mathfrak{A} \vdash \mathfrak{B}}{\mathfrak{A} \vdash \mathfrak{B}} \hat{\mathbf{I}}\vdash \qquad \frac{\mathfrak{A} \vdash \mathbf{I}}{\mathfrak{A} \vdash \mathfrak{B}} \vdash\mathbf{I}$$

$$\frac{\mathfrak{A}; (\mathfrak{B}; \mathfrak{C}) \vdash \mathfrak{D}}{\mathfrak{A}; \mathfrak{B}; \mathfrak{C} \vdash \mathfrak{D}} B\vdash \qquad \frac{\mathfrak{A}; \mathfrak{B} \vdash \mathfrak{C}}{\mathfrak{B}; \mathfrak{A} \vdash \mathfrak{C}} T\vdash$$

$$\frac{\mathfrak{A} \vdash \mathfrak{B}}{\mathfrak{A}; \mathfrak{C}^{\boxtimes} \vdash \mathfrak{B}} K^{\boxtimes}\vdash \qquad \frac{\mathfrak{A}^{\boxtimes}; \mathfrak{A}^{\boxtimes} \vdash \mathfrak{B}}{\mathfrak{A}^{\boxtimes} \vdash \mathfrak{B}} M^{\boxtimes}\vdash$$

The notions of a *proof* and of a *theorem* are as before.

Theorem 6.11. *The display cut rule is* admissible *in DL_L.*

Exercise* 6.2.1. Prove the theorem. [Hint: The proof of the theorem is outlined in Section 7.4.]

The real novelty of linear logic is the *modalization of the structural rules* (as we already pointed out in Section 4.5). Having displayed linear logic, Belnap introduced *punctual logic*, which goes all the way with respect to the modalization of all the structural rules.

The *display calculus* DL_P for punctual logic is defined like DL_L except that $(B\vdash)$ and $(T\vdash)$ are omitted. Instead, the following modalized permutation and associativity rules are postulated.[4]

$$\frac{\mathfrak{A};\mathfrak{B}^{\boxtimes}\vdash\mathfrak{C}}{\mathfrak{B}^{\boxtimes};\mathfrak{A}\vdash\mathfrak{C}}\;T_1^{\boxtimes}\vdash \qquad \frac{\mathfrak{A}^{\boxtimes};\mathfrak{B}\vdash\mathfrak{C}}{\mathfrak{B};\mathfrak{A}^{\boxtimes}\vdash\mathfrak{C}}\;t_1^{\boxtimes}\vdash \qquad \frac{\mathfrak{A}^{\boxtimes};(\mathfrak{B};\mathfrak{C})\vdash\mathfrak{D}}{\mathfrak{A}^{\boxtimes};\mathfrak{B};\mathfrak{C}\vdash\mathfrak{D}}\;B_1^{\boxtimes}\vdash$$

$$\frac{\mathfrak{A};(\mathfrak{B}^{\boxtimes};\mathfrak{C})\vdash\mathfrak{D}}{\mathfrak{A};\mathfrak{B}^{\boxtimes};\mathfrak{C}\vdash\mathfrak{D}}\;B_2^{\boxtimes}\vdash \qquad \frac{\mathfrak{A}^{\boxtimes};\mathfrak{B};\mathfrak{C}\vdash\mathfrak{D}}{\mathfrak{A}^{\boxtimes};(\mathfrak{B};\mathfrak{C})\vdash\mathfrak{C}}\;b_3^{\boxtimes}\vdash \qquad \frac{\mathfrak{A};\mathfrak{B}^{\boxtimes};\mathfrak{C}\vdash\mathfrak{D}}{\mathfrak{A};(\mathfrak{B}^{\boxtimes};\mathfrak{C})\vdash\mathfrak{D}}\;b_2^{\boxtimes}\vdash$$

Linear and punctual logics provide an opportunity to illustrate some of the difficulties that remain even in a powerful framework like display logics. Belnap says that "[t]he problem of finding a consecution calculus for **R** *without* boolean negation remains open." ([19, p. 397] and [8, p. 315].)

The * in DL_R stands for the structural equivalent of Boolean negation and allows the manipulation of the extensional structures, that is, structures of the form $\mathfrak{A},\mathfrak{B}$. As we have seen in the previous two chapters, the logistic sequent calculi such as LR_+ or the consecution calculi such as LT_+, have two kinds of structural connectives and two kinds of structures. Display logic does not eliminate the need for an additional structural connective if a logic contains $\wedge$ and $\vee$, which distribute over each other. Once there is a structural connective $,$, in order to display components of structures built by this connective, one needs some further structural connective.

It so happens that in a Boolean algebra, the *residual* of $\wedge$ is definable, because $a\wedge b\leq c$ iff $a\leq\neg b\vee c$, and dually, the residual of $\vee$ is definable too, because $c\leq a\vee b$ iff $c\wedge\neg a\leq c$. Even though $\wedge$, $\vee$ and $\neg$ do not form a family in the sense of gaggle theory, because families are defined in terms of residuation, the three connectives together are sufficient to express the residual of $\wedge$ and that of $\vee$, whereas $\neg$ is its own residual.[5] There is no similar relationship between $\sim$ and conjunction and disjunction. $\sim(a\vee b)=\sim a\wedge\sim b$ in the algebra of R, however, the residual of $\wedge$ or $\vee$ is not definable using $\sim$. Indeed, if it were, then $\sim$ would be a Boolean negation (or algebraically speaking, an orthonegation), rather than a De Morgan negation.

The intensional connectives of R do not run into a similar problem. When $\circ$, that is, fusion is included, $a\circ b\leq c$ iff $a\leq b\to c$. If fusion is not included, then it is still definable—with the help of $\sim$, namely, as $a\circ b=\sim(a\to\sim b)$. Fission, $+$, is also definable, which ensures that $;$ may be interpreted on both sides of a consecution. In the sense of gaggle theory, we have three complete families of intensional connectives, if we permit defined

[4]The modalized rules are labeled by letters that suggest a connection to combinators and dual combinators, and the subscripts refer to the argument place in which modalization is required. See also [38, Ch. 3].

[5]We use the term "family" in the sense as it is defined in [38].

connectives to be included. $\sim$ is its own residual, $\to$ is the residual of $\circ$ (which is commutative in R) and the negation of $\to$ is the residual of $+$.

Thus whether $\neg$, or $\supset$ and $\not\supset$ (the negation of implication) is added to R, it is paramount that such an the extension can be and has been proved to be *conservative*. One might think that it is sufficient to add the corresponding structural connectives—without the connectives themselves. However, if there are not enough connectives (primitive or definable) in the logic, then the interpretation of the consecutions becomes a problem.

No similar problems arise in the display logic in the next section, which starts with a complete family of connectives (like the one in the Lambek calculi) and also includes a complete dual family.

6.3 Display logic for symmetric gaggles

Section 5.5 showed how to formulate a consecution calculus with multiple right-hand sides so that the rules for the connectives that correspond to ; are in harmony with the cut rule. However, we left open the problem of finding suitable formulations for the implication-like connectives so that the cut remains admissible.

A way to start to consider the problem is to note that L_{HD} is well behaved. In $L_{HD?}$, we have four more connectives, and then, we might as well have *structural connectives* thatcorrespond to them. Formulating DL_G creates an opportunity to compare a consecution calculus and a display logic, as well as their cuts. We use the subscript $_G$, because the algebra of this calculus is like $\mathfrak{A}_\mathfrak{g}$ in Definition 5.42.

The *connectives* in DL_G are $\leftarrow, \circ, \to, \succ, +$ and $\prec$; they are all binary. The structural connectives are $\backslash$, ; , $/$, $\oslash$, $\oplus$ and $\oslash$. (We listed the structural connectives so that they match their pair in the list of connectives.)

The set of structures is as in Definition 6.2, but with the six binary structural connectives generating structures from the set of wff's.

The *display calculus* DL_G consists of an axiom, operational rules for the six connectives as well as four display equivalences.

$$\mathcal{A} \vdash \mathcal{A} \ \text{id}$$

$$\frac{\mathcal{A}; \mathcal{B} \vdash \mathfrak{A}}{\mathcal{A} \circ \mathcal{B} \vdash \mathfrak{A}} \circ\vdash \qquad \frac{\mathfrak{A} \vdash \mathcal{A} \qquad \mathfrak{B} \vdash \mathcal{B}}{\mathfrak{A}; \mathfrak{B} \vdash \mathcal{A} \circ \mathcal{B}} \vdash\circ$$

$$\frac{\mathcal{A} \vdash \mathfrak{A} \qquad \mathcal{B} \vdash \mathfrak{B}}{\mathcal{A} + \mathcal{B} \vdash \mathfrak{A} \oplus \mathfrak{B}} +\vdash \qquad \frac{\mathfrak{A} \vdash \mathcal{A} \oplus \mathcal{B}}{\mathfrak{A} \vdash \mathcal{A} + \mathcal{B}} \vdash+$$

$$\frac{\mathfrak{A} \vdash \mathcal{A} \qquad \mathcal{B} \vdash \mathfrak{B}}{\mathcal{A} \to \mathcal{B} \vdash \mathfrak{A} / \mathfrak{B}} \to\vdash \qquad \frac{\mathfrak{A}; \mathcal{A} \vdash \mathcal{B}}{\mathfrak{A} \vdash \mathcal{A} \to \mathcal{B}} \vdash\to$$

$$\dfrac{\mathfrak{A} \vdash \mathcal{A} \qquad \mathcal{B} \vdash \mathfrak{B}}{\mathcal{B} \leftarrow \mathcal{A} \vdash \mathfrak{B} \backslash \mathfrak{A}} \leftarrow\vdash \qquad \dfrac{\mathcal{A}; \mathfrak{A} \vdash \mathcal{B}}{\mathfrak{A} \vdash \mathcal{B} \leftarrow \mathcal{A}} \vdash\leftarrow$$

$$\dfrac{\mathcal{A} \vdash \mathcal{B} \oplus \mathfrak{A}}{\mathcal{A} \succ \mathcal{B} \vdash \mathfrak{A}} \succ\vdash \qquad \dfrac{\mathfrak{A} \vdash \mathcal{A} \qquad \mathcal{B} \vdash \mathfrak{B}}{\mathfrak{A} \obslash \mathfrak{B} \vdash \mathcal{A} \succ \mathcal{B}} \vdash\succ$$

$$\dfrac{\mathcal{A} \vdash \mathfrak{A} \oplus \mathcal{B}}{\mathcal{B} \prec \mathcal{A} \vdash \mathfrak{A}} \prec\vdash \qquad \dfrac{\mathfrak{A} \vdash \mathcal{A} \qquad \mathcal{B} \vdash \mathfrak{B}}{\mathfrak{B} \oslash \mathfrak{A} \vdash \mathcal{B} \prec \mathcal{A}} \vdash\prec$$

$$\dfrac{\mathfrak{A}; \mathfrak{B} \vdash \mathfrak{C}}{\mathfrak{A} \vdash \mathfrak{B} / \mathfrak{C}} \vdash/ \qquad \dfrac{\mathfrak{A}; \mathfrak{B} \vdash \mathfrak{C}}{\mathfrak{B} \vdash \mathfrak{C} \backslash \mathfrak{A}} \vdash\backslash \qquad \dfrac{\mathfrak{A} \vdash \mathfrak{B} \oplus \mathfrak{C}}{\mathfrak{C} \oslash \mathfrak{A} \vdash \mathfrak{B}} \oslash\vdash \qquad \dfrac{\mathfrak{A} \vdash \mathfrak{B} \oplus \mathfrak{C}}{\mathfrak{A} \obslash \mathfrak{B} \vdash \mathfrak{C}} \obslash\vdash$$

The notion of a *proof* is as before.

The operational rules may look like simplified versions of the operational rules in $L_{HD?}$. We mentioned that a problem in $L_{HD?}$ for the cut to be admissible is that implication-like rules do not allow all the component subformulas to be embedded within structures. We may consider different cut rules in DL_G; first of all, the display cut rule is the same as in Section 6.1. However, we may recall that $\oplus$ and $;$ are both denoted by $;$ in $L_{HD?}$. We may consider the single cut rule in DL_G in the following form.

$$\dfrac{\mathfrak{A} \vdash \mathfrak{B} \oplus \mathcal{C} \qquad \mathcal{C}; \mathfrak{C} \vdash \mathfrak{D}}{\mathfrak{A}; \mathfrak{C} \vdash \mathfrak{B} \oplus \mathfrak{D}} \text{cut}_1$$

The subscript suggests that we will look at more than one version of the cut rule, because with the abundance of structural connectives, we can no longer use the notation $\mathfrak{A}[\mathfrak{B}]$ or $\mathfrak{A}\langle\mathfrak{B}\rangle$. The location of a structural connective with respect to $\vdash$ does not determine which structural connective that is. We can still consider structures on the left and on the right of the turnstile, and if they have the desired structural connective at the *surface* level, then the cut formula may appear on the left or on the right of that structural connective. There are four potential variations in total.

$$\dfrac{\mathfrak{A} \vdash \mathcal{C} \oplus \mathfrak{B} \qquad \mathcal{C}; \mathfrak{C} \vdash \mathfrak{D}}{\mathfrak{A}; \mathfrak{C} \vdash \mathfrak{D}; \mathfrak{B}} \text{cut}_2 \qquad \dfrac{\mathfrak{A} \vdash \mathfrak{B} \oplus \mathcal{C} \qquad \mathfrak{C}; \mathcal{C} \vdash \mathfrak{D}}{\mathfrak{C}; \mathfrak{A} \vdash \mathfrak{B}; \mathfrak{D}} \text{cut}_3$$

$$\dfrac{\mathfrak{A} \vdash \mathcal{C} \oplus \mathfrak{B} \qquad \mathfrak{C}; \mathcal{C} \vdash \mathfrak{D}}{\mathfrak{C}; \mathfrak{A} \vdash \mathfrak{D}; \mathfrak{B}} \text{cut}_4$$

These cut rules are intimately connected to the hemi-distributivity principles relating $\circ$ and $+$, which were introduced in Definition 5.41.

Lemma 6.12. *The hemi-distributivity principles are* not provable *in DL_G. The cut$_i$ rules ($1 \leq i \leq 4$) are* not admissible *in DL_G.*

Proof: There is a certain symmetry between the four hemi-distributivity principles, as well as between the four latest cut rules. We give the details of one case, namely, of the hemi-distributivity principle (2).

$$\dfrac{\dfrac{\dfrac{\dfrac{\mathcal{B} \vdash \mathcal{B} \quad \mathcal{C} \vdash \mathcal{C}}{\mathcal{B} + \mathcal{C} \vdash \mathcal{B} \oplus \mathcal{C}}\,{+\vdash} \quad \dfrac{\mathcal{A} \vdash \mathcal{A} \quad \mathcal{C} \vdash \mathcal{C}}{\mathcal{A}; \mathcal{C} \vdash \mathcal{A} \circ \mathcal{C}}\,{\vdash\circ}}{\mathcal{A}; \mathcal{B} + \mathcal{C} \vdash \mathcal{B} \oplus \mathcal{A} \circ \mathcal{C}}\,\mathrm{cut}_3}{\mathcal{A}; \mathcal{B} + \mathcal{C} \vdash \mathcal{B} + (\mathcal{A} \circ \mathcal{C})}\,{\vdash +}}{\mathcal{A} \circ (\mathcal{B} + \mathcal{C}) \vdash \mathcal{B} + (\mathcal{A} \circ \mathcal{C})}\,{\circ\vdash}$$

The derivation shows that in the presence of cut_3, the consecution corresponding to $a \circ (b + c) \leq b + (a \circ c)$ is provable. Next, let us try to prove the end consecution *without* any application of cut rules.

The operational rules have a single wff on one side of the $\vdash$. In order to construct a proof, we have to be able to disassemble $\mathcal{B} + \mathcal{C}$ and $\mathcal{A} \circ \mathcal{C}$. We may start by reversing the $(\circ\vdash)$ and $(\vdash+)$ steps. The new bottom consecution is $\mathcal{A}; \mathcal{B} + \mathcal{C} \vdash \mathcal{B} \oplus \mathcal{A} \circ \mathcal{C}$. We may continue with $\mathcal{B} + \mathcal{C}$ or $\mathcal{A} \circ \mathcal{C}$.

$$\dfrac{\dfrac{?}{(\mathcal{A}; \mathcal{B} + \mathcal{C}) \oslash \mathcal{B} \vdash \mathcal{A} \circ \mathcal{C}}}{\mathcal{A}; \mathcal{B} + \mathcal{C} \vdash \mathcal{B} \oplus \mathcal{A} \circ \mathcal{C}} \qquad \dfrac{\dfrac{?}{\mathcal{B} + \mathcal{C} \vdash (\mathcal{B} \oplus \mathcal{A} \circ \mathcal{C}) \backslash \mathcal{A}}}{\mathcal{A}; \mathcal{B} + \mathcal{C} \vdash \mathcal{B} \oplus \mathcal{A} \circ \mathcal{C}}$$

The **?**'s indicate that we are stuck in our backward proof search. We have a single wff displayed, however, we do not have the suitable structural connective at the top level on the other side of the $\vdash$, which would result from an application of the $(\vdash\circ)$ or of the $(+\vdash)$ rule. Of course, the display equivalences allow us to continue building the tree upward, but in a futile way. We will never get any new consecutions beyond those already seen. (We tacitly assumed, as usual, that the meta-variables may stand for atomic wff's; hence, we cannot assume or use anything about their internal construction in a proof search.)

We have shown that the second hemi-distributivity principle is not provable without an application of a cut rule. But it is provable with cut; hence, the cut rule (cut_3, to be precise) is not admissible. qed

Exercise 6.3.1. Finish the proof of the lemma. [Hint: Once you have matched a hemi-distributivity principle with a cut, the proof is straightforward.]

We mentioned, in Section 5.5, twelve more inequations that could hold in $\mathfrak{A}_{\mathfrak{g}}$. We may ask the question whether, by any chance, some of the corresponding consecutions are provable in DL_G.

Let us consider the consecution $\mathcal{A} \prec (\mathcal{C} \leftarrow \mathcal{B}) \vdash (\mathcal{A} \prec \mathcal{C}) \leftarrow \mathcal{B}$. First, we do not suppose the availability of any cut rule. The consecution contains exactly one formula on each side, hence, there are two ways that we could have obtained the consecution in a proof.

$$\dfrac{\dfrac{\overset{\vdots}{\mathcal{B};\mathcal{A}\prec(\mathcal{C}\leftarrow\mathcal{B})\vdash\mathcal{C}}\qquad \overset{?}{\mathcal{A}\vdash}}{\mathcal{B};\mathcal{A}\prec(\mathcal{C}\leftarrow\mathcal{B})\vdash\mathcal{A}\prec\mathcal{C}}}{\mathcal{A}\prec(\mathcal{C}\leftarrow\mathcal{B})\vdash(\mathcal{A}\prec\mathcal{C})\leftarrow\mathcal{B}} \qquad \dfrac{\dfrac{\overset{?}{\vdash\mathcal{B}}\qquad \overset{\vdots}{\mathcal{C}\vdash(\mathcal{A}\prec\mathcal{C})\leftarrow\mathcal{B}\oplus\mathcal{A}}}{\mathcal{C}\leftarrow\mathcal{B}\vdash(\mathcal{A}\prec\mathcal{C})\leftarrow\mathcal{B}\oplus\mathcal{A}}}{\mathcal{A}\prec(\mathcal{C}\leftarrow\mathcal{B})\vdash(\mathcal{A}\prec\mathcal{C})\leftarrow\mathcal{B}}$$

The two **?**'s show that there is no way to continue the proof at that point, even if we would define and allow the empty structure. ($\mathcal{A}$ is not assumed to be a falsity or absurdity sort of wff, and $\mathcal{B}$ is not a triviality or truth formula.) Incidentally, the other branches of the tree will not end in instances of the axiom either. Roughly speaking, the missing formulas in the consecutions marked by **?** turn up as superfluous formulas on the other branch of the tree.

We may contemplate that with a cut rule, the consecution may still be derivable. Here is a derivation of the same consecution that involves a new version of the cut rule. As it should be expected, the cut allows a pair of consecutions to be interwoven, where the cut formula is within a complex structure. Just as $\oplus$ and ; were paired into a cut that was used in the proof of a hemi-distributivity principle, the structural connectives that correspond to $\prec$ and $\leftarrow$, that is, $\oslash$ and $\backslash$ are paired in the cut.

$$\dfrac{\dfrac{\dfrac{\dfrac{\dfrac{\dfrac{\mathcal{B}\vdash\mathcal{B}\quad \mathcal{C}\vdash\mathcal{C}}{\mathcal{C}\leftarrow\mathcal{B}\vdash\mathcal{C}\backslash\mathcal{B}}{\scriptstyle\leftarrow\vdash}\qquad \dfrac{\mathcal{A}\vdash\mathcal{A}\quad \mathcal{C}\vdash\mathcal{C}}{\mathcal{A}\oslash\mathcal{C}\vdash\mathcal{A}\prec\mathcal{C}}{\scriptstyle\vdash\prec}}{\mathcal{A}\oslash\mathcal{C}\leftarrow\mathcal{B}\vdash\mathcal{A}\prec\mathcal{C}\backslash\mathcal{B}}\,\text{cut}_5}{\mathcal{C}\leftarrow\mathcal{B}\vdash(\mathcal{A}\prec\mathcal{C}\backslash\mathcal{B})\oplus\mathcal{A}}\,{\scriptstyle\oslash\vdash}}{\mathcal{A}\prec(\mathcal{C}\leftarrow\mathcal{B})\vdash\mathcal{A}\prec\mathcal{C}\backslash\mathcal{B}}\,{\scriptstyle\prec\vdash}}{\mathcal{B};\mathcal{A}\prec(\mathcal{C}\leftarrow\mathcal{B})\vdash\mathcal{A}\prec\mathcal{C}}\,{\scriptstyle\vdash\backslash}}{\mathcal{A}\prec(\mathcal{C}\leftarrow\mathcal{B})\vdash(\mathcal{A}\prec\mathcal{C})\leftarrow\mathcal{B}}\,{\scriptstyle\vdash\leftarrow}$$

We could not have performed a cut earlier in the proof, and we could not have gotten to the end consecution otherwise.

The cut in the proof above is an instance of the following rule.

$$\dfrac{\mathfrak{A}\vdash\mathcal{C}\backslash\mathfrak{B}\qquad \mathfrak{C}\oslash\mathcal{C}\vdash\mathfrak{D}}{\mathfrak{C}\oslash\mathfrak{A}\vdash\mathfrak{D}\backslash\mathfrak{B}}\,\text{cut}_5$$

The rest of the proof, after the application of the cut rule, is slightly longer than the similar parts in the proofs of the hemi-distributivity principles. The reason behind the added length is that the operational rules, in their premises, involve only one or the other of the two structural connectives ; and $\oplus$, which correspond to the only structural connective in L_{HD}.

Exercise* 6.3.2. Define suitable versions of the cut rule that allow you to prove the eleven other consecutions that correspond to the inequations (5)–(15). [Hint: The display cut is admissible in DL_G, and a good inkling is that none of the consecutions mentioned are provable.]

To actually substantiate the claim that we repeated again and again about the connection between the structural connectives and the (object-language) connectives, we note that we could have defined the right rules for the implication connectives with the premises explicitly containing their matching structural connectives. This would also ensure that the only structural connective in the rules for $\rightarrow$ is $/$, and similarly, for $\leftarrow$, it is $\backslash$.

For example, the following rules are derivable in DL_G.

$$\frac{\mathfrak{A} \vdash \mathcal{B} \backslash \mathcal{A}}{\mathfrak{A} \vdash \mathcal{B} \leftarrow \mathcal{A}} \vdash\!\leftarrow_2 \qquad\qquad \frac{\mathfrak{A} \vdash \mathcal{A} / \mathcal{B}}{\mathfrak{A} \vdash \mathcal{A} \rightarrow \mathcal{B}} \vdash\!\rightarrow_2$$

Exercise 6.3.3. Prove that $(\vdash\!\leftarrow_2)$ and $(\vdash\!\rightarrow_2)$ are derivable in DL_G, and conversely, if we omit $(\vdash\!\leftarrow)$ and $(\vdash\!\rightarrow)$, but include these new rules, then $(\vdash\!\leftarrow)$ and $(\vdash\!\rightarrow)$ are derivable.

Exercise 6.3.4. Define new left rules for the connectives $\succ$ and $\prec$ in a similar fashion, and prove their equivalence to those that are in DL_G.

DL_G does not have any structural rules; the display equivalences capture the residuation patterns at the level of structures. Having solidified the relationship between connectives and their structural pair, we may contemplate adding structural rules to enable us to prove the four hemi-distributivity principles or the other inequations.

Following the idea of transforming inequations into rules that we outlined in Section 5.2, we could consider adding a structural rule of the following shape for the hemi-distributivity principle (1), that is, $(a+b)\circ c \leq (a \circ c)+b$.

$$\frac{(\mathfrak{A};\mathfrak{C}) \oplus \mathfrak{B} \vdash \mathfrak{D}}{(\mathfrak{A} \oplus \mathfrak{B});\mathfrak{C} \vdash \mathfrak{D}} \text{ hd}_1?$$

A problem with this rule is that it would not allow us to prove the consecution $(\mathcal{A}+\mathcal{B})\circ\mathcal{C} \vdash (\mathcal{A}\circ\mathcal{C})+\mathcal{B}$, because we cannot build up the complex structure in the upper consecution by applying the operational rules with some atomic wff's in place of $\mathcal{A}$, $\mathcal{B}$ and $\mathcal{C}$.

Exercise 6.3.5. Verify that the rule $(\text{hd}_1?)$ is not suitable to prove $(p+q)\circ r \vdash (p \circ r)+q$.

A way to discover an appropriate rule is to utilize some easy consequences of the residuation between structures. Since $\mathcal{A}\circ\mathcal{C}$ occurs on the right-hand side of the intended consecution, and similarly, $\mathcal{A}+\mathcal{B}$ occurs on the left-hand side, one of these two formulas has to be introduced in the consecution first. However, immediately afterward, the structures have to be rearranged for the other formula to be introduced. This suggests that the structural rule will involve a pair of structural connectives appearing on the same side, where they appear in the display equations. In other words, $;$, $\oslash$ and $\circledslash$ belong to the left-hand side, whereas $\oplus$, $/$ and $\backslash$ are native to the right-hand side. (In $(\text{hd}_1?)$, $\oplus$ appears on the wrong side, so to speak.)

These ideas lead to the following proof into which we already incorporated the new structural rule.

$$\dfrac{\dfrac{\dfrac{\dfrac{\dfrac{\dfrac{\dfrac{\mathcal{A}\vdash\mathcal{A}\qquad\mathcal{C}\vdash\mathcal{C}}{\mathcal{A};\mathcal{C}\vdash\mathcal{A}\circ\mathcal{C}}\,{\scriptstyle\vdash\circ}}{\mathcal{A}\vdash\mathcal{C}/\mathcal{A}\circ\mathcal{C}}\,{\scriptstyle\vdash/}\qquad\mathcal{B}\vdash\mathcal{B}}{\mathcal{A}+\mathcal{B}\vdash(\mathcal{C}/\mathcal{A}\circ\mathcal{C})\oplus\mathcal{B}}\,{\scriptstyle+\vdash}}{\mathcal{A}+\mathcal{B}\vdash\mathcal{C}/(\mathcal{A}\circ\mathcal{C}\oplus\mathcal{B})}\,{\scriptstyle\mathrm{hd}_1}}{\mathcal{A}+\mathcal{B};\mathcal{C}\vdash\mathcal{A}\circ\mathcal{C}\oplus\mathcal{B}}\,{\scriptstyle\vdash/}}{\mathcal{A}+\mathcal{B};\mathcal{C}\vdash(\mathcal{A}\circ\mathcal{C})+\mathcal{B}}\,{\scriptstyle\vdash+}}{(\mathcal{A}+\mathcal{B})\circ\mathcal{C}\vdash(\mathcal{A}\circ\mathcal{C})+\mathcal{B}}\,{\scriptstyle\circ\vdash}$$

Alternatively, we could have proceeded by switching our preference to $\mathcal{A}+\mathcal{B}$. Then we would have added a different structural rule.

$$\dfrac{\dfrac{\dfrac{\dfrac{\dfrac{\dfrac{\dfrac{\mathcal{A}\vdash\mathcal{A}\qquad\mathcal{B}\vdash\mathcal{B}}{\mathcal{A}+\mathcal{B}\vdash\mathcal{A}\oplus\mathcal{B}}\,{\scriptstyle+\vdash}}{\mathcal{B}\oslash\mathcal{A}+\mathcal{B}\vdash\mathcal{A}}\,{\scriptstyle\oslash\vdash}\qquad\mathcal{C}\vdash\mathcal{C}}{(\mathcal{B}\oslash\mathcal{A}+\mathcal{B});\mathcal{C}\vdash\mathcal{A}\circ\mathcal{C}}\,{\scriptstyle\vdash\circ}}{\mathcal{B}\oslash(\mathcal{A}+\mathcal{B};\mathcal{C})\vdash\mathcal{A}\circ\mathcal{C}}\,{\scriptstyle\mathrm{hd}'_1}}{\mathcal{A}+\mathcal{B};\mathcal{C}\vdash\mathcal{A}\circ\mathcal{C}\oplus\mathcal{B}}\,{\scriptstyle\oslash\vdash}}{(\mathcal{A}+\mathcal{B})\circ\mathcal{C}\vdash\mathcal{A}\circ\mathcal{C}\oplus\mathcal{B}}\,{\scriptstyle\circ\vdash}}{(\mathcal{A}+\mathcal{B})\circ\mathcal{C}\vdash(\mathcal{A}\circ\mathcal{C})+\mathcal{B}}\,{\scriptstyle\vdash+}$$

Exercise 6.3.6. The two proofs above used instances of the rules that we want to call (hd_1) and (hd'_1). (hd_1) is included below and (hd'_1) is obtained by replacing the wff's by structures in its concrete application above. Prove that (hd_1) and (hd'_1) are interderivable in DL_G.

We list the structural rules that are sufficient to prove analogs of the sixteen inequations that we considered in the context of $\mathfrak{A}_\mathfrak{g}$. Each rule given is applicable to structures on the right.

$$\dfrac{\mathfrak{D}\vdash(\mathfrak{C}/\mathfrak{A})\oplus\mathfrak{B}}{\mathfrak{D}\vdash\mathfrak{C}/(\mathfrak{A}\oplus\mathfrak{B})}\,\mathrm{hd}_1\qquad\qquad\dfrac{\mathfrak{D}\vdash\mathfrak{B}\oplus(\mathfrak{C}\backslash\mathfrak{A})}{\mathfrak{D}\vdash(\mathfrak{B}\oplus\mathfrak{C})\backslash\mathfrak{A}}\,\mathrm{hd}_2$$

$$\dfrac{\mathfrak{D}\vdash\mathfrak{A}\oplus(\mathfrak{C}/\mathfrak{B})}{\mathfrak{D}\vdash\mathfrak{C}/(\mathfrak{A}\oplus\mathfrak{B})}\,\mathrm{hd}_3\qquad\qquad\dfrac{\mathfrak{D}\vdash(\mathfrak{B}\backslash\mathfrak{A})\oplus\mathfrak{C}}{\mathfrak{D}\vdash(\mathfrak{B}\oplus\mathfrak{C})\backslash\mathfrak{A}}\,\mathrm{hd}_4$$

$$\dfrac{\mathfrak{D}\vdash\mathfrak{A}/(\mathfrak{C}/\mathfrak{B})}{\mathfrak{D}\vdash\mathfrak{C}/(\mathfrak{A}/\mathfrak{B})}\,\mathrm{iq}_5\qquad\qquad\dfrac{\mathfrak{D}\vdash\mathfrak{B}/(\mathfrak{C}\backslash\mathfrak{A})}{\mathfrak{D}\vdash(\mathfrak{B}/\mathfrak{C})\backslash\mathfrak{A}}\,\mathrm{iq}_6$$

$$\dfrac{\mathfrak{D}\vdash(\mathfrak{C}/\mathfrak{B})\backslash\mathfrak{A}}{\mathfrak{D}\vdash\mathfrak{C}/(\mathfrak{B}\backslash\mathfrak{A})}\,\mathrm{iq}_7\qquad\qquad\dfrac{\mathfrak{D}\vdash(\mathfrak{C}\backslash\mathfrak{A})\backslash\mathfrak{B}}{\mathfrak{D}\vdash(\mathfrak{C}\backslash\mathfrak{B})\backslash\mathfrak{A}}\,\mathrm{iq}_8$$

$$\frac{\mathfrak{D} \vdash (\mathfrak{C} \oplus \mathfrak{B}) \oplus \mathfrak{A}}{\mathfrak{D} \vdash \mathfrak{C} \oplus (\mathfrak{B} \oplus \mathfrak{A})}\ \mathrm{iq}_9 \qquad \frac{\mathfrak{D} \vdash \mathfrak{A} \oplus (\mathfrak{C} \oplus \mathfrak{B})}{\mathfrak{D} \vdash \mathfrak{C} \oplus (\mathfrak{A} \oplus \mathfrak{B})}\ \mathrm{iq}_{10}$$

$$\frac{\mathfrak{D} \vdash (\mathfrak{B} \oplus \mathfrak{A}) \oplus \mathfrak{C}}{\mathfrak{D} \vdash (\mathfrak{B} \oplus \mathfrak{C}) \oplus \mathfrak{A}}\ \mathrm{iq}_{11} \qquad \frac{\mathfrak{D} \vdash \mathfrak{B} \oplus (\mathfrak{C} \oplus \mathfrak{A})}{\mathfrak{D} \vdash (\mathfrak{B} \oplus \mathfrak{C}) \oplus \mathfrak{A}}\ \mathrm{iq}_{12}$$

$$\frac{\mathfrak{D} \vdash \mathfrak{A}/(\mathfrak{C} \oplus \mathfrak{B})}{\mathfrak{D} \vdash \mathfrak{C} \oplus (\mathfrak{A}/\mathfrak{B})}\ \mathrm{iq}_{13} \qquad \frac{\mathfrak{D} \vdash (\mathfrak{C} \oplus \mathfrak{B})\backslash\mathfrak{A}}{\mathfrak{D} \vdash \mathfrak{C} \oplus (\mathfrak{B}\backslash\mathfrak{A})}\ \mathrm{iq}_{14}$$

$$\frac{\mathfrak{D} \vdash \mathfrak{B}/(\mathfrak{C} \oplus \mathfrak{A})}{\mathfrak{D} \vdash (\mathfrak{B}/\mathfrak{C}) \oplus \mathfrak{A}}\ \mathrm{iq}_{15} \qquad \frac{\mathfrak{D} \vdash (\mathfrak{C} \oplus \mathfrak{A})\backslash\mathfrak{B}}{\mathfrak{D} \vdash (\mathfrak{C}\backslash\mathfrak{B}) \oplus \mathfrak{A}}\ \mathrm{iq}_{16}$$

Exercise* 6.3.7. Prove that if DL_G is extended with one of the listed structural rules, then the consecution corresponding to the identically labeled inequation is provable. [Hint: Each case is straightforward, but there are sixteen rule–inequation pairs and things can get confusing.]

DL_G contains plenty of structural connectives. We suggested that the most immediate idea to accommodate the hemi-distributivity principles does not work, and we went on to give other "one-sided" structural rules. The connection to the inequations may not be fully obvious everywhere, but in some cases, the structural rules reveal deep links between the inequations and some familiar properties of the operations.

For instance, (9) links $+$ and $\succ$, but the corresponding structural rule can be seen stating one direction of the associativity for $+$. The structural rule applies on the right-hand side of the turnstile, hence, it can be viewed as corresponding to moving the parentheses to the right, that is, to $(a+b)+c \leq a+(b+c)$.

Given the abundance of structural connectives, the rules that are sufficient to derive each of the inequations are not unique. We could have made each rule applicable on the left-hand side of the $\vdash$. Here is a handful of equivalent rules.

$$\frac{\mathfrak{A} \oslash (\mathfrak{C} \oslash \mathfrak{B}) \vdash \mathfrak{D}}{\mathfrak{C} \oslash (\mathfrak{A} \oslash \mathfrak{B}) \vdash \mathfrak{D}}\ \mathrm{iq}'_{11} \qquad \frac{\mathfrak{A} \oslash (\mathfrak{C} \obslash \mathfrak{B}) \vdash \mathfrak{D}}{(\mathfrak{A} \oslash \mathfrak{C}) \obslash \mathfrak{B} \vdash \mathfrak{D}}\ \mathrm{iq}'_{12}$$

$$\frac{\mathfrak{A} \oslash (\mathfrak{C}; \mathfrak{B}) \vdash \mathfrak{D}}{(\mathfrak{A} \oslash \mathfrak{C}); \mathfrak{B} \vdash \mathfrak{D}}\ \mathrm{iq}'_{15} \qquad \frac{\mathfrak{A} \oslash (\mathfrak{B}; \mathfrak{C}) \vdash \mathfrak{D}}{\mathfrak{B}; (\mathfrak{A} \oslash \mathfrak{C}) \vdash \mathfrak{D}}\ \mathrm{iq}'_{16}$$

Exercise 6.3.8. Perform a similar conversion of the remaining eleven rules. [Hint: We already mentioned (hd'_1).]

In sequent calculi, especially, in right-singular sequent calculi, there is a preference to turn axioms or inequations into rules that apply on the left-hand side. However, in display calculi, where we can move structures across the $\vdash$, this restriction may be abandoned. Not only right rules work, but also rules that disperse the structural effects between the two sides. This is, certainly possible in the case of the inequations we are interested in. Algebraically speaking, the next series of rules could be viewed to correspond to *quasi-inequational* formulations of the same principles. (The "qi" in the label of the rules stands for "quasi-inequation.")

$$\frac{\mathfrak{B}\oslash\mathfrak{D}\vdash\mathfrak{C}/\mathfrak{A}}{\mathfrak{D};\mathfrak{C}\vdash\mathfrak{A}\oplus\mathfrak{B}}\;\mathrm{qi}_1 \qquad \frac{\mathfrak{D}\obslash\mathfrak{B}\vdash\mathfrak{C}\backslash\mathfrak{A}}{\mathfrak{A};\mathfrak{D}\vdash\mathfrak{B}\oplus\mathfrak{C}}\;\mathrm{qi}_2$$

$$\frac{\mathfrak{D}\obslash\mathfrak{A}\vdash\mathfrak{C}/\mathfrak{B}}{\mathfrak{D};\mathfrak{C}\vdash\mathfrak{A}\oplus\mathfrak{B}}\;\mathrm{qi}_3 \qquad \frac{\mathfrak{C}\oslash\mathfrak{D}\vdash\mathfrak{B}\backslash\mathfrak{A}}{\mathfrak{A};\mathfrak{D}\vdash\mathfrak{B}\oplus\mathfrak{C}}\;\mathrm{qi}_4$$

$$\frac{\mathfrak{D};\mathfrak{A}\vdash\mathfrak{C}/\mathfrak{B}}{\mathfrak{D};\mathfrak{C}\vdash\mathfrak{A}/\mathfrak{B}}\;\mathrm{qi}_5 \qquad \frac{\mathfrak{D};\mathfrak{B}\vdash\mathfrak{C}\backslash\mathfrak{A}}{\mathfrak{A};\mathfrak{D}\vdash\mathfrak{B}/\mathfrak{C}}\;\mathrm{qi}_6$$

$$\frac{\mathfrak{A};\mathfrak{D}\vdash\mathfrak{C}/\mathfrak{B}}{\mathfrak{D};\mathfrak{C}\vdash\mathfrak{B}\backslash\mathfrak{A}}\;\mathrm{qi}_7 \qquad \frac{\mathfrak{B};\mathfrak{D}\vdash\mathfrak{C}\backslash\mathfrak{A}}{\mathfrak{A};\mathfrak{D}\vdash\mathfrak{C}\backslash\mathfrak{B}}\;\mathrm{qi}_8$$

$$\frac{\mathfrak{A}\oslash\mathfrak{D}\vdash\mathfrak{C}\oplus\mathfrak{B}}{\mathfrak{D}\obslash\mathfrak{C}\vdash\mathfrak{B}\oplus\mathfrak{A}}\;\mathrm{qi}_9 \qquad \frac{\mathfrak{D}\obslash\mathfrak{A}\vdash\mathfrak{C}\oplus\mathfrak{B}}{\mathfrak{D}\obslash\mathfrak{C}\vdash\mathfrak{A}\oplus\mathfrak{B}}\;\mathrm{qi}_{10}$$

$$\frac{\mathfrak{C}\oslash\mathfrak{D}\vdash\mathfrak{B}\oplus\mathfrak{A}}{\mathfrak{A}\oslash\mathfrak{D}\vdash\mathfrak{B}\oplus\mathfrak{C}}\;\mathrm{qi}_{11} \qquad \frac{\mathfrak{D}\obslash\mathfrak{B}\vdash\mathfrak{C}\oplus\mathfrak{A}}{\mathfrak{A}\oslash\mathfrak{D}\vdash\mathfrak{B}\oplus\mathfrak{C}}\;\mathrm{qi}_{12}$$

$$\frac{\mathfrak{D};\mathfrak{A}\vdash\mathfrak{C}\oplus\mathfrak{B}}{\mathfrak{D}\obslash\mathfrak{C}\vdash\mathfrak{A}/\mathfrak{B}}\;\mathrm{qi}_{13} \qquad \frac{\mathfrak{A};\mathfrak{D}\vdash\mathfrak{C}\oplus\mathfrak{B}}{\mathfrak{D}\obslash\mathfrak{C}\vdash\mathfrak{B}\backslash\mathfrak{A}}\;\mathrm{qi}_{14}$$

$$\frac{\mathfrak{D};\mathfrak{B}\vdash\mathfrak{C}\oplus\mathfrak{A}}{\mathfrak{A}\oslash\mathfrak{D}\vdash\mathfrak{B}/\mathfrak{C}}\;\mathrm{qi}_{15} \qquad \frac{\mathfrak{B};\mathfrak{D}\vdash\mathfrak{C}\oplus\mathfrak{A}}{\mathfrak{A}\oslash\mathfrak{D}\vdash\mathfrak{C}\backslash\mathfrak{B}}\;\mathrm{qi}_{16}$$

Example 6.13. Let us consider (10), that is, the consecution $(\mathcal{A}+\mathcal{B})\succ\mathcal{C}\vdash\mathcal{A}+(\mathcal{B}\succ\mathcal{C})$. The next proof shows how this consecution is proved using (qi_{10}).

$$\dfrac{\dfrac{\dfrac{\dfrac{\dfrac{\dfrac{\mathcal{A}\vdash\mathcal{A}\qquad\dfrac{\dfrac{\mathcal{B}\vdash\mathcal{B}\qquad\mathcal{C}\vdash\mathcal{C}}{\mathcal{B}\obslash\mathcal{C}\vdash\mathcal{B}\succ\mathcal{C}}\;\vdash\succ}{\mathcal{B}\vdash\mathcal{C}\oplus\mathcal{B}\succ\mathcal{C}}\;\obslash\vdash}{\mathcal{A}+\mathcal{B}\vdash\mathcal{A}\oplus(\mathcal{C}\oplus\mathcal{B}\succ\mathcal{C})}\;+\vdash}{\mathcal{A}+\mathcal{B}\obslash\mathcal{A}\vdash\mathcal{C}\oplus\mathcal{B}\succ\mathcal{C}}\;\obslash\vdash}{\mathcal{A}+\mathcal{B}\obslash\mathcal{C}\vdash\mathcal{A}\oplus\mathcal{B}\succ\mathcal{C}}\;\mathrm{qi}_{10}}{\mathcal{A}+\mathcal{B}\vdash\mathcal{C}\oplus(\mathcal{A}\oplus\mathcal{B}\succ\mathcal{C})}\;\obslash\vdash}{(\mathcal{A}+\mathcal{B})\succ\mathcal{C}\vdash\mathcal{A}\oplus\mathcal{B}\succ\mathcal{C}}\;\succ\vdash}{(\mathcal{A}+\mathcal{B})\succ\mathcal{C}\vdash\mathcal{A}+(\mathcal{B}\succ\mathcal{C})}\;\vdash+$$

Exercise 6.3.9. Prove the same end consecution using (iq_{10}), and compare the two proofs.

Exercise 6.3.10. Establish that the other "quasi-inequational" rules allow the proof of their respective consecutions.

Exercise 6.3.11. Verify that the "inequational" and the "quasi-inequational" rules are derivable from each other, if added to DL_G.

To motivate the "quasi-inequational" label for the last group of rules, we show how the fact that the quasi-inequation $d \succ a \leq c + e$ implies $d \succ c \leq a + e$ can be used to prove the inequation (10). We assume that we add this quasi-inequation to $\mathfrak{A}_{\mathfrak{g}}$.

Let us start with $a + b \leq a + b$, which holds by the reflexivity of $\leq$. By residuation, we obtain $(a + b) \succ a \leq b$. Similarly, from $b \succ c \leq b \succ c$, we get $b \leq c + (b \succ c)$. The transitivity of $\leq$ yields $(a + b) \succ a \leq c + (b \succ c)$, which is an instance of $d \succ a \leq c + e$. Therefore, $(a + b) \succ c \leq a + (b \succ c)$.

The quasi-inequational formulations of the rules also show, in an abstract way, the connection to the cut rule. The transitivity of the $\leq$ is an analog of the display cut rule, but that would not be sufficient to prove the target consecution. The swap of two structures—while retaining or modifying the structural connectives that attach them to other structures—is like applying two display equivalences simultaneously. The consecution calculus L_{HD} achieved a similar effect (i.e., the provability of sequents that correspond to the inequations (1)–(4)) by embedding structures into each other in the rules for $\circ$ and $+$.

Theorem 6.14. *If the structural rules corresponding to the inequations added to* DL_G *(in either form), then the display cut rule remains* admissible *in the expanded calculus.*

This is, essentially, Theorem 4.5 in [40], which may be proved like the cut theorems in Section 7.4. The cited theorem is not literally what we stated above, because in [40], we gave half of the inequational rules as left and the other half as right structural rules.

There is an abundance of structural connectives in DL_G. As long as there is one structural connective such as ; that can appear on either side of the $\vdash$, the formulation of the algebraic cut is unproblematic. An intuition is that the display cut is *weaker* than the algebraic cut, because the former is admissible DL_G, whereas the cut_1–cut_4 rules (that we considered at the beginning of this section) are not admissible. Those cut rules allow the hemi-distributivity principles to be proved, which were linked to the algebraic cut, and they can be seen to be special instances of the algebraic cut rule. However, it is not clear how to formulate the algebraic cut in DL_G (or in its extension with the (hd_i) rules), because an atomic structure occurring in the antecedent may not be an antecedent occurrence, and dually, an occurrence of a wff in the succedent does not need to be a succedent occurrence.

Exercise* 6.3.12. Formulate the algebraic cut rule for DL_G with (hd_1)–(hd_4) added, and prove its admissibility.

6.4 Hypersequent calculi

Rules in sequent calculi, typically, have one or two premises. In the latter case, both premises are necessary for the rule to be applicable, that is, they are conjoined with a meta-language *and*. The two premises may be two instances of one sequent. The placement of the premises as the left premise and the right premise is for convenient reference (as in talking about the cut rule) or for the ease of the recognition of a concrete application of a rule. We could define rules with a *multiset* of premises instead.

Hypersequent calculi expand the notion of a sequent. A hypersequent is a (finite) multiset of (usual) sequents, where the elements of the hypersequent are thought of as *alternatives*.[6] A *rule* in a hypersequent calculus looks like a rule in a sequent calculus—except that hypersequents replace the sequents. We will use Ψ, possibly with subscripts, as a meta-variable for hypersequents, and $\Psi[\Gamma \vdash \Delta]$ will indicate that $\Gamma \vdash \Delta$ is a component sequent in the hypersequent Ψ. Sequents are defined as in LK.

Example 6.15. $\Gamma \vdash \Delta$ is by itself a hypersequent, however, it may be a component of a more complex hypersequent. If Ψ consists of $k+1$ sequents, then it may look like the following, where the sequents are separated by $|$.

$$\Theta_1 \vdash \Lambda_1 \mid \ldots \mid \Theta_k \vdash \Lambda_k \mid \Gamma \vdash \Delta$$

We may denote this as $\Psi[\Gamma \vdash \Delta]$, especially when we are interested in pointing out that there is at least one sequent $\Gamma \vdash \Delta$ in Ψ, but the actual shape of the rest of the hypersequent is irrelevant. $\Psi[\Gamma \vdash \Delta]$ could be a different hypersequent (than the one above) such as the following.

$$\Gamma \vdash \Delta \mid \Theta \vdash \Lambda \mid \Gamma \vdash \Delta \mid \Theta \vdash \Lambda \mid \Gamma \vdash \Delta$$

We also use the bracket notation for the result of a replacement in a limited context. (The use of brackets in sequents in earlier chapters has been similar.)

At first, it may appear that hypersequents are merely a concise way to present rules, however, they genuinely increase expressibility. We illustrate this by two hypersequent calculi for $S5$.[7] $S5$ is not only one of the calculi originally proposed by Clarence I. Lewis as a logic of strict implication (though nowadays, it is usually presented with unary operations), it is one of the simplest normal modal logics. $S5$ does not have distinguishable stacked modalities (e.g., $\Box\Diamond\Box\mathcal{A}$ is equivalent to $\Box\mathcal{A}$), and its relational semantics

[6] There appears to be a duality here with respect to tableaux, in which each node contains a set of wff's, where the elements of the set are thought to be joined by conjunction.

[7] These calculi, with some variations, may be found in [83], [190] and [159]. We adjust the notation to ours.

can be defined from an equivalence relation. Even its informal interpretation is easy, because all worlds are accessible from each other, hence, the accessibility relation—as the total relation on the set of worlds—may be more or less ignored. Thus it is somewhat of a disappointment that $S5$ does not have a comparably simple sequent calculus formalization. This is remedied by hypersequent calculi.

The *language* of $S5$ contains the connectives $\neg$, $\wedge$, $\vee$, $\supset$ (as in classical logic) as well as the unary connectives $\Box$ (necessity) and $\Diamond$ (possibility). The set of well-formed formulas is generated from a set of propositional variables in the usual fashion.

We use the notations $\Gamma^{\Box}$ and $\Gamma^{\Diamond}$ to mean the set of wff's obtained by prefixing each formula in Γ—whether that wff is itself modalized or not—with $\Box$ and $\Diamond$, respectively.

A *hypersequent calculus* H_{S5} formalizing $S5$ comprises the following axiom and rules.

$$\Psi[\mathcal{A} \vdash \mathcal{A}] \text{ id}$$

$$\frac{\Psi[\mathcal{A}, \Gamma \vdash \Delta]}{\Psi[\mathcal{A} \wedge \mathcal{B}, \Gamma \vdash \Delta]} \quad \frac{\Psi[\mathcal{B}, \Gamma \vdash \Delta]}{\Psi[\mathcal{A} \wedge \mathcal{B}, \Gamma \vdash \Delta]} \wedge\vdash\text{'s} \qquad \frac{\Psi[\Gamma \vdash \Delta, \mathcal{A}] \quad \Psi[\Gamma \vdash \Delta, \mathcal{B}]}{\Psi[\Gamma \vdash \Delta, \mathcal{A} \wedge \mathcal{B}]} \vdash\wedge$$

$$\frac{\Psi[\mathcal{A}, \Gamma \vdash \Delta] \quad \Psi[\mathcal{B}, \Gamma \vdash \Delta]}{\Psi[\mathcal{A} \vee \mathcal{B}, \Gamma \vdash \Delta]} \vee\vdash \qquad \frac{\Psi[\Gamma \vdash \Delta, \mathcal{A}]}{\Psi[\Gamma \vdash \Delta, \mathcal{A} \vee \mathcal{B}]} \quad \frac{\Psi[\Gamma \vdash \Delta, \mathcal{B}]}{\Psi[\Gamma \vdash \Delta, \mathcal{A} \vee \mathcal{B}]} \vdash\vee\text{'s}$$

$$\frac{\Psi[\Gamma \vdash \Delta, \mathcal{A}] \quad \Psi[\mathcal{B}, \Gamma \vdash \Delta]}{\Psi[\mathcal{A} \supset \mathcal{B}, \Gamma \vdash \Delta]} \supset\vdash \qquad \frac{\Psi[\mathcal{A}, \Gamma \vdash \Delta, \mathcal{B}]}{\Psi[\Gamma \vdash \Delta, \mathcal{A} \supset \mathcal{B}]} \vdash\supset$$

$$\frac{\Psi[\Gamma \vdash \Delta, \mathcal{A}]}{\Psi[\neg\mathcal{A}, \Gamma \vdash \Delta]} \neg\vdash \qquad \frac{\Psi[\mathcal{A}, \Gamma \vdash \Delta]}{\Psi[\Gamma \vdash \Delta, \neg\mathcal{A}]} \vdash\neg$$

$$\frac{\Psi[\Gamma \vdash \Delta]}{\Psi[\mathcal{A}, \Gamma \vdash \Delta]} K\vdash \qquad \frac{\Psi[\Gamma, \mathcal{A}, \mathcal{B}, \Theta \vdash \Delta]}{\Psi[\Gamma, \mathcal{B}, \mathcal{A}, \Theta \vdash \Delta]} C\vdash \qquad \frac{\Psi[\mathcal{A}, \mathcal{A}, \Gamma \vdash \Delta]}{\Psi[\mathcal{A}, \Gamma \vdash \Delta]} W\vdash$$

$$\frac{\Psi[\Gamma \vdash \Delta]}{\Psi[\Gamma \vdash \Delta, \mathcal{A}]} \vdash K \qquad \frac{\Psi[\Gamma \vdash \Delta, \mathcal{A}, \mathcal{B}, \Theta]}{\Psi[\Gamma \vdash \Delta, \mathcal{B}, \mathcal{A}, \Theta]} \vdash C \qquad \frac{\Psi[\Gamma \vdash \Delta, \mathcal{A}, \mathcal{A}]}{\Psi[\Gamma \vdash \Delta, \mathcal{A}]} \vdash W$$

$$\frac{\Psi[\mathcal{A}, \Gamma, \Gamma^{\Box} \vdash \Delta^{\Diamond}, \Delta]}{\Psi[\Diamond\mathcal{A}, \Gamma^{\Box} \vdash \Delta^{\Diamond}]} \Diamond\vdash \qquad \frac{\Psi[\Gamma, \Gamma^{\Box} \vdash \Delta^{\Diamond}, \Delta, \mathcal{A}]}{\Psi[\Gamma^{\Box} \vdash \Delta^{\Diamond}, \Box\mathcal{A}]} \vdash\Box$$

$$\frac{\Psi[\mathcal{A}, \Gamma \vdash \Delta]}{\Psi[\Box\mathcal{A}, \Gamma \vdash \Delta]} \Box\vdash \qquad \frac{\Psi[\Gamma \vdash \Delta, \mathcal{A}]}{\Psi[\Gamma \vdash \Delta, \Diamond\mathcal{A}]} \vdash\Diamond$$

$$\frac{\Psi[\Gamma^{\triangle}, \Theta \vdash \Lambda, \Delta^{\triangle}]}{\Psi[\Gamma^{\triangle} \vdash \Delta^{\triangle} \mid \Theta \vdash \Lambda]} \text{ split} \qquad \frac{\Psi[\Gamma \vdash \Delta \mid \Gamma \vdash \Delta]}{\Psi[\Gamma \vdash \Delta]} M$$

The superscript $^{\triangle}$ indicates that each formula in the sequence is modalized, that is, of the form $\Box\mathcal{A}$ or $\Diamond\mathcal{A}$ (for some $\mathcal{A}$).

The first part of H_{S5} is a hypersequent version of LK, then come the modal rules. Among all the rules, the only rules that really utilize hypersequents are the *split* and the (M) rule. Split allows us to gather together some modalized formulas from the antecedent plus some modalized formulas from the succedent of a sequent, and create a new sequent comprising only those formulas, while leaving the remainder of the sequent as an alternative. This is a rule, which is specifically needed for $S5$. (M) is not a modal rule, however, it would be superfluous, if split were absent. This rule *contracts sequents*—unlike $(W\vdash)$ and $(\vdash W)$, which permit the contraction of formulas inside sequents.

The notion of a *proof* is similar to the notion of a proof in sequent calculi except that the nodes of a proof tree are occurrences of hypersequents. A formula $\mathcal{A}$ is a *theorem* if there is a proof of the hypersequent $\vdash \mathcal{A}$, that is, of the hypersequent comprising solely this sequent.

Obviously, any theorem of classical propositional logic or of $S4$ is a theorem, because one may ignore the Ψ in every rule save split (and (M)). Some other rules, which are derived and are of some interest, are the following. (We label them by "abs," because a modal formula *absorbs* a formula from another sequent or even the whole other sequent.)

$$\frac{\Psi[\Box\mathcal{A},\Gamma\vdash\Delta \mid \mathcal{A},\Theta\vdash\Lambda]}{\Psi[\Box\mathcal{A},\Gamma\vdash\Delta \mid \Theta\vdash\Lambda]}\ \text{abs}_1 \qquad \frac{\Psi[\Gamma\vdash\Delta,\Diamond\mathcal{A} \mid \Theta\vdash\Lambda,\mathcal{A}]}{\Psi[\Gamma\vdash\Delta,\Diamond\mathcal{A} \mid \Theta\vdash\Lambda]}\ \text{abs}_2$$

$$\frac{\Psi[\Diamond\mathcal{A},\Gamma\vdash\Delta \mid \mathcal{A}\vdash\ ]}{\Psi[\Diamond\mathcal{A},\Gamma\vdash\Delta]}\ \text{abs}_3 \qquad \frac{\Psi[\Gamma\vdash\Delta,\Box\mathcal{A} \mid\ \vdash\mathcal{A}]}{\Psi[\Gamma\vdash\Delta,\Box\mathcal{A}]}\ \text{abs}_4$$

Exercise 6.4.1. Prove that these rules are derived rules in H_{S5}.

The *characteristic axiom* of $S5$ is $\Diamond\mathcal{A}\supset\Box\Diamond\mathcal{A}$. Here is a proof showing that this wff is a theorem of H_{S5}.

$$\dfrac{\dfrac{\dfrac{\dfrac{\Diamond\mathcal{A}\vdash\Diamond\mathcal{A}}{\Diamond\mathcal{A}\vdash\Diamond\mathcal{A},\Box\Diamond\mathcal{A}}\ \vdash K}{\Diamond\mathcal{A}\vdash\Box\Diamond\mathcal{A} \mid\ \vdash\Diamond\mathcal{A}}\ \text{split}}{\Diamond\mathcal{A}\vdash\Box\Diamond\mathcal{A}}\ \text{abs}_4}{\vdash\Diamond\mathcal{A}\supset\Box\Diamond\mathcal{A}}\ \vdash\supset$$

Exercise 6.4.2. Prove that (K) $\Box(\mathcal{A}\supset\mathcal{B})\supset(\Box\mathcal{A}\supset\Box\mathcal{B})$, (T) $\Box\mathcal{A}\supset\mathcal{A}$, (4) $\Box\mathcal{A}\supset\Box\Box\mathcal{A}$ and (B) $\mathcal{A}\supset\Box\Diamond\mathcal{A}$ are also theorems of H_{S5}. [Hint: The labels for these wff's are often used in axiomatizations of some normal modal logics such as K, T, B and $S4$.]

The cut rule takes the following form.

$$\frac{\Psi_1[\Gamma\vdash\Delta[\mathcal{C}]] \quad \Psi_2[\Theta[\mathcal{C}]\vdash\Lambda]}{\Psi_1[\Psi_2[\Theta[\Gamma]\vdash\Delta[\Lambda]]]}\ \text{hyper cut}$$

Exercise* 6.4.3. Prove that the hyper cut rule is admissible in H_{S5}.

The *hypersequent calculus* $H_{S5}^{\Box}$ omits $\Diamond$, which is, of course, definable in $S5$; $\Diamond\mathcal{A}$ is $\neg\Box\neg\mathcal{A}$. $H_{S5}^{\Box}$ extends the hypersequent version of classical logic (as above) by the following two rules.

$$\frac{\Psi[\mathcal{A},\Gamma\vdash\Delta\mid\Theta\vdash\Lambda]}{\Psi[\Box\mathcal{A},\Theta\vdash\Lambda\mid\Gamma\vdash\Delta]}\;\Box\vdash \qquad\qquad \frac{\Psi[\;\vdash\mathcal{A}\mid\Gamma\vdash\Delta]}{\Psi[\Gamma\vdash\Delta,\Box\mathcal{A}]}\;\vdash\Box$$

The notions of a *proof* and of a *theorem* are as in H_{S5}. It may be interesting to note that both rules make genuine use of hypersequents. The rules are also appealing in the sense that there is just one left and one right introduction rule for $\Box$, and there are no side conditions (except the presence of another sequent in the hypersequent).

Example 6.16. We prove—once again—that $S5$'s characteristic axiom is a theorem.

$$\begin{array}{cl}
\mathcal{A}\vdash\mathcal{A}\mid\;\vdash\;\mid\;\vdash & \\
\hline
\vdash\mathcal{A}\mid\;\vdash\;\mid\Box\mathcal{A}\vdash & \Box\vdash \\
\hline
\vdash\Box\mathcal{A}\mid\Box\mathcal{A}\vdash & \vdash\Box \\
\hline
\neg\Box\mathcal{A}\vdash\;\mid\Box\mathcal{A}\vdash & \neg\vdash \\
\hline
\neg\Box\mathcal{A}\vdash\;\mid\;\vdash\neg\Box\mathcal{A} & \vdash\neg \\
\hline
\neg\Box\mathcal{A}\vdash\Box\neg\Box\mathcal{A} & \vdash\Box \\
\hline
\vdash\neg\Box\mathcal{A}\supset\Box\neg\Box\mathcal{A} & \vdash\supset
\end{array}$$

Exercise 6.4.4. Construct proofs of the modal formulas that are mentioned in Exercise 6.4.2.

Hypersequents have been used to formalize a range of non-classical logics, including many-valued logics—such as three-valued logics and R-mingle—as well as newer extensions of the basic normal modal logics.[8]

[8] For further hypersequent calculi and references, see [150], [190] and [159].

Chapter 7

Cut rules and cut theorems

The theorem that claims the admissibility (or rather, eliminability) of the single cut rule for LK and LJ was termed *Hauptsatz* by Gentzen. "Hauptsatz" simply means *main theorem* in German. However, in the English language literature, the term is used exclusively for the cut theorem. (Surely, there are other important or main theorems out there!) The cut rule is *vital* for any sequent calculus as we explain in detail in Section 7.8. It is much preferable to have the cut rule as an *admissible rule* rather than an explicit rule, because of the benefits of having the cut rule available when needed without always having to deal with it. Thus the Hauptsatz is indeed important.

The cut rule has many shapes beyond the original *single cut* in LK and LJ. Some versions of the cut rule were specifically invented to enable the proof of its admissibility—like the *mix rule*, which is an extreme form of the cut rule. However, not all forms of the cut rule are equivalent in all calculi; therefore, one has to carefully select a concrete cut rule so that it can be proved admissible, while it yields the outcomes expected from the cut rule.

Some proofs of the cut theorem are not proof-theoretic. But, perhaps, the most interesting ones are, and they are fascinating, because they rely on the examination and analysis of proofs and their structure.

The shape of the cut rule, the potential proofs of the cut theorem and the usefulness of the cut rule are tightly connected. In this chapter, we first move *from classical toward non-classical logics*, then we consider a semantic way of proving the cut theorem; lastly, we record some *consequences* and *uses* of the cut rule and of the cut theorem.

7.1 Uniform cut theorem

In Section 3.5, we described three *uniform sequent calculi* for classical first-order logic. LU_1 had the simplest underlying data type (i.e., sets) and the smallest number of rules, namely, four (one for each of α, β, γ and δ). This calculus allows us to construct a short and elegant proof of the cut theorem for a sequent calculus for classical first-order logic. LU_2 is the most suitable calculus to connect with tableaux (which we will introduce in Section 8.2), and LU_3 is the closest to the original LK.[1]

The rules (α) and (β) together with Table 3.1 undoubtedly suggest that there is a relationship between these rules. Similarly, there is a systematic way to relate (γ) and (δ) rules to each other. Although the α's, β's, etc. may be formulas signed with $\mathfrak{t}$ or $\mathfrak{f}$, that is, an α may start with either of the two signs, once the sign of a particular α is changed to the other sign, we have a β. Informally speaking, the De Morgan laws (in a somewhat generalized form) connect logical components, that is, negation is a dual isomorphism in the complete Boolean algebra of classical logic.

Definition 7.1. (Conjugation) The *conjugate* of $\mathfrak{t}\mathcal{A}$ is $\mathfrak{f}\mathcal{A}$, and vice versa. If X is a signed formula, then we denote the conjugate of X by $-X$.

As a turn of phrase, we will call $\mathfrak{t}\mathcal{A}$ and $\mathfrak{f}\mathcal{A}$ *conjugates* (of each other). Obviously, if X and Y are conjugates, then they are distinct, but the wff inside them is the same formula.

Lemma 7.2. *Conjugation has the following* duality-like *properties.*

(1) $--X$ *is* X;

(2) $-\alpha$ *is a* β *and* $-\beta$ *is an* α;

(3) $-\gamma$ *is a* δ *and* $-\delta$ *is a* γ;

(4) *if* $-\alpha$ *is* β, *then* $-\alpha_1$ *is* β_1 *and* $-\alpha_2$ *is* β_2;

(5) *if* $-\beta$ *is* α, *then* $-\beta_1$ *is* α_1, *and* $-\beta_2$ *is* α_2;

(6) *if* $-\gamma$ *is* δ, *then* $-\gamma(a)$ *is* $\delta(a)$;

(7) *if* $-\delta$ *is* γ, *then* $-\delta(a)$ *is* $\gamma(a)$.

Exercise 7.1.1. Convince yourself that the claims in the lemma are true. [Hint: Use Tables 3.1 and 3.2 from Section 3.5.]

If we try to extend this duality to the rules in LU_1, then we discover that in fact the two (α) rules in LU_3 are the duals of the two-premise (β) rule (in LU_1). Thus, we replace the (α) rule in LU_1 by the rules (α_1) and (α_2) (while still viewing Γ as a set).

Exercise 7.1.2. Prove that this modification to LU_1 yields an equivalent calculus. [Hint: Γ in (id) is arbitrary.]

Definition 7.3. If Γ has a proof such that in the last step of the proof X is the principal signed formula of the rule and the proof contains $n+1$ sequents, then Γ is provable *with weight* n. Briefly, Γ *is provable via* X *with weight* n.

[1]The uniform calculi were invented by Smullyan in [171]. We more or less follow [170] in this section.

This notion is related to the height of the proof tree and to the number of applications of rules in the proof of a sequent, but it is neither of those two. If Γ is an instance of (id), then the weight n is 0. If none of the rules applied is a (β) rule, then the number of rules applied is the weight, which is the height of the tree minus 1. From the above notion, we can obtain—in the usual way—the notion of Γ *is provable with weight* n and the notion of Γ *is provable.* That is, we existentially quantify the signed wff X, and then the weight n. Notice that in the proof of the cut theorem in Chapter 2, we started with the notion of *provable,* without further specifications, then for the proof we introduced ϱ, δ and μ and divided the cases according to which rule was applied last.

Lemma 7.4. (Provability properties)

(P1) (a) *If* Γ, α, α_i $(i \in \{1,2\})$ *is provable, so is* Γ, α;

(b) *if* Γ, β, β_1 *and* Γ, β, β_2 *are provable, so is* Γ, β;

(c) *if* $\Gamma, \gamma, \gamma(a)$ *is provable, so is* Γ, γ;

(d) *if* $\Gamma, \delta, \delta(a)$ *with* a *not occurring in* Γ, δ *is provable, then* Γ, δ *is provable.*

(P2) *If* Γ *is provable with weight* k *and* $\Gamma \subseteq \Delta$, *then* Δ *is provable with weight* k.

(P3) (a) *If* Γ, α *is provable via* α *with weight* k, *then either* Γ, α, α_1 *or* Γ, α, α_2 *is provable with weight* n *such that* $n < k$;

(b) *if* Γ, β *is provable via* β *with weight* k, *then both* Γ, β, β_1 *and* Γ, β, β_2 *are provable with weight* n *such that* $n < k$;

(c) *if* Γ, γ *is provable via* γ *with weight* k, *then for some* a, $\Gamma, \gamma, \gamma(a)$ *is provable with weight* n *such that* $n < k$;

(d) *if* Γ, δ *is provable via* δ *with weight* k, *then for some* a *that does not occur in* Γ, δ, $\Gamma, \delta, \delta(a)$ *is provable with weight* n *such that* $n < k$.

(P4) *If* $\Gamma, \delta(a)$ *is provable with weight* k, *where* a *has no occurrences in* Γ, δ, *then for any* b, $\Gamma, \delta(b)$ *is provable with weight* k.

Proof: **(P1)** This property is obvious from the shape of the rules in LU_1. For instance, if we have $\Gamma, \gamma, \gamma(a)$, then by the rule (γ), Γ, γ.
(P2) Δ is a finite set of signed wff's (by our convention), and Γ is a subset of Δ. It is easy to see (and to prove by induction on a proof of Γ), that any X's that are elements of Δ, but not of Γ, may be inserted into the sequents without destroying the proof.
(P3) This step requires only a little more proving. For example, if Γ, α is proved via α, then the last rule applied was (α_1) or (α_2). That is, Γ, α_1 or Γ, α_2 is provable, and by (P2), we know that Γ, α, α_1 or Γ, α, α_2 is also provable. The other cases are similar.

(P4) We first show that the replacement of name constants can be described systematically. Let $X[a/b]$ denote the signed formula in which all occurrences of a (if there are any) are replaced by b. This replacement operation is very similar to substitution, because it changes *all occurrences* of a; on the other hand, substitution is usually defined to affect free variables not *name constants*. The replacement of a name constant and the four types of rules have the following property.

(a) For $i \in \{1,2\}$, $(\alpha[a/b])_i$ is $\alpha_i[a/b]$;

(b) for $i \in \{1,2\}$, $(\beta[a/b])_i$ is $\beta_i[a/b]$;

(c) if c is not a, then $\gamma(c)[a/b]$ is $\gamma[a/b](c)$; otherwise, $\gamma(c)[a/b]$ is $\gamma(b)$;

(d) if c is not a, then $\delta(c)[a/b]$ is $\delta[a/b](c)$; otherwise, $\delta(c)[a/b]$ is $\delta(b)$.

Let us assume that $\Gamma, \delta(a)$ is provable with weight k, with a not in Γ, δ. That is, there is a proof of Γ, δ. We can use this proof to construct a proof of $\Gamma, \delta(b)$. First, we construct a proof in which all name constants that disappear from the proof by an application of a δ rule are distinct both from a and b. Then we replace all occurrences of a by b everywhere in the proof. The resulting tree is a proof tree.

First of all, replacing a name constant by another in any of the α, β or γ rules creates new sequents that form a new instance of the same rule. If $\delta(c)$ with a c distinct from a is modified to $\delta(c)[c/d]$ with d fresh, then the δ rule is applicable as before. If $\delta(c)$ is $\delta(a)$, then $\delta(c)[a/b]$ is $\delta(b)$, and since a does not occur in Γ, δ, neither does b. Thus, we indeed have a proof. qed

Exercise 7.1.3. Scrutinize and finish steps **(P1)** and **(P3)** in the above proof.

Theorem 7.5. (Uniform cut theorem) *For every Γ and for every X, if both Γ, X and $\Gamma, -X$ are provable, then Γ is provable.*

Proof: Notice that the statement in the theorem has a slightly different form than, for instance, the original cut rule—not only because we have signed formulas rather than formulas, but both assumptions contain the same Γ. Of course, *LK* has thinning and contraction, which means that antecedents and succedents could be enlarged so that they are the same in both premises. Further, using contractions, the duplicates could be weaned to single copies. In other words, we should not be concerned that this theorem is less general or more restricted than the one in Chapter 2.

The proof is by *double induction*, however, instead of ϱ, the rank of the cut, one parameter in the induction is the *sum of the weights* of the premises, that is, the weight of Γ, X and of $\Gamma, -X$. In order to give a meta-language formula on which the double induction is based, we will use $=$ and $<$ with their usual meaning, and E as a binary predicate. $E(k,d)$ is thought to mean that "a cut, in which the weight of the premises is k and the wff in the signed formula X has degree d, is eliminable (that is, the conclusion of the application of the rule, Γ is provable)."

$$\forall d \, \forall k \left[(\forall n \, (n < d \Rightarrow \forall l \, E(l,n)) \wedge \forall n \, (n = d \Rightarrow \forall l \, (l < k \Rightarrow E(l,n)))) \Rightarrow E(k,d) \right] \Rightarrow \forall n \, \forall l \, E(l,n)$$

Our aim is to prove $E(k,d)$ based on the two assumptions $\forall n \, (n < d \Rightarrow \forall l \, E(l,n))$ and $\forall n \, (n = d \Rightarrow \forall l \, (l < k \Rightarrow E(l,n)))$. (The latter may be written as $\forall l \, (l < k \Rightarrow E(l,d))$.) Then, by the above induction principle, it will follow that Γ is provable, provided that Γ, X and $\Gamma, -X$ are.

1. If Γ, X or $\Gamma, -X$ is provable with weight 0, then Γ is provable. To see this, and assuming that Γ, X is an instance of (id), we consider how this set can be an axiom. If $-X \notin \Gamma$, then there are Y and $-Y$ both of which are elements of Γ, hence, Γ itself is provable. On the other hand, if $-X \in \Gamma$, then $\Gamma, -X$ is Γ, and provable by assumption. Either way, Γ is provable. (Obviously, the argument is the same if we start with $\Gamma, -X$, because $-$ is a relative notation for conjugates.)

2. Now let us assume that Γ, X is provable via Y with weight l (where $l > 0$) and $\Gamma, -X$ is provable via Z with weight m (where $m > 0$), and further, either X is not Y or $-X$ is not Z. Let us assume that $-X$ is distinct from Z. (Again, choosing X or $-X$ does not impact the argument.) Then, $Z \in \Gamma$, and the assumption means that $\Gamma, -X, Z$ is provable via Z with weight m.

The α, γ and δ rules are one-premise rules. For these, the third provability property, (P3), guarantees that, given the assumption, there is a signed formula V with degree n less than the degree of Z, and $\Gamma, -X, Z, V$ is provable with weight $k < m$. For instance, if Z is of type α, then V is α_1 or α_2. We have that $\Gamma, -X, Z, V$ is provable with weight $k < m$. But Γ, X, Z, which is Γ, X, is provable with weight l, and hence, by (P2), so is Γ, X, Z, V. $k + l < m + l$, therefore, Γ, Z, V is provable by the second assumption of the double induction. By (P1), Γ, Z, V's provability implies Γ, Z's provability, because V is not an arbitrary signed formula, but it is appropriately related to Z. However, $Z \in \Gamma$, that is, Γ is provable.

The only two-premise rules are the β rules. Then (P3) gives that both $\Gamma, -X, Z, V$ and $\Gamma - X, Z, W$ are provable with weights $k < m$ and $n < m$, where V and W are β_1 and β_2, respectively. By (P2), we can conclude that both Γ, X, V and Γ, X, W are provable with weight l. We pair up the sets as Γ, X, Z, V and $\Gamma, -X, Z, V$, on one hand, and Γ, X, Z, W and $\Gamma, -X, Z, W$, on the other. $k + l < m + l$ and $n + l < m + l$, and by the second assumption in the double induction, we have that Γ, Z, V and Γ, Z, W are provable. However, V and W are exactly the signed formulas that we need in (P1) (b) to yield that Γ, Z, which is simply Γ, is provable.

3. Finally, we assume that Γ, X is provable via X with weight l (where $l > 0$) and $\Gamma, -X$ is provable via $-X$ with weight m (where $m > 0$). By (2)

and (3) from among the properties of conjugates, we know that either X or $-X$ is α or γ. Nothing hinges on whether it is X or $-X$, thus, we suppose that there are two possibilities for X.

If X is of type α, then, by (P3) (a), Γ, X, V or Γ, X, W is provable with weight $n < l$, where V and W are α_1 and α_2, respectively. Let us choose W, that is, let Γ, X, W be provable with weight $n < l$. We know that $\Gamma, -X, W$ is provable with weight m, by (P2). But $n + m < l + m$, which means by the second assumption of the double induction that Γ, W is provable. $-X$ is of type β, and so by (P3) (b), both $\Gamma, -X, -V$ and $\Gamma, -X, -W$ are provable with weight $k < m$. Γ, X is provable with weight l, hence, so is $\Gamma, X, -W$ by (P2). $k + l < m + l$, therefore, by the second assumption of the double induction, $\Gamma, -W$ is provable. Then, by the first assumption of the double induction, Γ is provable, because W has lower degree that X, and both Γ, W and $\Gamma, -W$ are provable.

If X is of type γ, then, by (P3) (c), Γ, X, V is provable with weight $n < l$, where V is $\gamma(a)$. Since $\Gamma, -X$ is provable with weight m, so is $\Gamma, -X, V$, by (P2). $n + m < l + m$, thus Γ, V is provable. $-X$ is of type δ, which gives us that $\Gamma, -X, W$ is provable with weight $k < m$. W is $\delta(c)$, where c has no occurrences in $\Gamma, -X$. From the provability of Γ, X with weight l, we obtain by (P2), that Γ, X, W is provable with the same weight. Then, Γ, W is provable, by the second assumption in the double induction. Now, V and W may or may not be conjugates. Although $-\gamma(a)$ is $\delta(a)$, it can differ from $\delta(c)$. However, the condition for (P4) is satisfied; therefore, from Γ, W being provable, it follows that $\Gamma, -V$ is provable too. The degree of V is less than the degree of X, and so we can conclude, by the first assumption of the double induction, that Γ is provable. qed

The uniform system LU_1 is obviously not Gentzen's original calculus. Nonetheless, it is remarkable how short the above proof is, and how easy it is to inspect the proof. Even with an unavoidable proof of equivalence to LK, LU_1 has undeniable advantages, with respect to comprehensibility, over either Gentzen's original proof via the mix rule, or our proof in Chapter 2.

7.2 Mix, multiple and single cuts

Gentzen must have realized that he did not have an inductive proof to show the eliminability of the cut rule. While he does not explicitly state this in [91], he introduces the *mix rule* to facilitate the proof of the Hauptsatz (i.e., the cut elimination theorem).

$$\frac{\Gamma \vdash \Delta \qquad \Theta \vdash \Lambda}{\Gamma, \Theta^* \vdash \Delta^*, \Lambda}\ \text{mix}$$

A condition of the applicability of this rule, which is somewhat obscured by the lack of good notation, is that there is a wff $\mathcal{M}$, which is called the *mix formula* that occurs both in Δ and Θ. The result of an application of mix eliminates *all the occurrences* of $\mathcal{M}$ from Δ and Θ, which is indicated by starring those two sequences of formulas.

Mix is a very radical rule, because it eradicates all occurrences of $\mathcal{M}$ from the succedent of the left premise and from the antecedent of the right premise. Importantly, "all occurrences" may be "more than one." In particular, a mix, which has been applied after a contraction that left one occurrence of a wff for mix to eliminate, can be replaced by a mix applied to the premise of the contraction, while retaining the other premise of the earlier mix. This is the situation where the double induction on the rank and degree of a formula fails to reduce the rank of the applications of the single cut rule.

The mix rule is unproblematic in LK and LJ, because anything can be thinned into a sequent, including whatever was "mixed away by accident," so to speak. The single cut rule is derivable if mix is a rule of LK or LJ, and the other way around.

Exercise 7.2.1. Demonstrate that each of the two rules can be derived from the other, in LK and LJ.

The mix rule is useless from the point of view of most of the non-classical sequent and consecution calculi, exactly because it indiscriminately deletes all occurrences of the mix formula. However, the mix rule resolves the difficulty in the double inductive proof for the original formulations of LK and LJ, as well as their variants that we considered in Chapter 3. As a result, there seems to have been no interest for a long time to provide a direct proof of the admissibility of the *single cut rule* in sequent calculi for classical or intuitionistic logics (or for non-classical logics).

Exercise 7.2.2. Modify the triple inductive proof in Section 2.3 for mix by omitting μ. That is, prove the admissibility of mix in LK, by double induction. [Hint: This is what Gentzen proved in [91], and his paper includes some details of the proof.]

Exercise 7.2.3. Prove the admissibility of the mix rule in LJ. [Hint: The proof from the previous exercise is useful here.]

7.2.1 Contraction and cut

The modifications to the original sequent calculi (i.e., LK and LJ) that were invented in the late 1950s led to sequent calculi that steered clear of the mix rule. The non-associative and associative Lambek calculi, which we mentioned in Sections 5.1 and 4.1, respectively, do not contain contraction rules in any form, nor do they have permutations. Then care has to be taken as to the proper placement of the cut formula in the single cut rule. The connective rules, as well as the single cut, are *properly formulated* in the sections mentioned above, with the rules referring to occurrences of formulas and replacing one of them by an occurrence of another formula or structure. Although LA can be emulated over LQ by structural rules, which provide associativity, LA itself has no structural rules, and neither does LQ. This allows a simplification of not only the triple inductive proof of the admissibility of the cut rule, but of the double inductive proof too.[2]

Definition 7.6. (Degree of a formula) The *degree* of a formula $\mathcal{A}$ is denoted by $\delta(\mathcal{A})$, and it is defined inductively by (1)–(4).

(1) $\delta(p) = 0$, where p is a propositional variable;

(2) $\delta(\mathcal{A} \to \mathcal{B}) = \delta(\mathcal{A}) + \delta(\mathcal{B}) + 1$;

(3) $\delta(\mathcal{A} \circ \mathcal{B}) = \delta(\mathcal{A}) + \delta(\mathcal{B}) + 1$;

(4) $\delta(\mathcal{A} \leftarrow \mathcal{B}) = \delta(\mathcal{A}) + \delta(\mathcal{B}) + 1$.

This definition of the degree of a formula differs from the one we gave in Section 2.3. Beyond the obvious difference in the set of the connectives, this definition counts the *number of occurrences of connectives* in a formula, whereas the degree of wff's in LK was the height of the formation tree of the formula.

Let us expand this notion to sequents (consecutions) by *summing up* the degrees of formulas in the sequent (consecution)—counting each occurrence of a formula separately. Then, we attach a pair of degrees to the rules. The *upper degree* is the degree of the premise (in a single-premise rule) and the sum of the degrees of the premises (in a two-premise rule). The *lower degree* is the degree of the lower sequent (consecution) in the rule.

Proposition 7.7. *Each rule in LA (and in LQ) is* degree increasing. *That is, the lower degree is* strictly greater *than the upper degree in each rule.*

Exercise 7.2.4. Verify that the proposition is true.

A bird's-eye view of the transformations in the triple inductive proof of the admissibility of the cut is that the top-most cut is moved all the way up. In that process, the cut may remain a cut on the same formula, but higher in the proof tree, or it may become one or two cuts applied to the same or to simpler formulas, and with at least one of those cuts is also higher in the proof tree.

The *degree of a cut* is identified with the upper degree of the cut rule.

Theorem 7.8. (Cut theorem for LA and LQ) *If $\Gamma \vdash \Delta$ is provable in LA (in LQ), then* there is a proof of $\Gamma \vdash \Delta$ *in LA (in LQ)* without *any applications of the* cut rule.

[2] Girard in [97, Ch. 11.1] claims that in non-associative calculi the cut rule is not admissible. This is, obviously, a misunderstanding. See Theorem 7.8 and its proof.

Proof: The proof is by complete induction on the degree of the cut. (We give the steps for the proof in LQ, and leave as an exercise the modifications needed for LA.)

1. If the degree of the cut is 0, then both premises are instances of the axiom. Obviously, one of the upper consecutions is sufficient as a proof.

We divide the rest of the cases into three groups.

2. If the two premises of the cut are by the rules for one and the same connective with their principal formulas being the cut formula, then there are three possibilities—according to which connective is introduced. We give the proof segments and their transformations as a pair.

$$\text{cut}\ \dfrac{\vdash\circ\ \dfrac{\mathfrak{A}\vdash\mathcal{A}\qquad\mathfrak{B}\vdash\mathcal{B}}{\mathfrak{A};\mathfrak{B}\vdash\mathcal{A}\circ\mathcal{B}}\qquad\dfrac{\mathfrak{C}[\mathcal{A};\mathcal{B}]\vdash\mathcal{C}}{\mathfrak{C}[\mathcal{A}\circ\mathcal{B}]\vdash\mathcal{C}}\ \circ\vdash}{\mathfrak{C}[\mathfrak{A};\mathfrak{B}]\vdash\mathcal{C}}\qquad\rightsquigarrow$$

$$\rightsquigarrow\qquad\dfrac{\mathfrak{B}\vdash\mathcal{B}\qquad\dfrac{\mathfrak{A}\vdash\mathcal{A}\qquad\mathfrak{C}[\mathcal{A};\mathcal{B}]\vdash\mathcal{C}}{\mathfrak{C}[\mathfrak{A};\mathcal{B}]\vdash\mathcal{C}}\ \text{cut}}{\mathfrak{C}[\mathfrak{A};\mathfrak{B}]\vdash\mathcal{C}}\ \text{cut}$$

$$\text{cut}\ \dfrac{\vdash\to\ \dfrac{\mathfrak{A};\mathcal{A}\vdash\mathcal{B}}{\mathfrak{A}\vdash\mathcal{A}\to\mathcal{B}}\qquad\dfrac{\mathfrak{B}\vdash\mathcal{A}\qquad\mathfrak{C}[\mathcal{B}]\vdash\mathcal{C}}{\mathfrak{C}[\mathcal{A}\to\mathcal{B};\mathfrak{B}]\vdash\mathcal{C}}\ \to\vdash}{\mathfrak{C}[\mathfrak{A};\mathfrak{B}]\vdash\mathcal{C}}\qquad\rightsquigarrow$$

$$\rightsquigarrow\qquad\dfrac{\text{cut}\ \dfrac{\mathfrak{B}\vdash\mathcal{A}\qquad\mathfrak{A};\mathcal{A}\vdash\mathcal{B}}{\mathfrak{A};\mathfrak{B}\vdash\mathcal{B}}\qquad\mathfrak{C}[\mathcal{B}]\vdash\mathcal{C}}{\mathfrak{C}[\mathfrak{A};\mathfrak{B}]\vdash\mathcal{C}}\ \text{cut}$$

$$\text{cut}\ \dfrac{\vdash\leftarrow\ \dfrac{\mathcal{A};\mathfrak{A}\vdash\mathcal{B}}{\mathfrak{A}\vdash\mathcal{B}\leftarrow\mathcal{A}}\qquad\dfrac{\mathfrak{B}\vdash\mathcal{A}\qquad\mathfrak{C}[\mathcal{B}]\vdash\mathcal{C}}{\mathfrak{C}[\mathfrak{B};\mathcal{B}\leftarrow\mathcal{A}]\vdash\mathcal{C}}\ \leftarrow\vdash}{\mathfrak{C}[\mathfrak{B};\mathfrak{A}]\vdash\mathcal{C}}\qquad\rightsquigarrow$$

$$\rightsquigarrow\qquad\dfrac{\text{cut}\ \dfrac{\mathfrak{B}\vdash\mathcal{A}\qquad\mathcal{A};\mathfrak{A}\vdash\mathcal{B}}{\mathfrak{B};\mathfrak{A}\vdash\mathcal{B}}\qquad\mathfrak{C}[\mathcal{B}]\vdash\mathcal{C}}{\mathfrak{C}[\mathfrak{B};\mathfrak{A}]\vdash\mathcal{C}}\ \text{cut}$$

In each of these three transformations, a cut is traded in for two cuts. However, both cuts have lower degrees.

3. If the left premise of the cut is obtained by a left rule, then the succedent of the upper and lower consecutions is unchanged. There are three left rules, and accordingly, three transformations.

$$\text{cut}\ \dfrac{\circ\vdash\ \dfrac{\mathfrak{A}[\mathcal{A};\mathcal{B}]\vdash\mathcal{C}}{\mathfrak{A}[\mathcal{A}\circ\mathcal{B}]\vdash\mathcal{C}}\qquad\mathfrak{B}[\mathcal{C}]\vdash\mathcal{D}}{\mathfrak{B}[\mathfrak{A}[\mathcal{A}\circ\mathcal{B}]]\vdash\mathcal{D}}\qquad\rightsquigarrow\qquad\dfrac{\dfrac{\mathfrak{A}[\mathcal{A};\mathcal{B}]\vdash\mathcal{C}\qquad\mathfrak{B}[\mathcal{C}]\vdash\mathcal{D}}{\mathfrak{B}[\mathfrak{A}[\mathcal{A};\mathcal{B}]]\vdash\mathcal{D}}\ \text{cut}}{\mathfrak{B}[\mathfrak{A}[\mathcal{A}\circ\mathcal{B}]]\vdash\mathcal{D}}\ \circ\vdash$$

$$\dfrac{\dfrac{\mathfrak{A}\vdash\mathcal{A}\qquad\mathfrak{B}[\mathcal{B}]\vdash\mathcal{C}}{\mathfrak{B}[\mathcal{A}\to\mathcal{B};\mathfrak{A}]\vdash\mathcal{C}}\,{\scriptstyle\to\vdash}\qquad\mathfrak{C}[\mathcal{C}]\vdash\mathcal{D}}{\mathfrak{C}[\mathfrak{B}[\mathcal{A}\to\mathcal{B};\mathfrak{A}]]\vdash\mathcal{D}}\,{\scriptstyle\mathrm{cut}}\quad\rightsquigarrow$$

$$\rightsquigarrow\quad\dfrac{\mathfrak{A}\vdash\mathcal{A}\qquad\dfrac{\mathfrak{B}[\mathcal{B}]\vdash\mathcal{C}\qquad\mathfrak{C}[\mathcal{C}]\vdash\mathcal{D}}{\mathfrak{C}[\mathfrak{B}[\mathcal{B}]]\vdash\mathcal{D}}\,{\scriptstyle\mathrm{cut}}}{\mathfrak{C}[\mathfrak{B}[\mathcal{A}\to\mathcal{B};\mathfrak{A}]]\vdash\mathcal{D}}\,{\scriptstyle\to\vdash}$$

$$\dfrac{\dfrac{\mathfrak{A}\vdash\mathcal{A}\qquad\mathfrak{B}[\mathcal{B}]\vdash\mathcal{C}}{\mathfrak{B}[\mathfrak{A};\mathcal{B}\leftarrow\mathcal{A}]\vdash\mathcal{C}}\,{\scriptstyle\leftarrow\vdash}\qquad\mathfrak{C}[\mathcal{C}]\vdash\mathcal{D}}{\mathfrak{C}[\mathfrak{B}[\mathfrak{A};\mathcal{B}\leftarrow\mathcal{A}]]\vdash\mathcal{D}}\,{\scriptstyle\mathrm{cut}}\quad\rightsquigarrow$$

$$\rightsquigarrow\quad\dfrac{\mathfrak{A}\vdash\mathcal{A}\qquad\dfrac{\mathfrak{B}[\mathcal{B}]\vdash\mathcal{C}\qquad\mathfrak{C}[\mathcal{C}]\vdash\mathcal{D}}{\mathfrak{C}[\mathfrak{B}[\mathcal{B}]]\vdash\mathcal{D}}\,{\scriptstyle\mathrm{cut}}}{\mathfrak{C}[\mathfrak{B}[\mathfrak{A};\mathcal{B}\leftarrow\mathcal{A}]]\vdash\mathcal{D}}\,{\scriptstyle\leftarrow\vdash}$$

In each case a cut is replaced by another cut of lower degree. Informally, the transformation may be viewed as swapping the cut rule with the left introduction rule of a connective.

4. If the left premise of the cut is by a right rule, then the principal formula of the rule determines the cut formula. In Case **1** above, we have considered what happens if the two premises of the cut result by rules that introduce the same connective. For example, if $(\vdash\to)$ is the rule in the left premise, then there are six possibilities for how the right premise came about.

If the right premise is by a right rule, then an insight behind the transformations is that none of the right rules moves a formula from the right-hand side of the $\vdash$ to its left; hence, if the cut could be applied after the connective rule, then it certainly can be applied before the connective rule. Here are the three possibilities.

$$\dfrac{\dfrac{\mathfrak{A};\mathcal{A}\vdash\mathcal{B}}{\mathfrak{A}\vdash\mathcal{A}\to\mathcal{B}}\,{\scriptstyle\vdash\to}\qquad\dfrac{\mathfrak{B}[\mathcal{A}\to\mathcal{B}];\mathcal{C}\vdash\mathcal{D}}{\mathfrak{B}[\mathcal{A}\to\mathcal{B}]\vdash\mathcal{C}\to\mathcal{D}}\,{\scriptstyle\vdash\to}}{\mathfrak{B}[\mathfrak{A}]\vdash\mathcal{C}\to\mathcal{D}}\,{\scriptstyle\mathrm{cut}}\quad\rightsquigarrow$$

$$\rightsquigarrow\quad\dfrac{\dfrac{\dfrac{\mathfrak{A};\mathcal{A}\vdash\mathcal{B}}{\mathfrak{A}\vdash\mathcal{A}\to\mathcal{B}}\,{\scriptstyle\vdash\to}\qquad\mathfrak{B}[\mathcal{A}\to\mathcal{B}];\mathcal{C}\vdash\mathcal{D}}{\mathfrak{B}[\mathfrak{A}];\mathcal{C}\vdash\mathcal{D}}\,{\scriptstyle\mathrm{cut}}}{\mathfrak{B}[\mathfrak{A}]\vdash\mathcal{C}\to\mathcal{D}}\,{\scriptstyle\vdash\to}$$

$$\dfrac{{\scriptstyle\vdash\to}\;\dfrac{\overset{\vdots}{\mathfrak{A};\mathcal{A}\vdash\mathcal{B}}}{\mathfrak{A}\vdash\mathcal{A}\to\mathcal{B}}\qquad\dfrac{\overset{\vdots}{\mathcal{C};\mathfrak{B}[\mathcal{A}\to\mathcal{B}]\vdash\mathcal{D}}}{\mathfrak{B}[\mathcal{A}\to\mathcal{B}]\vdash\mathcal{D}\leftarrow\mathcal{C}}\;{\scriptstyle\vdash\leftarrow}}{\mathfrak{B}[\mathfrak{A}]\vdash\mathcal{D}\leftarrow\mathcal{C}}\;{\scriptstyle\text{cut}}\qquad\rightsquigarrow$$

$$\rightsquigarrow\qquad\dfrac{\dfrac{{\scriptstyle\vdash\to}\;\dfrac{\overset{\vdots}{\mathfrak{A};\mathcal{A}\vdash\mathcal{B}}}{\mathfrak{A}\vdash\mathcal{A}\to\mathcal{B}}\qquad\overset{\vdots}{\mathcal{C};\mathfrak{B}[\mathcal{A}\to\mathcal{B}]\vdash\mathcal{D}}}{\mathcal{C};\mathfrak{B}[\mathfrak{A}]\vdash\mathcal{D}}\;{\scriptstyle\text{cut}}}{\mathfrak{B}[\mathfrak{A}]\vdash\mathcal{D}\leftarrow\mathcal{C}}\;{\scriptstyle\vdash\leftarrow}$$

$$\dfrac{{\scriptstyle\vdash\to}\;\dfrac{\overset{\vdots}{\mathfrak{A};\mathcal{A}\vdash\mathcal{B}}}{\mathfrak{A}\vdash\mathcal{A}\to\mathcal{B}}\qquad\dfrac{\overset{\vdots}{\mathfrak{C}[\mathcal{A}\to\mathcal{B}]\vdash\mathcal{C}}\quad\overset{\vdots}{\mathfrak{D}\vdash\mathcal{D}}}{\mathfrak{C}[\mathcal{A}\to\mathcal{B}];\mathfrak{D}\vdash\mathcal{C}\circ\mathcal{D}}\;{\scriptstyle\vdash\circ}}{\mathfrak{C}[\mathfrak{A}];\mathfrak{D}\vdash\mathcal{C}\circ\mathcal{D}}\;{\scriptstyle\text{cut}}\qquad\rightsquigarrow$$

$$\rightsquigarrow\qquad\dfrac{\dfrac{{\scriptstyle\vdash\to}\;\dfrac{\overset{\vdots}{\mathfrak{A};\mathcal{A}\vdash\mathcal{B}}}{\mathfrak{A}\vdash\mathcal{A}\to\mathcal{B}}\qquad\overset{\vdots}{\mathfrak{C}[\mathcal{A}\to\mathcal{B}]\vdash\mathcal{C}}}{\mathfrak{C}[\mathfrak{A}]\vdash\mathcal{C}}\;{\scriptstyle\text{cut}}\qquad\overset{\vdots}{\mathfrak{D}\vdash\mathcal{D}}}{\mathfrak{C}[\mathfrak{A}];\mathfrak{D}\vdash\mathcal{C}\circ\mathcal{D}}\;{\scriptstyle\vdash\circ}$$

(If $\mathcal{A}\to\mathcal{B}$ occurs in $\mathfrak{D}$, then cut is performed on $\mathfrak{D}[\mathcal{A}\to\mathcal{B}]\vdash\mathcal{D}$. The two premises of $(\vdash\circ)$ are symmetric in this respect.)

If the right premise is by a left rule, then we may assume (in view of Case **2**) that the principal formula of that rule is distinct from the cut formula. However, the rules are carefully formulated and there is no interference between the cut and the connective rules. Here are the three possibilities.

$$\dfrac{{\scriptstyle\vdash\to}\;\dfrac{\overset{\vdots}{\mathfrak{A};\mathcal{A}\vdash\mathcal{B}}}{\mathfrak{A}\vdash\mathcal{A}\to\mathcal{B}}\qquad\dfrac{\overset{\vdots}{\mathfrak{B}[\mathcal{C};\mathcal{D}]\,[\mathcal{A}\to\mathcal{B}]\vdash\mathcal{E}}}{\mathfrak{B}[\mathcal{C}\circ\mathcal{D}]\,[\mathcal{A}\to\mathcal{B}]\vdash\mathcal{E}}\;{\scriptstyle\circ\vdash}}{\mathfrak{B}[\mathcal{C}\circ\mathcal{D}]\,[\mathfrak{A}]\vdash\mathcal{E}}\;{\scriptstyle\text{cut}}\qquad\rightsquigarrow$$

$$\rightsquigarrow\qquad\dfrac{\dfrac{{\scriptstyle\vdash\to}\;\dfrac{\overset{\vdots}{\mathfrak{A};\mathcal{A}\vdash\mathcal{B}}}{\mathfrak{A}\vdash\mathcal{A}\to\mathcal{B}}\qquad\overset{\vdots}{\mathfrak{B}[\mathcal{C};\mathcal{D}]\,[\mathcal{A}\to\mathcal{B}]\vdash\mathcal{E}}}{\mathfrak{B}[\mathcal{C};\mathcal{D}]\,[\mathfrak{A}]\vdash\mathcal{E}}\;{\scriptstyle\text{cut}}}{\mathfrak{B}[\mathcal{C}\circ\mathcal{D}]\,[\mathfrak{A}]\vdash\mathcal{E}}\;{\scriptstyle\circ\vdash}$$

$$\dfrac{{\scriptstyle\vdash\to}\;\dfrac{\overset{\vdots}{\mathfrak{A};\mathcal{A}\vdash\mathcal{B}}}{\mathfrak{A}\vdash\mathcal{A}\to\mathcal{B}}\qquad\dfrac{\overset{\vdots}{\mathfrak{B}[\mathcal{A}\to\mathcal{B}]\vdash\mathcal{C}}\quad\overset{\vdots}{\mathfrak{C}[\mathcal{D}]\vdash\mathcal{E}}}{\mathfrak{C}[\mathcal{C}\to\mathcal{D};\mathfrak{B}[\mathcal{A}\to\mathcal{B}]]\vdash\mathcal{E}}\;{\scriptstyle\to\vdash}}{\mathfrak{C}[\mathcal{C}\to\mathcal{D};\mathfrak{B}[\mathfrak{A}]]\vdash\mathcal{E}}\;{\scriptstyle\text{cut}}\qquad\rightsquigarrow$$

$$\rightsquigarrow \quad \dfrac{\dfrac{\dfrac{\overset{\vdots}{\mathfrak{A}; \mathcal{A} \vdash \mathcal{B}}}{\mathfrak{A} \vdash \mathcal{A} \to \mathcal{B}}\,{\scriptstyle \vdash\to} \qquad \overset{\vdots}{\mathfrak{B}[\mathcal{A} \to \mathcal{B}] \vdash \mathcal{C}}}{\mathfrak{B}[\mathfrak{A}] \vdash \mathcal{C}}\,{\scriptstyle \mathrm{cut}} \qquad \overset{\vdots}{\mathfrak{C}[\mathcal{D}] \vdash \mathcal{E}}}{\mathfrak{C}[\mathcal{C} \to \mathcal{D}; \mathfrak{B}[\mathfrak{A}]] \vdash \mathcal{E}}\,{\scriptstyle \to\vdash}$$

$$\dfrac{\dfrac{\overset{\vdots}{\mathfrak{A}; \mathcal{A} \vdash \mathcal{B}}}{\mathfrak{A} \vdash \mathcal{A} \to \mathcal{B}}\,{\scriptstyle \vdash\to} \qquad \dfrac{\overset{\vdots}{\mathfrak{B} \vdash \mathcal{C}} \quad \overset{\vdots}{\mathfrak{C}[\mathcal{A} \to \mathcal{B}]\,[\mathcal{D}] \vdash \mathcal{E}}}{\mathfrak{C}[\mathcal{A} \to \mathcal{B}]\,[\mathfrak{B}; \mathcal{D} \leftarrow \mathcal{C}] \vdash \mathcal{E}}\,{\scriptstyle \leftarrow\vdash}}{\mathfrak{C}[\mathfrak{A}]\,[\mathfrak{B}; \mathcal{D} \leftarrow \mathcal{C}] \vdash \mathcal{E}}\,{\scriptstyle \mathrm{cut}} \quad \rightsquigarrow$$

$$\rightsquigarrow \quad \dfrac{\overset{\vdots}{\mathfrak{B} \vdash \mathcal{C}} \qquad \dfrac{\dfrac{\overset{\vdots}{\mathfrak{A}; \mathcal{A} \vdash \mathcal{B}}}{\mathfrak{A} \vdash \mathcal{A} \to \mathcal{B}}\,{\scriptstyle \vdash\to} \qquad \overset{\vdots}{\mathfrak{C}[\mathcal{A} \to \mathcal{B}]\,[\mathcal{D}] \vdash \mathcal{E}}}{\mathfrak{C}[\mathfrak{A}]\,[\mathcal{D}] \vdash \mathcal{E}}\,{\scriptstyle \mathrm{cut}}}{\mathfrak{C}[\mathfrak{A}]\,[\mathfrak{B}; \mathcal{D} \leftarrow \mathcal{C}] \vdash \mathcal{E}}\,{\scriptstyle \leftarrow\vdash}$$

(In the latter two transformations, we chose $\mathcal{A} \to \mathcal{B}$ to occur in different premises of the $(\to\vdash)$ and $(\leftarrow\vdash)$ rules. Again, the premises are alike with respect to the cut.)

The left premise may be by $(\vdash\circ)$ or $(\vdash\leftarrow)$, which do not cause difficulties either. We leave those cases for an exercise below. qed

The structure of the proof has an obvious similarity to the proof of the cut theorem in Chapter 2. The main difference is the absence of contraction and of its measure, and the combination of the degree of the cut formula and of the rank of the cut into the notion of *the degree of the cut*.

Exercise 7.2.5. Take a look at the above proof. Finish Case **4** and then note the changes that are needed to have a proof of the cut theorem for LA. [Hint: There are 12 transformations to consider; however, they fall into two groups, which share a general pattern with the subcases in Case **4**.]

The calculi LA and LQ are very special in that they do not have any structural rules. Many other logics have some structural rules in their sequent calculus formulation, but they do not have all of them. $LR_{\to}$ is an excellent example of such a logic, which is well-motivated, but omits only thinning. The absence of thinning means that the single cut rule and the mix rule are *not equivalent* in $LR_{\to}$. On the other hand, a double inductive argument along the lines of Gentzen's original proof of the admissibility of the single cut rule runs into a problem.

<u>*Example*</u> 7.9. Let us assume that we are considering the case when $\varrho_r \geq 2$, the right rank of the cut is greater than 2, and the right premise is by the $(W\vdash)$ rule. Moreover, the cut formula coincides with the formula that has

been contracted. The chunk of the proof and its transformation would look like the following.

$$\dfrac{\begin{matrix}\vdots \\ \Gamma \vdash \mathcal{A}\end{matrix} \qquad \dfrac{\begin{matrix}\vdots \\ \Delta[\mathcal{A};\mathcal{A}] \vdash \mathcal{B}\end{matrix}}{\Delta[\mathcal{A}] \vdash \mathcal{B}}\ {\scriptstyle W\vdash}}{\Delta[\Gamma] \vdash \mathcal{B}}\ {\scriptstyle \text{cut}} \qquad \dfrac{\dfrac{\begin{matrix}\vdots \\ \Gamma \vdash \mathcal{A}\end{matrix} \qquad \dfrac{\begin{matrix}\vdots \\ \Gamma \vdash \mathcal{A}\end{matrix} \quad \begin{matrix}\vdots \\ \Delta[\mathcal{A};\mathcal{A}] \vdash \mathcal{B}\end{matrix}}{\Delta[\Gamma;\mathcal{A}] \vdash \mathcal{B}}\ {\scriptstyle \text{cut}}}{\Delta[\Gamma;\Gamma] \vdash \mathcal{B}}\ {\scriptstyle \text{cut}}}{\Delta[\Gamma] \vdash \mathcal{B}}\ {\scriptstyle C\vdash\text{'s},\, W\vdash\text{'s}}$$

The step would need to be justified by a reduction in ϱ_r, but unfortunately, the right rank of the second cut is the same as that of the original cut.

A solution is to consider *multiple cut*. The problem with the double inductive proof is solved if the cut rule permits *several* occurrences of the cut formula to be replaced in one step, but it *does not force* all the occurrences to be replaced. Indeed, the following rule seems to be a straightforward solution to the problem in the double induction, whereas the mix rule appears to be overreaching.

$$\dfrac{\Gamma \vdash \mathcal{A} \qquad \Delta\langle\mathcal{A}\rangle \vdash \mathcal{B}}{\Delta\langle\Gamma\rangle \vdash \mathcal{B}}\ \text{multi-cut}$$

The angle brackets indicate *at least one*, but possibly *several* occurrences of the wff $\mathcal{A}$ being selected in the sequence Δ, each of which is replaced by Γ in the lower sequent.

This multiple cut rule is essentially the same that we introduced in Section 5.2 in structurally free logics. The notational difference is explained by the difference in the definition of a sequent in $LR_{\to}$ and of a consecution in a structurally free logic.

One might think that it might be useful to define a cut rule that is less permissive and more closely matches what is needed in the proof step illustrated above. Perhaps, the idea could be to merge the single cut rule and the $(W\vdash)$ rule.

$$\dfrac{\Gamma \vdash \mathcal{A} \qquad \Delta[\mathcal{A}^{(2)}] \vdash \mathcal{B}}{\Delta[\Gamma] \vdash \mathcal{B}}\ \text{cut}^{(2)}$$

By $\mathcal{A}^{(2)}$ we could mean $\mathcal{A}$ or $\mathcal{A};\mathcal{A}$. A proof of the admissibility of this rule would run into a similar problem as the single cut, because of the upper bound on the number of occurrences of the cut formula.

Exercise 7.2.6. Demonstrate that setting the number of occurrences of the cut formula to a particular natural number in a multiple cut rule is problematic for a double inductive proof (even without a built-in contraction). [Hint: If a sequent is by $(W\vdash)$, then no matter how many occurrences of the principal formula are selected, there is one more occurrence in the upper sequent.]

Exercise 7.2.7. Investigate whether the contraction in the cut$^{(2)}$ rule causes any problems. [Hint: Does it change the set of theorems? Does it prevent the double inductive argument to go through?]

Considering several occurrences of the cut formula in a sequent has certain drawbacks. One occurrence of a formula (in the lower sequent of a rule) is either the principal formula or parametric, but it cannot be both. (Recall that the typical transformation when a cut formula is principal, and when it is parametric are different, which means that we have to separate the multiple cut into more than one cut, while we have to ensure that each is justifiable in the induction.) In mix, all the occurrences of the mix formula are deleted, and so, in a sense, it does not matter whether they were principal or parametric. The multiple cut rule falls in between; it can include both kinds of occurrences, however, it does not require the elimination of all the occurrences. Furthermore, this rule is typically used when the thinning rules are not available, hence, the admissibility of mix is irrelevant.

A way around this problem is to consider an occurrence of a formula in a sequent with its history in the proof, so to speak. Curry's term, in [68], is *parametric ancestor* for an occurrence of a formula that is a precursor of an occurrence of a formula. We give the following definition for $LR_{\rightarrow}$.

Definition 7.10. (Parametric ancestor) An occurrence of $\mathcal{A}$ in the upper sequent of a rule is the *immediate ancestor* of an occurrence of $\mathcal{A}$ in the lower sequent of a rule according to one of (1)–(4), as appropriate. (In talking about sequences of wff's that are denoted by capital Greek letters, we assume that there is an obvious $1-1$ correspondence, namely, the identity function.)

(1) In the rule $(\rightarrow\vdash)$, the elements of Γ, Δ and the $\mathcal{C}$ in the upper sequents are immediate ancestors of the matching wff's in the lower sequent;

(2) in the rule $(\vdash\rightarrow)$, the elements of Γ in the upper sequent are immediate ancestors of the matching wff's in the lower sequent;

(3) in the rule $(C\vdash)$, $\mathcal{A}$, $\mathcal{B}$, the elements of Γ and $\mathcal{C}$ are immediate ancestors of $\mathcal{A}$, $\mathcal{B}$, the matching elements of Γ and $\mathcal{C}$ in the lower sequent;

(4) in the rule $(W\vdash)$, the two occurrences of $\mathcal{A}$, the elements of Γ, and $\mathcal{B}$ are immediate ancestors of $\mathcal{A}$, the matching elements of Γ and $\mathcal{B}$ in the lower sequent.

The *parametric ancestor* relation is the transitive closure of the immediate ancestor relation.

We gave this definition is great detail for clarity, and we will assume that similar definitions can be given for other calculi. We note that the notion of parametric ancestors may be defined for sequent calculi based on multisets too, for instance, by numbering or indexing the formula occurrences in the multisets. Such a maneuver is a step toward reinstating sequences of wff's, but it does not go all the way, because the indexing does not impose an order.

Exercise 7.2.8. In Section 4.3, we have mentioned that $LR_{\to}$ may be formulated with sequents where the antecedent is a multiset. Define $LR_{\to}$ with multisets, and give a precise definition of the notions of the immediate and of the parametric ancestors for that calculus.

The principal formula of a connective rule does not have any parametric ancestors, and the subalterns in it are not parametric ancestors of any formula (even if they have the same shape) in the lower sequent. Starting from a sequent upward, the parametric ancestors of a formula occur in sequents that form paths within a proof tree. Thus, we can talk about the length of those paths. (We assume that the paths contain all the parametric ancestors that they can, that is, the top element does not have a parametric ancestor in the tree.)

Definition 7.11. (Rank of the cut formula) The *rank of the cut formula* $\mathcal{C}$ is the sum of the *left rank* and the *right rank*. The left rank is the length of the path in the proof tree comprising sequents with $\mathcal{C}$ as the succedent of the sequent. The *right rank* is the maximal length of any paths containing parametric ancestors of the cut formula $\mathcal{C}$.

The left rank is easy to conceptualize in a single right-handed calculus, by counting how far up in the proof tree $\mathcal{C}$ occurs on the right-hand side of the $\vdash$. (Again, we assume that the sequences form a maximal path in the sense that the topmost $\mathcal{C}$ has no parametric ancestor above it.) The right rank is sensitive to how the occurrences of $\mathcal{C}$ in the right premise came about. In particular, if the principal formula of the rule that yielded the right premise of the cut is one of the occurrences of the cut formula, then the cut on that occurrence may be separated out, because other occurrences of that formula (if there are any), will not interfere with the right rank of the cut on the principal formula.

Theorem 7.12. *The multi-cut rule is* admissible *in* $LR_{\to}$.

Proof: The proof is by double induction on δ, the degree of the cut formula, and ϱ, the rank of the cut. The latter has been changed compared to the one in Definition 2.26. We present the proof concisely by grouping similar cases together.

1. If one of the premises of the cut is an instance of (id), then the application of the cut rule is immediately eliminable, by retaining the proof ending in the other premise of the cut.

2. If the cut formula is the principal formula in both premises, then we have the following chunk of derivation and its modification, which is justified by a reduction in the degree of the cut formula.

$$\dfrac{\dfrac{\begin{matrix}\vdots\\ \Gamma;\mathcal{A}\vdash\mathcal{B}\end{matrix}}{\Gamma\vdash\mathcal{A}\to\mathcal{B}}\,{\vdash\to}\qquad \dfrac{\begin{matrix}\vdots\\ \Delta\vdash\mathcal{A}\end{matrix}\quad\begin{matrix}\vdots\\ \Theta[\mathcal{B}]\vdash\mathcal{C}\end{matrix}}{\Theta[\mathcal{A}\to\mathcal{B};\Delta]\vdash\mathcal{C}}\,{\to\vdash}}{\Theta[\Gamma;\Delta]\vdash\mathcal{C}}\,\text{cut}\qquad\rightsquigarrow$$

$$\rightsquigarrow \quad \dfrac{\text{cut}\;\dfrac{\Delta \vdash \mathcal{A} \quad \Gamma;\mathcal{A} \vdash \mathcal{B}}{\Gamma;\Delta \vdash \mathcal{B}} \qquad \Theta[\mathcal{B}] \vdash \mathcal{C}}{\Theta[\Gamma;\Delta] \vdash \mathcal{C}}\;\text{cut}$$

3. If the left rank $\varrho_l > 1$, then we move the cut upward. The general shape of the proof chunk and the result of the transformation looks like the following.

$$\text{cut}\;\dfrac{\text{rule}\;\dfrac{\Gamma' \vdash \mathcal{C}}{\Gamma \vdash \mathcal{C}} \qquad \Delta\langle\mathcal{C}\rangle \vdash \mathcal{A}}{\Delta\langle\Gamma\rangle \vdash \mathcal{A}} \quad \rightsquigarrow \quad \dfrac{\dfrac{\Gamma' \vdash \mathcal{C} \quad \Delta\langle\mathcal{C}\rangle \vdash \mathcal{A}}{\Delta\langle\Gamma'\rangle \vdash \mathcal{A}}\;\text{cut}}{\Delta\langle\Gamma\rangle \vdash \mathcal{A}}\;\text{rule}$$

On the right-hand side, we indicated by a thicker line that several applications of the same rule may be required. This step is correct, because any left rule with Γ' as the antecedent is applicable when Γ' is embedded into a context, that is, into a sequence of wff's.

4. If the right rank $\varrho_r > 1$, then the right premise of the cut may be by a left rule or by $(\vdash\rightarrow)$. In the latter case, all the occurrences of the cut formula are parametric, and the cut and $(\vdash\rightarrow)$ may be swapped.

The three other rules may pose some complication, because their principal formula may be one of the occurrences of the cut formula. Otherwise, that is, when all the occurrences of the cut formula are parametric, we simply swap the cut rule and the left rule.

We consider $(W\vdash)$. The shape of the principal formula and its subalterns is the same, which means that we move the cut to the upper sequent.

$$\text{cut}\;\dfrac{\Gamma \vdash \mathcal{C} \qquad \dfrac{\Delta[\mathcal{C};\mathcal{C}]\langle\mathcal{C}\rangle \vdash \mathcal{A}}{\Delta[\mathcal{C}]\langle\mathcal{C}\rangle \vdash \mathcal{A}}\;W\vdash}{\Delta[\Gamma]\langle\Gamma\rangle \vdash \mathcal{A}} \quad \rightsquigarrow \quad \dfrac{\dfrac{\Gamma \vdash \mathcal{C} \quad \Delta[\mathcal{C};\mathcal{C}]\langle\mathcal{C}\rangle \vdash \mathcal{A}}{\Delta[\Gamma;\Gamma]\langle\Gamma\rangle \vdash \mathcal{A}}\;\text{cut}}{\Delta[\Gamma]\langle\Gamma\rangle \vdash \mathcal{A}}\;C\vdash\text{'s}, W\vdash\text{'s}$$

Depending on the shape of Γ, one or more applications of $(W\vdash)$ and none or one or more applications of $(C\vdash)$ are needed. (Incidentally, notice how useful it is to have a multiple cut in this case, which is applicable to one more occurrence in the upper sequent.)

The case of $(C\vdash)$ is similar, except that only $(C\vdash)$ is applied.

Lastly, if the right premise is by $(\rightarrow\vdash)$, and the principal formula and some parametric formulas are occurrences of the cut formula, then we replace the cut with two or three cuts (depending on whether the cut formula occurs in one or in both premises of the $(\rightarrow\vdash)$ rule).

$$\text{cut}\;\dfrac{\Gamma \vdash \mathcal{A} \rightarrow \mathcal{B} \qquad \dfrac{\Delta\langle\mathcal{A} \rightarrow \mathcal{B}\rangle \vdash \mathcal{A} \quad \Theta[\mathcal{B}]\langle\mathcal{A} \rightarrow \mathcal{B}\rangle \vdash \mathcal{D}}{\Theta[\mathcal{A} \rightarrow \mathcal{B};\Delta\langle\mathcal{A} \rightarrow \mathcal{B}\rangle]\langle\mathcal{A} \rightarrow \mathcal{B}\rangle \vdash \mathcal{D}}\;\rightarrow\vdash}{\Theta[\Gamma;\Delta\langle\Gamma\rangle]\langle\Gamma\rangle \vdash \mathcal{D}} \quad \rightsquigarrow$$

$$\rightsquigarrow \qquad \dfrac{\Gamma \vdash \mathcal{A} \to \mathcal{B} \qquad \dfrac{\text{cut}\;\dfrac{\Gamma \vdash \mathcal{C} \quad \Delta\langle \mathcal{C} \rangle \vdash \mathcal{A}}{\Delta\langle \Gamma \rangle \vdash \mathcal{A}} \qquad \dfrac{\Gamma \vdash \mathcal{C} \quad \Theta[\mathcal{B}]\langle \mathcal{C} \rangle \vdash \mathcal{D}}{\Theta[\mathcal{B}]\langle \Gamma \rangle \vdash \mathcal{D}}\;\text{cut}}{\Theta[\mathcal{A} \to \mathcal{B}; \Delta\langle \Gamma \rangle]\langle \Gamma \rangle \vdash \mathcal{D}}\;{\to}{\vdash}}{\Theta[\Gamma; \Delta\langle \Gamma \rangle]\langle \Gamma \rangle \vdash \mathcal{D}}\;\text{cut}$$

(In the resulting proof segment we abbreviated the cut formula in its parametric occurrences by $\mathcal{C}$, in order to squeeze the proof onto a page.) qed

In proofs of decidability based on sequent calculi, contraction is dispensed with by incorporating some contraction in the connective rules. As a result, the cut rule should also be amended. We consider the calculus $[LA_W]$ from Section 9.1.5, with the cut taking the following form.

$$\dfrac{\Gamma \vdash \mathcal{C} \qquad \Delta\langle \mathcal{C} \rangle \vdash \mathcal{A}}{[\Delta\langle \Gamma \rangle] \vdash \mathcal{A}}\;[\text{cut}]$$

The bracketed antecedent in the lower sequent is like $\Delta\langle \Gamma \rangle$, however, iteratively, $\Theta;\Theta$ may be replaced by Θ, if $\Theta;\Theta$ did not occur in the upper sequents.

Definition 7.13. (Height of cut) The *height of the cut*, denoted by ξ, is the sum of the left height, ξ_l, and of the right height, ξ_r, plus 1. The *left height* is the height of the proof tree ending in the left premise; the *right height* is the height of the proof tree ending in the right premise.

The least height of a cut is 3, when both premises are axioms. Although informally, we talk about "moving the topmost cut toward the top," the role of the cut height is not to replace the rank of the cut. For the rank, we use the definition that relies on the notion of parametric ancestors. There is no separate contraction rule, hence, we use the above rule. However, then we have to take care of cases where the principal formula of a rule is the cut formula. So we cannot abandon tracking the parametric ancestors of the cut formulas. In some cases, we have to add completely new cuts into the transformed proof in order to ensure that contractions that could be carried out before the transformation can be carried out after the transformation.

Theorem 7.14. *The* [cut] *rule is* admissible *in* $[LA_W]$.

Proof: We proceed by triple induction on the rank of the cut, and on the degree of the cut formula and on the height of the cut.

1. If one of the premises is the axiom, then the cut is eliminated by retaining the proof ending in the other premise. Notice that since the cut replaces a formula with itself, no additional contractions could have resulted from the cut. Therefore, nothing is lost by eliminating the cut.

2. There are three possibilities when the two premises are by the left and right introduction rules for a connective, that is, the rank of the cut, $\varrho = 2$.

$$\text{[cut]}\ \dfrac{\vdash\to\ \dfrac{\Gamma;\mathcal{A}\vdash\mathcal{B}}{\Gamma\vdash\mathcal{A}\to\mathcal{B}} \qquad \dfrac{\Delta\vdash\mathcal{A}\quad \Theta[\mathcal{B}]\vdash\mathcal{C}}{[\Theta[\mathcal{A}\to\mathcal{B};\Delta]]\vdash\mathcal{C}}\ \to\vdash}{[\Theta[\Gamma;\Delta]]\vdash\mathcal{C}} \quad \rightsquigarrow$$

$$\rightsquigarrow \quad \dfrac{\text{[cut]}\ \dfrac{\Delta\vdash\mathcal{A}\quad \Gamma;\mathcal{A}\vdash\mathcal{B}}{[\Gamma;\Delta]\vdash\mathcal{B}} \qquad \Theta[\mathcal{B}]\vdash\mathcal{C}}{[\Theta[\Gamma;\Delta]]\vdash\mathcal{C}}\ \text{[cut]}$$

The end sequent is identical to the one obtained earlier, provided that all the contractions that could have been carried out before, can be carried out now. It is so if the principal formula in $[\Theta[\mathcal{A}\to\mathcal{B};\Delta]]$ did not cause contractions. If $\mathcal{A}\to\mathcal{B}$ has no other occurrences, then the principal formula is not part of a sequence of formulas that is contracted. However, Δ and Θ may contain other occurrences of $\mathcal{A}\to\mathcal{B}$, which are not among the cut formulas. (The rank of the cut is tied to the parametric ancestors of the cut formula only, not to all the occurrences of the formula, which is like the cut formula.) The transformed proof avoids $\mathcal{A}\to\mathcal{B}$ as the principal formula, hence, it is possible that some contractions could not be carried out in it as before. This problem may be overcome by ensuring that the occurrences of $\mathcal{A}\to\mathcal{B}$, which are involved in contractions, are changed to Γ.

Let us assume that some occurrences of $\mathcal{A}\to\mathcal{B}$ are involved in contractions, and they have been selected in Δ and Θ. Then before performing cuts on formulas of lower degree, we make sure that the selected occurrences of $\mathcal{A}\to\mathcal{B}$ are turned into Γ's. (To keep the size of the proof reasonable, $\mathcal{A}\to\mathcal{B}$ is abbreviated as $\mathcal{D}$.)

$$\dfrac{\text{[cut]}\ \dfrac{\text{[cut]}\ \dfrac{\Gamma\vdash\mathcal{D}\quad \Delta\langle\mathcal{D}\rangle\vdash\mathcal{A}}{[\Delta\langle\Gamma\rangle]\vdash\mathcal{A}} \qquad \Gamma;\mathcal{A}\vdash\mathcal{B}}{[\Gamma;\Delta\langle\Gamma\rangle]\vdash\mathcal{B}} \qquad \dfrac{\Gamma\vdash\mathcal{D}\quad \Theta[\mathcal{B}]\langle\mathcal{D}\rangle\vdash\mathcal{C}}{[\Theta[\mathcal{B}]\langle\Gamma\rangle]\vdash\mathcal{C}}\ \text{[cut]}}{[\Theta[\Gamma;\Delta\langle\Gamma\rangle]\langle\Gamma\rangle]\vdash\mathcal{C}}\ \text{[cut]}$$

The cuts on $\mathcal{A}\to\mathcal{B}$ are justified by having lower cut height than the original cut on the same formula. The cuts on $\mathcal{A}$ and $\mathcal{B}$ are justified by having lower degree. The final cut can replicate all the contractions, because the selected occurrences of $\mathcal{A}\to\mathcal{B}$ have been replaced by Γ.

The case with $\mathcal{B}\leftarrow\mathcal{A}$ is symmetric, and we omit the details.

If the cut formulas are $\mathcal{A}\circ\mathcal{B}$, then we start with the following proof, and in the simplest case, we transform it as follows.

$$\dfrac{\vdash\circ\ \dfrac{\Gamma\vdash\mathcal{A}\quad \Delta\vdash\mathcal{B}}{[\Gamma;\Delta]\vdash\mathcal{A}\circ\mathcal{B}} \qquad \dfrac{\Theta[\mathcal{A};\mathcal{B}]\vdash\mathcal{C}}{[\Theta[\mathcal{A}\circ\mathcal{B}]]\vdash\mathcal{C}}\ \circ\vdash}{[\Theta[\Gamma;\Delta]]\vdash\mathcal{C}}\ \text{[cut]} \quad \rightsquigarrow$$

$$\rightsquigarrow \quad \dfrac{\Delta \vdash \mathcal{B} \qquad \dfrac{\Gamma \vdash \mathcal{A} \quad \Theta[\mathcal{A};\mathcal{B}] \vdash \mathcal{C}}{[\Theta[\Gamma;\mathcal{B}]] \vdash \mathcal{C}}\,[\text{cut}]}{[\Theta[\Gamma;\Delta]] \vdash \mathcal{C}}\,[\text{cut}]$$

It is possible that Θ has some occurrences of $\mathcal{A} \circ \mathcal{B}$ to start with. If this is the case, then we transform the proof as follows.

$$\dfrac{\Delta \vdash \mathcal{B} \qquad \dfrac{\Gamma \vdash \mathcal{A} \qquad \dfrac{[\Gamma;\Delta] \vdash \mathcal{A} \circ \mathcal{B} \quad [\Theta[\mathcal{A};\mathcal{B}]\langle \mathcal{A} \circ \mathcal{B} \rangle] \vdash \mathcal{C}}{[\Theta[\mathcal{A};\mathcal{B}]\langle \Gamma;\Delta \rangle] \vdash \mathcal{C}}\,[\text{cut}]}{[\Theta[\Gamma;\mathcal{B}]\langle \Gamma;\Delta \rangle] \vdash \mathcal{C}}\,[\text{cut}]}{[\Theta[\Gamma;\Delta]\langle \Gamma;\Delta \rangle] \vdash \mathcal{C}}\,[\text{cut}]$$

The top cut has the same degree as the original cut, but it has lower height. The two other cuts have lower degree than the original cut.

3. If the left rank, $\varrho_l > 1$, then we permute the cut rule with the left rule that resulted in the left premise of the cut. We note that each left rule has built-in contraction; therefore, if the original cut allowed some contraction due to the insertion of the antecedent of the left premise with the principal formula of the left rule resulting in the left premise, then those contractions may be carried out as part of the application of the connective rule after the cut. We consider one of the possibilities.

$$[\text{cut}]\,\dfrac{\circ\vdash\,\dfrac{\Gamma[\mathcal{A};\mathcal{B}] \vdash \mathcal{C}}{[\Gamma[\mathcal{A} \circ \mathcal{B}]] \vdash \mathcal{C}} \qquad \Delta\langle \mathcal{C} \rangle \vdash \mathcal{D}}{[\Delta\langle \Gamma[\mathcal{A} \circ \mathcal{B}] \rangle] \vdash \mathcal{D}} \quad \rightsquigarrow \quad \dfrac{\dfrac{\Gamma[\mathcal{A};\mathcal{B}] \vdash \mathcal{C} \quad \Delta\langle \mathcal{C} \rangle \vdash \mathcal{D}}{[\Delta\langle \Gamma[\mathcal{A};\mathcal{B}] \rangle] \vdash \mathcal{D}}\,[\text{cut}]}{[\Delta\langle \Gamma[\mathcal{A} \circ \mathcal{B}] \rangle] \vdash \mathcal{D}}\,\circ\vdash$$

4. If the left rank, $\varrho = 1$, but the right rank, $\varrho_r > 1$, then the right premise may be by a left or by a right rule. If the right premise is by a right rule, then the transformation is straightforward. If the right premise is by a left rule, then we have to consider whether the principal formula of the rule is one of the cut formulas. We consider six subcases in some detail, and leave the others for an exercise. In each subcase, we give a proof segment and how it is transformed.

4.1 The right premise is by the $(\vdash \circ)$ rule.

$$[\text{cut}]\,\dfrac{\Gamma \vdash \mathcal{C} \qquad \dfrac{\Delta\langle \mathcal{C} \rangle \vdash \mathcal{A} \quad \Theta\langle \mathcal{C} \rangle \vdash \mathcal{B}}{[\Delta\langle \mathcal{C} \rangle;\Theta\langle \mathcal{C} \rangle] \vdash \mathcal{A} \circ \mathcal{B}}\,\vdash\circ}{[\Delta\langle \Gamma \rangle;\Theta\langle \Gamma \rangle] \vdash \mathcal{A} \circ \mathcal{B}} \quad \rightsquigarrow$$

$$\rightsquigarrow \quad \dfrac{[\text{cut}]\,\dfrac{\Gamma \vdash \mathcal{C} \quad \Delta\langle \mathcal{C} \rangle \vdash \mathcal{A}}{[\Delta\langle \Gamma \rangle] \vdash \mathcal{A}} \qquad \dfrac{\Gamma \vdash \mathcal{C} \quad \Theta\langle \mathcal{C} \rangle \vdash \mathcal{B}}{[\Theta\langle \Gamma \rangle] \vdash \mathcal{B}}\,[\text{cut}]}{[\Delta\langle \Gamma \rangle;\Theta\langle \Gamma \rangle] \vdash \mathcal{A} \circ \mathcal{B}}\,\vdash\circ$$

4.2 The right premise is by the $(\vdash\to)$ rule.

$$[\text{cut}]\,\frac{\Gamma\vdash\mathcal{C} \qquad \dfrac{\Delta\langle\mathcal{C}\rangle;\mathcal{A}\vdash\mathcal{B}}{\Delta\langle\mathcal{C}\rangle\vdash\mathcal{A}\to\mathcal{B}}\,{\vdash\to}}{[\Delta\langle\Gamma\rangle]\vdash\mathcal{A}\to\mathcal{B}} \qquad\rightsquigarrow\qquad \frac{\dfrac{\Gamma\vdash\mathcal{C}\quad \Delta\langle\mathcal{C}\rangle;\mathcal{A}\vdash\mathcal{B}}{[\Delta\langle\Gamma\rangle;\mathcal{A}]\vdash\mathcal{B}}\,[\text{cut}]}{\Delta\langle\Gamma\rangle\vdash\mathcal{A}\to\mathcal{B}}\,{\vdash\to}$$

4.3 The right premise is by the $(\to\vdash)$ rule—without the principal formula being a cut formula.

$$[\text{cut}]\,\frac{\Gamma\vdash\mathcal{C} \qquad \dfrac{\Delta\langle\mathcal{C}\rangle\vdash\mathcal{A}\quad \Theta[\mathcal{B}]\langle\mathcal{C}\rangle\vdash\mathcal{D}}{[\Theta[\mathcal{A}\to\mathcal{B};\Delta\langle\mathcal{C}\rangle]\langle\mathcal{C}\rangle]\vdash\mathcal{D}}\,{\to\vdash}}{[\Theta[\mathcal{A}\to\mathcal{B};\Delta\langle\Gamma\rangle]\langle\Gamma\rangle]\vdash\mathcal{D}} \qquad\rightsquigarrow$$

$$\rightsquigarrow\qquad \frac{[\text{cut}]\,\dfrac{\Gamma\vdash\mathcal{C}\quad \Delta\langle\mathcal{C}\rangle\vdash\mathcal{A}}{[\Delta\langle\Gamma\rangle]\vdash\mathcal{A}} \qquad \dfrac{\Gamma\vdash\mathcal{C}\quad \Theta[\mathcal{B}]\langle\mathcal{C}\rangle\vdash\mathcal{D}}{[\Theta[\mathcal{B}]\langle\Gamma\rangle]\vdash\mathcal{D}}\,[\text{cut}]}{[\Theta[\mathcal{A}\to\mathcal{B};\Delta\langle\Gamma\rangle]\langle\Gamma\rangle]\vdash\mathcal{D}}\,{\to\vdash}$$

4.4 The right premise is by the $(\circ\vdash)$ rule—without the principal formula being a cut formula.

$$[\text{cut}]\,\frac{\Gamma\vdash\mathcal{C} \qquad \dfrac{\Delta[\mathcal{A};\mathcal{B}]\langle\mathcal{C}\rangle\vdash\mathcal{D}}{[\Delta[\mathcal{A}\circ\mathcal{B}]\langle\mathcal{C}\rangle]\vdash\mathcal{D}}\,{\circ\vdash}}{[\Delta[\mathcal{A}\circ\mathcal{B}]\langle\Gamma\rangle]\vdash\mathcal{D}} \qquad\rightsquigarrow\qquad \frac{\dfrac{\Gamma\vdash\mathcal{C}\quad \Delta[\mathcal{A};\mathcal{B}]\langle\mathcal{C}\rangle\vdash\mathcal{D}}{[\Delta[\mathcal{A};\mathcal{B}]\langle\Gamma\rangle]\vdash\mathcal{D}}\,[\text{cut}]}{[\Delta[\mathcal{A}\circ\mathcal{B}]\langle\Gamma\rangle]\vdash\mathcal{D}}\,{\circ\vdash}$$

4.5 The right premise is by the $(\circ\vdash)$ rule—with the principal formula being a cut formula.

$$[\text{cut}]\,\frac{\Gamma\vdash\mathcal{A}\circ\mathcal{B} \qquad \dfrac{\Delta[\mathcal{A};\mathcal{B}]\langle\mathcal{A}\circ\mathcal{B}\rangle\vdash\mathcal{D}}{[\Delta[\mathcal{A}\circ\mathcal{B}]\langle\mathcal{A}\circ\mathcal{B}\rangle]\vdash\mathcal{D}}\,{\circ\vdash}}{[\Delta[\Gamma]\langle\Gamma\rangle]\vdash\mathcal{D}} \qquad\rightsquigarrow$$

$$\rightsquigarrow\qquad \frac{\Gamma\vdash\mathcal{A}\circ\mathcal{B} \qquad \dfrac{\dfrac{\Gamma\vdash\mathcal{A}\circ\mathcal{B}\quad \Delta[\mathcal{A};\mathcal{B}]\langle\mathcal{A}\circ\mathcal{B}\rangle\vdash\mathcal{D}}{[\Delta[\mathcal{A};\mathcal{B}]\langle\Gamma\rangle]\vdash\mathcal{D}}\,[\text{cut}]}{[\Delta[\mathcal{A}\circ\mathcal{B}]\langle\Gamma\rangle]\vdash\mathcal{D}}\,{\circ\vdash}}{[\Delta[\Gamma]\langle\Gamma\rangle]\vdash\mathcal{D}}\,[\text{cut}]$$

4.6 The right premise is by the $(\leftarrow\vdash)$ rule—with the principal formula being a cut formula.

$$[\text{cut}]\,\frac{\Gamma\vdash\mathcal{B}\leftarrow\mathcal{A} \qquad \dfrac{\Delta\langle\mathcal{B}\leftarrow\mathcal{A}\rangle\vdash\mathcal{A}\quad \Theta[\mathcal{B}]\langle\mathcal{B}\leftarrow\mathcal{A}\rangle\vdash\mathcal{D}}{[\Theta[\Delta\langle\mathcal{B}\leftarrow\mathcal{A}\rangle;\mathcal{B}\leftarrow\mathcal{A}]\langle\mathcal{B}\leftarrow\mathcal{A}\rangle]\vdash\mathcal{D}}\,{\leftarrow\vdash}}{[\Theta[\Delta\langle\Gamma\rangle;\Gamma]\langle\Gamma\rangle]\vdash\mathcal{D}} \qquad\rightsquigarrow$$

$$\rightsquigarrow \quad \dfrac{\dfrac{\vdots}{\Gamma \vdash \mathcal{B} \leftarrow \mathcal{A}} \qquad \dfrac{\dfrac{\dfrac{\vdots}{\Gamma \vdash \mathcal{C}} \quad \dfrac{\vdots}{\Delta\langle \mathcal{C}\rangle \vdash \mathcal{A}}}{[\Delta\langle\Gamma\rangle] \vdash \mathcal{A}}\,[\text{cut}] \qquad \dfrac{\dfrac{\vdots}{\Gamma \vdash \mathcal{C}} \quad \dfrac{\vdots}{\Theta[\mathcal{B}]\langle \mathcal{C}\rangle \vdash \mathcal{D}}}{[\Theta[\mathcal{B}]\langle\Gamma\rangle] \vdash \mathcal{D}}\,[\text{cut}]}{[\Theta[\Delta\langle\Gamma\rangle; \mathcal{B} \leftarrow \mathcal{A}]\langle\Gamma\rangle] \vdash \mathcal{D}}\,{\leftarrow}\vdash}{[\Theta[\Delta\langle\Gamma\rangle; \Gamma]\langle\Gamma\rangle] \vdash \mathcal{D}}\,[\text{cut}]$$

(In the two top cuts, the cut formula is abbreviated by $\mathcal{C}$, in order to shrink the size of the chunk of the proof.)

We note that if in any of the cases contractions became possible because of combining two sequences specifically with the cut formula in it, then those occurrences (which are to be contracted) may be added to the cut (or cuts) that precede the connective rule in the transformed proof.

In the subcases, where the principal formula is one of the occurrences of the cut formula, the cut on the parametric formulas is justified by a decrease in rank, and so is the separate cut on the principal formula. qed

Exercise 7.2.9. Write out the details of the subcases that were mentioned but not included in the proof. [Hint: You may consider at least 6 cases.]

The admissibility of the [cut] rule is important for the decidability of LA_W, which we show in Section 9.1.5.

To conclude this section we return very briefly to the question of the relationship between mix and cuts.

In LK and LJ, the mix rule and the cut rule are interderivable; hence, the admissibility of one implies the admissibility of the other. Although we did not mention multi-cut in those calculi, the single and the multiple versions of the cut rule are, obviously, *interderivable* too.

In calculi where some of the structural rules are absent, especially when there is no thinning rule, the interderivability of the mix rule with a cut rule breaks down. The single and the multiple cut remain interderivable, however, not all proofs of the admissibility of the multi-cut are proofs of the admissibility of the single cut by a simple substitution of the second rule for the first.

The mix rule becomes a stand-alone rule, for example, in relevance logics. Arguably, it is not a very important rule either, because it is rather indiscriminate in eliminating all the occurrences of the mix formula. Whether expected, or perhaps unexpected, the mix rule may turn out to be admissible too. Furthermore, the admissibility of the mix rule is a consequence of the admissibility of the multiple cut rule in fragments of R. We state this result for $LR_{\overset{\sim}{\to}}$.[3]

Theorem 7.15. *The mix rule is* admissible *in* $LR_{\overset{\sim}{\to}}$.

[3]A similar claim was proved in [33], for a different fragment of R.

Proof: We will rely on the admissibility of the multi-cut in $LR_{\stackrel{\sim}{\to}}$, which takes the following form.

$$\dfrac{\Gamma \vdash \Delta; C \qquad \Theta\langle C\rangle \vdash \Lambda}{\Theta\langle\Gamma\rangle \vdash \Delta; \Lambda}\ \text{multi-cut}$$

We also know that this sequent calculus may be formulated in equivalent ways, by either postulating permutation rules and retaining sequences of formulas, or by having multisets of formulas and omitting the permutation rules. For the sake of transparency, we assume the latter formulation.

Let us assume that the mix formula is C. We use the * notation as in the mix rule. Let the premises of an application of the mix rule be $\Gamma \vdash \Delta^*; C; \ldots; C$ and $C; \ldots; C; \Theta^* \vdash \Lambda$, where the first premise contains n, and the second contains m occurrences of C. We perform n-many multiple cuts as follows. (C^n and Λ^m abbreviate n and m copies of C and Λ, respectively.)

$$\text{cut}\ \dfrac{\Gamma \vdash \Delta^*; C^n \quad C^m; \Theta^* \vdash \Lambda}{\Gamma^m; \Theta^* \vdash \Delta^*; \Lambda; C^{n-1}}$$

$$\vdots$$

$$\dfrac{\text{cut}\ \dfrac{\vdots}{(\ldots(\Gamma^m; \Theta^*)^m; \ldots; \Theta^*)^m; \Theta^* \vdash \Delta^*; \Lambda^{n-1}; C;} \qquad C^m; \Theta^* \vdash \Lambda}{(\ldots(\Gamma^m; \Theta^*)^m; \ldots; \Theta^*)^m; \Theta^* \vdash \Delta^*; \Lambda^n}\ \text{cut}$$

The bottom consecution contains the right components, but in way too many copies. However, $LR_{\stackrel{\sim}{\to}}$ contains the rules $(W\vdash)$ and $(\vdash W)$; therefore, sufficiently many applications of those rules lead to $\Gamma; \Theta^* \vdash \Delta^*; \Lambda$, which is the lower sequent that results by the mix rule. qed

Exercise 7.2.10. The admissibility of multi-cut in $LR_{\stackrel{\sim}{\to}}$ is known, and we have not proved it here. Construct a proof in detail. [Hint: The proof of the admissibility of cut for $LR_{\to}$ may serve as an example.]

7.3 Constants and the cut

Structurally free logics that we presented in Section 5.2 include combinators, which are zero-ary constants in the language of the logic. Moreover, they can be introduced by combinatory rules, which means that in an analysis, we want to declare the combinator that is introduced by such a rule to be the principal formula of the rule.

The combinatory rules create a situation in LC that is not paralleled in LK or LJ, or in the Lambek calculi, because the combinator does not become a subformula of a compound formula, while its presence is justified by a rule.

As it turns out, the cut theorem can be proved for multiple cut along the lines of the proof of the cut theorem for $LR_{\rightarrow}$.[4]

To start with, LC contains only some proper combinators. Often those are combinators that are better known and have been known to be related to similarly well-known implicational formulas. However, Y is also a well-known, and possibly, the most widely known combinator that is not proper. Y is added to LC in [31]. However, Y causes a new problem in the proof of the cut theorem, because if Y is the cut formula in the lower consecution of the $(\mathsf{Y}\vdash)$ rule, then the usual move in the transformation, that is, swapping the cut and the fixed point combinator rule does not work. Here is an illustration of the problem.

$$\dfrac{\mathfrak{A} \vdash \mathsf{Y} \qquad \dfrac{\mathfrak{B}[\mathfrak{C}; (\mathsf{Y}; \mathfrak{C})] \vdash \mathcal{A}}{\mathfrak{B}[\mathsf{Y}; \mathfrak{C}] \vdash \mathcal{A}}\,\mathsf{Y}\vdash}{\mathfrak{B}[\mathfrak{A}; \mathfrak{C}] \vdash \mathcal{A}}\,\text{cut} \qquad \not\rightsquigarrow \qquad \dfrac{\mathfrak{A} \vdash \mathsf{Y} \quad \mathfrak{B}[\mathfrak{C}; (\mathsf{Y}; \mathfrak{C})] \vdash \mathcal{A}}{\mathfrak{B}[\mathfrak{C}; (\mathfrak{A}; \mathfrak{C})] \vdash \mathcal{A}}\,\text{cut} \quad ??$$

The question marks are to indicate that there is no rule to allow a contraction-like move to get rid of the first occurrence of $\mathfrak{C}$.

To solve the problem, we have to notice that all the combinators are *special*. Namely, each of them comes with *one rule*—not two or three rules as some connectives or quantifiers. That is, there is no rule to introduce a combinator on the right-hand side of the $\vdash$. This does not imply that the situation illustrated above cannot emerge; rather, it means that Y must have been introduced above the left premise of the cut by an instance of (id). Then, from LC being a single right-handed consecution calculus, it follows that all the rules that have been applied to obtain the left premise, must have been left rules. Then, if all the left rules can be repeated in a context, then the application of the cut rule with the cut formula being Y is directly eliminable. Lemma 7.16 ensures that this is, indeed, so.

Y has a duplicative effect too, that is, it has a touch of similarity to the contraction rule. Of course, other (proper) combinators that duplicate one of their arguments may be included in an LC calculus. In Section 5.2, we included S, W and M. First, we consider the additional lemma.

Lemma 7.16. *Let us assume that* $\mathfrak{A} \vdash \mathsf{Y}$ *is provable together with* $\mathfrak{B}[\mathsf{Y}] \vdash \mathcal{A}$. *Then* $\mathfrak{B}[\mathfrak{A}] \vdash \mathcal{A}$ *is* provable *too.*

Proof: Let the height of the proof tree ending in $\mathfrak{A} \vdash \mathsf{Y}$ be χ.

1. If $\chi = 1$, then $\mathfrak{A} \vdash \mathsf{Y}$ is $\mathsf{Y} \vdash \mathsf{Y}$; hence, the claim is obvious.

2. The inductive step is divided into three groups of cases.

2.1 $\mathfrak{A} \vdash \mathsf{Y}$ is by $(\rightarrow\vdash)$. Then we have the following, where i.h. shows that the upper consecution on the right-hand side is guaranteed to be provable

[4]See [81] for the first proof of the cut theorem for an LC calculus.

by the hypothesis of the induction.

$$\to\vdash\ \dfrac{\mathfrak{C}\vdash\mathcal{B}\quad \mathfrak{D}[\mathcal{C}]\vdash\mathsf{Y}}{\mathfrak{D}[\mathcal{B}\to\mathcal{C};\mathfrak{C}]\vdash\mathsf{Y}} \qquad \overset{\text{i.h.}}{\rightsquigarrow} \qquad \dfrac{\mathfrak{B}[\mathfrak{D}[\mathcal{C}]]\vdash\mathcal{A}\quad \mathfrak{C}\vdash\mathcal{B}}{\mathfrak{B}[\mathfrak{D}[\mathcal{B}\to\mathcal{C};\mathfrak{C}]]\vdash\mathcal{A}}\ \to\vdash$$

2.2 $\mathfrak{A}\vdash\mathsf{Y}$ is by $(\circ\vdash)$. Then we have the next two chunks.

$$\circ\vdash\ \dfrac{\mathfrak{C}[\mathcal{B};\mathcal{C}]\vdash\mathsf{Y}}{\mathfrak{C}[\mathcal{B}\circ\mathcal{C}]\vdash\mathsf{Y}} \qquad \overset{\text{i.h.}}{\rightsquigarrow} \qquad \dfrac{\mathfrak{B}[\mathfrak{C}[\mathcal{B};\mathcal{C}]]\vdash\mathcal{A}}{\mathfrak{B}[\mathfrak{C}[\mathcal{B}\circ\mathcal{C}]]\vdash\mathcal{A}}\ \circ\vdash$$

Clearly, neither binary connective causes a problem. The last group comprises the combinatory rules. There is a general similarity between the combinatory rules—with a few subtleties. We consider $(\mathsf{S}\vdash)$ and $(\mathsf{Y}\vdash)$.
2.3 $\mathfrak{A}\vdash\mathsf{Y}$ is by $(\mathsf{S}\vdash)$.

$$\mathsf{S}\vdash\ \dfrac{\mathfrak{C}[\mathfrak{D};\mathfrak{G};(\mathfrak{E};\mathfrak{G})]\vdash\mathsf{Y}}{\mathfrak{C}[\mathsf{S};\mathfrak{D};\mathfrak{E};\mathfrak{G}]\vdash\mathsf{Y}} \qquad \overset{\text{i.h.}}{\rightsquigarrow} \qquad \dfrac{\mathfrak{B}[\mathfrak{C}[\mathfrak{D};\mathfrak{G};(\mathfrak{E};\mathfrak{G})]]\vdash\mathcal{A}}{\mathfrak{B}[\mathfrak{C}[\mathsf{S};\mathfrak{D};\mathfrak{E};\mathfrak{G}]]\vdash\mathcal{A}}\ \mathsf{S}\vdash$$

$\mathfrak{A}\vdash\mathsf{Y}$ is by $(\mathsf{Y}\vdash)$.

$$\mathsf{Y}\vdash\ \dfrac{\mathfrak{C}[\mathfrak{D};(\mathsf{Y};\mathfrak{D})]\vdash\mathsf{Y}}{\mathfrak{C}[\mathsf{Y};\mathfrak{D}]\vdash\mathsf{Y}} \qquad \overset{\text{i.h.}}{\rightsquigarrow} \qquad \dfrac{\mathfrak{B}[\mathfrak{C}[\mathfrak{D};(\mathsf{Y};\mathfrak{D})]]\vdash\mathcal{A}}{\mathfrak{B}[\mathfrak{C}[\mathsf{Y};\mathfrak{D}]]\vdash\mathcal{A}}\ \mathsf{Y}\vdash$$

qed

Exercise 7.3.1. Verify that the other combinatory rules are similar, in particular, proper combinators without a duplicative effect are completely unproblematic with single cut too.

Theorem 7.17. *The single cut rule is* admissible *in LC with* Y *included.*

Proof: The proof is by triple induction on the degree of the cut formula, on the rank of the cut and on its contraction measure, with an additional induction on χ the height of the proof tree of the left premise, when the cut formula is Y. (We have separated out the last induction into a lemma.) The following are some representative cases.
1. If either premise is an instance of the axiom, then the cut is eliminated, and the proof ending in the other premise is the cut-free proof.
2. If $\varrho = 2$ and the two premises are by a pair of matching rules, by $\langle(\vdash\to),(\to\vdash)\rangle$ or by $\langle(\vdash\circ),(\circ\vdash)\rangle$, then the cut is replaced by cuts of lower degree, like in the previous proofs.
3. If $\varrho_l > 1$, then the left premise is by a left rule. The cut is swapped with the left rule. Here is a combinatory rule that is switched with the cut, which reduces the left rank.

$$\text{cut}\ \dfrac{\mathsf{B}\vdash\ \dfrac{\mathfrak{A}[\mathfrak{B};(\mathfrak{C};\mathfrak{D})]\vdash\mathcal{C}}{\mathfrak{A}[\mathsf{B};\mathfrak{B};\mathfrak{C};\mathfrak{D}]\vdash\mathcal{C}}\qquad \mathfrak{C}[\mathcal{C}]\vdash\mathcal{A}}{\mathfrak{C}[\mathfrak{A}[\mathsf{B};\mathfrak{B};\mathfrak{C};\mathfrak{D}]]\vdash\mathcal{A}} \qquad \rightsquigarrow \qquad \dfrac{\dfrac{\mathfrak{A}[\mathfrak{B};(\mathfrak{C};\mathfrak{D})]\vdash\mathcal{C}\quad \mathfrak{C}[\mathcal{C}]\vdash\mathcal{A}}{\mathfrak{C}[\mathfrak{A}[\mathfrak{B};(\mathfrak{C};\mathfrak{D})]]\vdash\mathcal{A}}\ \text{cut}}{\mathfrak{C}[\mathfrak{A}[\mathsf{B};\mathfrak{B};\mathfrak{C};\mathfrak{D}]]\vdash\mathcal{A}}\ \mathsf{B}\vdash$$

4. If $\varrho_r > 1$, then the right premise may have been obtained by any of the rules. The four connective rules behave with respect to the cut as in some of the other systems, and there is no difficulty dealing with them. Most of the combinatory rules are straightforward too. Here is the proof segment and its transformation if the rule is $(\mathsf{C}\vdash)$.

$$\text{cut}\ \dfrac{\begin{array}{c}\vdots\\ \mathfrak{A}\vdash C\end{array}\qquad \dfrac{\mathfrak{B}[\mathfrak{C};\mathfrak{E};\mathfrak{D}]\,[C]\vdash\mathcal{A}}{\mathfrak{B}[\mathsf{C};\mathfrak{C};\mathfrak{D};\mathfrak{E}]\,[C]\vdash\mathcal{A}}\ \mathsf{C}\vdash}{\mathfrak{B}[\mathsf{C};\mathfrak{C};\mathfrak{D};\mathfrak{E}]\,[\mathfrak{A}]\vdash\mathcal{A}} \quad\rightsquigarrow\quad \dfrac{\dfrac{\mathfrak{A}\vdash C\qquad \mathfrak{B}[\mathfrak{C};\mathfrak{E};\mathfrak{D}]\,[C]\vdash\mathcal{A}}{\mathfrak{B}[\mathfrak{C};\mathfrak{E};\mathfrak{D}]\,[\mathfrak{A}]\vdash\mathcal{A}}\ \text{cut}}{\mathfrak{B}[\mathsf{C};\mathfrak{C};\mathfrak{D};\mathfrak{E}]\,[\mathfrak{A}]\vdash\mathcal{A}}\ \mathsf{C}\vdash$$

As long as a combinator is proper and it is without a duplicative effect, the exact place where C occurs in the antecedent of the right premise is of little interest for the proof of the cut theorem.

M, W and Y increase the contraction measure; therefore, we can add a cut as many times as needed to restore all the occurrences of a structure to some common shape before a duplicator is applied. Here is the case of M.

$$\text{cut}\ \dfrac{\begin{array}{c}\vdots\\ \mathfrak{A}\vdash C\end{array}\qquad \dfrac{\mathfrak{B}[\mathfrak{C}[C];\mathfrak{C}[C]]\vdash\mathcal{A}}{\mathfrak{B}[\mathfrak{C}[C]]\vdash\mathcal{A}}\ \mathsf{M}\vdash}{\mathfrak{B}[\mathfrak{C}[\mathfrak{A}]]\vdash\mathcal{A}} \quad\rightsquigarrow$$

$$\rightsquigarrow\quad \dfrac{\dfrac{\begin{array}{c}\vdots\\ \mathfrak{A}\vdash C\end{array}\qquad \dfrac{\mathfrak{A}\vdash C\qquad \mathfrak{B}[\mathfrak{C}[C];\mathfrak{C}[C]]\vdash\mathcal{A}}{\mathfrak{B}[\mathfrak{C}[\mathfrak{A}];\mathfrak{C}[C]]\vdash\mathcal{A}}\ \text{cut}}{\mathfrak{B}[\mathfrak{C}[\mathfrak{A}];\mathfrak{C}[\mathfrak{A}]]\vdash\mathcal{A}}\ \text{cut}}{\mathfrak{B}[\mathfrak{C}[\mathfrak{A}]]\vdash\mathcal{A}}\ \mathsf{M}\vdash$$

The transformation is justified by a reduction in the contraction measure of the cuts. The contraction measure was μ ($\mu \geq 1$), and the new cuts are above $(\mathsf{M}\vdash)$, which means that their contraction measure is $\leq \mu - 1$.

If the left rule leading to the right premise is $(\mathsf{Y}\vdash)$, then the cut formula may be Y itself or it might occur in the structure coming after the fixed point combinator, or it might occur somewhere else in the antecedent. In the first case, we appeal to Lemma 7.16. In the second case, we proceed as above with $(\mathsf{M}\vdash)$. The last case is completely usual, the cut and the $(\mathsf{Y}\vdash)$ rule may be simply switched. qed

Exercise 7.3.2. Write out the relevant proof segments and their transformations for some of the cases that we did not detail in the proof.

The proof may be structured differently than in [31] (with multi-cut) and than above (with single cut). Lemma 7.16 deals with the problem that arises when the cut formula is Y, which appears in both consecutions in the $(\mathsf{Y}\vdash)$ rule. But the lemma may be viewed as *reducing the left rank* of the cut. *LK* was designed to be symmetric, and has multiple left- and right-hand sides. But LC is not symmetric in a similar sense, and a consecution always has a single formula on the right-hand side of the $\vdash$. When Gentzen considered $\varrho > 2$, he first assumed $\varrho_r > 1$, and then $\varrho_r = 1$ and $\varrho_l > 1$.

If we impose an ordering on the cases under the rubric $\varrho > 2$, then we may require that if $\varrho_l > 1$ and $\varrho_r > 1$, then ϱ_l is reduced as much as possible, before ϱ_r is considered. If the cut formula is Y, or for that matter another combinator, then we will eliminate the cut without getting to ϱ_r, because the only way Y or another combinator can be the succedent is via the axiom. This way the lemma can be wholly integrated with the triple induction.

Proof: An alternative proof of Theorem 7.17 proceeds by triple induction on δ, μ and ϱ. If $\varrho > 2$ and both the left and the right rank are greater than 1, then the left rank is reduced first, as long as it can be reduced. qed

Exercise 7.3.3. Consider the proof of Lemma 7.16 and the first proof of the cut theorem, and round out those proofs so that you get the second proof.

Admittedly, Y may not be the most important combinator from the point of view of LC. The fixed point combinator is very important in combinatory logic and even in some functional programming languages. However, Y does not have a simple type, for instance. The impact of considerations about Y gave momentum to designing *consecution calculi* for certain *relevance logics*. In fact, the separate lemma's role was crucial, because it permitted cases of the cuts on combinators to be isolated from the rest of the proof of the cut theorem.

$\boldsymbol{t}$ has been introduced into a sequent calculus for the negation-free relevance logic R, and $\boldsymbol{t}$ turned out to be necessary for the algebraization of R too. $\boldsymbol{t}$ is identity for $\circ$ in R. However, in other relevance logics, notably, in E, T and B, $\boldsymbol{t}$ is not full identity. In some of these logics, special structural rules are needed to ensure that $\boldsymbol{t}$ can be moved around a little, and so the replacement theorem can be proved. However, when the general variant of that structural rule is not in the logic, then the usual proof of the cut theorem runs into a problem if $\boldsymbol{t}$ is the cut formula. $\boldsymbol{t}$ is very much like the identity combinator I. The $(\boldsymbol{t}\vdash)$ rule (that we occasionally also denote by $(\mathsf{KI}_t\vdash)$) is like the $(\mathsf{I}\vdash)$ rule. Although $\boldsymbol{t}$ may be included in some other rules, they are all left rules and they do not cause a problem in a lemma such as Lemma 7.16.

If we consider $LB^t_{\rightarrow}$ with $\circ$ and $\leftarrow$ also added, then $\boldsymbol{t}$ appears in six rules: $(\mathsf{KI}_t\vdash)$, $(\mathsf{M}_t\vdash)$, $(\mathsf{B}_t\vdash)$, $(\mathsf{B}'_t\vdash)$, $(\mathsf{b(cb)}_t\vdash)$ and $(\mathsf{b}_t\vdash)$. These rules were included in Sections 5.3 and 5.4; we repeat the two less obvious ones.

$$\frac{\mathfrak{A}[\,\boldsymbol{t};\mathfrak{B};\mathfrak{C}]\vdash\mathcal{A}}{\mathfrak{A}[\mathfrak{B};(\boldsymbol{t};\mathfrak{C})]\vdash\mathcal{A}}\ \mathsf{b(cb)}_t\vdash \qquad\qquad \frac{\mathfrak{A}[\,\boldsymbol{t};\mathfrak{B};\mathfrak{C}]\vdash\mathcal{A}}{\mathfrak{A}[\,\boldsymbol{t};(\mathfrak{B};\mathfrak{C})]\vdash\mathcal{A}}\ \mathsf{b}_t\vdash$$

We can consider the cut with the cut formula $\boldsymbol{t}$.

Lemma 7.18. *If* $\mathfrak{A} \vdash t$ *and* $\mathfrak{B}[t] \vdash A$ *are provable, then* $\mathfrak{B}[\mathfrak{A}] \vdash A$ *is* provable.

Proof: The proof is by induction on χ, the height of the proof of $\mathfrak{A} \vdash t$. (We consider some cases, and leave the details of some other cases as an exercise.)

1. If $\mathfrak{A} \vdash t$ is $t \vdash t$, then the claim is obvious.
2. $\mathfrak{A} \vdash t$ is by one of the three connective rules $(\leftarrow\vdash)$, $(\circ\vdash)$ or $(\rightarrow\vdash)$. We have the following for $(\leftarrow\vdash)$ (and we omit the rest of the details).

$$\leftarrow\vdash \frac{\mathfrak{C} \vdash B \quad \mathfrak{D}[C] \vdash t}{\mathfrak{D}[\mathfrak{C}; C \leftarrow B] \vdash t} \quad \overset{\text{i.h.}}{\rightsquigarrow} \quad \frac{\mathfrak{B}[\mathfrak{D}[C]] \vdash A \quad \mathfrak{C} \vdash B}{\mathfrak{B}[\mathfrak{D}[\mathfrak{C}; C \leftarrow B]] \vdash A} \leftarrow\vdash$$

3. $\mathfrak{A} \vdash t$ is by the $(\mathsf{Kl}_t\vdash)$ rule. Then we have the following proof segments.

$$\mathsf{Kl}_t\vdash \frac{\mathfrak{C}[\mathfrak{D}] \vdash t}{\mathfrak{C}[t; \mathfrak{D}] \vdash t} \quad \overset{\text{i.h.}}{\rightsquigarrow} \quad \frac{\mathfrak{B}[\mathfrak{C}[\mathfrak{D}]] \vdash A}{\mathfrak{B}[\mathfrak{C}[t; \mathfrak{D}]] \vdash A} \mathsf{Kl}_t\vdash$$

4. $\mathfrak{A} \vdash t$ is by $(\mathsf{b}(\mathsf{cb})_t\vdash)$.

$$\mathsf{b}(\mathsf{cb})_t\vdash \frac{\mathfrak{C}[t; \mathfrak{D}; \mathfrak{E}] \vdash t}{\mathfrak{C}[\mathfrak{D}; (t; \mathfrak{E})] \vdash t} \quad \overset{\text{i.h.}}{\rightsquigarrow} \quad \frac{\mathfrak{B}[\mathfrak{C}[t; \mathfrak{D}; \mathfrak{E}]] \vdash A}{\mathfrak{B}[\mathfrak{C}[\mathfrak{D}; (t; \mathfrak{E})]] \vdash A} \mathsf{b}(\mathsf{cb})_t\vdash$$

5. $\mathfrak{A} \vdash t$ is by $(\mathsf{b}_t\vdash)$.

$$\mathsf{b}_t\vdash \frac{\mathfrak{C}[t; \mathfrak{D}; \mathfrak{E}] \vdash t}{\mathfrak{C}[t; (\mathfrak{D}; \mathfrak{E})] \vdash t} \quad \overset{\text{i.h.}}{\rightsquigarrow} \quad \frac{\mathfrak{B}[\mathfrak{C}[t; \mathfrak{D}; \mathfrak{E}]] \vdash A}{\mathfrak{B}[\mathfrak{C}[t; \mathfrak{D}; \mathfrak{E}]] \vdash A} \mathsf{b}_t\vdash$$

qed

Exercise 7.3.4. Work out the details of the cases not included in the proof. [Hint: They are similar to the cases that we considered.]

This lemma also could be integrated into a double or a triple inductive proof of the cut theorem by requiring that, if it is possible, the left rank should be reduced first.

7.4 Display cut

H. B. Curry developed a proof of the admissibility of the cut rule based on the idea that certain *properties of rules* can guarantee that the cut rule is admissible. Gentzen's proof involves scrutinizing individual rules and making *local* changes to the given proof (though occasionally whole subtrees of the proof tree are duplicated). Curry's proof method involves more *global* changes, because it allows whole swaths to be modified at once, with an appeal to properties of the rules. N. D. Belnap adapted Curry's ideas to show the admissibility of the *display cut* rule in display calculi. We outline two versions of such proofs.

For each display logic, there is an *analysis*, which is like the one that we provided for LK in Definition 2.22. The correspondence between formulas in the upper consecutions and the lower consecution is called a *congruence*. There are eight conditions for the display calculi that utilize a Boolean negation.

Definition 7.19. The *conditions on the rules* in a display calculus with a star are (1)–(8).

(1) Every formula in a premise is a subformula in the conclusion.

(2) Congruent formulas and congruent structural constants are occurrences of one and the same formula and structural constant, respectively.

(3) A formula or a structural constant in an upper consecution is congruent to at most one formula or structural constant, respectively, in the lower consecution.

(4) Congruence respects $\mathfrak{a}$ and $\mathfrak{s}$ (from Definition 6.3), that is, congruent components have $\mathfrak{a}$ occurrences, or they have $\mathfrak{s}$ occurrences in their respective consecutions, but not both.

(5) The principal formula or structural constant is displayed in the lower consecution.

(6) An instance of a rule remains an instance of the same rule if a structure is substituted for occurrences of a parametric wff $\mathcal{A}$ that are congruent to $\mathfrak{s}(\mathcal{A})$. The substitution does not change the analysis of the unaffected components of the consecutions, and every component of the substituted structure is parametric.

(7) An instance of a rule remains an instance of the same rule if a structure $\mathfrak{A}$ is substituted for occurrences of a parametric wff $\mathcal{A}$ that are congruent to $\mathfrak{a}(\mathcal{A})$, provided that $\mathfrak{A} \vdash \mathcal{A}$ is provable with $\mathcal{A}$ being principal.

(8) The display cut rule is eliminable, in the following three cases, when the cut formula is principal in both premises. (i) The conclusion is the left premise. (ii) The conclusion is the right premise. (iii) The conclusion follows from the premises of the rules yielding the premises of the cut by applications of rules and by cuts on proper subformulas of the original cut formula.

The verification that the conditions (1)–(5) hold is easy. (6) and (7) are unproblematic, when no rule has side conditions. In the case of $(\check{\mathbf{I}}_e \vdash)$, in DL_E, (7) holds too, because the only way to prove $\mathfrak{A} \vdash p_e$ is as $p_e \vdash p_e$, that is, when $\mathfrak{A}$ is p_e itself. The last condition is the most complicated to check. Indeed, this corresponds to the core cases with $\varrho = 2$ in other proofs.

Theorem 7.20. (Display cut theorem) *If for a display calculus DL, there is an analysis and a congruence satisfying conditions* (2)–(8) *(above), then the* display cut rule is admissible.

Proof: The proof is divided into three stages.[5] The stages accomplish the following.
1. This stage shows that $\mathfrak{A} \vdash \mathfrak{B}$ is provable, given that (i) $\mathfrak{A} \vdash \mathcal{C}$ is provable, and that (ii) if $\mathfrak{D} \vdash \mathcal{C}$, where $\mathcal{C}$ is principal, is provable, then so is $\mathfrak{D} \vdash \mathfrak{B}$.
2. This stage demonstrates that $\mathfrak{A} \vdash \mathfrak{B}$ is provable, given that (i) $\mathcal{C} \vdash \mathfrak{B}$ is provable, and that (ii) if $\mathcal{C} \vdash \mathfrak{D}$ is provable with $\mathcal{C}$ being principal, then so is $\mathfrak{A} \vdash \mathfrak{D}$, and that (iii) $\mathfrak{A} \vdash \mathcal{C}$ is provable, where $\mathcal{C}$ is principal.
3. This stage shows that $\mathfrak{A} \vdash \mathfrak{B}$ is provable, given that (i) $\mathfrak{A} \vdash \mathcal{C}$, where $\mathcal{C}$ is principal, is provable, and that (ii) $\mathcal{C} \vdash \mathfrak{B}$, where $\mathcal{C}$ is principal, is provable, and that (iii) for any subformulas $\mathcal{D}$ of $\mathcal{C}$, $\mathfrak{C} \vdash \mathfrak{D}$ is provable, if so are $\mathfrak{C} \vdash \mathcal{D}$ and $\mathcal{D} \vdash \mathfrak{D}$. qed

Exercise* 7.4.1. Consider DL_R, and show how the details of the above proof can be filled out.

Exercise* 7.4.2. Recast the proof from the previous exercise so that you appeal to the conditions on the rules (rather that the rules themselves). [Hint: See [19] and [8, §62] for the details.]

For DL_L, a new condition is introduced and the proof of the cut theorem is made completely *symmetric* just as LL itself is symmetric.

Definition 7.21. (Regularity) A formula $\mathcal{A}$ is *regular* iff it is $\mathfrak{a}$- or $\mathfrak{s}$-regular.

$\mathcal{A}$ is $\mathfrak{a}$*-regular* iff (1) an instance of a rule remains an instance of the same rule, when a parametric $\mathfrak{a}(\mathcal{A})$ is replaced by $\mathfrak{B}$, and (2) an instance of a rule remains an instance of the same rule if a parametric $\mathfrak{s}(\mathcal{A})$ is replaced by $\mathfrak{C}$, provided that $\mathcal{A} \vdash \mathfrak{C}$ is provable with $\mathcal{A}$ being principal.

$\mathcal{A}$ is $\mathfrak{s}$*-regular* iff (1) an instance of a rule remains an instance of the same rule, when a parametric $\mathfrak{s}(\mathcal{A})$ is replaced by $\mathfrak{B}$, and (2) an instance of a rule remains an instance of the same rule if a parametric $\mathfrak{a}(\mathcal{A})$ is replaced by $\mathfrak{C}$, provided that $\mathfrak{C} \vdash \mathcal{A}$ is provable with $\mathcal{A}$ being principal.

Condition (9) for the rules is that *every formula is regular*.

Exercise 7.4.3. Verify that (9) holds for DL_L. [Hint: See [20, §5].]

Theorem 7.22. *The display cut rule is* admissible *in* DL_L.

Proof: First of all, condition (9) is satisfied by DL_L, that is, all formulas are regular. The proof of Theorem 7.20 proceeded via three stages. Now we add two *dual stages*, which we denote by $\mathbf{1}^{\eth}$ and $\mathbf{2}^{\eth}$. Then the admissibility of the display cut follows, because either **1**, **2** and **3**, or $\mathbf{1}^{\eth}$, $\mathbf{2}^{\eth}$ and **3** are applicable to any cut formula $\mathcal{C}$. We include the outline of the effect of the two new stages.

$\mathbf{1}^{\eth}$ This stage shows that $\mathfrak{A} \vdash \mathfrak{B}$ is provable, given that (i) $\mathcal{C} \vdash \mathfrak{B}$ is provable, and that (ii) if $\mathcal{C} \vdash \mathfrak{D}$, where $\mathcal{C}$ is principal, is provable, then so is $\mathfrak{A} \vdash \mathfrak{D}$.

[5] The structure of the proof resembles Curry's proof, who calls these general steps "stages."

2.$^{\eth}$ This stage demonstrates that $\mathfrak{A} \vdash \mathfrak{B}$ is provable, given that (i) $\mathfrak{A} \vdash C$ is provable, and that (ii) if $\mathfrak{D} \vdash C$ is provable with C being principal, then so is $\mathfrak{D} \vdash \mathfrak{B}$, and that (iii) $C \vdash \mathfrak{B}$ is provable, where C is principal. qed

Exercise* 7.4.4. Fill out the details of the proof of the cut theorem for DL_L. [Hint: The proof is outlined in [20].]

7.5 Cut theorem via normal proofs

Calculi with multiple right-hand sides—save LK—were not well understood for some time. In Section 5.5, we introduced a consecution calculus with multiple right-hand sides that we called L_{HD}. We also illustrated there that in $L_{HD?}$, which is an extension of L_{HD} with further connectives, the single algebraic cut rule is not admissible.

Exercise 7.5.1. The usual double inductive proof on the rank of the cut and on the degree of the cut formula does not work for L_{HD}. Try to find the place where there is a problem with such an attempted proof.

The proof of the admissibility of algebraic cut in [40] uses a completely new idea. The proof is not by double or triple induction; rather it uses *proofs in normal form*. The informal idea is that given a consecution, the structure of the antecedent and that of the succedent determines a proof, which may be different from the given one, but always exists.

We want to be able to turn a consecution into a pair of formulas and similarly, turn formulas into structures.

Definition 7.23. The operations of *fusing* and *fissioning* a structure are inductively defined by (1)–(2). They are denoted by $^\circ$ and $^+$, respectively.

(1) If $\mathfrak{A}$ is an atomic structure (i.e., a formula), then $\mathfrak{A}^\circ$ and $\mathfrak{A}^+$ are $\mathcal{A}$;

(2) if $\mathfrak{A}$ is $\mathfrak{B};\mathfrak{C}$, then $\mathfrak{A}^\circ$ is $\mathfrak{B}^\circ \circ \mathfrak{C}^\circ$, and $\mathfrak{A}^+$ is $\mathfrak{B}^+ + \mathfrak{C}^+$.

The operations of *unfusing* and *unfissioning* a structure are inductively defined by (3)–(6). They are denoted by $^{\underline{\circ}}$ and $^\pm$, respectively.

(3) If p is a propositional variable, then $p^{\underline{\circ}}$ and $p^\pm$ is p;

(4) if $\mathcal{A}$ is $\mathcal{B} \circ \mathcal{C}$, then $\mathcal{A}^{\underline{\circ}}$ is $\mathcal{B};\mathcal{C}$ and $\mathcal{A}^\pm$ is $\mathcal{B} \circ \mathcal{C}$;

(5) if $\mathcal{A}$ is $\mathcal{B} + \mathcal{C}$, then $\mathcal{A}^{\underline{\circ}}$ is $\mathcal{B} + \mathcal{C}$ and $\mathcal{A}^\pm$ is $\mathcal{B};\mathcal{C}$;

(6) if $\mathfrak{A}$ is $\mathfrak{B};\mathfrak{C}$, then $\mathfrak{A}^{\underline{\circ}}$ is $\mathfrak{B}^{\underline{\circ}};\mathfrak{C}^{\underline{\circ}}$ and $\mathfrak{A}^\pm$ is $\mathfrak{B}^\pm;\mathfrak{C}^\pm$.

The definition allows us to view a pair of formulas as a pair of structures, as much unfused and unfissioned, respectively, as possible. Conversely, a consecution may be viewed as a pair of formulas, when the antecedent is fused and the succedent is fissioned.

Lemma 7.24. *If* $\mathfrak{A} \vdash \mathfrak{B}$ *is provable in* L_{HD}, *then* $\mathfrak{A}^{\circ} \vdash \mathfrak{B}^{+}$ *and* $\mathfrak{A}^{\underline{\circ}} \vdash \mathfrak{B}^{\pm}$ *are* provable *too.*

Proof: **1.** The first claim is nearly obvious. $\mathfrak{A}$ can be turned into a formula by as many applications of the $(\circ\vdash)$ rule as there are ;'s in $\mathfrak{A}$. Similarly, for $\mathfrak{B}$, applications of the $(\vdash+)$ rule yield the result.
2. The second claim is proved by inductions on χ, the height of the proof of $\mathfrak{A} \vdash \mathfrak{B}$, and on the degree of a wff in an axiom in the proof.
2.1 If $\chi = 0$, then $\mathfrak{A} \vdash \mathfrak{B}$ is $\mathcal{A} \vdash \mathcal{A}$. If $d = 0$, then we have $p \vdash p$. Therefore, this proof itself is the proof of $\mathfrak{A}^{\circ} \vdash \mathfrak{B}^{+}$ and $\mathfrak{A}^{\underline{\circ}} \vdash \mathfrak{B}^{\pm}$.

If $d > 1$, then $\mathcal{A}$ is $\mathcal{B} \circ \mathcal{C}$ or $\mathcal{B} + \mathcal{C}$. By the hypothesis of induction, $\mathcal{B}^{\underline{\circ}} \vdash \mathcal{B}^{\pm}$ and $\mathcal{C}^{\underline{\circ}} \vdash \mathcal{C}^{\pm}$ are provable. The following two proofs suffice to show that $\mathcal{A}^{\underline{\circ}} \vdash \mathcal{A}^{\pm}$ are provable.

$$\vdash\circ\;\dfrac{\vdash+\text{'s}\;\dfrac{\mathcal{B}^{\underline{\circ}} \vdash \mathcal{B}^{\pm}}{\mathcal{B}^{\underline{\circ}} \vdash \mathcal{B}} \qquad \dfrac{\mathcal{C}^{\underline{\circ}} \vdash \mathcal{C}^{\pm}}{\mathcal{C}^{\underline{\circ}} \vdash \mathcal{C}}\;\vdash+\text{'s}}{\mathcal{B}^{\underline{\circ}};\mathcal{C}^{\underline{\circ}} \vdash \mathcal{B} \circ \mathcal{C}} \qquad\qquad \dfrac{\circ\vdash\text{'s}\;\dfrac{\mathcal{B}^{\underline{\circ}} \vdash \mathcal{B}^{\pm}}{\mathcal{B} \vdash \mathcal{B}^{\pm}} \qquad \dfrac{\mathcal{C}^{\underline{\circ}} \vdash \mathcal{C}^{\pm}}{\mathcal{C} \vdash \mathcal{C}^{\pm}}\;\circ\vdash\text{'s}}{\mathcal{B} + \mathcal{C} \vdash \mathcal{B}^{\pm};\mathcal{C}^{\pm}}\;+\vdash$$

The bottom consecutions are $\mathcal{A}^{\underline{\circ}} \vdash \mathcal{A}^{\pm}$, as we wanted to show.
3. If $\chi > 1$, then the last consecution is by one of the connective rules. We consider the rules for $\circ$. If the rule is $(\circ\vdash)$, then the resulting consecution is $\mathfrak{A}[\mathcal{A} \circ \mathcal{B}] \vdash \mathfrak{B}$. By the hypothesis of the induction, $\mathfrak{A}[\mathcal{A};\mathcal{B}]^{\underline{\circ}} \vdash \mathfrak{B}^{\pm}$ is provable. However, $\mathfrak{A}[\mathcal{A};\mathcal{B}]^{\underline{\circ}} \vdash \mathfrak{B}^{\pm}$ is $\mathfrak{A}[\mathcal{A} \circ \mathcal{B}]^{\underline{\circ}} \vdash \mathfrak{B}^{\pm}$.

If the rule is $(\vdash\circ_1)$, then by the hypothesis of the induction, $\mathfrak{A}^{\underline{\circ}} \vdash \mathfrak{B}[\mathcal{A}]^{\pm}$ and $\mathfrak{C}^{\underline{\circ}} \vdash \mathfrak{D}[\mathcal{B}]^{\pm}$ are provable. By an application of the rule, we obtain $\mathfrak{A}^{\underline{\circ}};\mathcal{C}^{\underline{\circ}} \vdash \mathfrak{D}[\mathfrak{B}[\mathcal{A} \circ \mathcal{B}]^{\pm}]^{\pm}$, which is $(\mathfrak{A};\mathfrak{C})^{\underline{\circ}} \vdash (\mathfrak{D}[\mathfrak{B}[\mathcal{A} \circ \mathcal{B}]])^{\pm}$.

If the rule is $(\vdash\circ_2)$, then the lower consecution is $\mathfrak{A};\mathfrak{C} \vdash \mathfrak{B}[\mathfrak{D}[\mathcal{A} \circ \mathcal{B}]]$, and by an application of the rule we obtain $(\mathfrak{A};\mathfrak{C})^{\underline{\circ}} \vdash (\mathfrak{B}[\mathfrak{D}[\mathcal{A} \circ \mathcal{B}]])^{\pm}$ from the hypotheses. qed

Exercise 7.5.2. Write out the details of the subcases (in step **3**) for the $+$ rules.

The lemma ensures that we can go back and forth between $\mathfrak{A}^{\circ} \vdash \mathfrak{B}^{+}$ and $\mathfrak{A}^{\underline{\circ}} \vdash \mathfrak{B}^{\pm}$, without losing provability.

A version of the cut rule that was introduced in display calculi, (hence, it is called *display cut*), requires the cut formula to be the whole succedent of the left premise and the whole antecedent of the right premise.

$$\dfrac{\mathfrak{A} \vdash \mathcal{C} \qquad \mathcal{C} \vdash \mathfrak{D}}{\mathfrak{A} \vdash \mathfrak{D}}\;\text{display cut}$$

It may be unexpected that this cut rule comes handy in no other calculus, but in L_{HD}, which permits multiple formulas to occur on both sides of the

the $\vdash$. However, we will use the display cut rule to prove the cut theorem for the genuine cut of L_{HD}, which has been called *single algebraic cut* (that we shortened to algebraic cut, in Chapter 5). This rule looks like the single cut rule that we used in LK, in Chapter 2, with structures replacing sequences of formulas.

$$\frac{\mathfrak{A} \vdash \mathfrak{B}[C] \qquad \mathfrak{C}[C] \vdash \mathfrak{D}}{\mathfrak{C}[\mathfrak{A}] \vdash \mathfrak{B}[\mathfrak{D}]} \text{ algebraic cut}$$

We show the admissibility of this cut rule by proving four lemmas. We first prove two lemmas about "embedding" a fused–fissioned consecution into a fused or a fissioned consecution. The bracket notation here is extended in an obvious way to formulas. For example, $\mathfrak{D}^\circ[\mathfrak{A}^\circ]$ is a formula obtained from the structure $\mathfrak{D}$, which has an occurrence of a formula $\mathfrak{A}^\circ$ as its atomic structure. Then, in a limited context, such as a consecution, we mean by $\mathfrak{D}^\circ[\mathfrak{B}^+[C]]$ a formula that is obtained by replacing $\mathfrak{A}^\circ$ with $\mathfrak{B}^+[C]$.

Lemma 7.25. *If the consecution $\mathfrak{A} \vdash \mathfrak{B}[C]$ is provable in L_{HD}, then the consecution $\mathfrak{D}^\circ[\mathfrak{A}^\circ] \vdash \mathfrak{D}^\circ[\mathfrak{B}^+[C]]$ is* provable *too.*

Proof: First of all, the consecution $\mathfrak{D}^\circ[\mathfrak{A}^\circ] \vdash \mathfrak{D}^\circ[\mathfrak{B}^+[C]]$ comprises a pair of wff's, which differ at the place where $\mathfrak{A}^\circ$ and $\mathfrak{B}^+[C]$ occur (if they are distinct). Let p be a new propositional variable (that does not occur in the consecution). Then $\mathfrak{D}^\circ[p] \vdash \mathfrak{D}^\circ[p]$ is provable, not only as an instance of the axiom (id), but by a proof in which the leaves of the proof tree instantiate the axiom with atomic formulas.

By assumption, $\mathfrak{A} \vdash \mathfrak{B}[C]$ has a proof, and so $\mathfrak{A}^\circ \vdash \mathfrak{B}^+[C]$ can be obtained from that proof by finitely many applications of the $(\circ\vdash)$ and $(\vdash+)$ rules. Replacing the leaf $p \vdash p$ with the latter proof, we can carry out the steps in the proof of $\mathfrak{D}^\circ[p] \vdash \mathfrak{D}^\circ[p]$ with $\mathfrak{A}^\circ$ and $\mathfrak{B}^+[C]$ replacing the subsequent occurrences of the p's. The concrete shape of a formula does not affect the applicability of a rule, hence all the proof steps remain proof steps, and we get a proof of $\mathfrak{D}^\circ[\mathfrak{A}^\circ] \vdash \mathfrak{D}^\circ[\mathfrak{B}^+[C]]$, as desired. qed

Lemma 7.26. *If the consecution $\mathfrak{A}[C] \vdash \mathfrak{B}$ is provable in L_{HD}, then the consecution $\mathfrak{D}^+[\mathfrak{A}^\circ[C]] \vdash \mathfrak{D}^+[\mathfrak{B}^+]$ is* provable *in L_{HD} too.*

Proof: The proof of this lemma is similar to the proof of the previous lemma, except that we start with $\mathfrak{D}^+[p] \vdash \mathfrak{D}^+[p]$. qed

The next lemma reflects the general distributivity property of $\circ$ and $+$ that we mentioned in Section 5.5.

Lemma 7.27. *The sequent $\mathfrak{A}^\circ[\mathfrak{B}^+[C]] \vdash \mathfrak{B}^+[\mathfrak{A}^\circ[C]]$ is* provable *in L_{HD}.*

Proof: Let p and q be new propositional variables that do not occur in $\mathfrak{A}^\circ[\mathfrak{B}^+[C]] \vdash \mathfrak{B}^+[\mathfrak{A}^\circ[C]]$. Both $\mathfrak{A}^\circ[p] \vdash \mathfrak{A}^\circ[p]$ and $\mathfrak{B}^+[q] \vdash \mathfrak{B}^+[q]$ are provable. By a previous lemma, $\mathfrak{A}^\circ_\circ[p] \vdash \mathfrak{A}^\circ[p]$ and $\mathfrak{B}^+[q] \vdash \mathfrak{B}^\pm[q]$ are also provable. In the proof tree of the latter, there is a leaf that is of the form

$q \vdash q$. We replace the leaf by the former proof tree with end consecution $\mathfrak{A}^{\underline{\circ}}[p] \vdash \mathfrak{A}^{\circ}[p]$, and restore the p's to $\mathcal{C}$. In building up $\mathfrak{B}^{+}[\mathcal{C}]$ on the left-hand side of the $\vdash$, we follow the steps in the proof of the consecution $\mathfrak{B}^{+}[q] \vdash \mathfrak{B}^{+}[q]$. Since $\mathcal{C}$ is an atomic structure in $\mathfrak{A}^{\underline{\circ}}[\mathcal{C}]$, $\mathfrak{B}^{+}[\mathcal{C}]$ remains an atomic structure in the process, whereas the matching structures on the right-hand side of the $\vdash$ are built around $\mathfrak{A}^{\circ}[\mathcal{C}]$. This blending of the proofs of $\mathfrak{A}^{\underline{\circ}}[\mathcal{C}] \vdash \mathfrak{A}^{\circ}[\mathcal{C}]$ and $\mathfrak{B}^{+}[\mathcal{C}] \vdash \mathfrak{B}^{\pm}[\mathcal{C}]$ together is unproblematic, as their versions with propositional variables were intended to show. The result is $\mathfrak{A}^{\underline{\circ}}[\mathfrak{B}^{+}[\mathcal{C}]] \vdash \mathfrak{B}^{\pm}[\mathfrak{A}^{\circ}[\mathcal{C}]]$, which may be turned into $\mathfrak{A}^{\circ}[\mathfrak{B}^{+}[\mathcal{C}]] \vdash \mathfrak{B}^{+}[\mathfrak{A}^{\circ}[\mathcal{C}]]$, by finitely many applications of the $(\circ\vdash)$ and $(\vdash+)$ rules. qed

As a final step toward the cut theorem, we show that the display cut rule is admissible in L_{HD}.[6]

Lemma 7.28. *The display cut rule is* admissible *in* L_{HD}.

Proof: We prove this lemma by complete induction on the degree of the cut, where the latter is defined as in LQ, by calculating the degree of $\mathcal{A}+\mathcal{B}$ similarly to how we did it for $\mathcal{A}\circ\mathcal{B}$.

1. If the degree is 0, then the lower sequent of the cut is identical to both premises, one of which is chosen to be retained.

2. If the two premises are by matching rules, that is, the cut formula is principal in both premises, then there are two possibilities, namely, when the cut formula is $\mathcal{A}\circ\mathcal{B}$ and when it is $\mathcal{A}+\mathcal{B}$.

We start with a proof segment of the following form.

$$\dfrac{\vdash\circ\ \dfrac{\begin{matrix}\vdots\\ \mathfrak{A}\vdash\mathcal{A}\end{matrix}\quad\begin{matrix}\vdots\\ \mathfrak{B}\vdash\mathcal{B}\end{matrix}}{\mathfrak{A};\mathfrak{B}\vdash\mathcal{A}\circ\mathcal{B}}\qquad\dfrac{\begin{matrix}\vdots\\ \mathcal{A};\mathcal{B}\vdash\mathfrak{C}\end{matrix}}{\mathcal{A}\circ\mathcal{B}\vdash\mathfrak{C}}\ \circ\vdash}{\mathfrak{A};\mathfrak{B}\vdash\mathfrak{C}}\ \text{cut}$$

From the provability of the right premise, $\mathcal{A}\circ\mathcal{B}\vdash\mathfrak{C}$, we know that $(\mathcal{A}\circ\mathcal{B})^{\underline{\circ}}\vdash\mathfrak{C}^{\pm}$ is provable too. Further, there is a proof ending in the following segment.

$$\dfrac{\dfrac{\begin{matrix}\vdots\\ \mathcal{A}\vdash\mathfrak{C}_1\end{matrix}\quad\begin{matrix}\vdots\\ \mathcal{B}\vdash\mathfrak{C}_2\end{matrix}}{\mathcal{A};\mathcal{B}\vdash\mathfrak{C}^{\pm}}\ \vdash\circ}{\mathcal{A};\mathcal{B}\vdash\mathfrak{C}}\ \vdash+\text{'s}$$

We consider this particular proof, even if this is not the proof, that is given.

[6] In [40], we gave a proof of the next lemma using a conservative extension of a display calculus, and soundness and completeness. Here we proceed within the boundaries of L_{HD} itself.

The cut on $\mathcal{A} \circ \mathcal{B}$ is replaced by two cuts, one on $\mathcal{A}$ and one on $\mathcal{B}$.

$$\dfrac{\dfrac{\dfrac{\dfrac{\vdots}{\mathfrak{A} \vdash \mathcal{A}} \quad \dfrac{\vdots}{\mathcal{A} \vdash \mathfrak{C}_1}}{\mathfrak{A} \vdash \mathfrak{C}_1}\,\text{cut} \qquad \dfrac{\dfrac{\vdots}{\mathfrak{B} \vdash \mathcal{B}} \quad \dfrac{\vdots}{\mathcal{B} \vdash \mathfrak{C}_2}}{\mathfrak{B} \vdash \mathfrak{C}_2}\,\text{cut}}{\mathfrak{A}; \mathfrak{B} \vdash \mathfrak{C}^{\pm}}\,\vdash\circ}{\mathfrak{A}; \mathfrak{B} \vdash \mathfrak{C}}\,\vdash +\text{'s}$$

The case when the cut formula is $\mathcal{A} + \mathcal{B}$ is very similar. We are given the following proof.

$$\dfrac{\dfrac{\dfrac{\vdots}{\mathfrak{A} \vdash \mathcal{A}; \mathcal{B}}}{\mathfrak{A} \vdash \mathcal{A} + \mathcal{B}}\,\vdash + \qquad \dfrac{\dfrac{\vdots}{\mathcal{A} \vdash \mathfrak{B}} \quad \dfrac{\vdots}{\mathcal{B} \vdash \mathfrak{C}}}{\mathcal{A} + \mathcal{B} \vdash \mathfrak{B}; \mathfrak{C}}\,+\vdash}{\mathfrak{A} \vdash \mathfrak{B}; \mathfrak{C}}\,\text{cut}$$

From the provability of the left premise, we get that $\mathfrak{A}^{\circ} \vdash (\mathcal{A} + \mathcal{B})^{\pm}$ is provable too. It has a proof ending with an application of $(+\vdash)$ to $\mathfrak{A}_1 \vdash \mathcal{A}$ and $\mathfrak{A}_2 \vdash \mathcal{B}$. We take this proof for the left premise of the cut, and replace the cut by two cuts of lower degree.

$$\dfrac{\dfrac{\dfrac{\dfrac{\vdots}{\mathfrak{A}_1 \vdash \mathcal{A}} \quad \dfrac{\vdots}{\mathcal{A} \vdash \mathfrak{B}}}{\mathfrak{A}_1 \vdash \mathfrak{B}}\,\text{cut} \qquad \dfrac{\dfrac{\vdots}{\mathfrak{A}_2 \vdash \mathcal{B}} \quad \dfrac{\vdots}{\mathcal{B} \vdash \mathfrak{C}}}{\mathfrak{A}_2 \vdash \mathfrak{C}}\,\text{cut}}{\mathfrak{A}^{\circ} \vdash \mathfrak{B}; \mathfrak{C}}\,+\vdash}{\mathfrak{A} \vdash \mathfrak{B}; \mathfrak{C}}\,\circ\vdash\text{'s}$$

3. If the left premise is by a left rule, that is, by $(\circ\vdash)$ or $(+\vdash)$, then we swap the left rule and the cut.

4. If the right premise is by a right rule, that is, by $(\vdash\circ)$ or $(\vdash+)$, then we swap the right rule and the cut. $\mathfrak{qed}$

Exercise 7.5.3. Work out the details of Cases **3** and **4**, and verify that each transformation is justified in the induction.

Now we are ready to prove the admissibility of the algebraic cut rule.

Theorem 7.29. *The single algebraic cut is* admissible *in* L_{HD}.

Proof: Let us assume that the consecutions that constitute the left and the right premise of the cut are provable. That is, $\mathfrak{A} \vdash \mathfrak{B}[C]$ and $\mathfrak{C}[C] \vdash \mathfrak{D}$ are provable. By Lemma 7.25, $\mathfrak{C}^{\circ}[\mathfrak{A}^{\circ}] \vdash \mathfrak{C}^{\circ}[\mathfrak{B}^{+}[C]]$, and by Lemma 7.26, $\mathfrak{B}^{+}[\mathfrak{C}^{\circ}[C]] \vdash \mathfrak{B}^{+}[\mathfrak{D}^{+}]$ are provable too. From Lemma 7.27, we have that $\mathfrak{C}^{\circ}[\mathfrak{B}^{+}[C]] \vdash \mathfrak{B}^{+}[\mathfrak{C}^{\circ}[C]]$ is provable. Then by the display cut rule, $\mathfrak{C}^{\circ}[\mathfrak{A}^{\circ}] \vdash \mathfrak{B}^{+}[\mathfrak{C}^{\circ}[C]]$, and further, $\mathfrak{C}^{\circ}[\mathfrak{A}^{\circ}] \vdash \mathfrak{B}^{+}[\mathfrak{D}^{+}]$ are provable. But, then $(\mathfrak{C}[\mathfrak{A}])^{\circ} \vdash (\mathfrak{B}[\mathfrak{D}])^{\pm}$ is provable, from which we get that $\mathfrak{C}[\mathfrak{A}] \vdash \mathfrak{B}[\mathfrak{D}]$ is provable by finitely many applications of the $(\circ\vdash)$ and $(\vdash+)$ rules. $\mathfrak{qed}$

7.6 Cut theorem via interpretations

The cut rule occupies a special place in sequent calculi, and this is highlighted by the two proofs that we describe in this section. Classical propositional logic is sound and complete for two-valued interpretations. Under the two-valued interpretation, $\mathcal{A}_1, \ldots, \mathcal{A}_n \vdash \mathcal{B}_1, \ldots, \mathcal{B}_m$ is satisfied by an interpretation when either one of the $\mathcal{A}$'s is false or one of the $\mathcal{B}$'s is true.

7.6.1 Partial interpretations

However, sequents may be given other interpretations, specifically, interpretations in which not all formulas have at least one of the two truth values T (true) or F (false), or in which some formulas are both T and F. The idea of partial interpretations has been introduced by Kleene in connection with recursive functions, and the idea of using them in a cut proof goes back to Schütte.

Kleene's strong three-valued logic, which we denote by Kl_3, is defined by the following matrices. The third truth value is denoted by U (undefined). We use the same connectives as in the language of LK—with an eye toward using the interpretation for wff's of classical logic. (The values of the first argument of a binary connective are listed vertically.)

	$\neg$
T	F
U	U
F	T

$\vee$	T	U	F
T	T	T	T
U	T	U	U
F	T	U	F

$\wedge$	T	U	F
T	T	U	F
U	U	U	F
F	F	F	F

$\supset$	T	U	F
T	T	U	F
U	T	U	U
F	T	T	T

Kleene, in [113, §64], introduces $\leftrightarrow$ too (as well as his weak three-valued logic), but we really need only his strong $\neg$ and $\vee$. A characteristic feature of Kl_3, which flows from the idea of a computation being not yet completed, is that a formula's truth or falsity cannot change by defining an undefined truth value.

Lemma 7.30. *If all atomic formulas that are subformulas of $\mathcal{A}$ are T or F according to an interpretation I, then $\mathcal{A}$'s value is T or F in Kl_3.*

Exercise 7.6.1. Prove the lemma. [Hint: It is an easy induction on wff's.]

Let I be an interpretation of atomic wff's in $\{T, U, F\}$, and let I be extended to compound formulas according to the Kl_3 matrices. I' is an *amelioration* of I iff for every atomic formula $\mathcal{A}$, if $I(\mathcal{A}) = T$, then $I'(\mathcal{A}) = T$ and if $I(\mathcal{A}) = F$, then $I'(\mathcal{A}) = F$.

Theorem 7.31. (Schütte's lemma) *If $I(\mathcal{A}) = T$ (or $I(\mathcal{A}) = F$) and I' is an amelioration of I, then $I'(\mathcal{A}) = T$ (or $I'(\mathcal{A}) = F$).*

Proof: We prove the claim by induction on the structure of formulas.
1. If $\mathcal{A}$ is atomic, then since by assumption $I(\mathcal{A}) = T$ or $I(\mathcal{A}) = F$, then any amelioration I' of I, assigns the same truth value to $\mathcal{A}$, which proves the claim for $\mathcal{A}$.
2. If $\mathcal{A}$ is $\neg\mathcal{B}$, then by hypothesis, $I'(\mathcal{B}) = T$, if $I(\mathcal{B}) = T$, and $I'(\mathcal{B}) = F$, if $I(\mathcal{B}) = F$. If $I(\neg\mathcal{B}) = T$, then $I(\mathcal{B}) = F$, hence, $I'(\mathcal{B}) = F$ and so $I'(\neg\mathcal{B}) = T$. If $I(\neg\mathcal{B}) = F$, then similar steps give $I'(\neg\mathcal{B}) = F$.
3 . If $\mathcal{A}$ is $\mathcal{B} \vee \mathcal{C}$, then $I(\mathcal{B}) = T$ implies $I'(\mathcal{B}) = T$, $I(\mathcal{B}) = F$ implies $I'(\mathcal{B}) = F$, and similarly for $\mathcal{C}$. We consider the possibility when $I(\mathcal{B} \vee \mathcal{C}) = T$, however, $I(\mathcal{B}) = U$ or $I(\mathcal{C}) = U$. I' may define the truth value of a previously undefined subformula. $\vee$ is commutative, hence, we assume $I(\mathcal{B}) = U$ and $I'(\mathcal{B}) \in \{T, F\}$. If $I(\mathcal{B} \vee \mathcal{C}) = T$, then $I(\mathcal{C}) = T = I'(\mathcal{C})$, hence, $I'(\mathcal{B} \vee \mathcal{C}) = T$. We leave the rest of the proof as an exercise. qed

Exercise 7.6.2. Finish Case 3, and the cases for $\wedge$ and $\supset$. [Hint: The latter are similar to 3.]

Obviously, if I assigns U to some atomic formulas, then there is a two-valued interpretation, which is an amelioration of I and can be conceptualized as turning *each* U into T or F. In short, a partial interpretation can be extended to a two-valued interpretation.

In order to be able to use the three-valued interpretations in the proof of the cut theorem, we want to show that any formula, which cannot be assigned the value F by a three-valued interpretation, is provable in $LK^{\curlyvee}$, the sequent calculus we introduced in Section 3.4.

First, we slightly reformulate the propositional part of $LK^{\curlyvee}$ into $LK^{\curlyvee'}$. Given the variations of LK that we considered in Chapter 3, the changes should not be surprising. We take a sequent to be a *set of formulas*. The *sequent calculus* $LK^{\curlyvee'}$ comprises the axiom $\Gamma, \mathcal{A}, \neg\mathcal{A}$ together with the rules $(\vee)$, (DM) and (DN).

Exercise 7.6.3. Prove that the calculi $LK^{\curlyvee}$ and $LK^{\curlyvee'}$ are equivalent. [Hint: Probably, the lengthiest part of the argument is to show that $LK^{\curlyvee'}$ can emulate (K).]

We note that in the case of the three connective rules, the application of a rule to the same subalterns always leads to the same principal formula, or looking at the rule upside down, a principal formula is always deconstructed into the same subalterns. We might build from, let us say, $\mathcal{A}, \mathcal{B}, \mathcal{C}$ both $(\mathcal{A} \vee \mathcal{B}) \vee \mathcal{C}$ and $(\mathcal{A} \vee \mathcal{B}) \vee (\mathcal{A} \vee \mathcal{C})$—using that a sequent is a set of formulas. However, once a wff is given and we are deconstructing the wff, there is no need to keep track of several occurrences of $\mathcal{A}$, because $\mathcal{A}, \mathcal{A}, \neg\mathcal{A}$ is "not more an axiom" than $\mathcal{A}, \neg\mathcal{A}$ is.

Theorem 7.32. *If for every interpretation I in $\{T, U, F\}$, $I(\mathcal{A}) \neq F$, then $\mathcal{A}$ is provable in $LK^{\curlyvee}$.*

Proof: We assume that $\mathcal{A}$ is not provable in $LK^{\curlyvee}$, hence, in $LK^{\curlyvee'}$. We construct a proof-search tree by putting $\mathcal{A}$ in the bottom sequent, and then we build a tree on top of it.

If Γ is a sequent, then we select each formula $\mathcal{B}$ such that $\mathcal{B}$ is not a literal and $\mathcal{B} \in \Gamma$, and apply a rule to it. In each case, the parametric part of the sequent is $\Gamma - \{\mathcal{B}\}$. If the rule is (DM) then we add two new nodes above Γ, otherwise we add one node. We repeat the step at each level, for each sequent and for each formula that has not yet been selected below that node. Some of the branches may lead to a sequent that is an instance of the axiom, and then we may discontinue building that branch further.

Since $\mathcal{A}$ is not provable, at least one of the branches will terminate in a sequent Δ that has a subset of literals, let us say, $\mathcal{C}_1, \ldots, \mathcal{C}_n$. The sequence is not an instance of the axiom, hence, no pair of wff's among the $\mathcal{C}$'s is an atomic formula and its negation.

We define an interpretation I to assign F to the atomic wff's among the $\mathcal{C}$'s and T to the atomic wff's the negations of which are among the $\mathcal{C}$'s. (We may assign arbitrary values to any other atomic wff's, because they do not occur in $\mathcal{A}$ at all.)

Next, we prove that stepping from the set of literals (at the top) toward $\mathcal{A}$ (at the bottom), all wff's in all sequents take the value F. First, we deal with the principal formulas of the rules.

1. Each element of $\mathcal{C}_1, \ldots, \mathcal{C}_n$ is assigned F, obviously.
2. If the rule is (DN) and $I(\mathcal{D}) = F$, then $I(\neg\neg\mathcal{D}) = F$.
3. If the rule is $(\vee)$ and $I(\mathcal{D}) = F = I(\mathcal{E})$, then $I(\mathcal{D} \vee \mathcal{E}) = F$.
4. If the rule is (DM) and $I(\neg\mathcal{D}) = F$, then $I(\neg(\mathcal{D} \vee \mathcal{E})) = F$ and $I(\neg(\mathcal{E} \vee \mathcal{D})) = F$, because $I(\mathcal{D}) = T$, hence $I(\mathcal{D} \vee \mathcal{E}) = I(\mathcal{E} \vee \mathcal{D}) = T$.

There are no other possibilities, which means that if all the elements of Δ are F, then for each sequent Γ, $I(\bigvee \Gamma) = F$, including the bottom sequent. That is, $I(\mathcal{A}) = F$, as we wanted to prove. qed

Theorem 7.33. (Soundness of $LK^{\curlyvee}$ with cut) *If $\mathcal{A}$ is a theorem of the sequent calculus $LK^{\curlyvee}$ together with the cut rule, then for all interpretations I in $\{T, F\}$, $I(\mathcal{A}) = T$.*

Proof: The sequents are interpreted by clauses (5)–(6) from Definition 2.18. We prove the claim by induction on the structure of the proof of $\mathcal{A}$.

1. If $\mathcal{A}$ is an instance of the axiom, then it is $\mathcal{B} \vee \neg\mathcal{B}$, which always takes the value T.
2. If the last rule is $(\vee)$, then the claim holds by the interpretation of the sequences and the hypothesis.
3. If the last rule is (DM), then $I(\bigvee(\Gamma, \neg\mathcal{A})) = T$ and $I(\bigvee(\Gamma, \neg\mathcal{B})) = T$. By the definition of $\bigvee$ and I, either one of the elements of Γ is T, or otherwise, $\neg\mathcal{A}$ and $\neg\mathcal{B}$ are both T. If the former, then $I(\bigvee(\Gamma, \mathcal{C})) = T$ with arbitrary

$\mathcal{C}$, hence, with $\neg(\mathcal{A} \vee \mathcal{B})$. If the latter, then $I(\mathcal{A}) = F = I(\mathcal{B})$, hence, $I(\mathcal{A} \vee \mathcal{B}) = F$. Then $I(\neg(\mathcal{A} \vee \mathcal{B})) = T$. This is sufficient for $I(\bigvee(\Gamma, \neg(\mathcal{A} \vee \mathcal{B}))) = T$.
4. If the last step is by the cut rule, then $I(\bigvee(\Gamma, \mathcal{A})) = T$ and $I(\bigvee(\Delta, \neg\mathcal{A})) = T$. By the definition of $\bigvee$ and I, at least one element of each sequent takes the value T. Clearly, either $\neg\mathcal{A}$ or $\mathcal{A}$ is F. Let us assume that $I(\mathcal{A}) = F$. (The other case is symmetric.) Then at least one element of Γ is T, which is sufficient for $I(\bigvee(\Gamma, \Delta)) = T$, because of the definition of I and $\bigvee$.

We leave the remaining cases as an exercise. qed

Exercise 7.6.4. Finish the proof of the theorem. [Hint: There are four rules left to consider, (DN), (C), (W) and (K).]

Now we can gather the bits and pieces together for a proof of the admissibility of the cut rule in LK^{γ}.

Theorem 7.34. (Cut theorem for LK^{γ}) *The cut rule is* admissible *in* LK^{γ}.

Proof: Let us assume that $\mathcal{A}$ is not a theorem of LK^{γ} (without the cut). Then by Theorem 7.32, there is an interpretation in $\{T, U, F\}$ such that $I(\mathcal{A}) = F$. By Schütte's lemma, every amelioration I' of I preserves F, including all ameliorations that are interpretations in $\{T, F\}$. However, by Theorem 7.33, if $I'(\mathcal{A}) = F$, then LK^{γ} with the cut rule included does not prove $\mathcal{A}$. qed

Exercise 7.6.5.** Extend the above treatment to the whole LK^{γ}.

7.6.2 Consequence relation and subformula property

The cut rule is paramount to establishing that LK can imitate the workings of an axiom system for classical logic, which is formulated with the inference rule modus ponens. On the other hand, $\mathcal{A} \supset \mathcal{B}$ is simply $\neg\mathcal{A} \vee \mathcal{B}$. Ackermann denoted the third rule of his logic of strong implication by γ, which is—viewed classically—simply modus ponens or the so-called disjunctive syllogism. The logic of entailment, E does not include γ as an inference rule, and Anderson, in [6], posed as the first open problem the question whether γ is admissible in E. The (positive) answer was given by Meyer and Dunn in [137]. Having noticed the similarity between γ and cut, Dunn and Meyer in [80] applied a version of a proof of the admissibility of γ to prove the admissibility of the cut for LK^{γ}.[7] It seems to us that this proof should be known more widely, hence, we present the proof—with some small modifications—from [80].

A sequent can be viewed as a consequence relation. However, a deducibility relation can also be defined for LK^{γ}, and that is a different relation. $\mathbb{G} \Vdash \Gamma$, that is, the sequent Γ is *deducible* from the set of sequents $\mathbb{G}$ iff

[7] They follow Schütte in calling $\mathbf{K}_1$ a sequent calculus, which is almost identical to LK^{γ}. We renamed a slightly different version into LK^{γ} to emphasize the remarkable connection to the proof of the admissibility of the rule γ in relevance logics. We harmonize their notation with ours.

there is a tree rooted in Γ, which is like a proof tree, but its leaves may be elements of $\mathbb{G}$ (in addition to being instances of the axiom), and any variable quantified by an application of (UG) may occur in a leaf only if that is an instance of the axiom. (We use a different symbol, $\Vdash$ to emphasize that this is *not* the $\vdash$ of a sequent.) First of all, while a sequent is always finite, in the deducibility relation, $\mathbb{G}$ may be infinite. Every sequent is interpreted as a disjunction of its elements, hence, under that interpretation, $\Vdash$ turns into a relation between a set of wff's and a wff.

Lemma 7.35. (Properties of $\Vdash$) *The deducibility relation* possesses properties (1)–(4).[8]

(1) *If* $\Gamma \in \mathbb{G}$, *then* $\mathbb{G} \Vdash \Gamma$.

(2) *If* Γ *is an instance of* (id), *then* $\mathbb{G} \Vdash \Gamma$.

(3) *If* $\mathbb{G} \Vdash \Gamma$, *then* $\mathbb{G}, \mathbb{D} \Vdash \Gamma$.

(4) *If* $\mathbb{G} \Vdash \Gamma$ *and* $\Gamma, \mathbb{D} \Vdash \Delta$, *then* $\mathbb{G}, \mathbb{D} \Vdash \Delta$.

Proof: First of all, we stress that (4) is *not* the cut rule in LK^{γ}. We prove only this property. By assumption there is a tree ending in Γ and one ending in Δ. If the latter does not have a leaf Γ, then the claim is obvious, because it is not required that all the elements of $\mathbb{G}, \mathbb{D}$ occur as leaves in a tree. If Γ is a leaf in the tree ending in Δ, then we replace that leaf with the first tree. The joined tree ends in Δ and its leaves are axioms or elements of $\mathbb{G}, \mathbb{D}$. qed

Exercise 7.6.6. Prove that $\Vdash$ has (1)–(3) from the above lemma.

It may be interesting to note that $\Vdash$ lacks one of the properties of Tarski's classical consequence relation. Classically, there is a formula that implies all formulas. Cut-free proofs, however, have the subformula property, which immediately excludes this property, whether for a formula or for a sequent.

Lemma 7.36. (Renaming free variables) *If* $\mathbb{G} \Vdash \Gamma, \mathcal{A}(x)$ *and all occurrences of* x *in the leaves of the derivation tree are in instances of the axiom, then* $\mathbb{G} \Vdash \Gamma, \mathcal{A}(y)$, by renaming x to y *in the derivation tree.*

Proof: The proof is by induction on the derivation tree. The only rule that may be affected by the renaming of a variable is (UG), which has a restriction. Any other application of a rule remains an application of the same rule after the renaming. A variable that is quantified in (UG) may occur in a leaf, if that is an axiom. If that variable happens to be y, then that can be renamed to z. x is not the quantified variable, if it occurs in $\mathcal{A}(x)$. qed

We have two uses of the comma now, as set union (like in Lemma 7.35) and as concatenation of sequences of wff's (like in Lemma 7.36). In order to avoid confusing these uses, the concatenation of the sequents Γ and Δ (each of which is a sequence of wff's) is denoted by $\langle \Gamma, \Delta \rangle$.

[8]The comma indicates the union of sets as in sequent calculi based on sets of formulas.

Lemma 7.37. (Monotonicity of ,) *If* $\mathbb{G},\Gamma \Vdash \Delta$*, then* $\mathbb{G},\langle\Gamma,\Theta\rangle \Vdash \langle\Delta,\Theta\rangle$ *as well as* $\mathbb{G},\langle\Theta,\Gamma\rangle \Vdash \langle\Theta,\Delta\rangle$.

Proof: The proof is by structural induction on the derivation of Δ from $\mathbb{G},\Gamma$. The second part of the claim reduces to the first part by applications of (C).
1. If Γ is Δ, then the claim is obvious, because $\langle\Gamma,\Theta\rangle$ is $\langle\Delta,\Theta\rangle$.
2. If Δ is an axiom or $\Delta \in \mathbb{G}$, then $\langle\Delta,\Theta\rangle$ is by applications of (K) and (C).
3. If Δ is by a rule, then there are eight possibilities. We consider three of them.
3.1 If Δ is by the (DM) rule, then Δ is $\langle\Delta',\neg(\mathcal{A}\vee\mathcal{B})\rangle$. By hypothesis, $\mathbb{G},\langle\Gamma,\Theta\rangle \Vdash \langle\Delta',\neg\mathcal{A},\Theta\rangle$ and $\mathbb{G},\langle\Gamma,\Theta\rangle \Vdash \langle\Delta',\neg\mathcal{B},\Theta\rangle$. Having applied (C) sufficiently many times, we may apply (DM) to obtain $\langle\Delta',\Theta,\neg(\mathcal{A}\vee\mathcal{B})\rangle$, and further, $\langle\Delta',\neg(\mathcal{A}\vee\mathcal{B}),\Theta\rangle$, by more (C)'s.
3.2 If Δ is by (UG), then it is $\langle\Delta',\neg\exists x\,\mathcal{A}(x)\rangle$. By hypothesis, $\mathbb{G},\langle\Gamma,\Theta\rangle \Vdash \langle\Delta',\neg\mathcal{A}(y),\Theta\rangle$. If y happens to occur in Θ, then we rename y in the derivation of $\langle\Delta',\neg\mathcal{A}(y)\rangle$, which is possible by the previous lemma. With applications of (C) interspersed, as needed, we get $\mathbb{G},\langle\Gamma,\Theta\rangle \Vdash \langle\Delta',\neg\exists x\,\mathcal{A}(x),\Theta\rangle$.
3.3 If Δ is by (K), and so Δ is $\langle\Delta',\mathcal{A}\rangle$, then by inductive hypothesis, $\mathbb{G},\langle\Gamma,\Theta\rangle \Vdash \langle\Delta',\Theta\rangle$. By (C)'s and (K), we can derive $\langle\Delta',\mathcal{A},\Theta\rangle$. qed

Exercise 7.6.7. Finish the proof. [Hint: There are five subcases left from 3.]

Corollary 7.37.1. *If* $\mathbb{G},\Gamma \Vdash \Delta$ *and* $\mathbb{G},\Theta \Vdash \Lambda$*, then* $\mathbb{G},\langle\Gamma,\Theta\rangle \Vdash \langle\Delta,\Lambda\rangle$.

Proof: By two applications of the previous lemma, we have $\mathbb{G},\langle\Gamma,\Theta\rangle \Vdash \langle\Delta,\Theta\rangle$ and $\mathbb{G},\langle\Delta,\Theta\rangle \Vdash \langle\Delta,\Lambda\rangle$, from which we get $\mathbb{G},\langle\Gamma,\Theta\rangle \Vdash \langle\Delta,\Lambda\rangle$ by a property of $\Vdash$. qed

Observation. If $\Gamma \Vdash \Delta$ and $\Delta \Vdash \Gamma$, then $\langle\Gamma,\Theta\rangle \Vdash \langle\Delta,\Theta\rangle$ and $\langle\Delta,\Theta\rangle \Vdash \langle\Gamma,\Theta\rangle$.

The observation may be seen to hold, if we notice that mutual derivability of a pair of sequents implies that they have the same elements, that is, they are the same as sets of wff's. If a sequent comprises one formula, then we identify the wff and the sequent.

Definition 7.38. (Theories) A *theory* is a set of sequents that is closed under $\Vdash$. (1)–(7) define several kinds of theories.

(1) A theory $\mathfrak{T}$ is *prime* iff $\langle\mathcal{A},\mathcal{B}\rangle \in \mathfrak{T}$ implies that $\mathcal{A} \in \mathfrak{T}$ or $\mathcal{B} \in \mathfrak{T}$;

(2) a theory $\mathfrak{T}$ is *consistent* iff $\mathcal{A} \notin \mathfrak{T}$ or $\neg\mathcal{A} \notin \mathfrak{T}$;

(3) a theory $\mathfrak{T}$ is *complete* iff $\mathcal{A} \in \mathfrak{T}$ or $\neg\mathcal{A} \in \mathfrak{T}$;

(4) a theory $\mathfrak{T}$ is *normal* iff $\mathfrak{T}$ is prime and consistent;

(5) a theory $\mathfrak{T}$ is *$\exists$-prime* iff $\exists x\,\mathcal{A}(x) \in \mathfrak{T}$ implies $\mathcal{A}(y) \in \mathfrak{T}$, for some y;

(6) a theory $\mathfrak{T}$ is *rich* iff for all y, $\mathcal{A}(y) \in \mathfrak{T}$ implies $\neg\exists x\,\mathcal{A}(x) \in \mathfrak{T}$;

(7) a theory $\mathfrak{T}$ is *opulent* iff $\mathfrak{T}$ is normal and rich.

Theories are typically defined as sets of wff's, and using sequents instead introduces additional objects. This may appear objectionable to those who worry about Ockham's razor. However, the set of wff's and the set of sequents have the same cardinality and every sequent implies the wff by which it is interpreted. Also, there are finitely many wff's derivable from a sequent by applications of $(\vee)$ alone.

Exercise 7.6.8. Prove that an opulent theory of $LK^{\curlyvee}$ has all the other properties listed. [Hint: (1), (2), (4) and (6) are immediate from the definition; (3) and (5) have to be established.]

Theorem 7.39. *If $\mathcal{A}$ is not a theorem of $LK^{\curlyvee}$, then there is a* prime *and* rich theory $\mathfrak{T}$ *such that* $\mathcal{A} \notin \mathfrak{T}$.

Proof: The proof is by alternating Lindenbaum constructions and expansions of the language with new sets of variables. The consequence relation is not classical, therefore, instead of not allowing F to be deduced, a growing set of sequents has to remain excluded.

We define two operations on sets of sequents, $\mathcal{T}$ and $\mathcal{I}$. $\mathcal{T}(\mathbb{G}) = \{\,\Gamma\colon \mathbb{G} \Vdash \Gamma\,\}$, that is, $\mathcal{T}$ forms the deductive closure of a set of sequents. $\mathcal{I}(\mathbb{G})$ contains $\mathbb{G}$, and if $\Gamma, \Delta \in \mathcal{I}(\mathbb{G})$, then $\langle \Gamma, \Delta \rangle \in \mathcal{I}(\mathbb{G})$ too. In words, $\mathcal{I}$ is the least set of sequents obtained by finitely many iterations of concatenation of sequents.

1. Let $\mathfrak{T}_{0,0} = \mathcal{T}(\{\,\Gamma\colon \emptyset \Vdash \Gamma\,\})$, that is, the set of sequents that are provable in $LK^{\curlyvee}$, including all wff's that are theorems. Let $\mathfrak{R}_0 = \mathcal{I}(\{\,\mathcal{A}\,\})$. Let $\mathcal{E}_1, \mathcal{E}_2, \ldots$ be an enumeration of wff's. $\mathfrak{T}_{n+1,0} = \mathfrak{T}_{n,0}$, if there is a $\Gamma \in \mathfrak{R}_0$ such that $\mathfrak{T}_{n,0}, \mathcal{E}_n \Vdash \Gamma$. Otherwise, $\mathfrak{T}_{n+1,0} = \mathcal{T}(\mathfrak{T}_{n,0}, \mathcal{E}_n)$. The theory $\mathfrak{T}_{\omega,0} = \bigcup_{i=0}^{i\in\omega} \mathfrak{T}_{i,0}$, which is a theory itself.
2. If $\neg\exists x\, \mathcal{B}(x) \notin \mathfrak{T}_{\omega,0}$, then $\neg\mathcal{B}(y)$, with a new y is added to $\mathfrak{R}_0$, which yields $\mathfrak{R}_1$ by an application of $\mathcal{I}$. $\mathfrak{T}_{\omega,0}$ is extended to $\mathfrak{T}_{0,1}$ by adding all provable sequents in the extended language and by taking the deductive closure of the set.
3. Step **1** is repeated with $\mathfrak{R}_1$ in place of $\mathfrak{R}_0$, and $\mathfrak{T}_{0,1}$ in place of $\mathfrak{T}_{0,0}$. The result is theory $\mathfrak{T}_{\omega,1}$. Step **2** is repeated with the latter set and a new set of variables, generating $\mathfrak{R}_2$ and $\mathfrak{T}_{0,2}$, respectively. $\mathfrak{T} = \bigcup_{i=0}^{i\in\omega} \mathfrak{T}_{\omega,i}$, which is a theory, because $\Vdash$ is defined by finite derivation trees.
4. We wish to show now that none of the $\mathfrak{T}$'s overlaps any $\mathfrak{R}$'s; furthermore, the $\mathfrak{T}_{\omega,i}$'s are prime.

$\mathfrak{T}_{0,0}$ does not overlap with $\mathfrak{R}_0$, because the latter comprises $\mathcal{A}$, which is not a theorem by assumption. The construction precludes the addition of a sequent into $\mathfrak{T}_{n+1,0}$, if that would cause $\mathcal{A}$ to become derivable. We note that once we know that $\mathfrak{T}_{0,m}$ has an empty intersection with $\mathfrak{R}_m$, the construction guarantees that $\mathfrak{T}_{n+1,m}$ cannot include a sequent from $\mathfrak{R}_m$ either.

Let us assume that $\langle \mathcal{B}, \mathcal{C} \rangle \in \mathfrak{T}_{\omega,i}$, whereas $\mathcal{B}, \mathcal{C} \notin \mathfrak{T}_{\omega,i}$. The latter may be the case only if there are some $\mathcal{D}, \mathcal{G} \in \mathfrak{R}_i$ such that $\mathfrak{T}_{\omega,i}, \mathcal{B} \Vdash \mathcal{D}$ and $\mathfrak{T}_{\omega,i}, \mathcal{C} \Vdash \mathcal{G}$. But then $\mathfrak{T}_{\omega,i}, \langle \mathcal{B}, \mathcal{C} \rangle \Vdash \langle \mathcal{D}, \mathcal{G} \rangle$, which means that $\langle \mathcal{B}, \mathcal{C} \rangle \notin \mathfrak{T}_{\omega,i}$ either.

If $\mathfrak{T}_{\omega,i}$ is prime and does not overlap with $\mathfrak{R}_i$, then $\mathfrak{T}_{0,i+1}$ does not overlap with $\mathfrak{R}_i$ either, because if it would contain a sequent $\langle \mathcal{A}, \mathcal{B}, \neg\mathcal{C}_1(y_1), \ldots, \neg\mathcal{C}_n(y_n)\rangle$ from $\mathfrak{R}_i$, then by applications of the (C) and (UG) rules, $\langle \mathcal{A}, \mathcal{B}, \neg\exists x_1\, \mathcal{C}_1(x_1), \ldots, \neg\exists x_n\, \mathcal{C}_n(x_n)\rangle$. The $\mathcal{C}$'s are the wff's that are added in expanding $\mathfrak{R}_{i-1}$ to $\mathfrak{R}_i$. If the latter sequent is an element of $\mathfrak{T}_{\omega,i}$, then the instances of the $\mathcal{C}$'s could not have been added to $\mathfrak{R}_{i-1}$, as stipulated.

Lastly, $\mathfrak{T}$ is rich, because of step **2**, which guarantees that if $\neg\exists x\, \mathcal{B}(x) \notin \mathfrak{T}$, then $\neg\mathcal{B}(y)$ is in $\mathfrak{R}_i$ at some stage i, hence, $\neg\mathcal{B}(y) \notin \mathfrak{T}$. qed

For every sequent, there is a formula that interprets that sequent.

Definition 7.40. (Extensional valuation) Let $\mathfrak{T}$ be a prime rich theory. An *extensional valuation* v maps formulas into $\{T, F\}$, where T and F are the truth values, and it satisfies (1)–(4).

(1) If $\mathcal{A}$ is atomic, then $v(\mathcal{A}) = T$ iff $\mathcal{A} \in \mathfrak{T}$;

(2) $v(\neg\mathcal{A}) = T$ iff $v(\mathcal{A}) = F$;

(3) $v(\mathcal{A} \vee \mathcal{B}) = T$ iff $v(\mathcal{A}) = T$ or $v(\mathcal{B}) = T$;

(4) $v(\exists x\, \mathcal{A}(x)) = T$ iff for some y, $\mathcal{A}(y) = T$.

Lemma 7.41. (Completeness) *If $\mathfrak{T}$ is a prime rich theory and v is an extensional valuation, then* (1) $v(\mathcal{A}) = T$ implies $\mathcal{A} \in \mathfrak{T}$, *and* (2) $v(\mathcal{A}) = F$ implies $\neg\mathcal{A} \in \mathfrak{T}$.

Proof: (1) and (2) are proved in parallel by induction on the structure of $\mathcal{A}$.
1. If $\mathcal{A}$ is atomic, then by (1) in the definition of v, the first part of the claim is immediate. Also, if $v(\mathcal{A}) = F$, then $\mathcal{A} \notin \mathfrak{T}$. However, $\langle \mathcal{A}, \neg\mathcal{A}\rangle$ is provable, hence, an element of $\mathfrak{T}$, which is prime. Then, $\neg\mathcal{A} \in \mathfrak{T}$.
2. If $\mathcal{A}$ is $\exists x\, \mathcal{B}(x)$, then if v assigns T to $\mathcal{A}$, then $v(\mathcal{B}(y)) = T$, hence, by hypothesis, $\mathcal{B}(y) \in \mathfrak{T}$. But $\mathfrak{T}$ is a theory, hence $\exists x\, \mathcal{B}(x)$, the consequence of $\mathcal{B}(y)$ by (EG), is also in $\mathfrak{T}$. If $v(\exists x\, \mathcal{B}(x)) = F$, then $v(\mathcal{B}(y)) = F$ for all y, hence, by hypothesis, $\neg\mathcal{B}(y) \in \mathfrak{T}$, for all y. However, $\mathfrak{T}$ is rich, therefore, $\neg\exists x\, \mathcal{B}(x) \in \mathfrak{T}$.

The two other cases are left for the next exercise. qed

Exercise 7.6.9. Finish the proof by considering when $\mathcal{A}$ is $\neg\mathcal{B}$, and also when $\mathcal{A}$ is $\mathcal{B} \vee \mathcal{C}$.

Lemma 7.42. (Soundness) *For any formula $\mathcal{A}$, if it is a theorem of $LK^{\curlyvee}$, then* for any first-order valuation, $\mathcal{A}$ is T.

Exercise 7.6.10. Prove this lemma. [Hint: A first-order valuation is v from Definition 7.40 when $\mathfrak{T}$ and (1) are ignored.]

Theorem 7.43. *$\mathcal{A}$ is a theorem of $LK^{\curlyvee}$ iff $v(\mathcal{A}) = T$, for every extensional valuation.*

Proof: A combination of Theorem 7.39, Lemma 7.41 and Lemma 7.42 suffices for a proof of this theorem. Notice that if $\mathfrak{T}$ in the definition of extensional valuation is taken to be a set of some atomic formulas, then v is simply a first-order valuation. qed

Theorem 7.44. (Cut theorem) *If* $\langle \Gamma, \mathcal{A} \rangle$ *and* $\langle \Delta, \neg\mathcal{A} \rangle$ *are provable, then* $\langle \Gamma, \Delta \rangle$ *is* provable *too.*

Proof: $v(\bigvee(\langle \Gamma, \mathcal{A} \rangle)) = T$ and $v(\bigvee(\langle \Delta, \neg\mathcal{A} \rangle)) = T$, by soundness. By the definitions of $\bigvee$ and that of the extensional valuation, some $\mathcal{B}$, which belongs to $\langle \Gamma, \Delta \rangle$, takes the value T, which means that $v(\bigvee(\langle \Gamma, \Delta \rangle)) = T$. However, LK^{γ} is complete, hence, $\langle \Gamma, \Delta \rangle$ is provable. qed

One might think about the second half of the construction on a par with proofs of the admissibility of the rule γ. This view may be summarized by saying that *every prime rich theory has an opulent subtheory.*

7.7 Analytic cut

We have seen various formulations of the cut rule and a range of proofs of cut theorems. We look at classical logic again, with a version of the cut rule that is called *analytic cut*, which is a restriction of the cut rule.[9]

The analytic cut rule, for the propositional part of LK, looks like the single cut rule (on p. 26), but with the side condition that $\mathcal{C}$, the cut formula, must be a subformula of a formula in the lower sequent of the cut rule (i.e., in $\Theta_1, \Gamma, \Theta_2 \vdash \Delta_1, \Lambda, \Delta_2$).

The side condition is clearly a restriction that is not satisfied by all applications of the single cut in LK. A very simple example goes as follows.

$$\dfrac{\dfrac{p \vdash p}{p \vdash p, r} \qquad \dfrac{q \vdash q}{r, q \vdash q}}{p, q \vdash p, q}\ \text{cut}$$

Of course, we constructed this proof specifically to serve as an example, and the whole proof might appear to be artificial. Whatever the meaning of "artificial," it is completely obvious that $p, q \vdash p, q$ is provable in LK without any sort of cut.

The analytic cut rule is a *special instance* of the (unrestricted) cut rule; hence, from the cut theorem for LK, it follows that for any sequent that is provable with some applications of the analytic cut rule, it is provable without it too.

One of the uses of the cut rule is in the emulation of applications of modus ponens. A little thought makes clear that it is not obvious that the cut, which

[9] This version of the cut rule was introduced by Smullyan in [172].

is used in obtaining $\vdash \mathcal{B}$ from $\vdash \mathcal{A}$ and $\vdash \mathcal{A} \supset \mathcal{B}$, can be replaced by analytic cut. Indeed, the difficulty here matches the difficulty one faces when proofs are constructed from axioms with the detachment rule. But the admissibility of the cut means that $\vdash \mathcal{B}$ itself has a proof (without a cut), and that proof has the *subformula property*. The analytic cut rule combines the best of two worlds: it is a cut rule, but it does not violate the subformula property.

As it is mentioned in [172], Gentzen *did* consider trading in some of the rules of *LK* for axioms.[10] The introduction of analytic cut (perhaps, together with insights from the uniform calculi) led to the completion of Gentzen's project—the elimination of all operational rules in lieu of axioms. On one hand, this is the only sequent calculus for classical logic (that we consider) with respect to which we will not be interested in the admissibility of cut (because it is obviously false). On the other hand, this is the only axiomatic calculus (that we consider) with the subformula property.

This calculus is closely related to the version of *LK* in Section 3.2, where there are sets of formulas on both sides of the $\vdash$. We denote the calculus by LK_{ac}, where $_{ac}$ is to remind us of the role of the *a*nalytic *c*ut rule. The *axioms* and the *rule* (for the propositional part of the calculus) are as follows.

$$\Gamma, \mathcal{A} \vdash \mathcal{A}, \Theta \ \text{id}$$

$$\mathcal{A} \wedge \mathcal{B} \vdash \mathcal{A} \ {\scriptstyle \wedge\vdash_1} \qquad \mathcal{A} \wedge \mathcal{B} \vdash \mathcal{B} \ {\scriptstyle \wedge\vdash_2} \qquad \mathcal{A}, \mathcal{B} \vdash \mathcal{A} \wedge \mathcal{B} \ {\scriptstyle \vdash\wedge}$$

$$\mathcal{A} \vdash \mathcal{A} \vee \mathcal{B} \ {\scriptstyle \vdash\vee_1} \qquad \mathcal{B} \vdash \mathcal{A} \vee \mathcal{B} \ {\scriptstyle \vdash\vee_2} \qquad \mathcal{A} \vee \mathcal{B} \vdash \mathcal{A}, \mathcal{B} \ {\scriptstyle \vee\vdash}$$

$$\mathcal{A}, \neg\mathcal{A} \vdash \quad {\scriptstyle \neg\vdash} \qquad \vdash \mathcal{A}, \neg\mathcal{A} \ {\scriptstyle \vdash\neg}$$

$$\vdash \mathcal{A}, \mathcal{A} \supset \mathcal{B} \ {\scriptstyle \vdash\supset_1} \qquad \mathcal{A} \vdash \mathcal{B} \supset \mathcal{A} \ {\scriptstyle \vdash\supset_2} \qquad \mathcal{A}, \mathcal{A} \supset \mathcal{B} \vdash \mathcal{B} \ {\scriptstyle \supset\vdash}$$

$$\frac{\Gamma \vdash \mathcal{C}, \Delta \qquad \Theta, \mathcal{C} \vdash \Lambda}{\Theta, \Gamma \vdash \Lambda, \Delta} \ \text{cut}^{\oslash}$$

The $^{\oslash}$ indicates that the cut rule is restricted, namely, it is the analytic version of the cut rule.

It is easy to see that the axioms are always true, when we use the translation τ from Definition 2.18 of sequents into formulas. (Some of the axioms, for instance, $(\vdash\wedge)$ and $(\vee\vdash)$ become instances of $\mathcal{C} \supset \mathcal{C}$.) The implicational axioms are obviously true too, though their shape may be somewhat unexpected. (For instance, $\tau(\vdash \mathcal{A}, \mathcal{A} \supset \mathcal{B})$ is $(\mathcal{A} \supset \mathcal{A}) \supset (\mathcal{A} \vee (\mathcal{A} \supset \mathcal{B}))$.)

[10] See [91, 2.2], where the conjunction, the disjunction and the negation rules, as well as $(\forall\vdash)$, $(\vdash\exists)$ and $(\supset\vdash)$ are replaced by axioms.

A question that has a less obvious answer is whether all (propositional) theorems of *LK* are provable. A way to establish that no theorems are lost is by demonstrating that all the connective rules of *LK* are *derived rules* in the above system.

For example, given $\mathcal{A}, \Gamma \vdash \Delta$, we want to obtain $\mathcal{A} \wedge \mathcal{B}, \Gamma \vdash \Delta$. By $(\wedge\vdash_1)$ and $(\text{cut}^{\oslash})$, we have the target sequent in one step. The application of the analytical cut rule is justified, because $\mathcal{A}$ is a subformula of $\mathcal{A} \wedge \mathcal{B}$.

Exercise 7.7.1. Check that all the connective rules of *LK* are derived rules. [Hint: Note the place and role of the implicational axioms.]

The notion of a subformula in the presence of the quantifiers is more complicated than in propositional logic, and this forces certain complications in extending the above calculus. In particular, a wff $\mathcal{A}(x)$ is a subformula of the quantified wff's $\forall y\, \mathcal{A}(y)$ and $\exists y\, \mathcal{A}(y)$. [172] adds a set of parameters to a sequent. In the formulation of *LK*, the role of parameters is played by variables; nonetheless, we will adopt the term "parameterized" here.

Definition 7.45. (Parameterized sequents) A *parameterized sequent* is a pair of (finite) sets of formulas and a (finite) set of variables.

The set of variables pertains to the whole sequent; hence, attaching it to the $\vdash$ seems to be a good notational choice. For example, if $\{x, y, z\}$ is the set of variables for the sequent $\Gamma \vdash \Delta$, then we will write this as $\Gamma \vdash^{x,y,z} \Delta$. As a notational convention, we omit $\emptyset$ or any similar decoration on the turnstile, when the set of variables is empty.

The *axioms* and the *rule* of LK_{ac} comprise the *axioms* for the propositional calculus above together with the following *four axioms* and a new version of the *analytic cut* rule.

$$\forall x\, \mathcal{A}(x) \vdash \mathcal{A}(y) \quad {\scriptstyle \forall\vdash} \qquad\qquad \mathcal{A}(y) \vdash^{y} \forall x\, \mathcal{A}(x) \quad {\scriptstyle \vdash\forall^{\oslash}}$$

$$\exists x\, \mathcal{A}(x) \vdash^{y} \mathcal{A}(y) \quad {\scriptstyle \exists\vdash^{\oslash}} \qquad\qquad \mathcal{A}(y) \vdash \exists x\, \mathcal{A}(x) \quad {\scriptstyle \vdash\exists}$$

$$\frac{\Gamma \vdash^{V_1} \mathcal{C}, \Delta \qquad \Theta, \mathcal{C} \vdash^{V_2} \Lambda}{\Theta, \Gamma \vdash^{V} \Lambda, \Delta}\ \text{cut}^{\oslash}$$

The axioms that are marked with $^{\oslash}$ as well as the cut rule have to satisfy certain side conditions. In the case of the axioms $(\exists\vdash)$ and $(\vdash\forall)$, the variable y cannot occur in $\mathcal{A}(x)$. For the cut rule to be applicable, either $V_1 = \emptyset$ or $V_2 = \emptyset$, and then, $V \subseteq V_2$ or $V \subseteq V_1$, respectively, where $z \in V$ only if z has a free occurrence in a formula in the lower sequent $\Theta, \Gamma \vdash \Lambda, \Delta$. (We also assume the usual stipulations about substitutions avoiding variable clashes.)

The soundness of LK_{ac} is straightforward. $\mathcal{A} \vdash^{x_1,\ldots,x_n} \mathcal{B}$ is interpreted as the wff $\exists x_1 \ldots \exists x_n\, (\mathcal{A} \supset \mathcal{B})$.

Some of the axioms of LK_{ac} are often taken to be axioms in axiomatic calculi with $\vdash$ traded for $\supset$. (For example, then $(\forall\vdash)$ is (A4) in K in Section 2.2.)

The question of completeness may be reduced again to the question of the derivability of the rules of LK. First of all, there is no change in demonstrating the derivability of the connective rules—despite the change in the notion of a sequent. We look at the two rules for $\forall$.

Let us assume that we are given $\mathcal{A}(y), \Gamma \vdash \Delta$. By the axiom $(\forall\vdash)$, and by $(\text{cut}^{\oslash})$, we obtain $\forall x\, \mathcal{A}(x), \Gamma \vdash \Delta$.

For $(\vdash\forall^{\oslash})$, let the starting sequent be $\Gamma \vdash \Delta, \mathcal{A}(y)$, with the condition that y does not occur in Γ, Δ. We select a suitable instance of the axiom $(\vdash\forall^{\oslash})$, $\mathcal{A}(y) \vdash^{y} \forall x\, \mathcal{A}(x)$. The set of variables in the left premise of the application of the cut is empty, and y does not occur in Γ, Δ; hence, we get $\Gamma \vdash^{\emptyset} \Delta, \forall x\, \mathcal{A}(x)$ (where $^{\emptyset}$ is added for emphasis).

Exercise 7.7.2. Show that $(\exists\vdash^{\oslash})$ and $(\vdash\exists)$ are derived rules in LK_{ac}. [Hint: These rules are duals of the rules for $\forall$, which we dealt with above.]

Exercise* 7.7.3. Investigate whether there are interesting applications of LK_{ac} in the meta-theory of FOL.

The analytic cut rule has proved useful in sequent calculus formalizations of modal logics, where the cut theorem is often not provable for natural formulations of well-known modal logics. We briefly mentioned some results about B and $S5$ in Section 4.7, and further applications of the analytic cut rule for normal modal logics may be found in [176].

7.8 Consequences of the cut theorem and uses of the cut rules

We have already mentioned some uses and applications of the cut theorem, and we detail some applications in Chapter 9. In this section, we summarize the consequences of the cut theorem. Sequent calculi, as well as cut rules, come in a great variety, hence, not all logics possess all the properties.

Consistency. If every formula or sequent is provable in a system, then that system is not good for much. The *existence of cut-free proofs* for provable sequents imparts consistency in the sense that there is a sequent that is not provable.

For instance, in LK and LJ, the sequent $\vdash$ is not provable. In many consecution calculi, where $\vdash$ is not a consecution, a consecution of the form $p \vdash q$ (where p and q are distinct propositional variables) can show that the calculus is consistent.

Exercise 7.8.1. Select a handful of calculi from earlier chapters and verify that a sequent that is not provable may be found.

Subformula property. The subformula property may sound less grandiose than consistency (or it may be thought of to be a component of that). However, this concept is deeply characteristic of sequent calculi and much liked. It plays into how consistency is shown, and it gives a constructive feel to proofs and proof search. This property is a manifestation of the *control over proofs* that we gain when we use a sequent calculus.

Conservative extensions. Logicians are often interested in full-scale logics with plenty of connectives and quantifiers. However, investigating fragments of logics provides a better understanding of the whole logic, and sometimes limiting the vocabulary is helpful for practical purposes. From the point of view of fragments, the whole logic is a *conservative extension*, because the extension does not produce new theorems or provable sequents in the old vocabulary.

Sequent calculi can facilitate establishing such relations between calculi. If one system is properly included in the other (with respect to its axioms and rules), and the cut theorem holds for the extension and has a reasonable proof, then it is conservative. By a reasonable proof, we mean that if a proof step concerns a certain logical particle, let us say, $\to$ (because the rule is $(\to\vdash)$), then other logical particles are not required or enter into that step.

Exercise 7.8.2. Find calculi in Chapters 4 and 5, for which such a relationship obtains.

Algebraization. Forming the *Lindenbaum algebra* of a logic, especially of quantifier-free logics, allows certain simplifications in the meta-theory as well as opens the door for applications of algebraic methods and theorems. This procedure goes through finding an equivalence relation that is a congruence, that is, formulas that are in the equivalence relation cannot be distinguished by the logical particles. A usual way to guarantee that this holds is to prove what is called the *replacement theorem*. In the case of sequent calculi, the proof of the latter typically requires applications of the cut rule.

Proof of equivalence to axiomatic calculi. K is formulated with one inference rule, modus ponens, and with a proof rule, universal generalization. Emulating the latter rule in LK is easy. However, the minor premise of detachment may be a wff that is completely disjoint from the conclusion. Thus, to prove that every theorem of K is a theorem of LK, we need to employ the cut rule. This situation is not unique to classical logic. If an axiom system contains the detachment rule—and many do—then it is nearly certain that the cut rule has to be utilized in showing that the sequent calculus is equivalent to its axiomatic counterpart. This is true for LJ in Chapter 3, as well as for many of the calculi in Chapters 4 and 5.

Disjunction property. Some calculi have disjunction as a connective and they allow exactly one or at most one formula on the right-hand side of the

$\vdash$. With the availability of cut-free proofs, the calculus is likely to have the disjunction property, which promises a *constructive meaning* for disjunction and sometimes allows a simplification in the relational semantics of the logic. We briefly considered this property for LJ in Section 3.6.

Exercise 7.8.3. Are there logics (other than intuitionistic logic), for example, in Chapters 4 and 5, that have this property?

Decidability. Some proofs of decidability for logics formalized by sequent calculi use that all provable sequents have cut-free proofs. However, decidability is not simply a consequence of the cut theorem, indeed, even LK itself is not decidable. We detail some decidability proofs in Section 9.1.
Interpolation. The so-called extended Hauptsatz is Theorem 2.1 in [92]. Roughly, it says that the proof of a wff in prenex normal form in LK may be transformed into a proof, in which the top part is purely propositional (and followed by the quantificational steps). William Craig in [65], used this theorem to introduce a calculus for classical logic, in which he could show that implicational theorems of classical logic (except in two special cases) have an *interpolant*. Theorem 5, in [65], is what has become known as the *Craig interpolation lemma*. If $\mathcal{A} \supset \mathcal{B}$ is a theorem of classical logic, where $\mathcal{A}$ and $\mathcal{B}$ have a predicate symbol in common, then there is a $\mathcal{C}$ such that $\mathcal{A} \supset \mathcal{C}$ and $\mathcal{C} \supset \mathcal{B}$ are theorems and $\mathcal{C}$ is a wff in the common vocabulary of $\mathcal{A}$ and $\mathcal{B}$.
To have *and* not to have cut. We typically formulated the cut theorem as stating the *admissibility* of one or another formulation of the cut rule. We started a listing in this section by praising the existence of cut-free proofs, which may leave one wondering why all the fuss about the cut rule and the cut theorem. Why not simply leave the cut rule out of all the calculi? Beyond the mere convenience of using the cut (as demonstrated by Boolos in [49]), the availability of the cut is almost always needed for the replacement theorem and to establish which logic is formalized by a sequent calculus. The importance of the cut theorem, to state it quickly, is to have our cut-free cake and to be able to eat (or at least, to cut) it too.

Chapter 8

Some other proof systems

Axiomatic systems are the oldest proof systems, but they are also the most onerous ones when it comes to finding a proof for a statement. Axioms have their appeal, especially if they are thought to encapsulate some essential, basic, self-evident truths about some area of investigation or some set of objects. Axiomatic systems are not close relatives of sequent calculi, but some other proof systems are.

This chapter briefly introduces some prototypical examples of proof systems that are readily seen to be connected to sequent calculi—at least if scrutinized from an appropriate angle. An obvious start is to consider *natural deduction systems.* Not only were they invented at the same time as sequent calculi were, but there is an extensive literature on defining and clarifying the relationship between sequent calculi and natural deduction systems.

We consider some *tableaux* and *resolution* systems too. These turn out to be the proof systems that are closest to sequent calculi; somebody even might contend that they are versions of sequent calculi. The connections between sequent calculi, tableaux and resolution are fairly easy to see in the case of classical propositional logic. However, resolution and tableaux gained an existence of their own, and especially the latter type of formalism turned out to be applicable to a wide range of logics. The connection between tableaux and sequent calculi for some non-classical logics is somewhat moot.

8.1 Natural deduction systems

Natural deduction systems, in particular NK and NJ, were invented by Gentzen in [89]. In [91, p. 288], he says that "I intended, first of all, to set up a formal system which came as close as possible to actual reasoning." Although he himself noted a certain complication in NK, the naturalness of the natural deduction systems appears to have convinced generations of people that these are the best-suited proof systems to be used and taught.[1]

[1]Some philosophers disagree, e.g., J. von Plato in [151] suggests that the natural deduction rules of NJ and NK are neither well motivated, nor in a sufficiently general form.

The calculi LK and LJ were introduced right after NK and NJ, with the idea that there would be *no assumptions* in them. This seems to explain why they are named *logistic* calculi. (Of course, as we mentioned earlier, provable sequents can be viewed as instances of the logical consequence relation with a finite set of premises.) Gentzen also found it pleasing that classical logic formulated as LK exhibits a *symmetry* with respect to the two sides of $\vdash$, unlike the set of rules in NK. Another advantage the sequent calculi have is that they separate the *structural rules* that are applied in an inference from the connective and quantifier rules.[2]

Natural deduction systems have a considerable pre-history before Gentzen as well as rich developments after Gentzen, including attempts to make precise his idea about a correspondence between the natural deduction and the sequent calculi, for one and the same logic. We will limit our consideration here to the natural deduction systems for classical and intuitionistic logic, and provide pointers for other natural deduction systems.

The *natural deduction systems* NJ (for intuitionistic) and NK (for classical logic) include the connectives $\neg$, $\wedge$, $\vee$ and $\supset$, the two quantifiers $\forall$ and $\exists$, as well as the propositional constant $\boldsymbol{F}$. (Unlike in the sequent calculi LK and LJ, the constant $\boldsymbol{F}$ enters into the formulation of the rules in NJ and NK.) The *rules* of NJ and NK are the following.

$$\frac{\mathcal{A} \quad \mathcal{B}}{\mathcal{A} \wedge \mathcal{B}}\ I\wedge \qquad \frac{\mathcal{A} \wedge \mathcal{B}}{\mathcal{A}}\ E\wedge_1 \qquad \frac{\mathcal{A} \wedge \mathcal{B}}{\mathcal{B}}\ E\wedge_2$$

$$\frac{\mathcal{A}}{\mathcal{A} \vee \mathcal{B}}\ I\vee_1 \qquad \frac{\mathcal{B}}{\mathcal{A} \vee \mathcal{B}}\ I\vee_2 \qquad \frac{\mathcal{A} \vee \mathcal{B} \quad \begin{matrix}[\mathcal{A}] \\ \mathcal{C}\end{matrix} \quad \begin{matrix}[\mathcal{B}] \\ \mathcal{C}\end{matrix}}{\mathcal{C}}\ E\vee$$

$$\frac{\begin{matrix}[\mathcal{A}] \\ \mathcal{B}\end{matrix}}{\mathcal{A} \supset \mathcal{B}}\ I\supset \qquad \frac{\mathcal{A} \quad \mathcal{A} \supset \mathcal{B}}{\mathcal{B}}\ E\supset$$

$$\frac{\begin{matrix}[\mathcal{A}] \\ \boldsymbol{F}\end{matrix}}{\neg\mathcal{A}}\ I\neg \qquad \frac{\neg\neg\mathcal{A}}{\mathcal{A}}\ E\neg \qquad \frac{\mathcal{A} \quad \neg\mathcal{A}}{\boldsymbol{F}}\ I\boldsymbol{F} \qquad \frac{\boldsymbol{F}}{\mathcal{A}}\ E\boldsymbol{F}$$

$$\frac{\mathcal{A}(y)}{\forall x\, \mathcal{A}(x)}\ I\forall^{\oslash} \qquad \frac{\forall x\, \mathcal{A}(x)}{\mathcal{A}(y)}\ E\forall$$

$$\frac{\mathcal{A}(y)}{\exists x\, \mathcal{A}(x)}\ I\exists \qquad \frac{\exists x\, \mathcal{A}(x) \quad \begin{matrix}[\mathcal{A}(y)] \\ \mathcal{C}\end{matrix}}{\mathcal{C}}\ E\exists^{\oslash}$$

NK has all the above rules, whereas NJ omits $(E\neg)$. (If $(E\boldsymbol{F})$ is also dropped, then the so-called *minimal logic* results.)

[2] [167] discusses the history of sequent calculi with respect to this question.

The $^{\oslash}$ indicates that the rules are *restricted*. In $(I\forall)$, the variable y may not occur in $\forall x\, \mathcal{A}(x)$ or in a formula that $\mathcal{A}(y)$ depends on. Similarly, $(E\exists)$ is applicable, if $\mathcal{C}$ does not have an occurrence of y, and y does not occur in any premises, which $\mathcal{A}(y)$ depends on. The dependency of $\mathcal{A}$ on $\mathcal{B}$ means that if applications of rules are traced upward starting from $\mathcal{A}$, then at some stage, $\mathcal{B}$ is a premise of a rule, and it is not bracketed. The brackets also indicate that the formula does not need to appear in the derivation leading to the wff below it.

In natural deduction systems, the primary objects are derivations. A *derivation* is a tree comprising formula occurrences in which the leaves are assumptions, and the other formulas in the tree are justified by applications of the rules to formulas that occur on branches above the newly introduced formula occurrence. From the point of view of an application of a rule which contains bracketed formulas, the application of the rule is what allows the insertion of the *brackets*, and some call the bracketing *discharging*. A derivation is a *proof* iff all the assumptions have been discharged. A formula $\mathcal{A}$ is a *theorem* of NK or NJ iff there is a proof in which the root is $\mathcal{A}$.

Both NK and NJ have certain peculiarities. The only rule that does not eliminate (or introduce) exactly one occurrence of a connective or quantifier is the $(E\neg)$ rule. Of course, this rule is intuitionistically unacceptable, which would suggest that NJ should be a more pleasing system. However, NJ does not have a $(E\neg)$ rule at all, despite the fact that all other logical particles have at least one *introduction* and at least one *elimination* rule. Also, $\boldsymbol{F}$ is a component in three rules. (Gentzen himself noted a certain dissatisfaction with his own natural deduction systems.)

Example 8.1. Here are two trees; one is a proof, the other is a derivation (but not a proof).

$$\dfrac{\dfrac{\dfrac{\dfrac{[\mathcal{A}]_1}{\mathcal{A}\vee\mathcal{B}}\quad \dfrac{[\mathcal{A}]_1}{\mathcal{A}\vee\mathcal{C}}}{(\mathcal{A}\vee\mathcal{B})\wedge(\mathcal{A}\vee\mathcal{C})}{\scriptstyle I\vee,\,2\times;\ I\wedge,\,2\times}\qquad \dfrac{\dfrac{\dfrac{[\mathcal{B}\wedge\mathcal{C}]_1}{\mathcal{B}}}{\mathcal{A}\vee\mathcal{B}}\quad \dfrac{\dfrac{[\mathcal{B}\wedge\mathcal{C}]_1}{\mathcal{C}}}{\mathcal{A}\vee\mathcal{C}}}{(\mathcal{A}\vee\mathcal{B})\wedge(\mathcal{A}\vee\mathcal{C})}{\scriptstyle E\wedge,\,2\times;\ I\vee,\,2\times}\qquad [(\mathcal{B}\wedge\mathcal{C})\vee\mathcal{A}]_2}{(\mathcal{A}\vee\mathcal{B})\wedge(\mathcal{A}\vee\mathcal{C})_1}\,E\vee}{((\mathcal{B}\wedge\mathcal{C})\vee\mathcal{A})\supset((\mathcal{A}\vee\mathcal{B})\wedge(\mathcal{A}\vee\mathcal{C}))_2}\,I\supset$$

We added subscripts to the brackets and some formulas. The assumptions are discharged when the corresponding formulas are added to the proof tree.

$$I\supset\ \dfrac{E\supset\ \dfrac{E\supset\ \dfrac{\mathcal{A}\supset\mathcal{A}\supset\mathcal{A}\qquad [\mathcal{A}]_1}{\mathcal{A}\supset\mathcal{A}}\qquad [\mathcal{A}]_1}{\mathcal{A}}}{\mathcal{A}\supset\mathcal{A}_1}$$

This example shows that $(I\supset)$ may discharge more than one assumption (which must be occurrences of the same formula). On the other hand, a proof of $\mathcal{A} \supset \mathcal{B} \supset \mathcal{A}$ would include the introduction of $\mathcal{B} \supset \mathcal{A}$ without the assumption of $\mathcal{B}$, that is, that proof would illustrate a vacuous discharge.

The structural rules are fully absorbed by the rules and their applications, because branches in a derivation are treated as sets of formulas.

Exercise 8.1.1. Prove (1) $((\mathcal{A} \vee \mathcal{B}) \wedge (\mathcal{A} \vee \mathcal{C})) \supset ((\mathcal{B} \wedge \mathcal{C}) \vee \mathcal{A})$ (the converse of the wff in the above example), (2) $\mathcal{A} \supset \mathcal{B} \supset \mathcal{A}$ (the prototypical wff requiring thinning on the left in LK), and (3) $((\mathcal{A} \supset \mathcal{B}) \supset \mathcal{A}) \supset \mathcal{A}$ (Peirce's law, which requires both thinning and contraction on the right in LK). [Hint: The proof of Peirce's law in NK is quite different from its proof in LK.]

NK and NJ structure proofs into trees in which every node is a formula occurrence. Many natural deduction systems—for classical and non-classical logics—put derivations and proofs into a more streamlined or linear form. However, assumptions and their conclusions have to be separated out. A way to do this is to add *subproofs*, and allow proofs to contain not only formula occurrences, but also subproofs. A natural deduction system of this sort is often called a *Fitch-style natural deduction system*—after Frederic Fitch.

Further enhancements include *dependency sets*, and *dependency terms* attached to formulas, which proved to be useful in natural deduction systems for relevance logics. Several such calculi may be found in [7].

Beyond formulating natural deduction systems for various logics, there has been a lot of interest in the *normalization* of proofs and in the *relationship* between natural deduction and sequent calculus systems. It is remarkable that Gentzen hit upon two kinds of formal systems and seemed to have perceived them as related to each other. The relationship, however, is not as clear cut as could have been expected. The sequent calculi LK^F and LJ^F, that we described in Section 3.3, were introduced for the purpose of getting a better match between sequent calculi for classical and intuitionistic logics and natural deduction systems for those logics.[3]

8.2 Tableau systems

The classic book on tableaux for classical logic is Smullyan's [170]. The approach originated in the work of Evert Beth as a formalism using tables of formulas, which are not unlike the two sides of a sequent based on multisets. [170] makes the connection clear and provides several formulations of *analytic tableaux*. Arguably, tableaux is the easiest proof system to learn, perhaps, because the shape of a formula allows only one rule to be applied

[3] [182] provides a detailed treatment of normalization and of the relationships between classical and intuitionistic natural deduction systems and sequent calculi.

to the formula. The lack of choice seems to confer pedagogical advantages. This situation is unlike the one in LK or LJ, where a connective has left and right introduction rules, and some of those come in multiple forms.

The *informal motivation* behind proofs in tableaux systems is that they show that a set of formulas has no model. The rules ensure that expansion steps for a set of formulas preserve satisfiability, while the formulas added are subformulas of the formula to which a rule has been applied. (Strictly speaking, the subformula property holds for tableaux with signed formulas as we explain later, but in general, only an extra negation symbol is added here and there.) Analytic tableaux do not construct formulas (or use arbitrary new formulas) like natural deduction proofs do.

Tableaux proofs are *trees*, but structural rules are completely avoided by incorporating them into the concepts used in tableaux proofs (such as conjugation). The complete absorption of the structural rules makes the tableaux system especially well suited for classical logic. Loosely speaking, a proof in a tableaux system is a proof in a left-handed sequent calculus, with its tree flipped upside down, and at each step, only the subalterns added. (There is a connection to the uniform sequent calculi from Section 3.5 too.) After all these informal remarks, we turn to formal descriptions of three variants of analytic tableaux for classical logic.

The *analytic tableau system* TK_1 for classical logic is defined by the following set of rules. (The rules are given in a "fraction format," with an explanation of their application to follow.)

$$\frac{\mathcal{A} \land \mathcal{B}}{\mathcal{A}, \mathcal{B}}\,\land \qquad \frac{\neg(\mathcal{A} \land \mathcal{B})}{\neg\mathcal{A} \mid \neg\mathcal{B}}\,\neg\land \qquad \frac{\mathcal{A} \lor \mathcal{B}}{\mathcal{A} \mid \mathcal{B}}\,\lor \qquad \frac{\neg(\mathcal{A} \lor \mathcal{B})}{\neg\mathcal{A}, \neg\mathcal{B}}\,\neg\lor$$

$$\frac{\mathcal{A} \supset \mathcal{B}}{\neg\mathcal{A} \mid \mathcal{B}}\,\supset \qquad \frac{\neg(\mathcal{A} \supset \mathcal{B})}{\mathcal{A}, \neg\mathcal{B}}\,\neg\supset \qquad \frac{\neg\neg\mathcal{A}}{\mathcal{A}}\,\neg\neg$$

$$\frac{\forall x\, \mathcal{A}(x)}{\mathcal{A}(a)}\,\forall \qquad \frac{\neg\forall x\, \mathcal{A}(x)}{\neg\mathcal{A}(n)}\,\neg\forall^{\oslash} \qquad \frac{\exists x\, \mathcal{A}(x)}{\mathcal{A}(n)}\,\exists^{\oslash} \qquad \frac{\neg\exists x\, \mathcal{A}(x)}{\neg\mathcal{A}(a)}\,\neg\exists$$

We arranged the rules so that the rules that apply to wff's without the main connective being a negation are on the left-hand side in a pair of rules. The two rules marked with $^{\oslash}$ are *restricted*, namely, the name constant labeled as n has to be *new* at the point where the rule is applied in a proof.

Tableaux proofs are trees in which the nodes are occurrences of formulas and every node has zero, one or two children. In order to characterize which trees comprising formulas are proofs, we need some further concepts.

The rules fall into two categories. Those with a pair of formulas separated by a vertical bar are *disjunctive* rules, the others are *conjunctive* rules.

Definition 8.2. (Application of a rule) If $\mathcal{A}$ occurs on a branch, then an *application* of its rule *on a branch* (below $\mathcal{A}$, i.e., at a higher level in the tree than where $\mathcal{A}$ occurs) is an expansion of the branch as follows.

(1) If $\mathcal{A}$'s rule is *disjunctive*, then the branch is split by adding two new nodes, with the two lower formulas in the rule, as children of the leaf ending the branch;

(2) if $\mathcal{A}$'s rule is *conjunctive*, then the branch is extended by one or two nodes, with the lower formula (or formulas) in the rule, as the child (and its child) of the leaf ending the branch.

If $\mathcal{A}$ occurs in a tree, then an *application* of its rule means applications of its rule on all the branches to which $\mathcal{A}$ belongs.

The shape of the rules might already suggest that the first seven rules (i.e., those that involve the connectives only) always result in the same formulas, no matter how many times the rules are applied. This is not so for the last four (i.e., the quantifier) rules, where picking a new name constant in an application of the rule results in a new (distinct) formula. Then, it turns out that the application of a rule in a wholesale fashion (i.e., on all branches) is typically more useful in the propositional fragment (or in certain tableaux designed to have a special structure for a completeness proof), whereas the application of a rule on a branch sometimes leads to more comprehensible and smaller trees when quantified formulas are involved.

Definition 8.3. (Conjugates and closures) A pair of formulas of the form $\mathcal{A}$ and $\neg\mathcal{A}$ are *conjugates* (of each other). If there is a pair of conjugates on a branch in a tree, then that *branch is closed*. A *tree is closed* iff all its branches are closed.

As a terminological convention, we will call a branch *open* if it is not closed. It is obvious that if a branch is closed, then adding more formulas to the branch—whether splitting it or not at the same time—will keep it closed. This justifies the practice according to which as soon as it is recognized that a branch is closed, one stops expanding a branch. As a reminder that a *branch is closed*, we will put a $*$ below the last formula that we added.

A *proof* is a tree (with its root situated at the top) comprising formula occurrences, which (save the root) result by applications of rules, with all branches closed. The formula $\mathcal{A}$ is a *theorem* of TK_1 iff there is a proof in which the root is $\neg\mathcal{A}$. The formula $\mathcal{B}$ is a *consequence* of $\{\mathcal{A}_1, \ldots, \mathcal{A}_n\}$ iff $(\mathcal{A}_1 \wedge \ldots \wedge \mathcal{A}_n) \supset \mathcal{B}$ is a theorem.[4]

We give an example of a proof of a formula, namely, of $(\mathcal{A} \supset (\mathcal{B} \vee \mathcal{C})) \supset ((\mathcal{A} \supset \mathcal{B}) \vee (\mathcal{A} \supset \mathcal{C}))$. Figure 8.1 shows the tree that has in its root the *negation* of this formula.

The tableau system TK_1 is sound and complete for the class of interpretations of classical first-order logic that we presented in Section 2.4.[5]

[4]The notion of provability can be extended to infinite sets of premises if the notion of the proof is slightly modified so as to permit the insertion of nodes into a proof tree, which are formulas from the premise set.

[5][170, Ch. V, §3] presents a proof of completeness in an elegant and clear way.

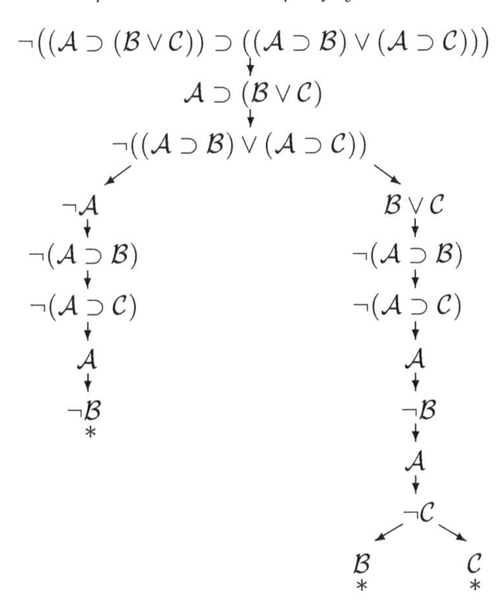

Figure 8.1. A proof in TK_1.

The definition of an application of a rule did not require that given two formulas on a branch, a rule should first be applied to the formula that is closer to the root. In Figure 8.1, on the branch that contains $\mathcal{B} \vee \mathcal{C}$, we did not apply a rule to this formula until the very end.

Exercise 8.2.1. Construct other tableaux proofs for $(\mathcal{A} \supset (\mathcal{B} \vee \mathcal{C})) \supset ((\mathcal{A} \supset \mathcal{B}) \vee (\mathcal{A} \supset \mathcal{C}))$. How many distinct proofs are there for this formula if to each formula in a proof its rule is applied at most once?

In general, applying disjunctive rules before conjunctive ones (if there is a choice) might lead to identical segments appearing on different branches—as the consecutive nodes $\neg(\mathcal{A} \supset \mathcal{B})$ and $\neg(\mathcal{A} \supset \mathcal{C})$ in Figure 8.1 illustrate. (More prosaically, one has to type or write the same formulas several times.) On the other hand, the rules in TK_1 do not require that at each step all the formulas be repeated, which have been accumulated by that point on the branch. In a sequent calculus, this would correspond to temporarily forgetting about the parametric formulas until they become principal formulas of a rule. (Recall that tableaux turn the proof tree upside down; hence in a sequent calculus we would be looking at the rules "applied" bottom-up, as in a proof-search tree.)

Exercise 8.2.2. Prove the following formulas by tableaux. (1) $(\mathcal{A} \supset \mathcal{B} \supset \mathcal{C}) \supset (\mathcal{D} \supset \mathcal{B}) \supset (\mathcal{A} \wedge \mathcal{D}) \supset \mathcal{C}$, (2) $\exists x\, \forall y\, R(x,y) \supset \forall x\, \exists y\, R(y,x)$ and (3) $\forall z\, (\forall x\, \forall y\, (R(x,y) \supset \neg R(y,x)) \supset \neg R(z,z))$.

The quantifier rules can be applied repeatedly to produce new formulas. It is advantageous to apply $(\exists)$ and $(\neg\forall)$ first (if possible). These rules are restricted, and the *name constant* that they introduce *must not have occurred* in any formula on the branch up to the point where the formula resulting from the application of the rule is added. In other words, it is not the place where $\exists x\,\mathcal{A}(x)$ or $\neg\forall x\,\mathcal{A}(x)$ occurs that determines the freshness of a name constant (i.e., whether it is new), rather what is decisive is the place where the rule is applied. (It is also unimportant if the name constant is used on another branch, as long as it is not in the overlapping section of the two branches.)

The distance between an occurrence of a formula and where its rule is applied may lead to some difficulties in seeing that a tree is a proof (unless the rules are applied in a prescribed fashion, for instance, in so-called *systematic tableaux*, which are useful in completeness proofs). A rule, in principle, can be applied even in the propositional case as many times as one wishes, hence, at any level in the proof tree, a rule could have been applied to any of the formulas on the branch, and as the branch grows so may the number of possibilities. Of course, this is a minor practical inconvenience (at most). It is *decidable* at any stage in the construction of a proof tree whether the expansion of the tree accords with the rules.

Exercise 8.2.3. Find some additional concepts for tableaux so that the rules can be used as a decision procedure for propositional classical logic.

Exercise 8.2.4. Use the concepts and procedure from the previous exercise to decide whether the following formulas are theorems of classical logic. (1) $(\mathcal{A}\wedge(\neg\neg\neg\mathcal{A}\vee(\neg\mathcal{A}\supset\mathcal{C})))\supset(\neg\mathcal{A}\vee\neg\mathcal{C})$ and (2) $(\mathcal{A}\supset\mathcal{B})\supset\mathcal{A}\supset\mathcal{B}\supset\mathcal{C}$.

Exercise* 8.2.5. Define formally a decision procedure based on Exercise 8.2.3, and prove that it is correct.

The tableau system TK_1, of course, is not a decision procedure for first-order logic. While in some cases (especially, after some practice of proving theorems with tableaux), we might conjecture that the necessary closure of all branches cannot be achieved, there is no effective procedure to make such a determination for arbitrary formulas. Informally, the culprits are the quantifier rules that introduce new name constants. The unrestricted rules can be used to introduce new name constants too, but as a tactic motivated by the goal to close a branch, the $(\forall)$ and $(\neg\exists)$ rules are hardly ever used in sensible proofs to introduce a new name constant. (An exception is when there are no name constants in the wff's on a branch and there are only universally quantified or negated existentially quantified formulas.) However, if an existentially quantified wff results after $(\forall)$ or $(\neg\exists)$ is applied, then by alternating the $(\forall)$ and $(\exists)$ rules, for instance, new and new instances of the same quantified formulas get added to the branch.

Exercise 8.2.6. Construct trees with the aim of proving (1) $\forall x\,\exists y\,R(x,y)\supset\exists y\,\forall x\,R(x,y)$ and (2) $\forall x\,\exists y\,(N(x)\supset(Q(y)\wedge x<y))\supset\exists x\forall y\,(Q(x)\wedge$

$(N(y) \supset y < x))$, where $<$ is used as a binary predicate in infix notation. [Hint: Apply the quantifier rules to each quantified formula a couple of times to get a feel for the difficulty that prevents closure.]

Of course, neither of the wff's in the exercise are theorems of first-order logic, hence there is no tableaux proof for them.

Exercise 8.2.7. Give first-order interpretations that make the two formulas above false. Can you define the models by lifting parts of the interpretation from the trees you constructed? [Hint: An easy interpretation for the wff in (2) can be constructed by taking N and Q to be "is a natural number" and "is a rational number," respectively.]

We likened TK_1 to a left-handed sequent calculus. The negation of classical logic facilitates moving all the formulas in a sequent to one or to the other side, but the extra negation symbols lead to a loss of the subformula property in the usual sense. A solution to this problem is to distinguish the negations that have the rest of the formula in their scope, and in effect, indicate that the wff (without the $\neg$) would appear on the left-hand side of the $\vdash$.

The *analytic tableaux system* TK_2 uses signed formulas. The signs are $\mathfrak{t}$ and $\mathfrak{f}$, and the set of signed formulas is obtained from the set of wff's by prefixing a sign to a wff, as in Definition 3.8 in Section 3.5. We include into the language four additional binary connectives, the *biconditional* ($\leftrightarrow$), the *exclusive disjunction* (∇), *Peirce's joint denial* ($\downarrow$) and *Sheffer's stroke* ($|$). We also give rules for the identity predicate.

$$\frac{\mathfrak{t}\neg\mathcal{A}}{\mathfrak{f}\mathcal{A}}\ \mathfrak{t}\neg \qquad \frac{\mathfrak{f}\neg\mathcal{A}}{\mathfrak{t}\mathcal{A}}\ \mathfrak{f}\neg$$

$$\frac{\mathfrak{t}(\mathcal{A}\land\mathcal{B})}{\mathfrak{t}\mathcal{A},\mathfrak{t}\mathcal{B}}\ \mathfrak{t}\land \qquad \frac{\mathfrak{f}(\mathcal{A}\land\mathcal{B})}{\mathfrak{f}\mathcal{A}\mid\mathfrak{f}\mathcal{B}}\ \mathfrak{f}\land \qquad \frac{\mathfrak{t}(\mathcal{A}\lor\mathcal{B})}{\mathfrak{t}\mathcal{A}\mid\mathfrak{t}\mathcal{B}}\ \mathfrak{t}\lor \qquad \frac{\mathfrak{f}(\mathcal{A}\lor\mathcal{B})}{\mathfrak{f}\mathcal{A},\mathfrak{f}\mathcal{B}}\ \mathfrak{f}\lor$$

$$\frac{\mathfrak{t}(\mathcal{A}\leftrightarrow\mathcal{B})}{\begin{array}{c|c}\mathfrak{t}\mathcal{A} & \mathfrak{f}\mathcal{A}\\ \mathfrak{t}\mathcal{B} & \mathfrak{f}\mathcal{B}\end{array}}\ \mathfrak{t}\leftrightarrow \qquad \frac{\mathfrak{f}(\mathcal{A}\leftrightarrow\mathcal{B})}{\begin{array}{c|c}\mathfrak{t}\mathcal{A} & \mathfrak{f}\mathcal{A}\\ \mathfrak{f}\mathcal{B} & \mathfrak{t}\mathcal{B}\end{array}}\ \mathfrak{f}\leftrightarrow \qquad \frac{\mathfrak{t}(\mathcal{A}\nabla\mathcal{B})}{\begin{array}{c|c}\mathfrak{t}\mathcal{A} & \mathfrak{f}\mathcal{A}\\ \mathfrak{f}\mathcal{B} & \mathfrak{t}\mathcal{B}\end{array}}\ \mathfrak{t}\nabla \qquad \frac{\mathfrak{f}(\mathcal{A}\nabla\mathcal{B})}{\begin{array}{c|c}\mathfrak{t}\mathcal{A} & \mathfrak{f}\mathcal{A}\\ \mathfrak{t}\mathcal{B} & \mathfrak{f}\mathcal{B}\end{array}}\ \mathfrak{f}\nabla$$

$$\frac{\mathfrak{t}(\mathcal{A}\downarrow\mathcal{B})}{\mathfrak{f}\mathcal{A},\mathfrak{f}\mathcal{B}}\ \mathfrak{t}\downarrow \qquad \frac{\mathfrak{f}(\mathcal{A}\downarrow\mathcal{B})}{\mathfrak{t}\mathcal{A}\mid\mathfrak{t}\mathcal{B}}\ \mathfrak{f}\downarrow \qquad \frac{\mathfrak{t}(\mathcal{A}\mid\mathcal{B})}{\mathfrak{f}\mathcal{A}\mid\mathfrak{f}\mathcal{B}}\ \mathfrak{t}\mid \qquad \frac{\mathfrak{f}(\mathcal{A}\mid\mathcal{B})}{\mathfrak{t}\mathcal{A},\mathfrak{t}\mathcal{B}}\ \mathfrak{f}\mid$$

$$\frac{\mathfrak{t}\forall x\,\mathcal{A}(x)}{\mathfrak{t}\mathcal{A}(a)}\ \mathfrak{t}\forall \qquad \frac{\mathfrak{f}\forall x\,\mathcal{A}(x)}{\mathfrak{f}\mathcal{A}(n)}\ \mathfrak{f}\forall^{\oslash} \qquad \frac{\mathfrak{t}\exists x\,\mathcal{A}(x)}{\mathfrak{t}\mathcal{A}(n)}\ \mathfrak{t}\exists^{\oslash} \qquad \frac{\mathfrak{f}\exists x\,\mathcal{A}(x)}{\mathfrak{f}\mathcal{A}(a)}\ \mathfrak{f}\exists$$

$$\frac{\begin{array}{c}\mathfrak{t}a=b\\ \mathfrak{t}\mathcal{A}(a)\end{array}}{\mathfrak{t}\mathcal{A}(b)}\ \mathfrak{t}{=_l} \qquad \frac{\begin{array}{c}\mathfrak{t}a=b\\ \mathfrak{t}\mathcal{A}(b)\end{array}}{\mathfrak{t}\mathcal{A}(a)}\ \mathfrak{t}{=_r} \qquad \frac{\begin{array}{c}\mathfrak{t}a=b\\ \mathfrak{f}\mathcal{A}(a)\end{array}}{\mathfrak{f}\mathcal{A}(b)}\ \mathfrak{f}{=_l} \qquad \frac{\begin{array}{c}\mathfrak{t}a=b\\ \mathfrak{f}\mathcal{A}(b)\end{array}}{\mathfrak{f}\mathcal{A}(a)}\ \mathfrak{f}{=_r}$$

In TK_2, tableaux are trees, however, the nodes are occurrences of *signed formulas*, rather than occurrences of formulas. A pair of formulas are *conjugates* of each other iff they are of the form $\mathfrak{t}\mathcal{A}$ and $\mathfrak{f}\mathcal{A}$. For any term a, $\mathfrak{f}a = a$ is a conjugate (of $\mathfrak{t}a = a$). Since $a = a$ is satisfied by all interpretations (unlike other atomic formulas), $\mathfrak{f}a = a$ alone suffices for the closure of a branch—even if $a = a$ does not occur on the branch (or in the whole tree). With these modifications, the notions of the *closure* of a branch and of the *closure* of a tableaux are as before.

The rules for $\leftrightarrow$ and ∇ combine binary branching and expanding a branch by two nodes. (An example below will illustrate the application of the $(\mathfrak{t} \leftrightarrow)$ rule.)

There are four identity rules, in each of which $a = b$ is signed with $\mathfrak{t}$. The two signed formulas above the horizontal line in the rules may occur in arbitrary order, however, they must occur on the same branch. Then the branch may be extended by the signed formula below the line.

$\mathcal{A}$ is a *theorem* of TK_2 iff it has a *proof*, that is, a closed tableau, which is a tableau with root $\mathfrak{f}\mathcal{A}$.

Example 8.4. Let us consider the example in Figure 8.2. We prove that $(1 \wedge 2) \supset 3$ is a theorem, where we use the numbers to abbreviate (1) $\forall x\,(\neg P(x) \leftrightarrow \neg Q(x))$, (2) $\exists x\,(Q(x) \wedge R(x))$ and (3) $\exists x\,((P(x) \wedge Q(x)) \wedge R(x))$.

The tree has four branches, each of which is closed. Hence, the closed tableau is a proof.

It is easy to see that if we would omit the first two nodes, then we would be left with a tree in which the root and the next two nodes can be viewed as the conclusion (i.e., (3)) prefixed with $\mathfrak{f}$ and the two premises (i.e., (1) and (2)) prefixed with $\mathfrak{t}$. This points toward how tableaux can be modified to define a syntactic consequence relation in addition to proving theorems. The elements of a premise set (with $\mathfrak{t}$) may be allowed to appear as nodes, though those nodes are, obviously, not obtained by applications of the rules.

Exercise 8.2.8. Scrutinize the proof in Figure 8.2 and determine the dependencies between the signed formula occurrences (i.e., annotate the tree).

Exercise 8.2.9. Prove, using TK_2, that (1) $\exists y\,\forall x\,R(x,y) \supset \forall x\,\exists y\,R(x,y)$ and (2) $\exists x\,\forall y\,(N(y) \supset (Q(x) \wedge y < x)) \supset \forall x\,\exists y\,(N(x) \supset (Q(y) \wedge x < y))$ are theorems. [Hint: Recall that $(\mathfrak{t}\exists)$ and $(\mathfrak{f}\forall)$ are restricted rules.]

Exercise 8.2.10. Prove that $=$ is reflexive, symmetric and transitive.

An advantage of having signed formulas is that the *subformula property* holds for TK_2 without identity with no side conditions. Another possibility would be to define subformulas so that if some occurrences of a in $\mathcal{A}(a)$ are replaced by b resulting in $\mathcal{A}(b)$, then the latter counts as a subformula of the former, and vice versa.

Exercise 8.2.11. Scrutinize the rules of TK_2, and try to determine if it would be sensible to extend the notion of subformulas.

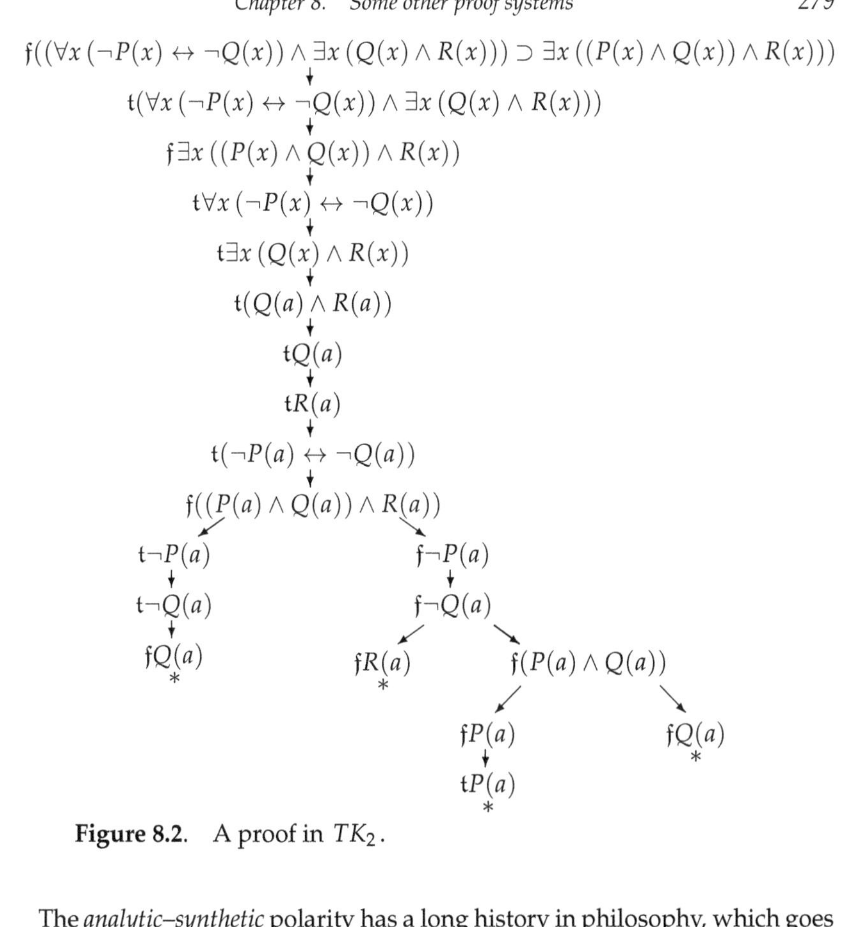

Figure 8.2. A proof in TK_2.

The *analytic–synthetic* polarity has a long history in philosophy, which goes back, at least, to I. Kant. The subformula property can be construed to mean that a tableaux system is *analytic*. *LK* has the subformula property—as long as the cut rule is not included. The parallel between tableaux and sequent calculi is completed by introducing *synthetic tableaux*. This version of TK_2 is obtained by adding the *split* rules.

$$\frac{\mathfrak{t}\mathcal{A}}{\mathfrak{f}\mathcal{B} \mid \mathfrak{t}\mathcal{B}}\ \text{split}_1 \qquad\qquad \frac{\mathfrak{f}\mathcal{A}}{\mathfrak{f}\mathcal{B} \mid \mathfrak{t}\mathcal{B}}\ \text{split}_2$$

The split rule can be formulated in different ways. We chose a form that allows the rule to be applied in a tree anywhere *after* the root. This allows us to emphasize that $\mathcal{B}$ is arbitrary with respect to $\mathcal{A}$. Since we explicitly mention the wff $\mathcal{A}$, the rule comes in two versions, depending on which sign $\mathcal{A}$ bears.

The claim that the synthetic version of TK_2 proves exactly the same formulas as TK_2 does mean that the split rules are admissible, which is the analog of the cut theorem in sequent calculi.

Exercise ** **8.2.12.** Prove that the split rules are admissible. [Hint: A way to go about such a proof is to translate TK_2 into a left-handed sequent calculus for classical logic with $=$ with signed formulas. Another route is to prove a soundness theorem for TK_2 with the split rules and then to show that if a wff is a logical validity with respect to the class of models for TK_2 with the split rules, then so it is in the class of models for TK_2.]

The benefits stemming from the use of signed formulas and from the generalization of rules can be combined into a *uniform* calculus. The connectives $\leftrightarrow$, ∇, $\downarrow$ and $|$ are enjoyable additions, and they are useful in some applications and even in certain meta-theorems. However, their rules deviate from the patterns exhibited by $\wedge$ and $\vee$.

We define the *uniform tableaux* TK_u to include the logical particles $\neg$, $\wedge$, $\vee$, $\forall$, $\exists$ and $=$. The rules fall into one of the following five types.

$$\frac{\alpha}{\alpha_1, \alpha_2}\ \alpha \qquad \frac{\beta}{\beta_1 \mid \beta_2}\ \beta \qquad \frac{\gamma}{\gamma(a)}\ \gamma \qquad \frac{\delta}{\delta(n)}\ \delta^{\oslash} \qquad \frac{\begin{array}{c}\varepsilon\\ \varepsilon(a)\end{array}}{\varepsilon(b)}\ \varepsilon$$

The *conjunction* and *disjunction* rules are easy to categorize. $(\mathfrak{t}\wedge)$ and $(\mathfrak{f}\vee)$ are (α) rules, whereas $(\mathfrak{f}\wedge)$ and $(\mathfrak{t}\vee)$ are (β) rules. In each of them, the subscripted letters stand for the immediate subformulas of the formula above with the same sign.

The *quantifier rules* are (γ) and (δ) type rules, the latter with a restriction. Namely, $(\mathfrak{t}\forall)$ and $(\mathfrak{f}\exists)$ are (γ) rules, and $(\mathfrak{f}\forall)$ and $(\mathfrak{t}\exists)$ are (δ) rules. Maybe, it is worth pointing out that, for instance, γ and $\gamma(a)$ have a slightly different relationship than $\mathcal{A}$ and $\mathcal{A}(a)$ have. Beyond the signs, γ includes a quantifier prefix, which is absent from $\gamma(a)$; similarly, for the δ's. (This notation is the same as in the uniform sequent calculi in Section 3.5.) The subformulas in these rules inherit the signs from the formula above the line.

It remains to classify the negation and identity rules. The *negation rules* could be viewed equivalently in either of the following forms.

$$\frac{\mathfrak{t}\neg\mathcal{A}}{\mathfrak{f}\mathcal{A}, \mathfrak{f}\mathcal{A}}\ \mathfrak{t}\neg_1 \qquad \frac{\mathfrak{f}\neg\mathcal{A}}{\mathfrak{t}\mathcal{A} \mid \mathfrak{t}\mathcal{A}}\ \mathfrak{f}\neg_1 \qquad \frac{\mathfrak{t}\neg\mathcal{A}}{\mathfrak{f}\mathcal{A} \mid \mathfrak{f}\mathcal{A}}\ \mathfrak{t}\neg_2 \qquad \frac{\mathfrak{f}\neg\mathcal{A}}{\mathfrak{t}\mathcal{A}, \mathfrak{t}\mathcal{A}}\ \mathfrak{f}\neg_2$$

Of course, these formulations are redundant in the sense that their application would expand a tree more that necessary, because duplicate signed formulas or even duplicate branches do not accelerate or facilitate the closing of a tableau. Choosing either the pair on the left or the pair on the right would allow us to place the rules into the (α) and (β) categories. It is also true that changing the sign, switches to another rule in the other category (as for $\wedge$ and $\vee$).

All the *identity rules* are (ε) rules. ε is the identity signed with $\mathfrak{t}$, whereas $\varepsilon(a)$ and $\varepsilon(b)$ are similar formulas, which differ in that some occurrences of a in the former are replaced by b in the latter.

The α–ε notation is particularly useful in concise proofs of meta-theorems. Neither the tableau systems, in general, nor their semantical adequacy is the focus of this book; hence, we leave the following as an exercise.

Exercise ** **8.2.13.** Prove the soundness and completeness of TK_u with respect to classical first-order interpretations. [Hint: The interpretations were defined in Chapter 2. [170, Ch. V] proves these results for the $=$-free fragment.]

Tableau systems have been designed for logics beyond classical logic. For example, Melvin Fitting adapted a version of tableaux, in which each node is labeled by a set of signed formulas, for intuitionistic logic. [130] gives tableaux for some relevance logics, and so does [138] for linear logic. We only mention here in some detail a version of analytic tableaux, which shows how tableaux for classical logic *integrate structural rules* such as thinning into the core concepts of tableaux.

In Section 4.6, we described a sequent calculus for **fde**, the logic of first-degree entailments. Now we present $T_{\mathbf{fde}}$, which is analytic tableaux comprising *a pair of trees*, and comes from [76].

First-degree entailments have only three connectives $\sim$, $\wedge$ and $\vee$, if we do not count $\rightarrow$, which is the main connective of a first-degree entailment formula. In other words, first-degree entailment formulas are of the form $\mathcal{A} \rightarrow \mathcal{B}$, where $\rightarrow$ does not occur in $\mathcal{A}$ or $\mathcal{B}$. The presence of these connectives matches perfectly the idea behind the (α)–(β) rules in TK_u. We use the signs and signed formulas as before.

The rules of $T_{\mathbf{fde}}$, a *tableau system for first-degree entailments*, are as follows.

$$\frac{\mathfrak{t}{\sim}\mathcal{A}}{\mathfrak{f}\mathcal{A}}\ \mathfrak{t}{\sim} \qquad\qquad \frac{\mathfrak{f}{\sim}\mathcal{A}}{\mathfrak{t}\mathcal{A}}\ \mathfrak{f}{\sim}$$

$$\frac{\mathfrak{t}(\mathcal{A}\wedge\mathcal{B})}{\mathfrak{t}\mathcal{A},\mathfrak{t}\mathcal{B}}\ \mathfrak{t}\wedge \qquad \frac{\mathfrak{f}(\mathcal{A}\wedge\mathcal{B})}{\mathfrak{f}\mathcal{A}\mid\mathfrak{f}\mathcal{B}}\ \mathfrak{f}\wedge \qquad \frac{\mathfrak{t}(\mathcal{A}\vee\mathcal{B})}{\mathfrak{t}\mathcal{A}\mid\mathfrak{t}\mathcal{B}}\ \mathfrak{t}\vee \qquad \frac{\mathfrak{f}(\mathcal{A}\vee\mathcal{B})}{\mathfrak{f}\mathcal{A},\mathfrak{f}\mathcal{B}}\ \mathfrak{f}\vee$$

$\mathcal{A} \rightarrow \mathcal{B}$ is a *theorem* of $T_{\mathbf{fde}}$ iff the trees started with $\mathfrak{t}\mathcal{A}$ and $\mathfrak{t}\mathcal{B}$ are such that every branch in the tree for $\mathfrak{t}\mathcal{A}$ *covers* some branch in the tree for $\mathfrak{t}\mathcal{B}$.

Definition 8.5. (Covering) If b is a branch in a tree in a tableaux, then b is *complete* iff for any signed formula on the branch, if there is a rule that is applicable to the signed formula, then the rule has been applied on the branch b.

A branch b_1 *covers* a branch b_2 iff they are both complete, and every signed atomic formula occurring on b_2 occurs on b_1.

To think of the covering relation informally, we can view b_1 as a conjunction of its signed atomic formulas, which implies b_2 as a conjunction of its signed atomic formulas. The requirement of completeness of the branches ensures that the covering does not happen for a trivial reason, namely, that the covered branch is unfinished (i.e., it has not yet been completed).

It may be interesting to note that conjugation and closure play no role in the definition of theoremhood. Dually, the splitting rules, are not allowed to be applied either.

It is convenient to turn the tree for $\mathrm{t}\mathcal{B}$ upside down when picturing a proof, and we will do this in our examples.

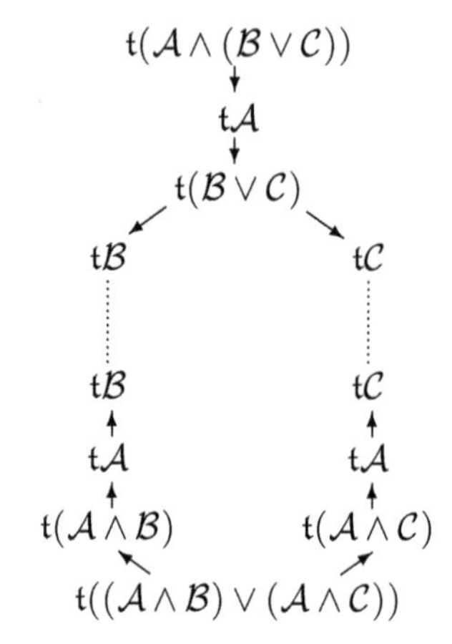

Figure 8.3. A proof in $T_{\mathbf{fde}}$.

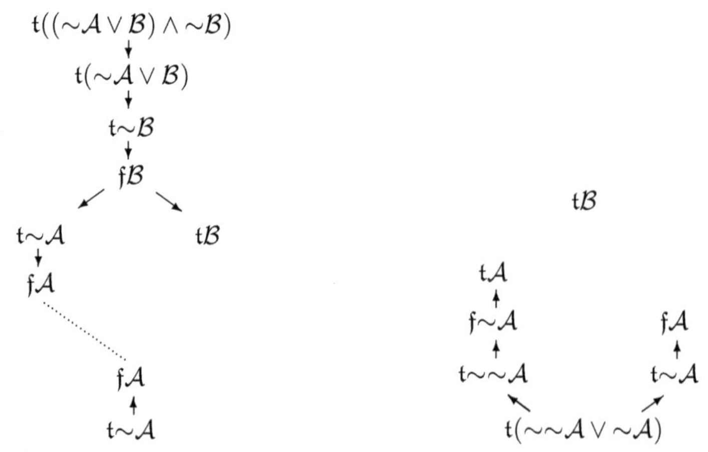

Figure 8.4. Two tableaux in $T_{\mathbf{fde}}$.

Example 8.6. Figure 8.3 depicts the proof of $(\mathcal{A} \wedge (\mathcal{B} \vee \mathcal{C})) \rightarrow ((\mathcal{A} \wedge \mathcal{B}) \vee (\mathcal{A} \wedge \mathcal{C}))$. The *dotted lines* indicate the covering between the branches.

The tableaux calculus $T_{\mathbf{fde}}$ is *sound* and *complete* for **fde**, and it is also a *decision procedure*.

Example 8.7. The two trees in Figure 8.4 are tableaux for $((\sim\mathcal{A} \vee \mathcal{B}) \wedge \sim\mathcal{B}) \to \sim\mathcal{A}$ and $\mathcal{B} \to (\sim\sim\mathcal{A} \vee \sim\mathcal{A})$, respectively. All the branches are complete, nonetheless, the tableaux are not proofs, because the covering relation fails to hold.

First-degree entailment can be formalized by tableaux comprising only one tree and utilizing a notion of conjugates and closures. However, we should be able to keep the information obtained from the antecedent and the consequent separate, which can be accomplished by *indexing the signs.* We do not go into the details here, which may be found in [32, §2].

8.3 Resolution systems

The *resolution calculus* for classical logic was invented by John A. Robinson in [161].[6] The *propositional* part of the calculus has a straightforward equivalent in the sequent calculus LK. The resolution calculus is based on applications of a single rule, called the *resolution rule,* which is applicable to a pair of *clauses.* A clause is a set of literals (i.e., atomic formulas or their negations), and it corresponds to a conjunct in a *conjunctive normal form* formula. A clause of the form $\{\neg\mathcal{A}_1, \ldots, \neg\mathcal{A}_n, \mathcal{B}_1, \ldots, \mathcal{B}_m\}$ may be thought of as the disjunction $\neg\mathcal{A}_1 \vee \ldots \vee \neg\mathcal{A}_n \vee \mathcal{B}_1 \vee \ldots \vee \mathcal{B}_m$.

Every wff of classical propositional logic has an equivalent wff, which is in *conjunctive normal form* (CNF), where the latter means that the wff is a conjunction of disjunctions of literals. If a formula is a disjunction of literals, then it is, by itself, considered to be a conjunction of disjunctions of literals. Similarly, a formula that is a literal is considered to be a disjunction, hence, being in CNF. For the purposes of CNF's (or disjunctive normal forms, DNF's), $\wedge$ and $\vee$ are viewed as *polyadic* connectives permitting the extremal cases of single conjunctions and single disjunctions (when desirable). This also justifies the omission of almost all of the parentheses.

Example 8.8. Let us assume that $\mathcal{A}$, $\mathcal{B}$ and $\mathcal{C}$ are atomic formulas. Then the following formulas are all in CNF.

(1) $(\mathcal{A} \vee \neg\mathcal{B} \vee \mathcal{C}) \wedge (\neg\mathcal{A} \vee \neg\mathcal{B} \vee \mathcal{C}) \wedge (\mathcal{A} \vee \mathcal{B} \vee \mathcal{C})$—This wff is in *complete* CNF, because each atomic wff occurs exactly once in each conjunct.

(2) $(\mathcal{A} \vee \neg\mathcal{A} \vee \mathcal{C} \vee \neg\mathcal{A}) \wedge (\mathcal{B} \vee \neg\mathcal{B})$—This wff is "overcomplicated," and obviously, it is equivalent to $\mathcal{A} \vee \neg\mathcal{A}$; nonetheless, it is in CNF.

(3) $\mathcal{A} \wedge \mathcal{B}$—Here $\mathcal{A}$ and $\mathcal{B}$ by themselves are considered to be disjunctions of literals.

[6] For a more contemporary treatment of the resolution calculus, see [123].

(4) $\mathcal{A} \vee \mathcal{B} \vee \neg\mathcal{C} \vee \mathcal{D}$—This disjunction of literals is itself a conjunction.

(5) $\mathcal{B}$—This atomic wff is also in CNF.

Exercise 8.3.1. Assume that $\mathcal{A}, \mathcal{B}, \mathcal{C}$ and $\mathcal{D}$ are atomic formulas. Find wff's in CNF that are equivalent to the given wff's. (a) $\neg(\mathcal{C} \wedge \neg\mathcal{A} \wedge \neg\neg\mathcal{B}) \vee (\neg\mathcal{A} \wedge \mathcal{B})$, (b) $\neg(((\mathcal{A} \wedge \neg\mathcal{B}) \vee (\mathcal{B} \wedge \neg\mathcal{C}) \wedge \neg\mathcal{C}) \vee \neg\mathcal{A})$, (c) $\neg(\neg(\mathcal{A} \wedge \mathcal{B}) \vee \neg(\neg\mathcal{A} \wedge \neg(\mathcal{B} \vee (\neg\mathcal{A} \wedge \neg\mathcal{B}))))$, (d) $(\neg(\mathcal{A} \wedge \mathcal{B}) \vee ((\mathcal{C} \wedge \neg\mathcal{A}) \wedge \neg(\neg\mathcal{A} \vee \neg\mathcal{B} \vee \mathcal{C} \vee \neg\neg\mathcal{D})))$ $\wedge \neg(\mathcal{A} \wedge \mathcal{C} \wedge \mathcal{B})$. [Hint: You may use any logical equivalences from classical logic, but you might try to narrow down the range of the equivalences to a handful of relatively simple ones.]

The resolution calculus is similar to tableaux in that a proof of a formula $\mathcal{A}$ starts with $\neg\mathcal{A}$. The next step is to generate a wff $\mathcal{B}$ that is logically equivalent to $\neg\mathcal{A}$ and is in conjunctive normal form. This step can be carried out fully algorithmically using some well-known logical equivalences, such as the De Morgan laws, in a predetermined order. The existence of CNF's (for any wff's) is provable; as is the fact that there are (deterministic) rewrite systems that terminate in a unique CNF for the formula. If no algorithm is specified, and repeated occurrences of a formula within a disjunction and within a conjunction are not excluded, then there is no unique equivalent wff in CNF for a given wff; moreover, there are infinitely many wff's.

Exercise 8.3.2. Suppose that a formula can contain only the connectives $\neg$, $\wedge$ and $\vee$. Define an algorithm, that is, a deterministic series of steps to rewrite a wff, in accordance with logical equivalences, to produce a wff in CNF.

Exercise* 8.3.3. Prove that the algorithm you defined in the previous exercise terminates with a unique wff in CNF that is equivalent to the starting wff.

A wff in CNF is turned into a *set of clauses* by taking the set of literals in a conjunct (which is a disjunction of literals) to be a clause (by omitting the disjunction symbols).

Example 8.9. The formulas in the previous example yield the following sets of clauses.

(1) $\{\{\mathcal{A}, \neg\mathcal{B}, \mathcal{C}\}, \{\neg\mathcal{A}, \neg\mathcal{B}, \mathcal{C}\}, \{\mathcal{A}, \mathcal{B}, \mathcal{C}\}\}$

(2) $\{\{\mathcal{A}, \neg\mathcal{A}, \mathcal{C}\}, \{\mathcal{B}, \neg\mathcal{B}\}\}$

(3) $\{\{\mathcal{A}\}, \{\mathcal{B}\}\}$

(4) $\{\{\mathcal{A}, \mathcal{B}, \neg\mathcal{C}, \mathcal{D}\}\}$

(5) $\{\{\mathcal{B}\}\}$

The *resolution rule* is as follows, where the $\mathcal{A}$'s and $\mathcal{B}$'s are literals.

$$\frac{\{\mathcal{A}_1, \ldots, \mathcal{A}_n, \mathcal{C}\} \qquad \{\mathcal{B}_1, \ldots, \mathcal{B}_m, \neg\mathcal{C}\}}{\{\mathcal{A}_1, \ldots, \mathcal{A}_n, \mathcal{B}_1, \ldots, \mathcal{B}_m\}}\ \text{res}$$

The rule allows taking a pair of clauses and generating a clause, when a pair of literals of the form $\mathcal{C}$ and $\neg\mathcal{C}$ occur in the clauses.

Notice that an application of this rule may lead to the *empty set* of literals. The *empty clause* is often denoted by $\Box$.[7] If the pair of clauses is of the form $\{\,\mathcal{C}\,\}$ and $\{\,\neg\mathcal{C}\,\}$, then the *resolvent*, that is, the resulting clause, is $\Box$. Indeed, having a pair of singleton clauses of this form is the only way to arrive at the empty clause. Given a wff in CNF, $\Box$ plays no role at the start, because no subformulas are empty. However, the aim of an attempted proof by resolution is to derive the empty clause.

A formula $\mathcal{A}$ is *provable* in the resolution calculus, when applications of the resolution rule to $\neg\mathcal{A}$ lead to the empty clause. A formula $\mathcal{B}$ is a *consequence* of the formulas $\mathcal{A}_1, \ldots, \mathcal{A}_n$ in the resolution calculus, when there is a derivation of the empty clause from the set of clauses obtained from $\mathcal{A}_1 \wedge \ldots \wedge \mathcal{A}_n \wedge \neg\mathcal{B}$.

Resolution proofs are often exhibited in a tree form—as the shape of the resolution rule already might have suggested. The resolution calculus can be thought of as a *right-handed sequent calculus* with sets (rather than sequences) of wff's. The resolution rule itself turns out to be the cut rule, with the cut formula restricted to be atomic.

<u>Example</u> 8.10. Let us assume that the negation of a formula yielded the next wff, which is in CNF (assuming again that the letters denote atomic wff's).

$$(\mathcal{A} \vee \mathcal{D} \vee \mathcal{A}) \wedge (\neg\mathcal{B} \vee \neg\mathcal{A}) \wedge (\mathcal{C} \vee \mathcal{B} \vee \mathcal{B}) \wedge (\neg\mathcal{D} \vee \neg\mathcal{B}) \wedge \neg\mathcal{C}$$

The series of applications of the resolution rule can be arranged in a tree with $\Box$ at the bottom.

$$\dfrac{\{\,\neg\mathcal{C}\,\} \qquad \dfrac{\dfrac{\{\,\neg\mathcal{B},\neg\mathcal{D}\,\} \qquad \dfrac{\{\,\mathcal{A},\mathcal{D}\,\} \quad \{\,\neg\mathcal{A},\neg\mathcal{B}\,\}}{\{\,\neg\mathcal{B},\mathcal{D}\,\}}}{\{\,\neg\mathcal{B}\,\}} \qquad \{\,\mathcal{B},\mathcal{C}\,\}}{\{\,\mathcal{C}\,\}}}{\Box}$$

Exercise 8.3.4. Prove the chain rule for implication, that is, prove that $\mathcal{A} \supset \mathcal{C}$ follows from $\mathcal{A} \supset \mathcal{B}$ and $\mathcal{B} \supset \mathcal{C}$. [Hint: This proof is special in the sense that each clause is used exactly once to obtain $\Box$.]

The example above also intended to emphasize that the clauses are sets of literals—unlike literals in a wff, where they may have multiple occurrences within a conjunct. A wff in CNF may contain conjuncts that yield the same set of literals, however, in the set of clauses it is recorded only once. (This is essentially the same sort of simplification that we already saw in Chapter 3 for *LK*.) The tree-form of a derivation may obscure the possibility of using a

[7]The same symbol is used in modal logic for necessity. However, it may be useful to think about something else here such as a box with nothing inside.

clause repeatedly. The above derivation could be presented equivalently as an expansion of the set of clauses.

$\{\{\mathcal{A},\mathcal{D}\},\{\neg\mathcal{A},\neg\mathcal{B}\},\{\neg\mathcal{B},\neg\mathcal{D}\},\{\mathcal{B},\mathcal{C}\},\{\neg\mathcal{C}\}\}$
$\{\{\mathcal{A},\mathcal{D}\},\{\neg\mathcal{A},\neg\mathcal{B}\},\{\neg\mathcal{B},\neg\mathcal{D}\},\{\mathcal{B},\mathcal{C}\},\{\neg\mathcal{C}\},\{\neg\mathcal{B},\mathcal{D}\}\}$
$\{\{\mathcal{A},\mathcal{D}\},\{\neg\mathcal{A},\neg\mathcal{B}\},\{\neg\mathcal{B},\neg\mathcal{D}\},\{\mathcal{B},\mathcal{C}\},\{\neg\mathcal{C}\},\{\neg\mathcal{B},\mathcal{D}\},\{\neg\mathcal{B}\}\}$
$\{\{\mathcal{A},\mathcal{D}\},\{\neg\mathcal{A},\neg\mathcal{B}\},\{\neg\mathcal{B},\neg\mathcal{D}\},\{\mathcal{B},\mathcal{C}\},\{\neg\mathcal{C}\},\{\neg\mathcal{B},\mathcal{D}\},\{\neg\mathcal{B}\},\{\mathcal{C}\}\}$
$\{\{\mathcal{A},\mathcal{D}\},\{\neg\mathcal{A},\neg\mathcal{B}\},\{\neg\mathcal{B},\neg\mathcal{D}\},\{\mathcal{B},\mathcal{C}\},\{\neg\mathcal{C}\},\{\neg\mathcal{B},\mathcal{D}\},\{\neg\mathcal{B}\},\{\mathcal{C}\},\Box\}$

Once $\Box$ has been added to the set, a proof has been constructed, which means that continued expansion of the set is unnecessary.

Exercise 8.3.5. Suppose that $\mathcal{A},\mathcal{B}$ and $\mathcal{C}$ are atomic wff's. Prove by resolution that $\mathcal{B}\supset(\mathcal{A}\vee\mathcal{C})$, $\neg\mathcal{B}\supset\mathcal{C}$, and $\mathcal{A}\supset(\mathcal{B}\supset\mathcal{C})$ imply $\mathcal{C}$. [Hint: Some clauses must be used more than once.]

The resolution rule amounts to concluding $\mathcal{A}_1\vee\ldots\vee\mathcal{A}_n\vee\mathcal{B}_1\vee\ldots\vee\mathcal{B}_m$ from $(\mathcal{A}_1\vee\ldots\vee\mathcal{A}_n\vee\mathcal{C})\wedge(\mathcal{B}_1\vee\ldots\vee\mathcal{B}_m\vee\neg\mathcal{C})$. This form of the premise may not be very suggestive, but it can be written as $(\neg(\mathcal{A}_1\vee\ldots\vee\mathcal{A}_m)\supset\mathcal{C})\wedge(\mathcal{C}\supset(\mathcal{B}_1\vee\ldots\vee\mathcal{B}_n))$, which is easily seen to imply $\neg(\mathcal{A}_1\vee\ldots\vee\mathcal{A}_m)\supset(\mathcal{B}_1\vee\ldots\vee\mathcal{B}_n)$. The latter wff corresponds to the resolvent clause.

Exercise 8.3.6. Show that the resolution rule is a one-way inference step, that is, the wff corresponding to the resolvent does not imply the wff's, which correspond to the initial pair of clauses.

The two implicational formulas, which we suggested in place of the clauses in the resolution rule, reiterate the idea that the resolution rule is related to the cut rule.

To establish this connection, let us start with a clause in which we separate the *positive* (unnegated) and the *negative* (negated) literals into $\mathcal{B}$'s and $\mathcal{A}$'s. By a De Morgan equivalence, we know that $\neg(\neg\mathcal{C}\vee\neg\mathcal{D})$ is equivalent to $\mathcal{C}\wedge\mathcal{D}$. Then $\{\neg\mathcal{A}_1,\ldots,\neg\mathcal{A}_n,\mathcal{B}_1,\ldots,\mathcal{B}_m\}$ may be viewed as the sequent $\mathcal{A}_1,\ldots,\mathcal{A}_n\vdash\mathcal{B}_1,\ldots,\mathcal{B}_m$ in the classical sequent calculus LK (or in $LK^{\{\ \}}$). The resolution rule applies to a pair of clauses, one with $\mathcal{C}$ and the other with $\neg\mathcal{C}$. The following is a slightly different formulation of the same rule as before, because now we assume that all the letters stand for atomic formulas rather than literals.

$$\frac{\{\neg\mathcal{A}_1,\ldots,\neg\mathcal{A}_n,\mathcal{B}_1,\ldots,\mathcal{B}_m,\mathcal{C}\}\qquad\{\neg\mathcal{C},\neg\mathcal{D}_1,\ldots,\neg\mathcal{D}_l,\mathcal{E}_1,\ldots,\mathcal{E}_k\}}{\{\neg\mathcal{A}_1,\ldots,\neg\mathcal{A}_n,\mathcal{B}_1,\ldots,\mathcal{B}_m,\neg\mathcal{D}_1,\ldots,\neg\mathcal{D}_l,\mathcal{E}_1,\ldots,\mathcal{E}_k\}}\ \text{res}$$

Now it is easy to see that in LK we would have—if we disregarded permutations—a rule of the following form.

$$\frac{\mathcal{A}_1,\ldots,\mathcal{A}_n\vdash\mathcal{B}_1,\ldots,\mathcal{B}_m,\mathcal{C}\qquad\mathcal{C},\mathcal{D}_1,\ldots,\mathcal{D}_l\vdash\mathcal{E}_1,\ldots,\mathcal{E}_k}{\mathcal{A}_1,\ldots,\mathcal{A}_n,\mathcal{D}_1,\ldots,\mathcal{D}_l\vdash\mathcal{B}_1,\ldots,\mathcal{B}_m,\mathcal{E}_1,\ldots,\mathcal{E}_k}$$

This is Gentzen's single cut rule from Section 2.3.1, except that the formulas here are all atomic. So, the resolution rule itself can be justified as a *special instance* of the single cut rule.

The whole procedure of proof by resolution requires us to show that a formula in CNF may be broken up into clauses. In LK, we can easily obtain from $\Gamma \vdash \mathcal{A} \wedge \mathcal{B}$ either $\Gamma \vdash \mathcal{A}$ or $\Gamma \vdash \mathcal{B}$.

$$\dfrac{\begin{matrix}\vdots \\ \Gamma \vdash \mathcal{A} \wedge \mathcal{B}\end{matrix} \qquad \dfrac{\mathcal{A} \vdash \mathcal{A}}{\mathcal{A} \wedge \mathcal{B} \vdash \mathcal{A}}\ \wedge\vdash}{\Gamma \vdash \mathcal{A}}\ \text{cut}$$

The Ketonen versions of the $(\wedge \vdash)$ and $(\vdash \vee)$ rules are derivable, and they contain both subalterns of the principal formulas. Those rules are invertible; hence, by using the cut rule, a disjunction occurring on the right-hand side of the $\vdash$ can be turned into a succedent comprising the subalterns in the disjunction.

Conceptually, a propositional resolution proof of $\mathcal{A}$ may be viewed as deducing from sequents that correspond to clauses obtained from $\neg\mathcal{A}$ the empty sequent $\vdash$. In the empty sequent, the left- and right-hand sides stand for two different constants, namely, $\boldsymbol{T}$ and $\boldsymbol{F}$. That is, the provability of $\vdash$ means that $\boldsymbol{T} \vdash \boldsymbol{F}$ is provable, which means that the initial sequents lead to inconsistency. (We have already seen in Section 7.8 that $\vdash$ is not provable in LK.)

The resolution proof in Example 8.10 turns into the proof below in a right-handed sequent calculus for classical logic. (In Section 3.4, we gave a right-handed sequent calculus for classical logic in a reduced vocabulary, in which sequents comprised sequences of formulas. In Section 3.2, we defined $LK^{\{\,\}}$, in which sets of wff's replaced sequences of formulas.)

Exercise 8.3.7. Formulate a right-handed sequent calculus, $LK_r^{\{\,\}}$, with a set of wff's with all the connectives of LK. [Hint: The cut rule should be admissible in $LK^{\{\,\}}$.]

We assume that the $\wedge$'s and $\vee$'s have been disassembled.

$$\dfrac{\vdash \neg\mathcal{C} \qquad \dfrac{\dfrac{\vdash \neg\mathcal{B}, \neg\mathcal{D} \qquad \dfrac{\vdash \mathcal{A}, \mathcal{D} \qquad \vdash \neg\mathcal{A}, \neg\mathcal{B}}{\vdash \neg\mathcal{B}, \mathcal{D}}}{\vdash \neg\mathcal{B}} \qquad \vdash \mathcal{B}, \mathcal{C}}{\vdash \mathcal{C}}}{\vdash}$$

Exercise 8.3.8. We argued that in a right-handed version of LK we can get from the CNF formula in Example 8.10 to the sequents that are the leaves in the above proof tree. Complete the proof tree by adding those steps. Then prove that such steps are available in general. [Hint: While completing the proof tree, you will notice that certain kinds of steps tend to appear one after another.]

$\boldsymbol{T}$ and $\boldsymbol{F}$ play a somewhat marginal role in classical logic, because they are definable. They were not included in *LK* itself. We appealed to those constants in our informal explanation above for the sake of transparency.

Exercise 8.3.9. The propositional part of the language that we introduced in Section 2.1 contains $\boldsymbol{T}$ and $\boldsymbol{F}$ too. Extend the notion of conjunctive normal forms to include these logical constants.

Exercise 8.3.10. Modify the notion of clauses and of the resolution rule to incorporate the logical constants $\boldsymbol{T}$ and $\boldsymbol{F}$.

Exercise* 8.3.11. Extend the right-handed sequent calculus $LK_r^{\{\,\}}$ (from Exercise 8.3.7) to include $\boldsymbol{T}$ and $\boldsymbol{F}$, and prove that the cut rule is admissible.

The *resolution calculus* has been defined for first-order classical logic. The match with sequent calculi is less obvious. We give a brief outline of the main components.

Definition 8.11. (Prenex normal form, PNF) A formula $\mathcal{A}$ is in *prenex normal form* iff it is of the shape $Q_1x_1 \ldots Q_nx_n\,\mathcal{B}$, where $n \in \mathbb{N}$ and the Q's are $\forall$ or $\exists$, and no quantifiers occur in $\mathcal{B}$.

Any formula that does not have any quantifiers is in PNF. The set of wff's in PNF is clearly a proper subset of wff's. However, the following claim, which is widely known, ensures that wff's in PNF form a sufficiently large and useful set.

Proposition 8.12. *For every wff* $\mathcal{A}$ *of classical logic,* there is a wff $\mathcal{B}$ *such that* $\mathcal{B}$ *is* in PNF, *and the two wff's are* logically equivalent.

Definition 8.13. (Skolemization) Given a wff $\mathcal{A}$ in PNF, the *Skolemization of* $\mathcal{A}$ is the wff obtained by setting $\mathcal{A}$ to be $\mathcal{A}'$, and then applying (1)–(3), in that order, as long as (1) is applicable.[8]

(1) Find the first (from left) $\exists$ in the quantifier prefix of $\mathcal{A}'$, let us say, $\exists x$;

(2) pick a new function symbol of arity, which equals the number of $\forall$'s to the left from this $\exists$, let us say, f^n;

(3) if the n $\forall$'s bind, let us say, $y_1, \ldots, y_n$, then omit $\exists x$ and replace all occurrences of x in $\mathcal{A}'$ with $f(y_1, \ldots, y_n)$; let the result be the new $\mathcal{A}'$.

Zero-ary function symbols are identified with name constants.

The finiteness of formulas is easily seen to imply that Skolemization terminates. We cannot talk about the starting $\mathcal{A}$ and the final $\mathcal{A}'$ being logically equivalent, because in the Skolemization process, we are constantly changing the language, hence, the interpretation should change too. However, if

[8]The noun "Skolemization" is derived from the name of T. Skolem.

$\mathcal{A}$ can be made true in a model, then that model can be modified so that $\mathcal{A}'$ becomes true.

Skolemization eliminates existential quantifiers, and the universal quantifiers will also be omitted. But a variable can denote variably—just as its name suggests. This is captured by the idea of the unification of terms.

Definition 8.14. (Unification) Let t_1 and t_2 be terms. They are *unifiable* iff there is a substitution σ such that $\sigma(t_1) = \sigma(t_2)$.

A substitution is a function assigning a term to each variable in the language. Terms are finite, and if a variable z does not occur in t_1 or t_2, then it is unimportant for the unification of t_1 and t_2 which terms σ assigns to z. The following facts are also well known.

Proposition 8.15. *A pair of terms is either* not unifiable *or unifiable, in which case, there are* one *or* $\aleph_0$-many *common instances. A pair of terms can be unified* in more than one way *iff they are unifiable and their common instance contains a variable. If a pair of terms is unifiable, then they have* a most general common instance (mgci), *which is unique up to renaming variables.*

Exercise 8.3.12. Consider the following pairs of terms. Decide whether they can be unified. If a pair is unifiable, then find the mgci. (1) $\langle f(x,y,z), g(z)\rangle$, (2) $\langle h(y_1,y_2), h(f(z), g(x_1,x_2,x_3))\rangle$, (3) $\langle x, f(f(f(h(g(y,g(z,z))))))\rangle$.

First-order resolution starts with negating the purported theorem, and the aim is to derive the empty clause.

Definition 8.16. (First-order resolution) The wff $\mathcal{A}$ is *theorem by resolution* iff starting with $\neg\mathcal{A}$, the steps (1)–(7) lead to a set of clauses that contains $\Box$.

(1) Find a $\mathcal{B}$ such that $\mathcal{B}$ is in PNF, and $\mathcal{B}$ is logically equivalent to $\neg\mathcal{A}$;

(2) Skolemize $\mathcal{B}$, let us say, producing $\forall x_1 \ldots \forall x_n\, \mathcal{C}$;

(3) turn $\mathcal{C}$ into CNF, giving $\forall x_1 \ldots \forall x_n\, (\mathcal{D}_1 \wedge \ldots \wedge \mathcal{D}_m)$;

(4) distribute the $\forall$'s over the $\wedge$'s, resulting in $\forall x_1 \ldots \forall x_n\, \mathcal{D}_1 \wedge \ldots \wedge \forall x_1 \ldots \forall x_n\, \mathcal{D}_m$;

(5) rename bound variables so that no pair of conjuncts has a common bound variable;

(6) omit all the quantifiers and form a set of clauses;

(7) iteratively apply the resolution rule to expand the set of clauses, if necessary, then via unifying some terms in a pair of clauses.

The addition of $\Box$ concludes the proof.

To complete our brief overview of first-order resolution, we give one detailed example.

Example 8.17. Let P and Q be unary, and R and S be binary predicates. The next formula is a theorem of classical first-order logic.

$$\exists x\, \forall y\, (P(x) \land (Q(y) \supset (R(x,y) \lor S(y,x)))) \supset \forall y\, \exists x\, (Q(y) \supset (\neg R(x,y) \supset (P(x) \land S(y,x))))$$

We take the negation of this formula and provide results of the steps in the definition above. (Some steps do not yield a unique wff.)

(1) $\exists x\, \forall y\, \exists z\, \forall v\, ((P(x) \land (Q(y) \supset (R(x,y) \lor S(y,x)))) \land (Q(z) \land \neg R(v,z) \land (\neg P(v) \lor \neg S(z,v))))$

(2) $\forall y\, \forall v\, ((P(a) \land (Q(y) \supset (R(a,y) \lor S(y,a)))) \land (Q(f(y)) \land \neg R(v,f(y)) \land (\neg P(v) \lor \neg S(f(y),v))))$

(3) $\forall y\, \forall v\, (P(a) \land (\neg Q(y) \lor R(a,y) \lor S(y,a)) \land Q(f(y)) \land \neg R(v,f(y)) \land (\neg P(v) \lor \neg S(f(y),v)))$

(4) $\forall y\, \forall v\, P(a) \land \forall y\, \forall v\, (\neg Q(y) \lor R(a,y) \lor S(y,a)) \land \forall y\, \forall v\, Q(f(y)) \land \forall y\, \forall v\, \neg R(v,f(y)) \land \forall y\, \forall v\, (\neg P(v) \lor \neg S(f(y),v)))$

(5) $P(a) \land \forall y\, (\neg Q(y) \lor R(a,y) \lor S(y,a)) \land \forall x\, Q(f(x)) \land \forall z\, \forall v\, \neg R(v,f(z)) \land \forall u\, \forall w\, (\neg P(w) \lor \neg S(f(u),w))$
(We omitted the vacuous quantifiers to simplify the wff.)

(6) $\{\, \{\, P(a) \,\}, \{\, \neg Q(y), R(a,y), S(y,a) \,\}, \{\, Q(f(x)) \,\}, \{\, \neg R(v,f(z)) \,\}, \{\, \neg P(w), \neg S(f(u),w) \,\} \,\}$

(7) A substitution σ that provides a suitable unification for the terms in the clauses at once is $x \mapsto x,\ y \mapsto f(x),\ z \mapsto x,\ v \mapsto a,\ w \mapsto a,\ u \mapsto x$. The clauses $\{\, P(a) \,\}$ and $\{\, \neg P(a), \neg S(f(x),a) \,\}$ yield $\{\, \neg S(f(x),a) \,\}$. Then,

$$\dfrac{\{\, \neg S(f(x),a) \,\} \qquad \dfrac{\dfrac{\{\, \neg Q(f(x)), R(a,f(x)), S(f(x),a) \,\} \quad \{\, Q(f(x)) \,\}}{\{\, R(a,f(x)), S(f(x),a) \,\}} \qquad \{\, \neg R(a,f(x)) \,\}}{\{\, S(f(x),a) \,\}}}{\Box}.$$

Chapter 9

Applications and applied calculi

An important area of application for the cut theorem is in proofs of the *decidability* of logics. Strictly speaking, decidability is never the consequence of the cut theorem per se. Limiting the search space to *cut-free proofs* is a big step forward, but there might be too many—even infinitely many—cut-free proofs for a sequent or a formula. The desire to prove decidability for various logics led to many interesting innovations and new results. We start the first section with an overview of earlier results concerning classical logic, intuitionistic logic and the logic of pure relevant implication. Until 2010, the last major open problem in the field of relevance logics was the *decidability of the logic of pure ticket entailment*. We briefly outline the solution that was originally published in [42], and then we add some further decidability results.

Calculi may be thought to be *applied* in at least two ways. First, we may expand the vocabulary by components that have some intended interpretation or meaning outside of logic. An obvious area for applied calculi in this sense is calculi that *formalize mathematical theories*. To give the flavor of existing and potential developments in this area, we give a sample theory in Section 9.2.

Some other applied calculi not only expand the vocabulary, but they make formulas part of a new kind of statement. *Typed calculi* are the oldest and most widely known method of combining some sort of computational information (such as terms, functions or proofs) with formulas. The last section deals with *simply typed calculi* and some initial steps toward extending the *Curry–Howard isomorphism* to combinators and classes of sequent calculus proofs. We conclude the section by mentioning other applied calculi—so-called *labeled calculi*—along the lines of pairing pieces of information from different sources within a sequent calculus framework.

9.1 Decidability

Sequent calculi may appear less natural than the natural deduction systems. Gentzen called NK and NJ *natural* for a reason. However, the *control* that a sequent calculus provides over proofs is unmatched by any other proof system, as this section superbly illustrates.

A typical question, which reflects the desire to have assurance that answers exist, is the problem of the decidability of a logic. A logic is *decidable* iff there is an effective procedure (or algorithm) that given a well-formed formula of the logic returns a "yes" or a "no" answer, after finitely many steps, to the question whether the formula is a *theorem*. "Effective procedure" and "algorithm" are technical terms here; any notion capturing full-fledged *computability* can stand in for them. Some of the most widely recognized names for computable functions are *partial recursive functions*, λ-*computable functions*, *Turing machines*, *Markov algorithms* and *abstract state machines*.[1] The notion of a step is dependent on the particulars of a concrete model, and a step in one formalism may equal a hundred in another or vice versa. Unless we consider the *complexity* of an algorithm, such differences are disregarded, and we are only concerned with *termination*. In other words, the number of steps taken must be *finite* before an answer is returned (which, of course, should be correct).

The problem of decidability loomed large in the early 20th century. On one hand, there was a desire to make mathematics more formal and precise. On the other hand, there were no exact definitions for a formal system or for an algorithm. Combined with the perpetual aspiration of mathematicians to solve problems, or as Hilbert so eloquently put it, to solve *all* problems, the lack of concepts and the lack of understanding of potential complications created a lot of interest in decidability.

To distinguish between the versions of the decidability problem concerning *theorems* and concerning *logically valid wff's*, the former is sometimes labeled as the syntactic and the latter as the semantic decision problem. If a logic is sound and complete for some interpretation, then the notions of logical validity and theoremhood are equivalent. Then, it is very well possible that an algorithm checks models of a logic, and it does not examine proofs at all. Of course, occasionally, the construction of models may very closely resemble how a search for proofs would proceed. Or the other way around, a proof system might yield not only completed proofs, but also blueprints for building models. For example, tableaux for classical propositional logic are often perceived (or even taught in introductory courses) as "truth trees" with the explicit suggestion that it is a method to find a truth assignment (if there is one) that satisfies a set of formulas.

Classical propositional logic has been known to be decidable for a long time. Indeed, it is difficult to pinpoint who formulated and answered the question. Ch. S. Peirce invented and used truth tables around the turn of the 19th and 20th centuries, but his work remained mostly unpublished until the 1930s. E. L. Post's work around 1920, published as [154], provided the first completeness proof, in a contemporary sense, for classical propositional logic with respect to its two-valued interpretation. In any case, by the mid-1930s,

[1] For formal definitions of some of these notions see, for example, [169], [132, Ch. 5], [68, Ch. 2], [62] and [36, Ch. 3], including references to their origins.

it was known that classical propositional logic is decidable, as it was noted by Gentzen in his [92, p. 204].

There is an inherent *asymmetry* hidden in the decidability problem. On one hand, if a logic is claimed to be decidable, then it is sufficient to define *one particular procedure* to prove the claim. Even if there is no agreed-upon notion of an algorithm (as there was none in the 1920s), a concrete procedure can be scrutinized, and possibly, criticized if deemed non-effective. In short, a positive answer to the decidability problem has an easy proof, in the sense that it requires only that an algorithm is presented. On the other hand, to show that there is no procedure to decide whether a formula is a theorem, one has to prove that *all* the algorithms, procedures and machines that could be thought suitable, are in fact, incapable of solving the problem. (To put it informally, the claim to decidability is an existential claim, whereas undecidability is a universal statement, moreover, in the 1930s, it was unclear what the domain of quantification is for this universal quantifier.) Given this asymmetry, it should not be surprising that, chronologically, decidability results preceded both the invention of precise notions of computability and the first undecidability result by Church [60].

9.1.1 Intuitionistic and classical propositional logics

Intuitionistic propositional logic is not a finitely valued logic (like its classical counterpart is)—as it was proved by Gödel in [100]. Thus, Gentzen's proof that propositional intuitionistic logic is decidable was not simply a new proof of a known result, but a proof of a new result. However, because of the relationship between LK and LJ, his proof applies to both calculi. Moreover, some of the underlying ideas motivated certain later decidability proofs based on sequent calculi.

We start with Gentzen's proof for propositional LJ (and LK). We consider separately the number of occurrences of formulas in the antecedent and in the succedent.

Definition 9.1. (Reduced sequents) The sequent $\Gamma \vdash \Delta$ is a *reduced sequent* iff no $\mathcal{A}$ that occurs in Γ occurs more than three times in Γ and no $\mathcal{B}$ that occurs in Δ occurs more than three times in Δ.

"More than three" means strictly greater than three. Some of the $\mathcal{A}$'s and $\mathcal{B}$'s may be the same wff, but it is an artifact that no formula can occur more than six times in the whole sequent. (Nonetheless, $\mathcal{A}, \mathcal{A}, \mathcal{A}, \mathcal{A} \vdash \mathcal{A}$ is not a reduced sequent.) When we are concerned with LJ only, then the provisions concerning the $\mathcal{B}$'s are superfluous.

Lemma 9.2. (Reduced proofs) *Given a cut-free proof with a reduced end sequent* $\Gamma \vdash \Delta$, there is a cut-free proof *of the same end sequent in which* all sequents are reduced. *Furthermore, such a proof may be obtained effectively from the proof that is given.*

Proof: **1.** It can be easily determined whether a sequent is reduced or not. However, typically, given a reduced sequent, there are other reduced sequents that contain the same wff's, but in different quantities. (The only exception is a sequent of the form $\vdash \mathcal{A}$ in LJ, because $\mathcal{A}$ cannot have fewer or more occurrences on the right-hand side of the $\vdash$.) All the other reduced sequents that are built from the same $\mathcal{A}$'s and $\mathcal{B}$'s can be obtained from each other by applications of the structural rules.

2.1 Given a proof ending in a reduced sequent, the leaves and the root in the proof remain as they are. The instances of the connective rules—with the exception of $(\supset\vdash)$—are modified so that each parametric formula in the antecedent and each parametric formula in the succedent has only one occurrence. In other words, in the other connective rules, all the Γ's and Δ's are shrunk so that neither Γ nor Δ has any duplicate wff's. This guarantees that the upper sequent of a rule is reduced with at most two occurrences of each formula, because a subaltern and a parametric formula may be the same formula. An application of the same rule yields a lower sequent, which is certainly reduced, and may differ from the earlier lower sequent only with respect to the number of the occurrences of wff's.

2.2 In the case of the $(\supset\vdash)$ rule, if Θ and Λ are also shrunk, then the rule is applicable as before, moreover, the resulting sequent is reduced too. This rule is the motivation for defining reduced sequents with the "magic number" *three*. Nothing prevents Γ and Θ from each having an occurrence of $\mathcal{A} \supset \mathcal{B}$. Due to the shape of the rule, namely, that Γ and Θ are not required to be the same sequences of wff's, the lower sequent "sums" the number of occurrences in the two sequences (unlike the $(\vdash\wedge)$ and $(\vee\vdash)$ rules do). Thus, it can happen that an application of $(\supset\vdash)$ results in a sequent with three occurrences of $\mathcal{A} \supset \mathcal{B}$ (and two occurrences of some other wff's).

2.3 In case of applications of thinning and contraction, reducing the number of occurrences of parametric formulas works exactly as in the general case for connective rules, because all structural rules have a single premise. According to the requirement that parametric formulas have one occurrence, the upper sequent of a contraction may have three occurrences of a formula (in the antecedent or in the succedent), because the contracted occurrences are not considered to be parametric. In the case of $(C\vdash)$ and $(\vdash C)$, we have both Γ and Δ in the antecedent or succedent, hence, Γ, Δ is contracted (if necessary) until each formula occurs once; Θ is shrunk similarly. Then a wff has three occurrences only if $\mathcal{A}$ and $\mathcal{B}$ are the same wff's, which renders the application of the permutation rule superfluous. qed

The role of the number three is that a given finite set of wff's there are *finitely many reduced sequents* that contain those wff's. If there are n (distinct) wff's in Γ, and m wff's in Δ, then there are 3^{n+m} different reduced sequents in $LK^{[\,]}$, and 3^n different sequents in $LJ^{[\,]}$ (where $m = 0$ or $m = 1$, by the definition of a sequent).

Definition 9.3. (Decision procedure for (propositional) LJ and LK) Let $\Gamma' \vdash \Delta'$ be a sequent, and let $\Gamma \vdash \Delta$ be a *reduced sequent*, such that Γ and Γ', and Δ and Δ', respectively, comprise the same formulas, though the number of formula occurrences may differ. Γ and Δ are finite, and the set of subformulas of formulas in them is a finite set of wff's, let us say, Φ. We may focus our attention on subformulas, because any provable sequent has a cut-free proof; hence, by the previous lemma, $\Gamma \vdash \Delta$ has a cut-free proof comprising reduced sequents only.

There are *finitely many* reduced sequents that contain no other wff's than those in Φ, and one of them is $\Gamma \vdash \Delta$. We partition the set of these reduced sequents into Ξ and Ψ. To start with, Ξ contains only those sequents that are instances of the axiom (id). $\Gamma \vdash \Delta$ may be an instance of the axiom, but in more interesting cases, it is not, hence, it is in Ψ.

Repeatedly, we consider the elements of Ξ, the connective and the structural rules, and if a sequent in Ψ would result, then we remove it from Ψ and add it to Ξ. If at any point $\Gamma \vdash \Delta$ is in Ξ, then it is provable. $\Xi \cup \Psi$ is finite, and there are finitely many rules to consider. Therefore, after finitely many steps, all the sequents that could have been transferred from Ψ to Ξ, in fact, have been transferred from Ψ to Ξ. If, at that point, $\Gamma \vdash \Delta$ is not an element of Ξ, then it is not provable.

What we gave here is a description of how the decision procedure would proceed. Indeed, Gentzen does not state separately that the procedure is *correct*, that is, it does achieve what it is claimed to achieve. In the case of LK and LJ, once the cut theorem (Theorem 2.27) and Lemma 9.2 have been proved, the correctness depends on the *finiteness* of various sets of objects generated from finitely many elements. So it must be obvious, indeed.

In practice, given a sequent, one is likely to try to figure out how that sequent could have been obtained, then how the premises could have been obtained, and so on. This way of thinking about the discovery of a proof makes clear that contraction is problematic, because the premises of successive contractions can become arbitrarily large. However, if a reduced sequent has a proof comprising reduced sequents only, then contraction has been curbed.

Many decidability proofs via sequent calculi appeal to a "backward" *proof-search tree* that results from what we have just outlined. That is, given a wff or a sequent that is claimed to be provable, we want to construct a proof (if there is one) by building a proof tree from bottom up. If we have more than one possibility at a step, then the search tree forks. A proof is a tree itself and the search for a proof also creates a tree, which—if the proof search is successful—contains a subtree that is a proof.

The decision procedure for LJ and LK can be recast so that it appears to be more practical. The two core components remain Theorem 2.27 and Lemma 9.2.

Definition 9.4. (Decision procedure for LJ and LK, 2) Given a sequent $\Gamma' \vdash \Delta'$, we consider the *reduced sequent* $\Gamma \vdash \Delta$, in which every formula that occurs in the antecedent Γ' has exactly one occurrence in Γ, and similarly for wff's in the succedents Δ' and Δ. Of course, the latter sequent differs from the former one at most with respect to the multiplicities of the wff's. We already know that one of these two sequents is provable if and only if so is the other. The *root* of the proof-search tree is the sequent $\Gamma \vdash \Delta$.

For each leaf in the proof-search tree, we repeat the following steps **(1)**–**(2)** to extend the tree (if possible). We may assume that, at each stage, there is a list of nodes that have to be considered as in **(1)**. Once **(1)** has been applied to a node n, n is deleted from the list and finitely many (possibly, zero) nodes are added to the list.

(1) Given a node n, which is a sequent, there are finitely many premise sequents and rules that can yield that sequent. We know that only reduced sequents need to be considered. That is, potential premises that are not reduced sequents are excluded from consideration. We also omit any sequent that is already on the path from the root to the given node. (Alternating applications of contraction and thinning could generate an infinite path comprising reduced sequents—with some sequents appearing infinitely many times on the path.)

It may happen that although the given node n is not an instance of (id), no extension is possible. For instance, it is possible that none of the potential premises is reduced, or that there are no potential premises at all.

(2) We may stop either when we have *a proof* constructed, or when we have *the full proof-search tree* constructed (possibly, already containing finished proofs). The first termination condition is helpful if we simply want to know if a sequent is provable or not (though finding the proof in a proof-search tree requires additional work). The second termination condition is useful when we want to have all the proofs (if there are any) comprising reduced sequents. Of course, the full proof-search tree will be constructed under the first termination condition too, when the starting sequent is not provable. If the proof search has not terminated with a proof, by the first condition, then after finitely many steps, we have a full proof-search tree, because it cannot be further expanded.

This description of the decision procedure provides more particulars about the concrete steps, and it brings into being a new object—a *proof-search tree.* The presence of details may make it less obvious—in an abstract sense—that the procedure works, but it yields a proof (which may be checked), if the starting sequent is provable.

Lemma 9.5. *The decision procedure above is* correct.

We leave piecing together the correctness of the procedure for the next exercise. However, it may be useful to emphasize that imposing a tree structure on the proof search does not guarantee by itself that we have a decision procedure. The proof-search tree has to be proved *finite,* which is usually carried out by an application of König's lemma (Lemma A.4). We applied the lemma tacitly by claiming that at each node the tree can be extended by finitely many premise nodes (i.e., the tree has the *finite fork* property), and by restricting paths to contain reduced sequents without repeating any sequent (i.e., the tree has the *finite branch* property too).

Exercise 9.1.1. Scrutinize the second decision procedure and show that it is correct. [Hint: Prove that if $\Gamma' \vdash \Delta'$ is provable, then the procedure will yield a proof, and that if $\Gamma' \vdash \Delta'$ is not provable, then the procedure will not result in a proof. Furthermore, show that either conclusion can be reached after finitely many steps.]

Exercise 9.1.2. Consider LK or LJ with quantification included, and a sequent $\Gamma' \vdash \Delta'$ given, which includes quantified formulas. Pinpoint where decision procedures along the lines of the propositional cases would "break down." [Hint: The monadic fragment of LK is decidable.]

9.1.2 $R^t_{\overset{\sim}{\to}}$ is decidable

Relevance logics are interesting and well-motivated logics. Such qualifications are somewhat subjective. However, the mathematics used in the metatheory of relevance logics is more intricate than what is needed for classical and intuitionistic logic, and this view is objective, or at least hard to dispute.

So far we have looked at the decidability of classical and intuitionistic propositional logics, which include a variety of connectives. In particular, there are connectives beyond $\supset$ in both. Decidability results for certain propositional relevance logics are not readily forthcoming. We have seen in Chapters 4 and 5 that there is no logistic sequent or consecution calculus for R, E or T. Of course, there are decidability proofs that do not utilize a sequent calculus or any proof system (if a logic is sound and complete with respect to a semantics). Nonetheless, the complications that entered into sequent calculi for some relevance logics in Chapters 4 and 5 may be viewed as a forewarning against overblown expectations about the decidability of propositional relevance logics, in general.

The logic $R_{\to}$ was proved *decidable* by Kripke in [115], and he used a sequent calculus. However, Kripke's proof is not a simple restriction or adaptation of Gentzen's proof to a one-connective case. The proof of the decidability of $R_{\to}$ uses several *new concepts* and *new lemmas*. That proof provided a template for decidability proofs for related logics—including $E_{\to}$, $R_{\overset{\sim}{\to}}$, $E_{\overset{\sim}{\to}}$, $R^t_{\to}$, and even for $T^t_{\to}$ and $T_{\to}$.[2]

We present a series of decidability results in this section, which we start by proving that $R^t_{\overset{\sim}{\to}}$ is *decidable*. This is the implication–negation fragment of R with intensional truth, which has not been proved decidable before. But it is not difficult to extend the result for $R_{\overset{\sim}{\to}}$ to $R^t_{\overset{\sim}{\to}}$ along the lines of [42],

[2]Some versions of these proofs may be found in [7, §13], [77, §3], [23] and [42]. See also [8, §63], [160] and [136].

which showed $R^t_{\rightarrow}$ decidable given that so is $R_{\rightarrow}$. $R_{\overset{\sim}{\rightarrow}}$ can be axiomatized by adding (A5)–(A6) to the axioms of $R_{\rightarrow}$ (listed in Exercise 4.3.1 on p. 104), with modus ponens as the only rule. Then, adding (A7) and (A8) yields $R^t_{\overset{\sim}{\rightarrow}}$.

(A5) $\sim\sim\mathcal{A} \rightarrow \mathcal{A}$

(A6) $(\mathcal{A} \rightarrow \sim\mathcal{B}) \rightarrow \mathcal{B} \rightarrow \sim\mathcal{A}$

(A7) $\boldsymbol{t}$

(A8) $\boldsymbol{t} \rightarrow \mathcal{A} \rightarrow \mathcal{A}$

A sequent calculus formulation for $R^t_{\overset{\sim}{\rightarrow}}$ can be obtained by extending $LR_{\overset{\sim}{\rightarrow}}$ from Section 4.3 by the axiom $(\vdash \boldsymbol{t})$ and the rule $(\boldsymbol{t}\vdash)$, where the latter is the following multiple right-hand side version of the rule.

$$\frac{\Gamma[\Delta] \vdash \Theta}{\Gamma[\boldsymbol{t};\Delta] \vdash \Theta}\ \boldsymbol{t}\vdash$$

The same addition to $[LR_{\overset{\sim}{\rightarrow}}]$ yields $[LR^t_{\overset{\sim}{\rightarrow}}]$.

Exercise 9.1.3. Prove that if $\mathcal{A}$ is a theorem of $R^t_{\overset{\sim}{\rightarrow}}$, then $\vdash \mathcal{A}$ is provable in $LR^t_{\overset{\sim}{\rightarrow}}$. [Hint: $LR_{\overset{\sim}{\rightarrow}}$ is the core of the sequent calculus $LR^t_{\overset{\sim}{\rightarrow}}$, for which Theorem 4.20 stated the admissibility of the cut.]

Exercise* 9.1.4. Prove the converse of the claim in the previous exercise. [Hint: Notice that $\sim$ allows the interpretation of sequents by formulas.]

Exercise 9.1.5. Prove that the cut rule is admissible in $[LR^t_{\overset{\sim}{\rightarrow}}]$ and that this calculus is equivalent to $LR^t_{\overset{\sim}{\rightarrow}}$. [Hint: The cut rule must be appropriately formulated, for instance, like for $[LR^t_{\rightarrow}]$, in Section 4.3.]

Using reduced sequents was a suitable way to restrict the number of sequents that had to be taken into account in LJ and LK. We have seen in Section 3.2 that a sequent calculus for classical logic can be based on *sets*; but for relevance logics, sets are not an appropriate data type, because thinning is not among the structural rules. (Only $\boldsymbol{t}$ can be thinned into a sequent, and only on the left-hand side of the $\vdash$.) However, contraction is a rule, and *multisets* can replace sequences of wff's, as in $LR^t_{\overset{\sim}{\rightarrow}}$.

The following definition is applicable to sequents both with multisets and with sequences of wff's.

Definition 9.6. (Cognate sequents) $\Gamma \vdash \Delta$ and $\Theta \vdash \Lambda$ are *cognate sequents* iff Γ and Θ yield the same set of wff's when multiple occurrences (and the order) of the wff's are disregarded, and so do Δ and Λ.

If $\Gamma \vdash \Delta$ and $\Theta \vdash \Lambda$ are cognate sequents, then $\Gamma;\Theta \vdash \Delta;\Lambda$ is a sequent that is cognate with both, and from which either can be obtained by contractions (and by permutations, if Γ, Δ, Θ and Λ are sequences). $\Gamma;\Theta \vdash$

$\Delta; \Lambda$ is not the smallest sequent that has both these properties. Let us assume that the distinct wff's in Γ (hence, in Θ) are $\mathcal{A}_1, \ldots, \mathcal{A}_n$. Similarly, let $\mathcal{B}_{n+1}, \ldots, \mathcal{B}_{n+m}$ be the distinct wff's in Δ (hence, in Λ). Let $\mathcal{A}_i^{\mathfrak{m}_i}$ be a multiset (or a sequence) of $\mathfrak{m}_i$-many $\mathcal{A}_i$'s, where $\mathfrak{m}_i$ is the *maximum* of the number of occurrences of $\mathcal{A}_i$ in Γ and in Θ; similarly, for the $\mathcal{B}$'s. The multiset (or sequence) that is cognate with both $\Gamma \vdash \Delta$ and $\Theta \vdash \Lambda$, and has the fewest number of occurrences of wff's among those from which both sequents can be obtained by finitely many (including zero) contractions (and permutations) is $\mathcal{A}_1^{\mathfrak{m}_1}; \ldots; \mathcal{A}_n^{\mathfrak{m}_n} \vdash \mathcal{B}_{n+1}^{\mathfrak{m}_{n+1}}; \ldots; \mathcal{B}_{n+m}^{\mathfrak{m}_{n+m}}$. On the other hand, $\mathcal{A}_1; \ldots; \mathcal{A}_n \vdash \mathcal{B}_{n+1}; \ldots; \mathcal{B}_{n+m}$ is a multiset (or a sequent) that is cognate with both $\Gamma \vdash \Delta$ and $\Theta \vdash \Lambda$, and contains the least possible number of occurrences of the wff's.

The decidability proof for $LR^t_{\underset{\to}{\sim}}$ combines *three lemmas*—Curry's, Kripke's and König's. Together they ensure that there is a *finite proof-search tree*. *König's lemma* (Lemma A.4) is not specific to a search for proofs; rather it justifies the conclusion that the proof-search tree is finite, provided that no branch is infinite and every splitting (or forking) is finite. The finite fork property is more or less obvious, however, the finite branch property is not, and that is what *Kripke's lemma* will guarantee. Finally, *Curry's lemma* ensures that the elimination of the explicit contraction rules, that is, hiding the applications of the $(W\vdash)$ and $(\vdash W)$ rules in the connective rules is successful.

Lemma 9.7. (Curry's lemma for $[LR^t_{\underset{\to}{\sim}}]$) *If the sequent $\Gamma \vdash \Delta$ is provable, and the height of the proof P is $\chi(P) = n$, and $\Gamma' \vdash \Delta'$ can be obtained from $\Gamma \vdash \Delta$ by finitely many contractions (i.e., by zero or more applications of the rules $(W\vdash)$ and $(\vdash W)$), then* there is a proof *P' of $\Gamma' \vdash \Delta'$ such that $\chi(P') \leq n$.*

Proof: We give the proof of Curry's lemma for this calculus in some detail. (Proofs of Curry's lemma for other calculi may be found, for example, in [77] and [41].) The proof is by induction on the height of the proof of $\Gamma \vdash \Delta$.

1. *Base case.* If $n = 1$, then the proof is an instance of one of the two axioms, (id) or $(\vdash t)$. Γ' is Γ and Δ' is Δ, because neither Γ nor Δ can be contracted; hence, the claim is obviously true.

2. *Inductive case.* $\Gamma \vdash \Delta$ is by an application of one of the five rules in $[LR^t_{\underset{\to}{\sim}}]$.

2.1 If the last rule is $(t\vdash)$, then there are three cases to consider. Some combinations of these cases are also possible, however, it is straightforward to combine their proofs. There may be parametric formulas, like $\mathcal{A}$, in Γ or in Δ that could be contracted. The premise of the new proof on the right-hand side exists, because of the hypothesis of the induction (i.e., by i.h.).

$$t\vdash \frac{\Gamma \vdash \mathcal{A}; \mathcal{A}; \Delta}{t; \Gamma \vdash \mathcal{A}; \mathcal{A}; \Delta} \quad \overset{\text{i.h.}}{\rightsquigarrow} \quad \frac{\Gamma \vdash \mathcal{A}; \Delta}{t; \Gamma \vdash \mathcal{A}; \Delta}\, t\vdash$$

The following is completely analogous.

$$t\vdash \frac{\Gamma; \mathcal{A}; \mathcal{A} \vdash \Delta}{t; \Gamma; \mathcal{A}; \mathcal{A} \vdash \Delta} \quad \overset{\text{i.h.}}{\rightsquigarrow} \quad \frac{\Gamma; \mathcal{A} \vdash \Delta}{t; \Gamma; \mathcal{A} \vdash \Delta}\, t\vdash$$

If the contracted formula includes the principal formula of the $(t\vdash)$ rule, then we can simply omit the application of the rule.

$$t\vdash\ \frac{t;\Gamma\vdash\Delta}{t;t;\Gamma\vdash\Delta} \quad\rightsquigarrow\quad t;\Gamma\vdash\Delta$$

2.2 If the rule is $([\vdash\rightarrow])$, then again there are three cases to consider. First of all, some contractions may involve parametric formulas on the left or on the right without involving the principal formula of the rule.

$$[\vdash\rightarrow]\ \frac{\Gamma;\mathcal{A};\mathcal{A};\mathcal{B}\vdash\mathcal{C};\Delta}{\Gamma;\mathcal{A};\mathcal{A}\vdash\mathcal{B}\rightarrow\mathcal{C};\Delta} \quad\overset{\text{i.h.}}{\rightsquigarrow}\quad \frac{\Gamma;\mathcal{A};\mathcal{B}\vdash\mathcal{C};\Delta}{\Gamma;\mathcal{A}\vdash\mathcal{B}\rightarrow\mathcal{C};\Delta}\ [\vdash\rightarrow]$$

$$[\vdash\rightarrow]\ \frac{\Gamma;\mathcal{B}\vdash\mathcal{C};\mathcal{A};\mathcal{A};\Delta}{\Gamma\vdash\mathcal{B}\rightarrow\mathcal{C};\mathcal{A};\mathcal{A};\Delta} \quad\overset{\text{i.h.}}{\rightsquigarrow}\quad \frac{\Gamma;\mathcal{B}\vdash\mathcal{C};\mathcal{A};\Delta}{\Gamma\vdash\mathcal{B}\rightarrow\mathcal{C};\mathcal{A};\Delta}\ [\vdash\rightarrow]$$

If the contraction became possible, because the principal formula created a double, then we appeal to the contraction that is part of the $([\vdash\rightarrow])$ rule.

$$[\vdash\rightarrow]\ \frac{\Gamma;\mathcal{B}\vdash\mathcal{C};\mathcal{B}\rightarrow\mathcal{C};\Delta}{\Gamma\vdash\mathcal{B}\rightarrow\mathcal{C};\mathcal{B}\rightarrow\mathcal{C};\Delta} \quad\rightsquigarrow\quad \frac{\Gamma;\mathcal{B}\vdash\mathcal{C};\mathcal{B}\rightarrow\mathcal{C};\Delta}{\Gamma\vdash\mathcal{B}\rightarrow\mathcal{C};\Delta}\ [\vdash\rightarrow]$$

Again, these possibilities may blend together. We give one example of how all the above can happen at once.

$$\frac{\Gamma;\mathcal{A};\mathcal{A};\mathcal{D};\mathcal{D};\mathcal{B}\vdash\mathcal{C};\mathcal{B}\rightarrow\mathcal{C};\mathcal{E};\mathcal{E};\Delta}{\Gamma;\mathcal{A};\mathcal{A};\mathcal{D};\mathcal{D}\vdash\mathcal{B}\rightarrow\mathcal{C};\mathcal{B}\rightarrow\mathcal{C};\mathcal{E};\mathcal{E};\Delta} \quad\overset{\text{i.h.}}{\rightsquigarrow}\quad \frac{\Gamma;\mathcal{A};\mathcal{D};\mathcal{B}\vdash\mathcal{C};\mathcal{B}\rightarrow\mathcal{C};\mathcal{E};\Delta}{\Gamma;\mathcal{A};\mathcal{D}\vdash\mathcal{B}\rightarrow\mathcal{C};\mathcal{E};\Delta}\ [\vdash\rightarrow]$$

2.3–2.4 If the rule is $([\vdash\sim])$, then there are two cases concerning parametric formulas, and one case when $\sim\mathcal{B}$ is affected by the contraction. Here are two of these three cases.

$$[\vdash\sim]\ \frac{\Gamma;\mathcal{A};\mathcal{A};\mathcal{B}\vdash\Delta}{\Gamma;\mathcal{A};\mathcal{A}\vdash\sim\mathcal{B};\Delta} \quad\overset{\text{i.h.}}{\rightsquigarrow}\quad \frac{\Gamma;\mathcal{A};\mathcal{B}\vdash\Delta}{\Gamma;\mathcal{A}\vdash\sim\mathcal{B};\Delta}\ [\vdash\sim]$$

$$[\vdash\sim]\ \frac{\Gamma;\mathcal{B}\vdash\sim\mathcal{B};\Delta}{\Gamma\vdash\sim\mathcal{B};\sim\mathcal{B};\Delta} \quad\rightsquigarrow\quad \frac{\Gamma;\mathcal{B}\vdash\sim\mathcal{B};\Delta}{\Gamma\vdash\sim\mathcal{B};\Delta}\ [\vdash\sim]$$

The cases with the last rule being $([\sim\vdash])$ are completely symmetric and we omit the details.

2.5 Lastly, the most interesting case is when the rule is $([\rightarrow\vdash])$. This is the only two-premise rule, moreover, the contexts are not shared. There are four situations that parallel the previous cases, when the contracted formulas are parametric. We omit the details of these.

There are two parametric cases, where the possibility for contraction is created by joining two multisets. An application of $([\rightarrow\vdash])$ suffices.

$$[\rightarrow\vdash]\ \frac{\Gamma;\mathcal{A}\vdash\mathcal{B};\Delta \qquad \Theta;\mathcal{A};\mathcal{C}\vdash\Lambda}{\Gamma;\Theta;\mathcal{A};\mathcal{A};\mathcal{B}\rightarrow\mathcal{C}\vdash\Delta;\Lambda} \quad\rightsquigarrow\quad \frac{\Gamma;\mathcal{A}\vdash\mathcal{B};\Delta \qquad \Theta;\mathcal{A};\mathcal{C}\vdash\Lambda}{\Gamma;\Theta;\mathcal{A};\mathcal{B}\rightarrow\mathcal{C}\vdash\Delta;\Lambda}\ [\rightarrow\vdash]$$

$$[\to\vdash]\ \frac{\Gamma\vdash\mathcal{A};\mathcal{B};\Delta\quad\Theta;\mathcal{C}\vdash\mathcal{A};\Lambda}{\Gamma;\Theta;\mathcal{B}\to\mathcal{C}\vdash\mathcal{A};\mathcal{A};\Delta;\Lambda}\quad\rightsquigarrow\quad\frac{\Gamma\vdash\mathcal{A};\mathcal{B};\Delta\quad\Theta;\mathcal{C}\vdash\mathcal{A};\Lambda}{\Gamma;\Theta;\mathcal{B}\to\mathcal{C}\vdash\mathcal{A};\Delta;\Lambda}\ [\to\vdash]$$

Finally, the principal formula can be part of one or two contractions. The latter situation looks like the following.

$$[\to\vdash]\ \frac{\Gamma;\mathcal{B}\to\mathcal{C}\vdash\mathcal{B};\Delta\quad\Theta;\mathcal{B}\to\mathcal{C};\mathcal{C}\vdash\Lambda}{\Gamma;\Theta;\mathcal{B}\to\mathcal{C};\mathcal{B}\to\mathcal{C};\mathcal{B}\to\mathcal{C}\vdash\Delta;\Lambda}\quad\rightsquigarrow$$

$$\frac{\Gamma;\mathcal{B}\to\mathcal{C}\vdash\mathcal{B};\Delta\quad\Theta;\mathcal{B}\to\mathcal{C};\mathcal{C}\vdash\Lambda}{\Gamma;\Theta;\mathcal{B}\to\mathcal{C}\vdash\Delta;\Lambda}\ [\to\vdash]$$

There are no other rules; hence, the proof is completed. qed

Exercise 9.1.6. Scrutinize the proof above and write out the few details that were omitted. Convince yourself that all the possible combinations of the subcases can be dealt with.

Before stating Kripke's lemma, we introduce a property that may (or may not) hold for a sequence of cognate sequents.

Definition 9.8. (Irredundancy) Let $\sigma = \langle\Gamma_1\vdash\Delta_1,\ldots,\Gamma_i\vdash\Delta_i,\ldots\rangle$ be a (possibly infinite) sequence of *cognate sequents*. σ is *irredundant* iff the cognate sequents are distinct and whenever $i < j$, $\Gamma_i\vdash\Delta_i$ is not a result of contractions applied to $\Gamma_j\vdash\Delta_j$.

We can take another look at Curry's lemma now. In searching for a proof, proofs that contain a sequent that could be contracted to a sequent appearing lower in the proof can be disregarded. In other words, no proof in which a branch (or more precisely, some of the sequents on that branch) would form an irredundant sequence has to be considered. However, we still have no assurance that the branch is not infinite. One source of infinity can be excluded like in the case of LJ and LK, namely, the set of subformulas of all the formulas in a given sequent is finite. Kripke's lemma excludes another source of infinity.

Lemma 9.9. (Kripke's lemma) *If σ is an irredundant sequence of (distinct) cognate sequents, then σ is* finite.[3]

Proof: First of all, we note that we based the notion of sequents in $LR^t_{\underset{\to}{\sim}}$ on a pair of (finite) multisets.[4] In other words, there are finitely many wff's that occur in each sequent in σ, and this set remains finite even if we distinguish between the same wff in the antecedent and in the succedent. Let us assume

[3]This proof is, essentially, the one due to Kripke as presented in [23] and [7, §13].

[4]Kripke's sequents comprised two sequences of wff's, however, the difference has no significant impact on the proof. Though the proof is more transparent with multisets.

that with the latter distinction between antecedent and succedent formulas there are k wff's in the sequents in σ. (We denote the wff's by $\mathcal{A}_1, \ldots, \mathcal{A}_k$.) The proof is by induction on k.

1. *Base case.* If $k = 1$, then $\mathcal{A}_1$ can have n occurrences. Recall that the sequent is either

$$\underbrace{\mathcal{A}_1; \ldots; \mathcal{A}_1}_{n} \vdash \qquad \text{or} \qquad \vdash \underbrace{\mathcal{A}_1; \ldots; \mathcal{A}_1}_{n},$$

because $\mathcal{A}_1$ is a wff that occurs on only one side of the $\vdash$. We have not determined on which side $\mathcal{A}_1$ occurs, because that is not important. What is important is that any cognate sequent, with let us say m occurrences of $\mathcal{A}_1$ is either identical to the given sequent, or can be obtained by sufficiently many contractions from it, or conversely, sufficiently many contractions applied to the new sequent give the old one. In this simple case, we could have said that either $m = n$, or $m < n$ or $m > n$.

2. *Inductive case.* Let us assume that for k, any σ is finite. We will show that for $k+1$ formulas any irredundant sequence σ must be finite too. Given σ, an irredundant sequence, in which the sequents have $\mathcal{A}_1, \ldots, \mathcal{A}_{k+1}$ as their distinct formulas, there is an $\mathcal{A}_i$ and a sequent $\Gamma_n \vdash \Delta_n$ in σ such that for all $m > n$, the number of occurrences of $\mathcal{A}_i$ in $\Gamma_m \vdash \Delta_m$ is not greater (i.e., less than or equal to) the number of occurrences of $\mathcal{A}_i$ in $\Gamma_n \vdash \Delta_n$. If there were no such formula and sequent, then σ has to be infinite, and after a finite initial segment a sequent that can be contracted to an earlier one or identical to an earlier one must occur. Then the sequence is not irredundant, either because the sequents are not distinct or because later sequents can be contracted to earlier ones.

We modify each sequent in σ by deleting all occurrences of $\mathcal{A}_i$. The modified sequents are cognate and contain k distinct formulas. However, it may no longer be true that they are all distinct or that earlier sequents do not result by contractions from later ones. Thus we define a new sequence σ' by starting it with $(\Gamma_n \vdash \Delta_n)'$, which is the $\mathcal{A}_i$-less $\Gamma_n \vdash \Delta_n$ with least n. Among the modified sequents that had higher index than n, there is a sequent, let us say $\Gamma_j \vdash \Delta_j$, such that for some $\mathcal{A}_l$, the number of occurrences of $\mathcal{A}_l$ are not increasing after that sequent. We stipulate that $(\Gamma_j \vdash \Delta_j)'$ is the next element of σ', and so on. Then σ' satisfies the conditions of the lemma and contains k formulas, hence, by the hypothesis of induction, it is finite. But σ' is irredundant if σ is, and σ is finite, if σ' is. Therefore, if σ is irredundant, then it is finite. qed

Now we present a more detailed way to think about cognate sequents and their irredundant sequences. We hope that this will provide insights into what is at the core of Kripke's lemma. We will deal again with multisets, which allows a more elegant presentation here. However, we note that each finite sequence has finitely many permutations, thus, using sequences instead of multisets would create only finitely many more cognate sequents.

Definition 9.10. (Distant sequents) Let $\Gamma \vdash \Delta$ and $\Theta \vdash \Lambda$ be cognate sequents. They are *distant* iff they are distinct and neither the first can be obtained from the second by contractions nor can the second be so obtained from the first.

We want to count the *maximal number* of mutually *distant sequents* in the class of cognate sequents with n wff's. Then we will show that instantiating a pair of the parameters with concrete positive integers gives us an *upper bound*, let us say m, on the possible distant sequents with respect to a given sequent in the set of cognate sequents. Lastly, in an irredundant sequence of sequents, we can add only finitely many further sequents, which are no longer distant from the m sequents, but can follow them—without making the sequence redundant.

If we consider only one wff $\mathcal{A}$, then obviously, we can have one sequent to start with, but no other cognate sequent will be distant. We hasten to note that the size of σ (an irredundant sequence of cognate sequents) may be arbitrarily large in this case, because we may consider an arbitrarily large sequent—like a sequent containing a "zillion" occurrences of $\mathcal{A}$. To keep things to manageable size, let us say we start with $\mathcal{A}$ having 10 occurrences, for short, $\mathcal{A}^{10}$. Then the following sequence of sequents is irredundant. (We omit $\vdash$, because we do not know on which side $\mathcal{A}$ is, and the location of $\vdash$ is unimportant for irredundant sequences.)

$$\langle \mathcal{A}^{10}, \mathcal{A}^{7}, \mathcal{A}^{6}, \mathcal{A}^{4}, \mathcal{A}^{3}, \mathcal{A} \rangle$$

Obviously, if we start with $\mathcal{A}^{10}$, then the *maximal length* of a σ is 10. But we would like to think of all these sequents as of the same kind or of one *type*, because choosing any one of them, there is no other distant sequent.

In order to form the right abstraction, let us consider what happens when we have two formulas $\mathcal{A}$ and $\mathcal{B}$. Then we can have $\mathcal{A}^n, \mathcal{B}^m$ to start with. (We continue to think about sequents for the purposes of calculations as multisets, and we omit $\vdash$.) If $l < n$, then $\mathcal{A}^l, \mathcal{B}^m$ can be obtained by contractions from $\mathcal{A}^n, \mathcal{B}^m$. However, if we increase the number of $\mathcal{B}$'s at the same time that we decrease the number of $\mathcal{A}$'s, then the two sequents are distant. The $\mathcal{A}$'s and $\mathcal{B}$'s matter, so to speak, that is, increasing the number of $\mathcal{A}$'s and decreasing the number of $\mathcal{B}$'s yields yet another sequent, which is distant from both earlier ones. In general, we may be able—depending on the concrete numbers—to further decrease an already decreased number while further increasing an already increased number. For instance, given $\mathcal{A}^6, \mathcal{B}^5$, both $\mathcal{A}^5, \mathcal{B}^7$ and $\mathcal{A}^2, \mathcal{B}^8$ could be obtained by a decrease in the number of $\mathcal{A}$'s and an increase in the number of $\mathcal{B}$'s, and so we want to think of them as being of the same type. The relation of the third sequent to the second is exactly that of the second to the first.

At the moment, we want to disregard that it may be possible to repeat the increase–decrease moves. We will return to that later. Then, it also does not matter what the concrete numbers are—as long as they are sufficiently large.

From $\mathcal{A}^{101}, \mathcal{B}^{1300}$ we could get $\mathcal{A}^{8}, \mathcal{B}^{9631}$ and $\mathcal{A}^{102}, \mathcal{B}^{13}$, but we could make the initial $\mathcal{A}$'s and $\mathcal{B}$'s equal in number, by adding 1199 to the $\mathcal{A}$'s. This does not change the relationship between the three sequents. That is, the following three sequents—separated by semi-colons—are pairwise distant: $\mathcal{A}^{1300}, \mathcal{B}^{1300}$; $\mathcal{A}^{1207}, \mathcal{B}^{9631}$; $\mathcal{A}^{1301}, \mathcal{B}^{13}$. For classifying distant sequents that can be obtained from a given sequent, the *sufficiently large number* is 2. The three sequents above fall into three types: $\mathcal{A}^{2}, \mathcal{B}^{2}$; $\mathcal{A}^{1}, \mathcal{B}^{3}$; $\mathcal{A}^{3}, \mathcal{B}^{1}$.

One can think of the above notation as ^{1} representing *decrease*, whereas ^{3} representing *increase*. For two sequents not to be contractible to one another, a *contraction* (decrease in the number of occurrences of a formula) must be counterbalanced with at least one *expansion* (or increase in the number of occurrences of another formula), and vice versa.

Definition 9.11. Given a sequent $\mathcal{A}_1^{m_1}, \ldots, \mathcal{A}_n^{m_n}$, the *type* of the sequent is the n-tuple $\langle 2, \ldots, 2 \rangle$. The *kind* of a distant sequent is an n-tuple built of 1's, 2's and 3's. A distant sequent has a *type* iff its kind contains as many 1's as 3's.

Every distant sequent has a kind, however, some kinds subsume several types in the sense that other distant sequents become obtainable by contractions. (For instance, $\langle 3,3,1,3,3 \rangle$ is a kind of a distant sequent (relative to $\langle 2,2,2,2,2 \rangle$), but then $\langle 2,2,1,2,3 \rangle$, $\langle 2,2,1,3,2 \rangle$, $\langle 2,3,1,2,2 \rangle$ and $\langle 3,2,1,2,2 \rangle$ are no longer distant from $\langle 3,3,1,3,3 \rangle$.)

Given a sequent, every cognate sequent could be assigned an n-tuple of 1's, 2's and 3's. However, we are interested in the types, because they represent a subclass of cognate sequents, which are not only distant from the originally given sequent, but they are *pairwise distant* too.

To count the number of types, we look at the number of *ordered sums* with summands limited to 1, 2 and 3, and adding up to $2 \cdot n$, if there are n distinct formulas. It is obvious, that there are finitely many ordered sums for a positive integer; hence, also for $2 \cdot n$. Therefore, there are *finitely many ones* that involve only 1, 2 and 3. In fact, we can calculate how many types there are, because the number of types depends only on the number of distinct formulas in a sequent. We do not allow a decrease to be balanced by several increases, because that would merge several types into one type.

<u>*Example*</u> 9.12. The above modeling of the problem allows us to crunch some numbers. The cases when $n = 1$ and $n = 2$ are really easy. 2 can be written in one way (with 2 as the single summand), and there is just one type $\langle 2 \rangle$. $2 \cdot 2 = 2 + 2$, but also $4 = 1 + 3$ and $4 = 3 + 1$. Combinatorially, we can calculate by taking $2 \cdot 1$—there are *two* choices (1 or 3) for $\mathcal{A}$, but then $\mathcal{B}$ is determined, that is, there is *one* choice. The all-2's type, $\langle 2, 2 \rangle$ adds 1 to the number of types. In total, $2 + 1$ gives us three as the number of types of distant sequents composed of two formulas.

Every pair of 2's can be traded for $1 + 3$ and $3 + 1$. For $n = 5$, the ordered sums must contain 5, 3 or 1 2's. The former is possible in one way. The second is possible in $5 \cdot 4 = 20$ ways, because there are 5 places to situate

3, and then 4 places for 1. All the 2's are indistinguishable, hence their choice drops out. (A bit more formally, $5 \cdot 4$ would be multiplied by $3 \cdot 2 \cdot 1$ and the result divided by 3!, that is, by 6.) Lastly, we have 5 places for a 3, 4 places for another 3 and 3 places for a 2. However, the two 3's are indistinguishable, hence there are $(5 \cdot 4 \cdot 3)/2 = 30$ possibilities for ordered sums with just one 2. The total is 51, that is, there are 51 types of sequents that are distant with respect to a sequent with five distinct formulas.

Exercise 9.1.7. Calculate the number of types of distant sequents when there are 3, 4 and 6 distinct wff's in a fixed sequent.

Exercise* 9.1.8. Find a closed formula that for any n ($n > 2$), gives the number of types of distant sequents. [Hint: It may be useful to consider odd and even integers separately.]

Now we want to add back some of the details that we left out of the concept of types of distant sequents. In particular, we no longer assume that each formula has the same number of occurrences in our starting sequent or that any wff occurs twice in that sequent. Then some types (for $n > 1$) may have many sequents in them.

<u>Example</u> 9.13. To return to our earlier example, given $\mathcal{A}^{101}, \mathcal{B}^{1300}$ (from which we obtained $\mathcal{A}^{8}, \mathcal{B}^{9631}$ and $\mathcal{A}^{102}, \mathcal{B}^{13}$), we could also have $\mathcal{A}^{7}, \mathcal{B}^{9633}$; $\mathcal{A}^{5}$, $\mathcal{B}^{9634}$; $\mathcal{A}^{3}, \mathcal{B}^{9635}$; $\mathcal{A}^{2}, \mathcal{B}^{9663}$; $\mathcal{A}^{1}, \mathcal{B}^{9999}$ as well as $\mathcal{A}^{103}, \mathcal{B}^{11}$; $\mathcal{A}^{104}, \mathcal{B}^{8}$; $\mathcal{A}^{105}$, $\mathcal{B}^{5}$; $\mathcal{A}^{106}, \mathcal{B}^{2}$ and $\mathcal{A}^{107}, \mathcal{B}^{1}$. All these sequents are pairwise distant.

Exercise 9.1.9. Scrutinize whether all the sequents in the example are pairwise distant. Are there sequents that are distant from all the above sequents (and from each other)? [Hint: The increase–decrease moves in the example are not quite systematic, and there are some gaps left.]

Clearly, even if we have tight control over the number of types (e.g., only three types for two formulas), huge numbers of occurrences may give rise to large sets of distant sequents. However, *finiteness* is ensured. If we have 101 occurrences, then at most 100 decreases are possible. The corresponding increases may take as large jumps as one wants to have, because there is no largest positive integer. But to generate as many distant sequents as possible, the most economical decrease is by 1, and even then, after finitely many steps we run out of positive integers, that is, of a potential for further decreases.

<u>Example</u> 9.14. Let us consider a sequent with $\mathcal{A}$ and $\mathcal{B}$, as before, but with much smaller superscripts. Let $\mathcal{A}^{3}, \mathcal{B}^{2}$ be given. This is the sequent of type $\langle 2, 2 \rangle$. There is one sequent of type $\langle 3, 1 \rangle$, namely, $\mathcal{A}^{n}, \mathcal{B}^{1}$ (where $n \geq 4$). There are two sequents of type $\langle 1, 3 \rangle$. With minimal increases, they are $\mathcal{A}^{2}, \mathcal{B}^{3}$, and $\mathcal{A}^{1}, \mathcal{B}^{4}$. In general, the sequents are of the form $\mathcal{A}^{2}, \mathcal{B}^{m}$ and $\mathcal{A}^{1}, \mathcal{B}^{k}$ where $3 \leq m < k$.

Exercise 9.1.10. Let $\mathcal{A}^2, \mathcal{B}^3, \mathcal{C}^4$ be the given sequent. Generate the largest set of mutually distant sequents with minimal increases (i.e., with $+1$). [Hint: There are 7 types, but not all of them can be inhabited by multiple sequents.]

Now let us turn back to the *irredundant sequences* of sequents. First of all, we should note that if $\Gamma \vdash \Delta$ and $\Theta \vdash \Lambda$ are distant sequents, then they are cognate, and they can appear in any order without making the sequence redundant. However, two cognate sequents $\Gamma \vdash \Delta$ and $\Theta \vdash \Lambda$ can appear in an irredundant sequence even if one of them, let us say, $\Theta \vdash \Lambda$ can be obtained by contractions from $\Gamma \vdash \Delta$. Simply, $\Gamma \vdash \Delta$ must come *before* $\Theta \vdash \Lambda$. This means that we have to take into account sequents that may be obtained by one or more contractions. The number of sequents that can be obtained via contractions is *finite* too, because to obtain another sequent, the number of occurrences of at least one formula must be decreased by at least one.

Example 9.15. Let us assume that given $\mathcal{A}^3, \mathcal{B}^2$ we obtained a set of mutually distant sequents via increases by $+1$. Our set of distant sequents contains three more sequents $\mathcal{A}^4, \mathcal{B}$; $\mathcal{A}^2, \mathcal{B}^3$; $\mathcal{A}, \mathcal{B}^4$ in addition to the original $\mathcal{A}^3, \mathcal{B}^2$. Via contractions, we can get six more sequents, which are $\mathcal{A}^2, \mathcal{B}^2$; $\mathcal{A}^2, \mathcal{B}^1$; $\mathcal{A}^3, \mathcal{B}^1$; $\mathcal{A}^1, \mathcal{B}^3$; $\mathcal{A}^1, \mathcal{B}^2$; $\mathcal{A}^1, \mathcal{B}^1$. The binary relation on the set of sequents in which the first yields the second via contractions defines a (strict) *partial order* on the set of these sequents as shown in Figure 9.1.

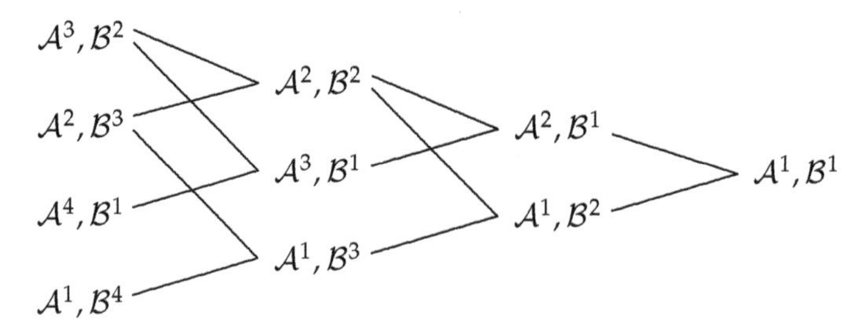

Figure 9.1. Partial order on a set of cognate sequents.

The sequents that are listed in a column are not ordered with respect to each other. But some of the sequents in different columns are not ordered by the relation either. For instance, $\mathcal{A}^1, \mathcal{B}^4$ and $\mathcal{A}^2, \mathcal{B}^1$ are distant, though the latter is not distant from the given sequent $\mathcal{A}^3, \mathcal{B}^2$. This means that there are many possibilities to *linearize* this partially ordered set of cognate sequents so that the resulting sequence is irredundant. We can simply go from the top to the bottom and from left to right. (For readability, we separate the sequents from each other by inserting ; between them.)

$$\langle \mathcal{A}^3, \mathcal{B}^2; \mathcal{A}^2, \mathcal{B}^3; \mathcal{A}^4, \mathcal{B}^1; \mathcal{A}^1, \mathcal{B}^4; \mathcal{A}^2, \mathcal{B}^2; \mathcal{A}^3, \mathcal{B}^1; \mathcal{A}^1, \mathcal{B}^3; \mathcal{A}^2, \mathcal{B}^1; \mathcal{A}^1, \mathcal{B}^2; \mathcal{A}^1, \mathcal{B}^1 \rangle$$

Exercise 9.1.11. Create other (or more ambitiously, create all the possible) irredundant sequences from the above set.

If we take as our starting sequent another sequent cognate with $\mathcal{A}^3, \mathcal{B}^2$, then the three types may have many more inhabitants. Depending on how large the numbers are that we pick for the increases, we may get way more sequents by contractions of the distant sequents inhabiting the three types.

Example 9.16. To illustrate the effect of the upward jumps on the maximal length of irredundant sequences, let us assume that we start with $\mathcal{A}^3, \mathcal{B}^2$, but we make slightly bigger upward jumps. Figure 9.2 shows the partial order on the set of sequents we generated.

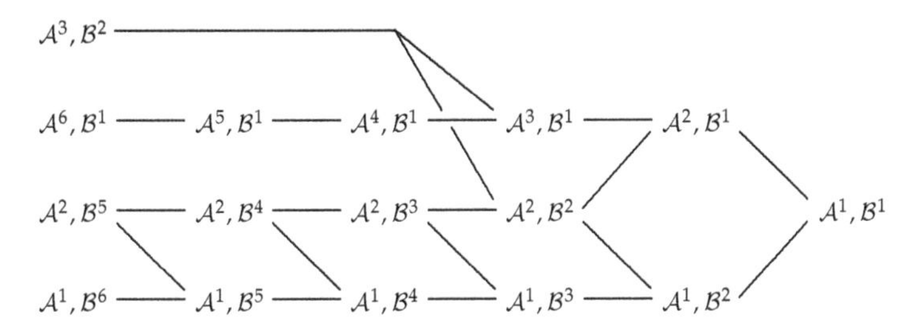

Figure 9.2. Partial order on a larger set of cognate sequents.

To sum up our findings, *finiteness* is guaranteed at each step. Given a concrete sequent, there are finitely many *types*. Each type can be *inhabited* by finitely many sequents, though some of the superscripts can be arbitrarily large (compared to the superscript on the same formula in the starting sequent). Once a maximal set of mutually distant sequents is fixed (i.e., the superscripts are concretized), there are finitely many sequents that can be obtained via *contractions* from those sequents.

We could conceptualize the gist of what happens in irredundant sequences by saying that as the sequence progresses, contractions must happen to ensure the distinctness of sequents (since no expansion can happen without making the sequence redundant). Taking away a small finite amount (e.g., 1) in one place (e.g., $\mathcal{A}^{n-1}$) opens up the possibility of adding a large finite amount in another place (e.g., $\mathcal{B}^{l+M}$), but these are *irreversible* and *asymmetric* moves. Once we have carried this trick out, the taking away a little in the second place (e.g., $\mathcal{B}^{l+M-1}$) while magnifying in the first place (e.g., $\mathcal{A}^{n-1+G}$) is no longer an option. (M and G are intended to suggest mega- and giga-size numbers.)

Dunn explains Kripke's lemma in [77, §3.6] and illustrates cognate sequents with two distinct formulas in a two-dimensional diagram. We enrich the idea of using the top right quadrant of the coordinate system with the concept of distant sequents. We will locate the sequents from Figure 9.2 in the diagram below.

The $\bullet$ is the given sequent, and the $*$'s are the three mutually distant sequents chosen in Example 9.16. The $\circ$'s are the sequents that can potentially

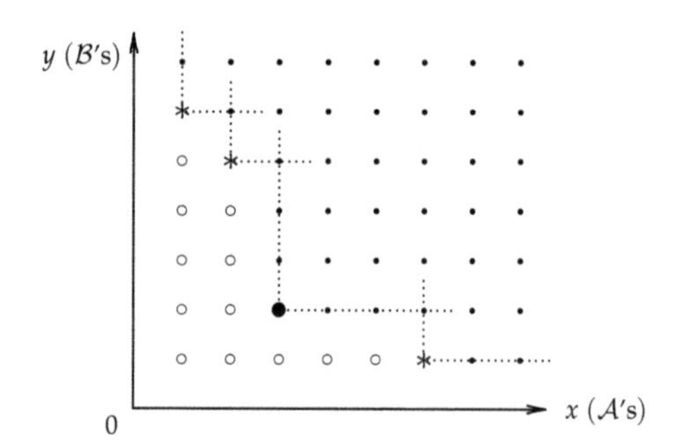

Figure 9.3. Finite enclosure by distant sequents.

be added into an irredundant sequence—if they are carefully ordered. The dotted lines indicate how each of the four mutually distant sequents exclude infinitely many sequents, some of which are shown as •'s. Although not everything generalizes easily (or at all) from two to three or to higher dimensions even in Euclidean space, the above modeling does. Although drawing a three-dimensional coordinate system is feasible, higher dimensional spaces are more difficult to visualize. However, we hope that this detailed treatment of the notion of irredundant sequences is convincing, and the truth of Kripke's lemma is now obvious.

Let us now return to the proof-search tree. The *cut theorem* for $[LR^t_{\underset{\sim}{\rightarrow}}]$ guarantees the subformula property. All the formulas in $R^t_{\underset{\sim}{\rightarrow}}$ have finitely many subformulas, and their set can be obtained in an effective way from a wff. *Curry's lemma* guarantees that no contraction is missed by building contraction into the connective rules, and that no branch in a proof contains sequents that would form an irredundant sequence of cognate sequents. The finite fork property is guaranteed by there being finitely many rules and each sequent containing finitely many formulas. The connective rules typically do not lead to a sequent that is cognate with its premises. If we look at a connective rule in a proof-search tree (bottom-up), then we see that the principal formula is replaced with shorter formulas, which is a component in the termination of the search. However, even if the upper and lower sequents in an application of a rule are cognate, *Kripke's lemma* guarantees that there are at most finitely many such sequents in a proof. (In fact, due to the shape of the rules, the big swings in the numbers of occurrences of wff's that we illustrated above are irrelevant here, even if they are not excluded by the definition of irredundant sequences.) The finiteness of the proof-search tree guarantees that if a formula is not provable, and so no proof can be found, the whole procedure, nonetheless, terminates.

Corollary 9.16.1. *The logics* $R^t_{\rightarrow}$ *and* $R_{\rightarrow}$ *are* decidable.

In Section 4.3, we pointed out that multiple right-hand sides do no harm when there is no way to introduce multiple wff's on the right in a calculus. This is the case for $LR_{\rightarrow}$ and $LR^t_{\rightarrow}$. A definition of a sequent calculus with multiple right-hand sides for these logics may look a little overindulgent. However, as we see now, this allows us to obtain the decidability of $R_{\rightarrow}$ and $R^t_{\rightarrow}$ as a corollary.

9.1.3 Pure ticket entailment is decidable

The decidability problem of $T_{\rightarrow}$, of the *implicational fragment of ticket entailment*, turned out to be even more recalcitrant than the decidability problem of R and E. It is likely that more attention was paid to the latter problem, though it seems that everybody who worked on the latter problem, did work on the former too. The problem appeared not only in [7, p. 69], but some twenty years later in [108, p. 106].[5]

$T_{\rightarrow}$ has a short and elegant formulation, and the Curry–Howard isomorphism extends to this logic straightforwardly. Thus it should not be surprising that this decidability problem captured the attention of the typed-λ-community. Keeping up with the practices of the time, the problem appeared on the website of the TLCA (Typed Lambda Calculi and Applications) around 2006.[6]

Originally, $T_{\rightarrow}$ was defined as a proof system, in particular, it was first carved out from Ackermann's strong implication logic very much like $E_{\rightarrow}$ was. Anderson in [5] characterized it by omitting not only Ackermann's γ rule, but also the δ rule (which is included in E). He also described a natural deduction system. Anderson's intuition about T somehow "lacking modality" (that E possesses) proved to be correct in the sense that the diminished amount of permutation cannot be characterized in the same way in T and E. Arguably, this is a reason why there were no suitable sequent calculi for T_{+} or even for $T_{\rightarrow}$ for a long time. Anderson and Belnap explicitly ask the question (in [7, p. 69]) whether decidability can be based on the merge systems. However, the complicated descriptions of the merge systems, and perhaps the absence of an independent source for understanding the mathematical properties of merged sequents seem to have left this question ignored altogether. $LT^t_{\rightarrow}$ and the decision procedure described here gained more inspiration from structurally free logics, which are presented in Section 5.2, and the decision procedure for $R_{\rightarrow}$ than from the ideas behind the merge calculi. $LT^t_{\rightarrow}$—which has a cut theorem and the subformula property—was created in 2005. It took a few more years (until 2010, to be precise) to find a way to

[5] It is also mentioned in [56, p. 373], and it is mistakenly claimed in [86, p. 227] to having been solved by Urquhart in [186]. The latter mistake may have resulted from a misunderstanding of English sentences like "There is no primitive recursive decision procedure for $\mathbf{R}_{\rightarrow\wedge}$, $\mathbf{E}_{\rightarrow\wedge}$ or $\mathbf{T}_{\rightarrow\wedge}$." in Urquhart's article [185, p. 1797], which was also published in 1999, like [186].

[6] The URL is `tlca.di.unito.it/opltlca`—as of August 2013.

use an extension of this calculus to reduce the decidability problem for $T^t_{\rightarrow}$ to the decidability problem of $R^t_{\rightarrow}$.[7]

The branching out into the computer science community via simple types paved another route toward the solution of the problem—via long normal inhabitants (which are certain λ-terms) of types.[8] Being a fan of combinatory logic, I tend to think that type-assignment systems involving implicational wff's are more elegant for combinators (in the sense of combinatory logic) than for λ-terms (including combinators in the sense of λ-calculi).[9] In any case, it would be too big of a diversion from the topic of this book to attempt to explain how the counting of certain λ-terms (so-called long normal form inhabitants) could solve the decidability problem for $T_{\rightarrow}$. We describe here the *proof-theoretic solution*.[10]

We defined a consecution calculus for $T^t_{\rightarrow}$ in Section 5.3, where we also described the consecution calculus $LR_{\rightarrow;}$ (together with three of its variants, which were distinguished by numerical subscripts). However, in Section 4.3, we have already seen $LR_{\rightarrow}$ and $LR^t_{\rightarrow}$, which are fragments of $LR^t_{\overset{\sim}{\rightarrow}}$, the contraction-free modification of which featured in the previous section. The only sequent calculus for $T_{\rightarrow}$ that is not a consecution calculus (i.e., it uses an associative structural connective) is the merge calculus $L_\mu T_{\rightarrow}$. For merge calculi, it is easy to get from $L_\mu T_{\rightarrow}$ to $L_\mu R_{\rightarrow}$—one simply omits the so-called ticket restriction (which, roughly speaking, amounts to the non-emptiness of left-hand side of the left premise in $(\rightarrow\vdash)$; see Section 4.8). Unfortunately, as we pointed out above, there is no known decision procedure based on merge calculi, which include an explicit contraction rule, and have not been formulated without a contraction rule. (We note though that $L_\mu R_{\rightarrow}$, probably, would not be much of a problem, because one could dispense with merging sequences, and in effect, simply return to $LR_{\rightarrow}$.)

We can easily obtain an axiomatic formulation of $R_{\rightarrow}$ from that of $T_{\rightarrow}$. Modus ponens is the rule in each of these calculi. (A1)–(A4) on page 167 is $T_{\rightarrow}$, and adding (A5) or (A6) yields $R_{\rightarrow}$. The problem is that if we have the axioms (A1)–(A2) plus (A4)–(A5), then (A3) is a theorem, whereas (A1)–(A2) and (A4) do not imply (A3). Similarly, as we have shown in Section 5.3, if we add a permutation rule like $(\mathsf{C}\vdash)$ or $(\mathsf{T}\vdash)$ to $LT^t_{\rightarrow}$, then we not only obtain a formalization of $R^t_{\rightarrow}$, but the $(\mathsf{B}'\vdash)$ rule becomes admissible. To put it yet another way, all the permutation licensed by (A3) is covered by (A5); therefore, adding (A5) or its equivalent in a consecution calculus obliterates any trace of a permutation that could have come from or was sanctioned by (A3). We may have a proof in $LR^t_{\rightarrow;}$ of a formula, but we may not be able to

[7]The results we have just mentioned may be found in [32], [41] and [42].

[8]See [108], [56] and [148].

[9]My fondness for combinators is evidenced by a bunch of my writings such as [27], [28], [29], [36] and [37].

[10]This section presents the proof of the decidability of $T_{\rightarrow}$ from [41] and [42] in a much shorter and less detailed form.

tell whether some permutation in the proof is a permutation in addition to those that could happen in $LT^t_{\rightarrow}$.

The presence of t may be thought to add some complications to the consecution calculi. As we have seen in Sections 4.2 and 5.3, in some cases, special structural rules are needed and their presence may render the usual double inductive proof of the cut theorem unavailable. The combinatorial investigations into $R_{\rightarrow}$, $E_{\rightarrow}$ and $T_{\rightarrow}$ led to the idea that their extensions with t can be viewed as *essentially* $\mathsf{BB'IW}$ *logics*, in which t has different properties. In $T^t_{\rightarrow}$, thought of as the axiom system $T_{\rightarrow}$ extended with t and $t \rightarrow \mathcal{A} \rightarrow \mathcal{A}$ as axioms, t is a left identity for the operation $\circ$ (which is not in the language, but can be conservatively added and of which $\rightarrow$ is a residual). In $E^t_{\rightarrow}$, t is a left identity and an upper right identity, whereas in $R^t_{\rightarrow}$, t is the identity.

The *core idea* behind the calculus $LT^{\textcircled{t}}_{\rightarrow}$ is to make more use of t—if we already have to contend with its presence in $LT^t_{\rightarrow}$ anyway. If we add a rule to $LT^t_{\rightarrow}$ that allows us to introduce t on the right-hand side of a structure, then we make t a lower right identity. We already know from Section 5.3 that the rule $(\hat{t}_r \vdash)$ makes t an upper-right identity. But a disadvantage of the latter rule is that it ruins the subformula property if we look at proofs that may not be proofs of a theorem (but only of a consecution). However, experience with special structural rules that involve t instigates the insight that a structural rule that is specific to t may be added. Since $R^t_{\rightarrow}$ has full permutation, we can add arbitrary permutation rules with t. However, the following proof suggests one particular rule. The proof assumes the usual rules for $\circ$ in consecution calculi, which in turn permit the straightforward conceptualization of the differences between identities as we have explained.

$$\dfrac{\dfrac{\dfrac{\mathcal{A} \vdash \mathcal{A} \qquad t \vdash t}{\mathcal{A}; t \vdash \mathcal{A} \circ t}\ {\vdash\circ}}{t; \mathcal{A} \vdash \mathcal{A} \circ t}\ \mathsf{T}_t{\vdash}}{t \vdash \mathcal{A} \rightarrow (\mathcal{A} \circ t)}\ {\vdash\rightarrow}$$

The step that we labeled with $(\mathsf{T}_t \vdash)$ is an instance of our proposed new rule. Of course, we have to make sure that the general form of the rule is properly structuralized and contextualized. The above proof also clarifies that this rule does not interact with any of the t introduction rules. In comparison to a blanket permutation rule like $(\mathsf{T}\vdash)$, the restricted rule $(\mathsf{T}_t \vdash)$ accomplishes very little. But that is exactly what we want, so that we can track what results from the extra rules added to $LT^t_{\rightarrow}$ to obtain $LT^{\textcircled{t}}_{\rightarrow}$.

The *consecution calculus* $LT^{\textcircled{t}}_{\rightarrow}$ is the extension of $LT^t_{\rightarrow}$ by the following two rules. (The axiom and the rules of $LT^t_{\rightarrow}$ are on page 178.)

$$\dfrac{\mathfrak{A}[\mathfrak{B}] \vdash \mathcal{A}}{\mathfrak{A}[\mathfrak{B}; t] \vdash \mathcal{A}}\ \mathsf{K}_t{\vdash} \qquad\qquad \dfrac{\mathfrak{A}[\mathfrak{B}; t] \vdash \mathcal{A}}{\mathfrak{A}[t; \mathfrak{B}] \vdash \mathcal{A}}\ \mathsf{T}_t{\vdash}$$

Before we make claims about $LT^{\textcircled{t}}_{\rightarrow}$, we should point out that the *circle* around t in the label is intended. $LT^t_{\rightarrow}$ and $LT^{\textcircled{t}}_{\rightarrow}$ are in the same language, but the latter is not a conservative extension of the former. The circle alludes to the ability of $\boldsymbol{t}$ to move around: it can move toward the left by applications of the $(\mathsf{B}'\vdash)$ or $(\mathsf{T}_t\vdash)$ rules, where it can be absorbed by other $\boldsymbol{t}$'s through applications of the $(\mathsf{M}_t\vdash)$ rule; but a $\boldsymbol{t}$ can re-appear on the right-hand side, by the $(\mathsf{K}_t\vdash)$ rule.

The single cut rule is *admissible* in $LT^{\textcircled{t}}_{\to}$. The two new rules are sufficient—together with $LT^{t}_{\to}$, of course—to capture $R^{t}_{\to}$. We know that $R^{t}_{\to}$ is decidable. It was first proved in [41], but the result also follows from the decidability of $LR^{t}_{\tilde{\to}}$, as stated in corollary 9.16.1 above.

Exercise 9.1.12. Prove the cut theorem for $LT^{\textcircled{t}}_{\to}$. [Hint: The basic structure of the proof with some details included may be found in [41, §5].]

Exercise 9.1.13. Give proofs for the axioms of $R_{\to}$ in $LT^{\textcircled{t}}_{\to}$. Show that modus ponens and the $\boldsymbol{t}$ rules (i.e., $\vdash \mathcal{A}$ iff $\vdash \boldsymbol{t} \to \mathcal{A}$) can be emulated by this consecution calculus. [Hint: A set of axioms is listed on page 104.]

Exercise 9.1.14. In Section 5.3, we listed some further axioms as (A7)–(A14). Find proofs for these in $LT^{\textcircled{t}}_{\to}$. [Hint: Try not to use the $(\mathsf{K}_t\vdash)$ or $(\mathsf{T}_t\vdash)$ rules, whenever it is possible not to use them.]

Exercise* 9.1.15. Prove that $LT^{\textcircled{t}}_{\to}$ is exactly $R^{t}_{\to}$. [Hint: A way to prove this is by adding $\circ$ to $LT^{\textcircled{t}}_{\to}$, and by establishing that the addition is conservative. Then, from the equivalence of $LT^{\textcircled{t}}_{\to\circ}$ and $R^{\circ t}_{\to}$, one can conclude the desired result, with the help of the conservativeness of $\circ$ over $R^{t}_{\to}$.]

Once we have created $LT^{\textcircled{t}}_{\to}$, and we know that it formalizes $R^{t}_{\to}$, the blueprint of a decision procedure is straightforward. We, of course, also know that every theorem of $T^{t}_{\to}$ is a theorem of $R^{t}_{\to}$. Given a wff $\mathcal{A}$ we determine using $[LR^{t}_{\to}]$ from Section 4.3, if the wff is a theorem of $R^{t}_{\to}$. If it is not, then we are finished, because $\mathcal{A}$ is surely not a theorem of $T^{t}_{\to}$.

If $\mathcal{A}$ is a theorem of $R^{t}_{\to}$, then it has *finitely many proofs* in $[LR^{t}_{\to}]$. We take the finitely many proofs and perform a series of transformations on each of them until we end up with a corresponding *finite set of proofs* in $LT^{\textcircled{t}}_{\to}$.

The *first transformation* makes all the contractions (if there are any in the proof) explicit.[11] A proof in $[LR^{t}_{\to}]$ has a unique annotation. The labels that we often attach to proofs are not part of the proof, but in $[LR^{t}_{\to}]$, if we are presented with a concrete proof, then there is no ambiguity as to which rules are applied at each step. This is true for $([\to\vdash])$ too; moreover, simple counting can determine which formulas have been contracted and how many times a formula has been contracted. The given proof is unaltered except possibly where $([\to\vdash])$ was applied; if there is no contraction in $([\to\vdash])$, then there is no change at this step; if $\mathcal{A}_1, \ldots, \mathcal{A}_n$ were contracted, then contractions are inserted after $(\to\vdash)$ and above the next node. Only the principal formula

[11] We assume here that $LR^{t}_{\to}$ is based on multisets—the possibility we mentioned in Section 4.3. See also [41] and [42]. Using sequences of wff's would add a small, but insignificant complication to the procedure.

of the rule can be contracted more than once, and if so, then two contractions are added on that formula. It is obvious that we get a proof in $LR^t_{\to}$. However, if there are at least two contractions on different formulas, then we collect all the distinct $LR^t_{\to}$ proofs (of which there are finitely many) that result by permutations of the contractions.

The *second transformation* massages a proof in $LR^t_{\to}$ into a proof (typically, another proof) in the same calculus. Recall that in $LR^t_{\to}$ there is an axiom that has no matching pair in $LT^{\textcircled{t}}_{\to}$, namely, $(\vdash t)$. Correspondingly, $\mathcal{A}$ is a theorem of $LT^{\textcircled{t}}_{\to}$ iff $t \vdash \mathcal{A}$ is provable, whereas, in $LR^t_{\to}$, $\vdash \mathcal{A}$ suffices. But these differences are cosmetic, because within $LR^t_{\to}$, $t \vdash \mathcal{A}$ is immediate from $\vdash \mathcal{A}$, and conversely, if $t \vdash \mathcal{A}$ is provable, then so is $\vdash \mathcal{A}$. We modify a given proof as follows. Each axiom $(\vdash t)$ is replaced by an instance of (id), that is, by $t \vdash t$. These new t's (if any) are propagated everywhere downward. We note that the addition of a t on the left of the turnstile leaves every instance of an application of a rule intact, that is, it remains an instance of an application of the same rule with an extra t. Now we scrutinize the proof once more, and if there is a sequent with an empty left-hand side (by necessity, after an application of $(\vdash\to)$), then we add a $(t\vdash)$ step before the $\to$ introduction rule; we carry this new t downward too. These modifications result in a proof. Furthermore, all the axioms are instances of (id), and there is no sequent in the proof with an empty left-hand side. However, the bottom sequent has changed, because it contains as many t's as there were $(\vdash t)$ axioms and sequents with an empty left-hand side. However, by finitely many contractions, we can obtain $t \vdash \mathcal{A}$ instead of the original $\vdash \mathcal{A}$. The contractions are unproblematic, and we have a proof to use in the last transformation.

The *third transformation* takes a proof of the latest kind and turns it into a finite set of proofs in $LT^{\textcircled{t}}_{\to}$. Typically, there will be more than one proof in the resulting set. We know for sure that having started with a proof ending with $\vdash \mathcal{A}$ in $LR^t_{\to}$, we have a proof ending with $t \vdash \mathcal{A}$ in $LT^{\textcircled{t}}_{\to}$, because all the rules in $LR^t_{\to}$ are rules in $LT^{\textcircled{t}}_{\to}$ modulo some permutation and regrouping of the formulas with respect to their listing as a multiset. In fact, $(\mathsf{W}\vdash)$ in $LT^{\textcircled{t}}_{\to}$ is a more powerful rule than $(W\vdash)$ in $LR^t_{\to}$, because it allows the contraction of several formulas in one step. The rules for t, importantly including $(\mathsf{K}_t\vdash)$ and $(\mathsf{T}_t\vdash)$—together with $(\mathsf{B}\vdash)$ and $(\mathsf{B}'\vdash)$—are sufficient to generate all the permutations and groupings.

Exercise* 9.1.16. Prove this last claim. [Hint: Extra t's may temporarily appear in a consecution, but as long as they are gathered in a suitable place, they do not cause a problem. (See also [42, §3].)]

Our goal is to generate at each step in a proof all the possible ways in which a rule could be applied. We do not aim at finding "the proof" of a formula in $LT^{\textcircled{t}}_{\to}$. Actually, every theorem of $R^t_{\to}$ has infinitely many proofs in $LT^{\textcircled{t}}_{\to}$ (just like in $LR^t_{\to}$). For instance, the presence of t in every proof of a formula—together with the $(t\vdash)$ and $(\mathsf{M}_t\vdash)$ rules—guarantees the multiplicity of proofs. Similar moves in $LR^t_\to$ yield cut-free proofs, when the end sequent has an occurrence of t, and proofs with cut, otherwise. However, we are not transforming infinitely many $LR^t_\to$ proofs, but a finite subset of the them. Permutations and regroupings (may) increase the number of proofs too, but a finite sequence has finitely many groupings and permutations.

We proceed systematically in the proof tree, from the top to the bottom, and from left to right. We also create a correspondence between consecutions in the generated proofs in $LT^{\textcircled{t}}_\to$ and sequents in the given proof in $LR^t_\to$. We number the cases for clarity.

1. If the node is a leaf, that is, an instance of (id) in $LR^t_\to$, then it is an instance of (id) in $LT^{\textcircled{t}}_\to$. (Recall that we are past the second transformation.)
2. If the node is by an application of the rule $(t\vdash)$, then we make as many copies of each corresponding consecution as they have substructures, and insert a t on the left of the substructures. If a corresponding consecution has a substructure t, then we also retain a copy of the consecution. All these consecutions correspond to the sequent that was obtained by $(t\vdash)$. (We note that the retained copies could be thought of as a result of $(t\vdash)$ and $(\mathsf{M}_t\vdash)$ applied successively, in which case there are two consecutions inserted into each of these proofs and the lower one corresponds to a sequent obtained by $(t\vdash)$. But we do not really need the extra steps—except as an explanation.)
3. If the node is by an application of the $(\vdash\to)$ rule, then the subalterns and the principal formula are determined, let us say, $\mathcal{A}$, $\mathcal{B}$ and $\mathcal{A}\to\mathcal{B}$. From the consecution that corresponds to the upper sequent of the rule, we generate all the antecedents of the form $\mathfrak{A};\mathcal{A}$ (where $\mathfrak{A}$ is the rest of the antecedent of the starting consecution, possibly, with some t's added). Then we apply the $(\vdash\to)$ rule. The resulting consecutions correspond to the original lower sequent.

The *generation of all antecedents* may involve the rules $(\mathsf{B}\vdash)$, $(\mathsf{B}'\vdash)$, $(\mathsf{T}_t\vdash)$, $(\mathsf{K}_t\vdash)$ and $(\mathsf{M}_t\vdash)$. We could allow the $(t\vdash)$ rule to be applied too, however, due to the shape of the three first rules mentioned, $(t\vdash)$ cannot contribute to permutations and regroupings.

Exercise 9.1.17. Verify that the $(t\vdash)$ rule, though it can add a t, and thereby, add new substructures, cannot facilitate getting $\mathfrak{B};\mathfrak{A}$ from $\mathfrak{A};\mathfrak{B}$ using the $(\mathsf{B}\vdash)$, $(\mathsf{B}'\vdash)$ and $(\mathsf{T}_t\vdash)$ rules.

The role of the $(\mathsf{M}_t\vdash)$ rule is to reduce duplicate t's occurring next to each other to a single t. Multiple adjacent t's can always be restored, if necessary, by the $(t\vdash)$ rule without jeopardizing the proof remaining within $LT^t_\to$.

There are finitely many ways to permute and regroup the atomic structures in each starting consecution. Since all the t's that may be inserted to facilitate permutations and regroupings are the same formulas, their permutations and groupings have no importance, and do not need to be considered.

From now on, we assume that the generation of all possible structures of some form or of all possible antecedents is understood in the described way.

The rules involved in a concrete proof justify the proof steps that lead to the consecutions that are inserted between the starting consecution and the one where the matching rule of $LT^{\textcircled{t}}_{\rightarrow}$ is applied.

4. If the sequent is by $(\rightarrow\vdash)$, then the subalterns and the principal formula of the rule are determined. Let these be $\mathcal{A}$, $\mathcal{B}$ and $\mathcal{A} \rightarrow \mathcal{B}$ again. Then from the consecutions that correspond to the premise sequents of the rule, we generate all the consecutions of the form $\mathfrak{A} \vdash \mathcal{A}$ and $\mathfrak{B}[\mathcal{B}] \vdash \mathcal{C}$, where $\mathfrak{A}$ and $\mathfrak{B}$ are all the possible permutations and groupings (with extra $\boldsymbol{t}$'s, if necessary). Then an application of $(\rightarrow\vdash)$ in each pairing results in $\mathfrak{B}[\mathcal{A} \rightarrow \mathcal{B}; \mathfrak{A}] \vdash \mathcal{C}$, and these are the consecutions that correspond to the lower sequent.

5. The last possibility is that the node is by $(W\vdash)$. This is a delicate situation, because $(\mathrm{W}\vdash)$ in $LT^{\textcircled{t}}_{\rightarrow}$ is more powerful than $(W\vdash)$ is in $LR^{\boldsymbol{t}}_{\rightarrow}$. On the other hand, the structural connective of $LR^{\boldsymbol{t}}_{\rightarrow}$ is more permissive than the one in $LT^{\textcircled{t}}_{\rightarrow}$. In order to make sure that we do not miss the opportunity to use $(\mathrm{W}\vdash)$ on a complex structure, if we encounter an application of the $(W\vdash)$ rule, then we look further down in the $LR^{\boldsymbol{t}}_{\rightarrow}$ proof to see whether other contractions follow next. If there are no other contractions, then we proceed similarly, as with other rules. That is, if the principal formula of $(W\vdash)$ is $\mathcal{A}$, then we generate from the consecution corresponding to the premise sequent of $(W\vdash)$ all the consecutions, possibly, with some extra $\boldsymbol{t}$'s, that are of the form $\mathfrak{A}[\mathfrak{B}; \mathcal{A}; \mathcal{A}] \vdash \mathcal{B}$. The lower consecutions are by $(\mathrm{W}\vdash)$, and they are of the form $\mathfrak{A}[\mathfrak{B}; \mathcal{A}] \vdash \mathcal{B}$. These consecutions correspond to the lower sequent of the $(W\vdash)$ rule.

If there are other applications of $(W\vdash)$, then we consider all those contractions at once. We divide the series of contractions into groups of contractions. To count the number of the latter possibilities, we turn to ordered sums again. This time, we generate all the ordered sums for n (if there are n consecutive contractions) without limitation on the size of positive integers. There are 2^{n-1} ordered sums for $n > 1$—counting n itself as a sum. For each summand, the principal formulas are given by the respective contractions in the $LR^{\boldsymbol{t}}_{\rightarrow}$ proof. There are finitely many (complex) structures that can be formed from them, let us say, $\mathfrak{B}_1, \ldots, \mathfrak{B}_m$. Then for each i $(1 \leq i \leq m)$, we generate all the possible consecutions with occurrences of these, as $\mathfrak{A}[\mathfrak{C}; \mathfrak{B}_i; \mathfrak{B}_i] \vdash \mathcal{A}$. When we have $\mathfrak{A}[\mathfrak{C}; \mathfrak{B}_i] \vdash \mathcal{A}$ for each, we proceed to the next summand. Clearly, there is a lot of bookkeeping in this step; notwithstanding, everything is finite. n elements (without repetitions) have $n!$ permutations; with $m_1, \ldots, m_l$ identical elements, there are

$$\frac{n!}{m_1! \cdot \ldots \cdot m_l!}$$

permutations. The number of all complex structures that arise from n distinct formulas can be calculated by multiplying $n!$ with the $n-1$th Catalan number. The nth Catalan number is given by the following formula.

$$C_n = \frac{1}{n+1}\binom{2n}{n}$$

Exercise 9.1.18. Choose a couple of theorems of $[LR^{t}_{\to}]$ and work out some parts of some of the transformations described. [Hint: You might try to pursue a thread toward a complete proof—instead of considering all the possibilities at each stage, which may quickly become overwhelming, even if there are only finitely many possibilities.]

Of course, having multiple proofs of an $R^{t}_{\to}$ theorem is pleasing, but it is not our goal. Given the set of proofs of $\mathcal{A}$ in $LT^{\textcircled{t}}_{\to}$ that are derived through these three transformations above from the set of proofs of $\mathcal{A}$ in $[LR^{t}_{\to}]$, we want to find out if any of the proofs is a proof in $LT^{t}_{\to}$. The proofs can be almost uniquely annotated in $LT^{\textcircled{t}}_{\to}$.

Exercise 9.1.19. Imagine a proof, and describe how the annotation algorithm would proceed. Where can ambiguity arise? What would be a good strategy for disambiguation?

We may assume that we have the annotation set up so that if a proof *could* belong to $LT^{t}_{\to}$, then it *does*. The last step in the decision procedure is the *inspection of the proofs* of the given formula. We have a finite set of proofs we derived from the proofs of the formula in $[LR^{t}_{\to}]$, thus we can go through the proofs in finitely many steps. (We can take some shortcuts too, e.g., if we find an application of $(\mathsf{K}_t\vdash)$ or $(\mathsf{T}_t\vdash)$, then we can skip checking the rest of the proof, because it is surely not an $LT^{t}_{\to}$ proof.) If *none of the proofs* is an $LT^{t}_{\to}$ proof, then the formula is not a theorem of $T^{t}_{\to}$. However, if we have found *at least one proof* that belongs to $LT^{t}_{\to}$, then the formula is a theorem of $T^{t}_{\to}$, because the sets of theorems of $LT^{t}_{\to}$ and $T^{t}_{\to}$ coincide.

The decision procedure is applicable to $T_{\to}$ *theorems* too. This places a restriction on the formula itself, since it cannot contain occurrences of $\boldsymbol{t}$. If a $\boldsymbol{t}$-free formula is given, then the above procedure can be carried out unchangeably. The role of the $\boldsymbol{t}$'s in a proof of a $\boldsymbol{t}$-free formula is limited to structural modifications, and to indicating the theoremhood of the wff in the consecution in the root of the proof tree.

9.1.4 Pure entailment and relevant implication with $\circ$ and t

The view that the main relevance logics are $\mathsf{BB'IW}$ logics, suggests that we may try to formulate a consecution calculus for $E^{t}_{\to}$ differently than $LE^{t}_{\to}$, which we presented in Section 5.3. We add $\circ$, the fusion operation, too.

The *consecution calculus* $LE^{\circ t}_{\to 2}$ is defined by the following axiom and rules.[12]

$$\mathcal{A} \vdash \mathcal{A} \ \text{id}$$

$$\frac{\mathfrak{A} \vdash \mathcal{A} \quad \mathfrak{B}[\mathcal{B}] \vdash \mathcal{C}}{\mathfrak{B}[\mathcal{A} \to \mathcal{B}; \mathfrak{A}] \vdash \mathcal{C}} \to\vdash \qquad \frac{\mathfrak{A}; \mathcal{A} \vdash \mathcal{B}}{\mathfrak{A} \vdash \mathcal{A} \to \mathcal{B}} \vdash\to$$

[12] We added a subscript $_2$ in the label to distinguish this calculus from others, for instance, from a fragment of LE^{t}_{+} in [34].

$$\frac{\mathfrak{A}[\mathcal{A};\mathcal{B}] \vdash \mathcal{C}}{\mathfrak{A}[\mathcal{A}\circ\mathcal{B}] \vdash \mathcal{C}}\ {\circ\vdash} \qquad \frac{\mathfrak{A} \vdash \mathcal{A} \quad \mathfrak{B} \vdash \mathcal{B}}{\mathfrak{A};\mathfrak{B} \vdash \mathcal{A}\circ\mathcal{B}}\ {\vdash\circ}$$

$$\frac{\mathfrak{A}[\mathfrak{B};(\mathfrak{C};\mathfrak{D})] \vdash \mathcal{A}}{\mathfrak{A}[\mathfrak{B};\mathfrak{C};\mathfrak{D}] \vdash \mathcal{A}}\ \mathsf{B}\vdash \qquad \frac{\mathfrak{A}[\mathfrak{C};(\mathfrak{B};\mathfrak{D})] \vdash \mathcal{A}}{\mathfrak{A}[\mathfrak{B};\mathfrak{C};\mathfrak{D}] \vdash \mathcal{A}}\ \mathsf{B}'\vdash \qquad \frac{\mathfrak{A}[\mathfrak{B};\mathfrak{C};\mathfrak{C}] \vdash \mathcal{A}}{\mathfrak{A}[\mathfrak{B};\mathfrak{C}] \vdash \mathcal{A}}\ \mathsf{W}\vdash$$

$$\frac{\mathfrak{A}[\mathfrak{B}] \vdash \mathcal{A}}{\mathfrak{A}[\boldsymbol{t};\mathfrak{B}] \vdash \mathcal{A}}\ \mathsf{Kl}_t\vdash \qquad \frac{\mathfrak{A}[\mathfrak{B};\boldsymbol{t}] \vdash \mathcal{A}}{\mathfrak{A}[\boldsymbol{t};\mathfrak{B}] \vdash \mathcal{A}}\ \mathsf{T}_t\vdash$$

$$\frac{\mathfrak{A}[\boldsymbol{t};\mathfrak{B};\mathfrak{C}] \vdash \mathcal{A}}{\mathfrak{A}[\boldsymbol{t};(\mathfrak{B};\mathfrak{C})] \vdash \mathcal{A}}\ \mathsf{b}_t\vdash \qquad \frac{\mathfrak{A}[\boldsymbol{t};\boldsymbol{t}] \vdash \mathcal{A}}{\mathfrak{A}[\boldsymbol{t}] \vdash \mathcal{A}}\ \mathsf{M}_t\vdash$$

The notions of a *proof* and of a *theorem* are as in other consecution calculi.

Exercise 9.1.20. Prove a suitable cut theorem for this calculus.

Exercise 9.1.21. Prove that $\mathcal{A}$ is a theorem of $E^{\circ t}_{\rightarrow}$ iff $\boldsymbol{t} \vdash \mathcal{A}$ is provable in $LE^{\circ t}_{\rightarrow 2}$. [Hint: (R1) and (R2) from Section 5.4 can be taken to be the rules for $\circ$ in an axiomatic formulation.]

In Section 9.1.2, we gave a proof of the decidability of $LR^{t}_{\overset{\sim}{\rightarrow}}$, which extended $LR_{\rightarrow}$ with $\boldsymbol{t}$ and $\sim$. Now we have not only $\boldsymbol{t}$, but also $\circ$ in the language, and so we first prove that $LR^{\circ t}_{\rightarrow}$ is decidable. (We will present this proof much more concisely, assuming the earlier more detailed discussion for $LR^{t}_{\overset{\sim}{\rightarrow}}$.)

The *sequent calculus* $LR^{\circ t}_{\rightarrow}$ is based on *multisets*, and comprises the following axioms and rules.

$$\mathcal{A} \vdash \mathcal{A}\ \text{id} \qquad \vdash \boldsymbol{t}\ \vdash t$$

$$\frac{\Gamma;\mathcal{A};\mathcal{B} \vdash \mathcal{C}}{\Gamma;\mathcal{A}\circ\mathcal{B} \vdash \mathcal{C}}\ {\circ\vdash} \qquad \frac{\Gamma \vdash \mathcal{A} \qquad \Delta \vdash \mathcal{B}}{\Gamma;\Delta \vdash \mathcal{A}\circ\mathcal{B}}\ {\vdash\circ}$$

$$\frac{\Gamma \vdash \mathcal{A} \qquad \Delta;\mathcal{B} \vdash \mathcal{C}}{\Gamma;\Delta;\mathcal{A}\rightarrow\mathcal{B} \vdash \mathcal{C}}\ {\rightarrow\vdash} \qquad \frac{\Gamma;\mathcal{A} \vdash \mathcal{B}}{\Gamma \vdash \mathcal{A}\rightarrow\mathcal{B}}\ {\vdash\rightarrow}$$

$$\frac{\Gamma \vdash \mathcal{A}}{\Gamma;\boldsymbol{t} \vdash \mathcal{A}}\ t\vdash \qquad \frac{\Gamma;\mathcal{A};\mathcal{A} \vdash \mathcal{B}}{\Gamma;\mathcal{A} \vdash \mathcal{B}}\ \mathsf{W}\vdash$$

$\mathcal{A}$ is a theorem of $LR^{\circ t}_{\rightarrow}$ iff $\vdash \mathcal{A}$ is provable. Next, we define a *contraction-free* version of this calculus, which is denoted by $[LR^{\circ t}_{\rightarrow}]$.

$$\mathcal{A} \vdash \mathcal{A}\ \text{id} \qquad \vdash \boldsymbol{t}\ \vdash t$$

$$\frac{\Gamma;\mathcal{A};\mathcal{B} \vdash \mathcal{C}}{[\Gamma;\mathcal{A}\circ\mathcal{B}] \vdash \mathcal{C}}\ [{\circ\vdash}] \qquad \frac{\Gamma \vdash \mathcal{A} \qquad \Delta \vdash \mathcal{B}}{[\Gamma;\Delta] \vdash \mathcal{A}\circ\mathcal{B}}\ [{\vdash\circ}]$$

$$\frac{\Gamma \vdash \mathcal{A} \qquad \Delta; \mathcal{B} \vdash \mathcal{C}}{[\Gamma; \Delta; \mathcal{A} \to \mathcal{B}] \vdash \mathcal{C}} \ [\to\vdash] \qquad\qquad \frac{\Gamma; \mathcal{A} \vdash \mathcal{B}}{\Gamma \vdash \mathcal{A} \to \mathcal{B}} \ \vdash\to$$

$$\frac{\Gamma \vdash \mathcal{A}}{\Gamma; t \vdash \mathcal{A}} \ t\vdash$$

The potential reductions in the number of occurrences of wff's are as listed below. The rules do not permit a loss of a wff, and they do not necessitate a contraction—with those constraints, (1)–(3) apply.

(1) In $[\Gamma; \mathcal{A} \circ \mathcal{B}]$, $\mathcal{A} \circ \mathcal{B}$ may have one fewer occurrences than in $\Gamma; \mathcal{A} \circ \mathcal{B}$; all other formulas have the same number of occurrences;

(2) in $[\Gamma; \Delta]$, a formula that is common to Γ and Δ, may have one fewer occurrences than in $\Gamma; \Delta$; all other formulas have the same number of occurrences;

(3) in $[\Gamma; \Delta; \mathcal{A} \to \mathcal{B}]$, $\mathcal{A} \to \mathcal{B}$ may have two or one fewer occurrences than in $\Gamma; \Delta; \mathcal{A} \to \mathcal{B}$; other formulas, which are common to Γ and Δ, may have one fewer occurrences than in $\Gamma; \Delta; \mathcal{A} \to \mathcal{B}$; all other formulas have the same number of occurrences.

Exercise 9.1.22. Prove that the cut rule is admissible in $[LR^{\circ t}_{\to}]$. [Hint: The cut rule will have to permit some contractions too.]

The definition of cognate sequents is as in Definition 9.6. The following version of Curry's lemma, which is adjusted to be appropriate for $[LR^{\circ t}_{\to}]$, is slightly different from Lemma 9.7.

Lemma 9.17. (Curry's lemma for $[LR^{\circ t}_{\to}]$) *If the sequent $\Gamma \vdash \mathcal{A}$ is provable, and the height of the proof P is $\chi(P) = n$, and furthermore, $\Gamma' \vdash \mathcal{A}$ can be obtained from $\Gamma \vdash \mathcal{A}$ by finitely many contractions, (i.e., by zero or more applications of the rule $(W\vdash)$), then* there is a proof *P' of $\Gamma' \vdash \mathcal{A}$ such that $\chi(P') \leq n$.*

Exercise 9.1.23. Prove the theorem. [Hint: Although the steps in the proof of the Lemma 9.7 cannot be reused verbatim, the steps here are similar.]

We defined cognate sequents for multiple right-handed sequents, but of course, we can now consider the formula on the right-hand side of the $\vdash$ as a singleton. Kripke's lemma is applicable as before, and of course, König's lemma is applicable too.

Theorem 9.18. *The logic $R^{\circ t}_{\to}$ is* decidable.

Exercise 9.1.24. Prove this theorem. [Hint: The general structure of the proof is similar to the structure of the proof for $LR^{t}_{\tilde{\to}}$. See also [42, §1].]

We know that every theorem of $E^{\circ t}_{\to}$ is a theorem of $R^{\circ t}_{\to}$. Having $LE^{\circ t}_{\to}$ defined so that the *subformula property* holds without exceptions, we can ask the question if we can use the idea that we used to prove $T^t_\to$ decidable. That is, can we add a rule to $LE^{\circ t}_\to$ so that the rule adds nothing more than what is missing from $LE^{\circ t}_\to$ from the point of view of $R^{\circ t}_\to$? The answer is "yes."

The *consecution calculus* $LE^{\textcircled{t}}_{\to\circ}$ is defined by adding $(\mathsf{K}_t\vdash)$ to $LE^{\circ t}_\to$. We use a notation that is somewhat similar to $LT^{\textcircled{t}}_\to$. This calculus formalizes $LR^{\circ t}_\to$, not $LE^{\circ t}_\to$. Here is a proof of the permutation axiom, as an example.

$$\dfrac{\dfrac{\dfrac{\dfrac{\dfrac{\mathcal{A}\vdash\mathcal{A}\qquad \dfrac{\dfrac{\dfrac{\mathcal{B}\vdash\mathcal{B}}{\mathcal{B};t\vdash\mathcal{B}}\,\mathsf{K}_t\vdash \qquad \mathcal{C}\vdash\mathcal{C}}{\mathcal{B}\to\mathcal{C};(\mathcal{B};t)\vdash\mathcal{C}}\,\to\vdash}{\mathcal{B};\mathcal{B}\to\mathcal{C};t\vdash\mathcal{C}}\,\mathsf{B}'\vdash}{\mathcal{B};(\mathcal{A}\to\mathcal{B}\to\mathcal{C};\mathcal{A});t\vdash\mathcal{C}}\,\to\vdash}{\mathcal{A}\to\mathcal{B}\to\mathcal{C};\mathcal{B};\mathcal{A};t\vdash\mathcal{C}}\,\mathsf{B}'\vdash}{\dfrac{t;(\mathcal{A}\to\mathcal{B}\to\mathcal{C};\mathcal{B};\mathcal{A})\vdash\mathcal{C}}{\dfrac{t;(\mathcal{A}\to\mathcal{B}\to\mathcal{C};\mathcal{B});\mathcal{A}\vdash\mathcal{C}}{t;\mathcal{A}\to\mathcal{B}\to\mathcal{C};\mathcal{B};\mathcal{A}\vdash\mathcal{C}}\,\mathsf{B}\vdash}\,\mathsf{B}\vdash}\,\mathsf{T}_t\vdash}{t\vdash(\mathcal{A}\to\mathcal{B}\to\mathcal{C})\to\mathcal{B}\to\mathcal{A}\to\mathcal{C}}\,\vdash\to\text{'s}$$

Exercise 9.1.25. Prove that the addition of the $(\mathsf{K}_t\vdash)$ rule leaves the cut rule admissible. Then prove that $LE^{\textcircled{t}}_{\to\circ}$ is equivalent to $LR^{\circ t}_\to$. [Hint: The presence of fusion makes it easier to prove the equivalence.]

Notice that $LE^{\textcircled{t}}_{\to\circ}$ is $LT^{\textcircled{t}}_\to$ with three additional rules that are needed to add $\circ$. These rules are also sufficient to add fusion to $T^t_\to$ as it was established in [32, §3.2].

Theorem 9.19. *The logics $E^{\circ t}_\to$, $E^{\circ}_\to$, $T^{\circ t}_\to$ and $T^{\circ}_\to$ are decidable.*

Exercise* 9.1.26. Describe a decision procedure for $T^{\circ t}_\to$. [Hint: $T^{\circ t}_\to$ is an extension of $T^t_\to$, which we already dealt with.]

Exercise* 9.1.27. Modify the solution to the previous exercise to obtain decision procedures for the two fragments of E mentioned in the theorem. [Hint: The crucial part is to define a transformation of proofs in $LR^{\circ t}_\to$ into proofs in $LE^{\textcircled{t}}_{\to\circ}$. However, the $(\circ\vdash)$ and $(\vdash\circ)$ rules are unproblematic.]

Exercise 9.1.28. You have seen two applications of the idea of using the decision procedure for $R^t_\to$ to create a decision procedure for another logic using a suitable formulation of a consecution calculus for $R^t_\to$. Is there an extension of $LE^t_\to$ from Section 5.3 that is suitable for a decidability proof for $E^t_\to$? [Hint: Recall how and why we added two rules to $LT^t_\to$.]

Exercise* 9.1.29. Having solved the previous exercise, prove the cut theorem and the equivalence of your calculus to $R^t_\to$.

Exercise 9.1.30. A crucial fact about $LT^{\textcircled{t}}_\to$ is that we can obtain arbitrary permutations and groupings (possibly, with some added t's). Formulate appropriate lemmas for your calculus and prove them. [Hint: The task may sound a tad open-ended, but you may find sample claims and proofs for $LT^{\textcircled{t}}_{\rightarrow}$ in [42, §3.3].]

Exercise* 9.1.31. Given the calculus you created for $R^t_{\rightarrow}$ from $LE^t_{\rightarrow}$, describe a decision procedure based on the calculus.

9.1.5 Lambek calculus with contraction

Now we turn to the decidability of an extension of LA that we considered in Section 4.2.[13] Recall that in the absence of the permutation rule, contraction of formulas is not sufficient for the cut rule to be admissible. By LA_W, we denote (as before) LA with the structural version of the contraction rule that we denoted by $(W' \vdash)$ on page 91. (Alternatively, it is LA^t_W, but without the rule $(t \vdash)$ and without t in the language.)

In order to prove the decidability of this logic, we formulate this logic without an explicit contraction rule. The *sequent calculus* $[LA_W]$ comprises the following axiom and rules. (The meta-variables $\Gamma, \Delta, \ldots$ range over non-empty sequences of wff's.)

$$\mathcal{A} \vdash \mathcal{A} \ \text{id}$$

$$\frac{\Gamma \vdash \mathcal{A} \qquad \Delta[\mathcal{B}] \vdash \mathcal{C}}{[\Delta[\mathcal{A} \rightarrow \mathcal{B}; \Gamma]] \vdash \mathcal{C}} \ [\rightarrow\vdash] \qquad \frac{\Gamma; \mathcal{A} \vdash \mathcal{B}}{\Gamma \vdash \mathcal{A} \rightarrow \mathcal{B}} \ \vdash\rightarrow$$

$$\frac{\Gamma \vdash \mathcal{A} \qquad \Delta[\mathcal{B}] \vdash \mathcal{C}}{[\Delta[\Gamma; \mathcal{B} \leftarrow \mathcal{A}]] \vdash \mathcal{C}} \ [\leftarrow\vdash] \qquad \frac{\mathcal{A}; \Gamma \vdash \mathcal{B}}{\Gamma \vdash \mathcal{B} \leftarrow \mathcal{A}} \ \vdash\leftarrow$$

$$\frac{\Gamma\,[\mathcal{A}; \mathcal{B}] \vdash \mathcal{C}}{[\Gamma\,[\mathcal{A} \circ \mathcal{B}]] \vdash \mathcal{C}} \ [\circ\vdash] \qquad \frac{\Gamma \vdash \mathcal{A} \qquad \Delta \vdash \mathcal{B}}{[\Gamma; \Delta] \vdash \mathcal{A} \circ \mathcal{B}} \ [\vdash\circ]$$

A *proof* is defined as usual, and there are no theorems in this calculus, rather we have *provable sequents* only.

In $[LR_{\rightarrow}]$, in the $([\rightarrow\vdash])$ rule, which has contraction built in, we could specify the number of occurrences of wff's in the lower sequent with respect to the number of occurrences in the lower sequent in the $(\rightarrow\vdash)$ rule in $LR_{\rightarrow}$. We cannot do that in $[LA_W]$, because $(W' \vdash)$ permits a sequence of wff's to be contracted. The contractions that one of the bracketed rules allows to be carried out is specified as a *finite iteration*.

(1) If $\Theta; \Theta$ occurs in the lower sequent (without the brackets) and it did not occur in the upper sequents of a bracketed rule, then it may be replaced by Θ;

[13] This problem is mentioned in [174], as an open problem. As it turns out, a proof-theoretic approach provides an answer.

(2) if $\Theta;\Theta$ results by (1) or (2) (in an application of a bracketed rule), then it may be replaced by Θ.

We hasten to note that applications of (1) and (2) certainly come to an end after finitely many rounds, simply because the antecedent of a sequent is a *finite sequence*. However, unlike in $[LR_{\rightarrow}]$, it is not possible to specify—independently of the length of the antecedent—an upper bound on how many contractions become possible as a result of the application of a rule.

Example 9.20. Let us assume that the $([\circ\vdash])$ rule is applied to the following sequent, with the principal wff $\mathcal{A}\circ\mathcal{B}$.

$$\mathcal{G};\mathcal{F};\mathcal{D};\mathcal{A}\circ\mathcal{B};\mathcal{E};\mathcal{C};\mathcal{H};\mathcal{G};\mathcal{F};\mathcal{D};\mathcal{A}\circ\mathcal{B};\mathcal{E};\mathcal{C};\mathcal{F};\mathcal{D};\mathcal{A}\circ\mathcal{B};\mathcal{E};\mathcal{D};\mathcal{A};\mathcal{B};\mathcal{E};$$
$$\mathcal{D};\mathcal{A}\circ\mathcal{B};\mathcal{E};\mathcal{C};\mathcal{F};\mathcal{D};\mathcal{A}\circ\mathcal{B};\mathcal{E};\mathcal{C};\mathcal{H}\vdash\mathcal{I}$$

If we assume that distinct letters stand for distinct formulas, then there is no possibility for contraction before the rule is applied. Once we have introduced $\circ$ (between $\mathcal{A}$ and $\mathcal{B}$), the following series of contractions becomes possible. (We underline the pairs of subsequences that are to be contracted.)

$$\mathcal{G};\mathcal{F};\mathcal{D};\mathcal{A}\circ\mathcal{B};\mathcal{E};\mathcal{C};\mathcal{H};\mathcal{G};\mathcal{F};\mathcal{D};\mathcal{A}\circ\mathcal{B};\mathcal{E};\mathcal{C};\mathcal{F};\underline{\mathcal{D};\mathcal{A}\circ\mathcal{B};\mathcal{E};}\underline{\mathcal{D};\mathcal{A}\circ\mathcal{B};\mathcal{E};}$$
$$\mathcal{D};\mathcal{A}\circ\mathcal{B};\mathcal{E};\mathcal{C};\mathcal{F};\mathcal{D};\mathcal{A}\circ\mathcal{B};\mathcal{E};\mathcal{C};\mathcal{H}\vdash\mathcal{I}$$
$$\mathcal{G};\mathcal{F};\mathcal{D};\mathcal{A}\circ\mathcal{B};\mathcal{E};\mathcal{C};\mathcal{H};\mathcal{G};\mathcal{F};\mathcal{D};\mathcal{A}\circ\mathcal{B};\mathcal{E};\mathcal{C};\mathcal{F};\underline{\mathcal{D};\mathcal{A}\circ\mathcal{B};\mathcal{E};}$$
$$\underline{\mathcal{D};\mathcal{A}\circ\mathcal{B};\mathcal{E};}\mathcal{C};\mathcal{F};\mathcal{D};\mathcal{A}\circ\mathcal{B};\mathcal{E};\mathcal{C};\mathcal{H}\vdash\mathcal{I}$$
$$\mathcal{G};\mathcal{F};\mathcal{D};\mathcal{A}\circ\mathcal{B};\mathcal{E};\mathcal{C};\mathcal{H};\mathcal{G};\underline{\mathcal{F};\mathcal{D};\mathcal{A}\circ\mathcal{B};\mathcal{E};\mathcal{C};}\underline{\mathcal{F};\mathcal{D};\mathcal{A}\circ\mathcal{B};\mathcal{E};\mathcal{C};}$$
$$\mathcal{F};\mathcal{D};\mathcal{A}\circ\mathcal{B};\mathcal{E};\mathcal{C};\mathcal{H}\vdash\mathcal{I}$$
$$\mathcal{G};\mathcal{F};\mathcal{D};\mathcal{A}\circ\mathcal{B};\mathcal{E};\mathcal{C};\mathcal{H};\mathcal{G};\underline{\mathcal{F};\mathcal{D};\mathcal{A}\circ\mathcal{B};\mathcal{E};\mathcal{C};}\underline{\mathcal{F};\mathcal{D};\mathcal{A}\circ\mathcal{B};\mathcal{E};\mathcal{C};}\mathcal{H}\vdash\mathcal{I}$$
$$\underline{\mathcal{G};\mathcal{F};\mathcal{D};\mathcal{A}\circ\mathcal{B};\mathcal{E};\mathcal{C};\mathcal{H};}\underline{\mathcal{G};\mathcal{F};\mathcal{D};\mathcal{A}\circ\mathcal{B};\mathcal{E};\mathcal{C};\mathcal{H}}\vdash\mathcal{I}$$
$$\mathcal{G};\mathcal{F};\mathcal{D};\mathcal{A}\circ\mathcal{B};\mathcal{E};\mathcal{C};\mathcal{H}\vdash\mathcal{I}$$

The lettering of the formulas may reveal how we can construct lengthy sequences of wff's with the property that after the introduction of just one new formula, a series of contractions becomes possible. We place the subalterns of an application of the $([\circ\vdash])$ rule in the middle of two similar sequences, but with the principal formula replacing the subalterns. Then we add separating formulas, and add similar sequences on the left and on the right. It should be clear informally, that this process of *triplication* can be continued indefinitely, but it will require further increases in the length of the antecedent. Or, conversely, if there are n formulas in a sequence, then at most $n-1$ contractions are possible, and if we require that no contraction is possible before the application of some rule, then the number, in general, decreases.

Exercise* 9.1.32. Determine the largest number of contractions possible given the length of sequences of wff's in each of the four bracketed rules.

The cut rule has to allow a series of contractions that become possible as a result of inserting the antecedent of the left premise into the location of the cut formula in the antecedent of the right premise.

Exercise 9.1.33. Find the greatest number of possible contractions given the lengths of the antecedents of the premises of the cut rule.

Theorem 9.21. *The cut rule is* admissible *in* $[LA_W]$.

We outlined the proof of the theorem in Section 7.3.
We want to make sure now that we have not lost anything provable through the reformulation of the calculus.

Lemma 9.22. (Curry's lemma for $[LA_W]$) *If the sequent* $\Gamma \vdash \mathcal{A}$ *is provable, and the height of the proof* P *is* $\chi(P) = n$, *and furthermore,* $\Gamma' \vdash \mathcal{A}$ *can be obtained from* $\Gamma \vdash \mathcal{A}$ *by finitely many contractions (i.e., by zero or more applications of the rule* $(W' \vdash)$*), then* there is a proof P' *of* $\Gamma' \vdash \mathcal{A}$ *such that* $\chi(P') \leq n$.

Proof: **1.** If the proof is an instance of the axiom (id), then the claim is obviously true.
2. In the inductive case, we have the six connective rules, and within each of those cases, we distinguish between contractions that were possible before and those that became possible after the application of the rule.
2.1 Let the last rule be $\vdash\rightarrow$. The antecedent cannot be contracted as part of an application of the rule, hence, the upper sequent must contain at least a pair of sequences of wff's that can be contracted. (We use lowercase Greek letters for sequences that may be empty.)

$$\vdash\rightarrow \frac{\alpha;\Gamma;\Gamma;\beta;\mathcal{A}\vdash\mathcal{B}}{\alpha;\Gamma;\Gamma;\beta\vdash\mathcal{A}\rightarrow\mathcal{B}} \quad \overset{\text{i.h.}}{\leadsto} \quad \frac{\alpha;\Gamma;\beta;\mathcal{A}\vdash\mathcal{B}}{\alpha;\Gamma;\beta\vdash\mathcal{A}\rightarrow\mathcal{B}}\vdash\rightarrow$$

2.2 If the last rule is $([\rightarrow\vdash])$, then the contracted sequences may be fully parametric or they may become contractible as the result of the blending of two sequences and the introduction of the principal formula. In proofs of Curry's lemma for other logics, we maintained that the different kinds of contractions can be combined without difficulty. In the parametric case, we may have contractible sequences in either or both premises.

$$[\rightarrow\vdash]\frac{\alpha;\Gamma;\Gamma;\beta\vdash\mathcal{A} \quad \gamma;\Delta;\Delta;\mathcal{B};\delta\vdash\mathcal{C}}{\gamma;\Delta;\Delta;\mathcal{A}\rightarrow\mathcal{B};\alpha;\Gamma;\Gamma;\beta\vdash\mathcal{C}} \quad \overset{\text{i.h.}}{\leadsto}$$

$$\frac{\alpha;\Gamma;\beta\vdash\mathcal{A} \quad \gamma;\Delta;\mathcal{B};\delta\vdash\mathcal{C}}{\gamma;\Delta;\mathcal{A}\rightarrow\mathcal{B};\alpha;\Gamma;\beta\vdash\mathcal{C}}[\rightarrow\vdash]$$

When the contractions emerge because of the newly introduced $\mathcal{A}\rightarrow\mathcal{B}$, then the sequents may be of the following shape.

$$[\rightarrow\vdash]\frac{\Gamma;\mathcal{A}\rightarrow\mathcal{B};\alpha\vdash\mathcal{A} \quad \beta;\Gamma;\mathcal{A}\rightarrow\mathcal{B};\Gamma;\mathcal{B}\vdash\mathcal{C}}{\beta;\Gamma;\mathcal{A}\rightarrow\mathcal{B};\Gamma;\mathcal{A}\rightarrow\mathcal{B};\Gamma;\mathcal{A}\rightarrow\mathcal{B};\alpha\vdash\mathcal{C}} \quad \leadsto$$

$$\rightsquigarrow \quad \frac{\Gamma\,;\mathcal{A}\rightarrow\mathcal{B};\alpha\vdash\mathcal{A}\quad \beta;\Gamma\,;\mathcal{A}\rightarrow\mathcal{B};\Gamma\,;\mathcal{B}\vdash\mathcal{C}}{\beta;\Gamma\,;\mathcal{A}\rightarrow\mathcal{B};\alpha\vdash\mathcal{C}}\;[\rightarrow\vdash]$$

Depending on the shape of α and β, further contractions may be possible. However, no difficulty arises, because of the $([\rightarrow\vdash])$ rule's specification. We leave the other cases for the next exercise. qed

Exercise 9.1.34. Finish the proof of Curry's lemma. [Hint: The $(\vdash\leftarrow)$ and $([\leftarrow\vdash])$ rules are very similar to Cases **2.1** and **2.2**, and the $([\circ\vdash])$ and $([\vdash\circ])$ are not much more difficult either.]

In $LR_{\rightarrow}$, contraction is applied to formulas, but the $(W'\vdash)$ rule is easily shown to be a derived rule in that calculus; on the other hand, $(W\vdash)$ is a special instance of $(W'\vdash)$. But in LA_W, the situation is different, because the two rules are not equipotent. However, a finite sequence has only finitely many (non-empty) subsequences, and the set of subsequences is determined by a sequence of wff's—just as the set of subformulas of a formula is. Given a sequent, we take the subsequences in place of formulas in Kripke's lemma, which guarantees that there are only finitely many distinct sequences that cannot yield the given sequent through applications of the $(W'\vdash)$ rule.

Theorem 9.23. $[LA_W]$ *is* decidable.

Proof: The structure of the proof is as in earlier cases. We generate a proof-search tree. The subformula property is guaranteed by the cut theorem, the finite fork property is guaranteed by the finite number of bracketed rules, in each of which the number of possible upper sequents is bounded by the size of the given lower sequent. Lastly, the finite branch property is guaranteed by the subformula property and Kripke's lemma. qed

9.1.6 LR, TW_+ and RW_+ are decidable

Now we look at some decidability results for relevance logics that contain $\wedge$ and $\vee$. LR was formulated and proved decidable in [135]. The proof of the decidability follows the same overall pattern as the decidability proof for $R_{\rightarrow}$. On the other hand, consecution formulations of TW_+ and RW_+ were given by Giambrone in [94] who proved these logics decidable by using some new concepts in order to gain control over the two kinds of structures.[14] (The W in the labels of these logics indicates that contraction for intensional structures is *excluded*, rather than included.)

The *sequent calculus* LR is described in Section 4.4. The main feature of this calculus is that it adds the *structure-free rules* for conjunction and disjunction. Although this yields a relatively elegant calculus, the distributivity of $\wedge$ and $\vee$ is not provable; hence, it is not quite R. First, we want to give an alternative formulation of LR, which discards the two contraction rules, but allows

[14]Our presentation of results in this section is based on [131], and [95] and [8, §67].

the same wff's to be proved. We denote by $[LR]$ the variant of LR that has no explicit contraction rules.

$$\mathcal{A} \vdash \mathcal{A} \text{ id}$$

$$\frac{\Gamma \vdash \mathcal{A};\Delta \quad \Theta;\mathcal{B} \vdash \Lambda}{[\Gamma;\Theta;\mathcal{A} \to \mathcal{B}] \vdash [\Delta;\Lambda]} [\to\vdash] \qquad \frac{\Gamma;\mathcal{A} \vdash \mathcal{B};\Delta}{\Gamma \vdash [\mathcal{A} \to \mathcal{B};\Delta]} [\vdash\to]$$

$$\frac{\Gamma;\mathcal{A};\mathcal{B} \vdash \Delta}{[\Gamma;\mathcal{A} \circ \mathcal{B}] \vdash \Delta} [\circ\vdash] \qquad \frac{\Gamma \vdash \mathcal{A};\Delta \quad \Theta \vdash \mathcal{B};\Lambda}{[\Gamma;\Theta] \vdash [\mathcal{A} \circ \mathcal{B};\Delta;\Lambda]} [\vdash\circ]$$

$$\frac{\Gamma;\mathcal{A} \vdash \Delta}{[\Gamma;\mathcal{A} \wedge \mathcal{B}] \vdash \Delta} [\wedge\vdash_1] \qquad \frac{\Gamma;\mathcal{B} \vdash \Delta}{[\Gamma;\mathcal{A} \wedge \mathcal{B}] \vdash \Delta} [\wedge\vdash_2] \qquad \frac{\Gamma \vdash \mathcal{A};\Delta \quad \Gamma \vdash \mathcal{B};\Delta}{\Gamma \vdash [\mathcal{A} \wedge \mathcal{B};\Delta]} [\vdash\wedge]$$

$$\frac{\Gamma;\mathcal{A} \vdash \Delta \quad \Gamma;\mathcal{B} \vdash \Delta}{[\Gamma;\mathcal{A} \vee \mathcal{B}] \vdash \Delta} [\vee\vdash] \qquad \frac{\Gamma \vdash \mathcal{A};\Delta}{\Gamma \vdash [\mathcal{A} \vee \mathcal{B};\Delta]} [\vdash\vee_1] \qquad \frac{\Gamma \vdash \mathcal{B};\Delta}{\Gamma \vdash [\mathcal{A} \vee \mathcal{B};\Delta]} [\vdash\vee_2]$$

$$\frac{\Gamma \vdash \mathcal{A};\Delta}{[\Gamma;{\sim}\mathcal{A}] \vdash \Delta} [{\sim}\vdash] \qquad \frac{\Gamma;\mathcal{A} \vdash \Delta}{\Gamma \vdash [{\sim}\mathcal{A};\Delta]} [\vdash{\sim}]$$

The number of occurrences of wff's in the bracketed multisets are specified according to (1)–(3). If a wff occurs in a multiset, then it has at least one occurrence in its bracketed version too. But no contraction is forced in bracketed multisets, that is, Γ and $[\Gamma]$ may be the same.

(1) If $\mathcal{A}$ occurs in Γ and in Θ, then $\mathcal{A}$ occurs as many times or one time less in $[\Gamma;\Theta]$ than in $\Gamma;\Theta$. Similarly, for $[\Delta;\Lambda]$;

(2) if a principal wff, let us say $\mathcal{C}$, is joined with one parametric multiset, let us say Γ, then every wff other than $\mathcal{C}$ occurs as many times in $[\Gamma;\mathcal{C}]$ as in $\Gamma;\mathcal{C}$, but $\mathcal{C}$ may occur one time less, provided that $\mathcal{C}$ occurs in Γ;

(3) if a principal wff and two parametric multisets are joined, then in the bracketed multiset, every non-principal formula that occurs in both parametric multisets may have one fewer occurrence, and the principal formula may have at most two fewer occurrences than in the non-bracketed multiset.

(1) and (3) are applicable to $([\to\vdash])$ and $([\vdash\circ])$, whereas (2) applies to all the other rules.

The notion of a *proof* is as usual. Recall that $\Gamma, \Delta, \ldots$ range over *multisets* of wff's, including the empty multiset (like in LR). Thus, $\mathcal{A}$ is a *theorem* iff the sequent $\vdash \mathcal{A}$ is provable.

Theorem 9.24. *The cut rule is* admissible *in* $[LR]$.

Exercise 9.1.35. Prove the theorem. [Hint: There is a certain similarity to the proof of the cut theorem in $[LR^t_{\underset{\to}{\sim}}]$.]

Having seen the detailed proof of the decidability of $[LR^{t}_{\underset{\sim}{\rightarrow}}]$ earlier, it should be clear that we have to prove Curry's lemma for this calculus. We can take it as having been formulated exactly as in Lemma 9.7.

Proof: We proceed now as if we were continuing the proof of Lemma 9.7, because we can re-use Cases **2.2–2.5** without any change, and we can use half of the base Case **1** too. (Case **2.1** is obviously superfluous, because $\boldsymbol{t}$ is not in the language of $[LR]$.) We deal with the rules for disjunction, and leave the cases of the two other connectives (i.e., $\circ$ and $\wedge$) as an exercise.

2.6 Let us assume that the rule applied is $([\vee\vdash])$. We consider three possibilities separately, but they may be mixed together. Multiple formulas may be parametric on either side.

$$[\vee\vdash]\ \frac{\Gamma;\mathcal{C};\mathcal{C};\mathcal{A}\vdash\Delta\quad \Gamma;\mathcal{C};\mathcal{C};\mathcal{B}\vdash\Delta}{\Gamma;\mathcal{C};\mathcal{C};\mathcal{A}\vee\mathcal{B}\vdash\Delta}\quad\overset{\text{i.h.}}{\rightsquigarrow}\quad\frac{\Gamma;\mathcal{C};\mathcal{A}\vdash\Delta\quad \Gamma;\mathcal{C};\mathcal{B}\vdash\Delta}{\Gamma;\mathcal{C};\mathcal{A}\vee\mathcal{B}\vdash\Delta}\ [\vee\vdash]$$

$$[\vee\vdash]\ \frac{\Gamma;\mathcal{A}\vdash\mathcal{C};\mathcal{C};\Delta\quad \Gamma;\mathcal{B}\vdash\mathcal{C};\mathcal{C};\Delta}{\Gamma;\mathcal{A}\vee\mathcal{B}\vdash\mathcal{C};\mathcal{C};\Delta}\quad\overset{\text{i.h.}}{\rightsquigarrow}\quad\frac{\Gamma;\mathcal{A}\vdash\mathcal{C};\Delta\quad \Gamma;\mathcal{B}\vdash\mathcal{C};\Delta}{\Gamma;\mathcal{A}\vee\mathcal{B}\vdash\mathcal{C};\Delta}\ [\vee\vdash]$$

Another subcase is when Γ may have an occurrence of $\mathcal{A}\vee\mathcal{B}$ already, which creates room for contraction. Then the sequent with one fewer occurrences of $\mathcal{A}\vee\mathcal{B}$ is by the same rule, but exploiting the built-in contraction.

$$[\vee\vdash]\ \frac{\Gamma;\mathcal{A}\vee\mathcal{B};\mathcal{A}\vdash\Delta\quad \Gamma;\mathcal{A}\vee\mathcal{B};\mathcal{B}\vdash\Delta}{\Gamma;\mathcal{A}\vee\mathcal{B};\mathcal{A}\vee\mathcal{B}\vdash\Delta}\quad\rightsquigarrow$$

$$\frac{\Gamma;\mathcal{A}\vee\mathcal{B};\mathcal{A}\vdash\Delta\quad \Gamma;\mathcal{A}\vee\mathcal{B};\mathcal{B}\vdash\Delta}{\Gamma;\mathcal{A}\vee\mathcal{B}\vdash\Delta}\ [\vee\vdash]$$

qed

Exercise 9.1.36. Complete the proof by considering the rules for $\circ$ and $\wedge$. [Hint: Notice that the concrete shape of the principal formula is not very important in the lemma. Rather, similarities between the rules that are characterized by (1)–(3) above are what matter.]

Exercise 9.1.37. Using the accumulated results, prove that LR is decidable.

The potentially *devastating effect* of contraction from the point of view of decidability should, certainly, be clear by now, since all the decision procedures we looked at so far circumvented explicit contraction. TW_{+} and RW_{+} are the *contraction-free* versions of the positive fragments of ticket entailment and relevant implication. Isn't their decidability immediately obvious? Of course, the level of obviousness can vary between individual people, but it should not be forgotten that in the sequent calculus formulation of TW_{+} and RW_{+}—at least if it is given along the lines in Sections 4.6 and 5.4—*there is a contraction rule* in the context of extensional sequences (or extensional structures). Moreover, the contraction rule is structuralized.

There are two structural connectives as before, however, we assume that , forms *multisets*, whereas ; is *binary* (and its associativity is not stipulated). We assume that extensional multisets are parenthesized when they become components of an intensional structure. (We will use $\Gamma, \Delta, \ldots$ as metavariables (like we did in LR_+) except that here they cannot be *empty*.)

The only *axiom* of LTW_+ and LRW_+ is (id). The *rules* for $\circ$, $\rightarrow$ and $\vee$ are as in LR_+. The rules for $\wedge$ and the structural rules are as follows, where $(\mathsf{T}\vdash)$ belongs only to LRW_+.

$$\frac{\Gamma[\mathcal{A}] \vdash \mathcal{C}}{\Gamma[\mathcal{A} \wedge \mathcal{B}] \vdash \mathcal{C}}\ {\wedge\vdash_1} \qquad \frac{\Gamma[\mathcal{B}] \vdash \mathcal{C}}{\Gamma[\mathcal{A} \wedge \mathcal{B}] \vdash \mathcal{C}}\ {\wedge\vdash_2} \qquad \frac{\Gamma \vdash \mathcal{A} \quad \Gamma \vdash \mathcal{B}}{\Gamma \vdash \mathcal{A} \wedge \mathcal{B}}\ {\vdash\wedge}$$

$$\frac{\Gamma[\Delta] \vdash \mathcal{A}}{\Gamma[\boldsymbol{t}; \Delta] \vdash \mathcal{A}}\ \check{\boldsymbol{t}}_l\vdash \qquad \frac{\Gamma[\boldsymbol{t}; \Delta] \vdash \mathcal{A}}{\Gamma[\Delta] \vdash \mathcal{A}}\ \hat{\boldsymbol{t}}_l\vdash$$

$$\frac{\Gamma[\Delta; (\Theta; \Lambda)] \vdash \mathcal{A}}{\Gamma[\Delta; \Theta; \Lambda] \vdash \mathcal{A}}\ \mathsf{B}\vdash \qquad \frac{\Gamma[\Delta; (\Theta; \Lambda)] \vdash \mathcal{A}}{\Gamma[\Theta; \Delta; \Lambda] \vdash \mathcal{A}}\ \mathsf{B}'\vdash \qquad \frac{\Gamma[\Delta; \Theta] \vdash \mathcal{A}}{\Gamma[\Theta; \Delta] \vdash \mathcal{A}}\ \mathsf{T}\vdash$$

$$\frac{\Gamma[\Delta] \vdash \mathcal{A}}{\Gamma[\Delta, \Theta] \vdash \mathcal{A}}\ \mathsf{K}\vdash \qquad \frac{\Gamma[\Delta, \Delta] \vdash \mathcal{A}}{\Gamma[\Delta] \vdash \mathcal{A}}\ \mathsf{W}\vdash$$

A *proof* is defined as usual, and $\mathcal{A}$ is a *theorem* iff $\boldsymbol{t} \vdash \mathcal{A}$ has a proof. $\boldsymbol{t}$ is sort of both in and out in these calculi, because the $(\hat{\boldsymbol{t}}_l\vdash)$ rule allows $\boldsymbol{t}$ to disappear as a proof grows, which means that the subformula property in its full scope does not hold.

Exercise 9.1.38. Prove a cut theorem for both calculi. [Hint: Compared to LR_+, the left-hand side cannot be empty, which simplifies the proof.]

Giambrone in [95, pp. 243–245] shows that it is possible to slightly modify these calculi so that $\boldsymbol{t}$ can be altogether omitted. The changes relate to whether the left-hand side of the $\vdash$ may become empty.

(1) $\boldsymbol{t}$ is omitted from the language together with $(\check{\boldsymbol{t}}_l\vdash)$ and $(\hat{\boldsymbol{t}}_l\vdash)$;

(2) a sequent may have an empty left-hand side;

(3) in the rules Γ may be empty with the restrictions (i)–(ii),

  (i) Γ is not empty in $(\rightarrow\vdash)$;

  (ii) Γ is not empty in $(\vdash\circ)$ unless Δ is.

The fact that the provability of $\boldsymbol{t}$-free formulas is as before is called the *vanishing* $\boldsymbol{t}$ *lemma*. (For the precise statement of this lemma and its proof, see [95, p. 244.])

A structure is *reduced* if no structure has more than two occurrences in an extensional multiset in it, and it is *super reduced* if no structure has more than one occurrence in an extensional multiset in it. Reduced and super reduced are defined for sequents in the obvious way, that is, by requiring the antecedent of the sequent to be reduced and super reduced, respectively. The core theorem is that a sequent is provable iff its super-reduced version has a proof in which every sequent is reduced.

Giambrone defines the notion of the *degree* of a structure and sequent as follows. (We denote the degree by d. This notion of degree is related to, but different than, the customarily used notion of the degree of a formula.)

(1) $d(p) = 1$, where p is a propositional variable;

(2) $d(\mathcal{A} \wedge \mathcal{B}) = d(\mathcal{A} \vee \mathcal{B}) = d(\mathcal{A}) + d(\mathcal{B})$;

(3) $d(\mathcal{A} \to \mathcal{B}) = d(\mathcal{A} \circ \mathcal{B}) = d(\mathcal{A}) + d(\mathcal{B}) + 1$;

(4) $d(\Gamma_1, \ldots, \Gamma_n) = \max(d(\Gamma_1), \ldots, d(\Gamma_n))$;

(5) $d(\Gamma; \Delta) = d(\Gamma) + d(\Delta) + 1$;

(6) $d(\Gamma \vdash \mathcal{A}) = d(\Gamma) + d(\mathcal{A})$.

In the modified (i.e., t-less) calculi LTW_+ and LRW_+, the rules do not decrease the degree of the sequents, that is, the degree of the conclusion is greater than or equal to the degree of either premise. Subformulas of a formula are defined as usual.

Exercise 9.1.39. Prove the claim that the rules do not decrease the degree of sequents. [Hint: The t rules have been dropped.]

Lemma 9.25. *For any* n *(where* $n \geq 0$*), and for any wff* $\mathcal{A}$*, there are* finitely many reduced structures Γ *built from subformulas of* $\mathcal{A}$ *such that* $d(\Gamma) \leq n$.

Exercise 9.1.40. Prove the lemma. [Hint: Use complete induction on n.]

Exercise* 9.1.41. Design a decision procedure given the modified LTW_+ and LRW_+ calculi together with already known results (such as König's lemma).

9.1.7 Some good and bad news

So far in this section we talked about decidability results—mainly for various relevance logics. There were positive fragments, implication–negation fragments, fusion and truth added, so one might wonder if finally, we are going to put all this together. Of course, we won't, for the simple reason that it is known (and it has been known for nearly thirty years by now) that the main relevance logics, T, E and R are *undecidable*.[15] The undecidability proofs

[15] See Urquhart's publications: [183], [8, §65], [54, §11.5] and [188].

are not less exciting that the decidability proofs above. However, it would be quite a stretch to include them—beyond a quick mention—in a book on sequent calculi, because they do not involve any such proof system at all. Of course, we did not present a sequent or a consecution calculus for T or R in Chapters 4 or 5, because there are none that would fit into those frameworks.

At the other edge of the spectrum, there are some decidability results that are very easy and straightforward to prove.[16]

Theorem 9.26. (LA **and** LQ **are decidable)** *Given a sequent (a consecution)* $\Gamma \vdash \mathcal{A}$*, it is* decidable *whether* $\Gamma \vdash \mathcal{A}$ *is provable in* LA *(in* LQ*).*

Proof: Both calculi enjoy the cut theorem.

1. Let us start with LQ and let us assume that $\Gamma \vdash \mathcal{A}$ is given. There are six rules; notably, none of them is a structural rule. We can generate a backward proof-search tree. The root of the tree is the given sequent. There are at most four rules that could have been applied, because the right rules are not interchangeable with each other. (That is, the main connective of $\mathcal{A}$ precludes two of the right rules to be applicable.) We add the potential premises to the proof-search tree as nodes. We note that each rule decreases the degree of a formula; therefore, after finitely many steps we reach leaves. The sequent is provable if the proof-search tree contains a proof, and it is not provable if the proof-search tree is complete but contains no proof.

2. The case of LA is similar, except that a sequence, in general, has more subsequences than a structure (of the same size) has substructures. Notwithstanding, the number of possibilities is finite. Hence, the finite fork property is maintained, and the finiteness of the whole proof-search tree is guaranteed. qed

Exercise 9.1.42. Try your hand at searching for proofs with some consecutions in LQ and with some sequents in LA. [Hint: Start with simpler consecutions or sequents and generate the whole proof-search tree, rather than using heuristics toward finding proofs.]

The *implicational fragment* of the minimal relevance logic B with intensional truth can be formalized as $LB^{\boldsymbol{t}}_{\to}$ in Section 5.3. Since only the number of $\boldsymbol{t}$'s can increase when we build a proof-search tree, the decidability seems to be guaranteed. But we wish to make this obvious decidability somewhat more rigorous.

We formulate the *consecution calculus* $[LB^{\boldsymbol{t}}_{\to}]$. Our goal is, as before, to eliminate explicit contraction rules. We omit the $(\mathsf{M}_t \vdash)$ rule, and we replace the $(\mathsf{B}_t \vdash)$ rule by the $([\mathsf{B}_t \vdash])$ rule.

$$\frac{\mathfrak{A}[\boldsymbol{t}; (\mathfrak{B}; \mathfrak{C})] \vdash \mathcal{A}}{[\mathfrak{A}[\boldsymbol{t}; \mathfrak{B}; \mathfrak{C}]] \vdash \mathcal{A}}\ [\mathsf{B}_t \vdash]$$

[16] The decidability of LQ and LA is due to J. Lambek.

In this rule, $[\mathfrak{A}[\,t;\mathfrak{B};\mathfrak{C}]]$ is either $\mathfrak{A}[\,t;\mathfrak{B};\mathfrak{C}]$, or $\mathfrak{A}[\,t;\mathfrak{C}]$, or $\mathfrak{A}[\,t\,]$, but the latter two are permitted only when $\mathfrak{B}$ or both $\mathfrak{B}$ and $\mathfrak{C}$ are t.

Exercise 9.1.43. Scrutinize the rules of $LB^t_\rightarrow$ to see that $(\rightarrow\vdash)$ and $(\vdash\rightarrow)$ cannot result in a consecution in which $(\mathsf{M}_t\vdash)$ becomes applicable due to the application of the rule. Further, $(\mathsf{B}'_t\vdash)$ yields such a consecution when $(\mathsf{B}'_t\vdash)$ is indistinguishable from $(\mathsf{B}_t\vdash)$.

Theorem 9.27. *The cut rule is* admissible *in* $[LB^t_\rightarrow]$.

Proof: The key components of the proof are the same as in the proof of the cut theorem for $LB^t_\rightarrow$. In particular, the induction that ensures that cuts, in which the cut formula is t, are eliminable is not affected by the modification of the $(\mathsf{B}_t\vdash)$ rule. We omit the details. qed

Exercise 9.1.44. Verify the details of the proof of the cut theorem.

Theorem 9.28. (Curry's lemma) *If the sequent* $\Gamma\vdash\mathcal{A}$ *is provable, and the height of the proof* P *is* $\chi(P)=n$, *and furthermore,* $\Gamma'\vdash\mathcal{A}$ *can be obtained from* $\Gamma\vdash\mathcal{A}$ *by finitely many* t*-contractions, (i.e., by zero or more applications of the* $(\mathsf{M}_t\vdash)$ *rule), then* there is a proof P' *of* $\Gamma'\vdash\mathcal{A}$ *such that* $\chi(P')\leq n$.

Proof: The proof is by induction on the height of the proof tree.
1. The case of (id) is obvious.
2. In the inductive step we consider whether the t's are parametric or principal. The applicability of the rules does not depend on the presence or the shape of the parametric formulas.
2.1 If the t's are parametric, then we use the inductive hypothesis, and apply the same rule. ($\mathfrak{A}$ results from $\mathfrak{A}'$ by one of the five rules, and $\mathcal{A}$ and $\mathcal{B}$ are the same formula except in the case when the $(\vdash\rightarrow)$ rule is applied.)

$$\frac{\mathfrak{A}'[\,t;t\,]\vdash\mathcal{A}}{\mathfrak{A}[\,t;t\,]\vdash\mathcal{B}} \quad\overset{\text{i.h.}}{\rightsquigarrow}\quad \frac{\mathfrak{A}'[\,t\,]\vdash\mathcal{A}}{\mathfrak{A}[\,t\,]\vdash\mathcal{B}}$$

2.2 If the last rule is $(\mathsf{KI}_t\vdash)$, then we can simply omit the application of the rule.

$$\mathsf{KI}_t\vdash\frac{\mathfrak{A}[\,t\,]\vdash\mathcal{A}}{\mathfrak{A}[\,t;t\,]\vdash\mathcal{A}} \quad\rightsquigarrow\quad \mathfrak{A}[\,t\,]\vdash\mathcal{A}$$

2.3 If the last rule is $([\mathsf{B}_t\vdash])$, then an application of the same rule gives the desired result. (If $\mathfrak{B}$ is also t, then the $([\mathsf{B}_t\vdash])$ rule can yield either $\mathfrak{A}[\,t;t\,]\vdash\mathcal{A}$ or $\mathfrak{A}[\,t\,]\vdash\mathcal{A}$, as desired.)

$$[\mathsf{B}_t\vdash]\frac{\mathfrak{A}[\,t;(t;\mathfrak{B})]\vdash\mathcal{A}}{\mathfrak{A}[\,t;t;\mathfrak{B}]\vdash\mathcal{A}} \quad\rightsquigarrow\quad \frac{\mathfrak{A}[\,t;(t;\mathfrak{B})]\vdash\mathcal{A}}{[\mathfrak{A}[\,t;t;\mathfrak{B}]]\vdash\mathcal{A}}[\mathsf{B}_t\vdash]$$

2.4 Finally, the last rule may be $(\mathsf{B}'_t\vdash)$, which can give rise to $t;t$ as a new substructure if $\mathfrak{B}$ is t. However, in this case, an application of $(\mathsf{B}'_t\vdash)$ is indistinguishable from an application of $([\mathsf{B}_t\vdash])$. We simply use the latter with the built-in t-contraction.

$$\mathsf{B}'_t\vdash\ \frac{\mathfrak{A}[t;(t;\mathfrak{C})]\vdash\mathcal{A}}{\mathfrak{A}[t;t;\mathfrak{C}]\vdash\mathcal{A}}\quad\rightsquigarrow\quad\frac{\mathfrak{A}[t;(t;\mathfrak{C})]\vdash\mathcal{A}}{[\mathfrak{A}[t;t;\mathfrak{C}]]\vdash\mathcal{A}}\ [\mathsf{B}_t\vdash]$$

The $(\to\vdash)$ and $(\vdash\to)$ rules cannot lead to a new substructure of the form $t;t$; hence, their applications fall under Case **2.1.** qed

Given a consecution $t\vdash\mathcal{A}$, we create a proof-search tree for proofs of this consecution. We look at all the rules "applied backward," that is, in building the tree we move from lower consecutions to upper consecutions. We may assume that all the applications of the $(\vdash\to)$ rule to the antecedents of $\mathcal{A}$ come toward the end of a proof of a theorem.

Exercise 9.1.45. Prove that in a proof of a theorem $(\mathcal{A}_1\to\ldots(\mathcal{A}_n\to a)\ldots)$, where ($a$ is an atomic formula), the instances of $(\vdash\to)$ applied to the $\mathcal{A}$'s may be permuted with applications of other rules toward the bottom in the proof tree.

Thus, we may assume that the proof-search tree starts with $t;\mathcal{A}_1;\ldots;\mathcal{A}_n\vdash a$. The two connective rules, $(\vdash\to)$ and $(\to\vdash)$, introduce a principal formula of strictly greater degree than its subalterns; hence, from bottom to top, they decrease the degree. There is no way to contract a formula unless it is t, which means that there are finitely many possible applications of the implication rules.

We extend the proof-search tree as usual, by adding all the upper consecutions that could be premises of a rule. Of course, we may stop building the tree as soon as it contains a subtree that is a proof. But we also want to stop expanding the search tree when we are sure that no proof can be found.

If the only rule that could have resulted in a structure of the form $\mathfrak{B}[\mathcal{D}\to\mathcal{E};(t;\mathfrak{C})]$ is $(\to\vdash)$, then if $(\vdash\to)$ yielded $\mathcal{D}$, and $(\to\vdash)$ resulted in $\mathcal{E}$, then we may find a proof. However, first "applying backward" rules to $t;\mathfrak{C}$ will not result in a proof, if another strategy will not.

There are three rules beyond the implication rules, namely, $(\mathsf{KI}_t\vdash)$, $([\mathsf{B}_t\vdash])$ and $(\mathsf{B}'_t\vdash)$. By combinations of these rules, we can obtain $t;\mathfrak{C}$ from one of the following structures that we put above the line.

$$\frac{\mathfrak{C}\quad t;t;\mathfrak{C}\quad t;t;t;\mathfrak{C}\quad t;(t;\mathfrak{C})\quad t;t;(t;\mathfrak{C})\quad t;(t;t);\mathfrak{C}\quad t;(t;(t;\mathfrak{C}))\quad t;(t;t;\mathfrak{C})}{t;\mathfrak{C}}$$

Let us denote one of the structures above the line by $\mathfrak{D}$. Then, it is easy to see that one of the following derivation segments is a proof if the other is, and vice versa. (We denote by $(3^*\vdash)$ some combination of the three rules mentioned above.)

$$\dfrac{3^*\vdash\ \dfrac{\mathfrak{D}\vdash\mathcal{D}}{t;\mathfrak{C}\vdash\mathcal{D}}\qquad\mathfrak{B}[\mathcal{E}]\vdash\mathcal{A}}{\mathfrak{B}[\mathcal{D}\to\mathcal{E};(t;\mathfrak{C})]\vdash\mathcal{A}}\to\vdash\qquad\qquad\dfrac{\dfrac{\mathfrak{D}\vdash\mathcal{D}\qquad\mathfrak{B}[\mathcal{E}]\vdash\mathcal{A}}{\mathfrak{B}[\mathcal{D}\to\mathcal{E};\mathfrak{D}]\vdash\mathcal{A}}\to\vdash}{\mathfrak{B}[\mathcal{D}\to\mathcal{E};(t;\mathfrak{C})]\vdash\mathcal{A}}\ 3^*\vdash$$

This means that subtrees in which such applications of $(\rightarrow\vdash)$ are permuted with $(3^*\vdash)$ will produce a proof too.

Definition 9.29. (t-free structures) The set of *t-free structures* is a set of structures in which none of the atomic structures (i.e., formulas) is t.

An important insight in the consecution calculi for relevance logics is that *all the t's are the same*, so to speak. Of course, t may have several occurrences in a consecution, which can be, and should be, distinguished.

Definition 9.30. (t variants) The t *variants* of the structure $\mathfrak{B}$ are recursively generated by applications of (1) and (2).

(1) The structure $\mathfrak{B}$ is a t-variant of itself;

(2) if $\mathfrak{A}$ is a substructure of a t-variant of the structure $\mathfrak{B}$, then $\mathfrak{B}[t;\mathfrak{A}]$ is a t-variant of $\mathfrak{B}$.

The t-variants of a structure capture the idea that the $(\mathsf{KI}_t\vdash)$ rule (together with the connective rules) ensures that t is a left identity for $\circ$ or for its structural alter ego ; . The t-variants of a structure can be obtained by finitely many applications of the $(\mathsf{KI}_t\vdash)$ rule.

Structures, in general, and especially the structures in $LB^t_{\rightarrow}$, cannot be considered to be multisets—without loss of important information about the relationship between substructures. These structures are not even sequences, because the structural connective ; is not associative.

Definition 9.31. (Cognate consecutions) $\mathfrak{A}\vdash\mathcal{A}$ and $\mathfrak{B}\vdash\mathcal{B}$ are *cognate consecutions* when $\mathcal{A}$ is $\mathcal{B}$, and $\mathfrak{A}$ and $\mathfrak{B}$ are *concomitant*, that is, (1) the multisets of the atomic structures of $\mathfrak{A}$ and $\mathfrak{B}$ coincide, or (2) there are concomitant structures $\mathfrak{A}'$ and $\mathfrak{B}'$ such that $\mathfrak{A}$ and $\mathfrak{B}$ are t-variants of $\mathfrak{A}'$ and $\mathfrak{B}'$, respectively.

Example 9.32. Concrete structures can be inspected, and the shape of the atomic structures can be determined. In this example, we assume that $\mathcal{A},\mathcal{B}$ and $\mathcal{C}$ are not identical to t, but otherwise, they are arbitrary. The structures $\mathcal{A};(\mathcal{B};\mathcal{B});\mathcal{C}$ and $\mathcal{B};(\mathcal{C};\mathcal{B});\mathcal{A}$ are concomitant, and they are both t-free. Neither of them is concomitant with $\mathcal{A};\mathcal{C};\mathcal{B}$, simply because the number of $\mathcal{B}$'s differs between the first two and the third structure. Incidentally, they are not concomitant even if some of the $\mathcal{A},\mathcal{B}$ and $\mathcal{C}$ are the same formula.

The structure $t;\mathcal{A};(t;t;\mathcal{B};\mathcal{B});(t;\mathcal{C})$ is concomitant with the first structure above, because the latter is a t-variant of the former. It is also concomitant with the second structure, as well as with $t;\mathcal{A};(t;\mathcal{B};(t;\mathcal{B};(t;\mathcal{C})))$. However, it is not concomitant with $\mathcal{A};\mathcal{B};\mathcal{B};\mathcal{C};t;t;t;t;t$. This latest structure is not a t-variant of $\mathcal{A};\mathcal{B};\mathcal{B};\mathcal{C}$, neither is $t;\mathcal{A};(t;t;\mathcal{B};\mathcal{B});(t;\mathcal{C})$ a t-variant of the structure $\mathcal{A};\mathcal{B};\mathcal{B};\mathcal{C};t;t;t;t;t$.

Definition 9.33. (Irredundancy) Let $\sigma = \langle \mathfrak{A}_1 \vdash \mathcal{A}, \ldots, \mathfrak{A}_i \vdash \mathcal{A}, \ldots \rangle$ be a (possibly infinite) sequence of *cognate consecutions*. σ is *irredundant* iff the cognate sequents are distinct and whenever $i < j$, $\mathfrak{A}_i \vdash \mathcal{A}$ cannot be obtained by applications of $(\mathsf{M}_t \vdash)$, $(\mathsf{B}_t \vdash)$ and $(\mathsf{B}'_t \vdash)$.

Lemma 9.34. *If σ is an irredundant sequence of (distinct) cognate consecutions, then σ is* finite.

Proof: Given a consecution $\mathfrak{A} \vdash \mathcal{A}$, there are finitely many cognate consecutions $\mathfrak{B} \vdash \mathcal{A}$ such that the multisets of atomic structures of $\mathfrak{A}$ and $\mathfrak{B}$ coincide. (This is obvious, because there are at most $n!$ permutations for n-element multisets, and each sequence of wff's representing a permutation can be parenthesized in finitely many ways.) Let $\mathfrak{C}$ be one of these structures. Of course, there are infinitely many t-variants of $\mathfrak{C}$. However, there are only finitely many t-variants in which there is no occurrence of $t; t$, but the structure does not contain a substructure of the form $t; (t; \mathfrak{C})$. qed

Theorem 9.35. *The logic $LB^t_{\rightarrow}$ is* decidable.

Proof: We can return to the proof-search tree now. The finite branching is obvious, and by the previous lemma, we do not need to construct an infinite branch. qed

Exercise* 9.1.46. Add conjunction and disjunction to $LB^t_{\rightarrow}$. Prove that the resulting calculus is decidable. [Hint: You may choose if $\wedge$ and $\vee$ should or should not distribute over each other.]

Exercise 9.1.47. Now choose the other possibility (than what you chose in the previous exercise) with respect to distributivity, and prove decidability.[17]

9.2 Sequent calculi for mathematical theories

Gentzen in [92] added certain *axioms* to LK, and proved the consistency of the resulting theory, moreover, he claimed that further axioms of certain very restricted shape could be added too—without leading to inconsistency. Alas, his axioms did not include the induction schema, and he remarked that the added axioms create a theory, which is of little practical significance.

Identity may be viewed as a purely logical predicate, and this view coincides with the use of identity (or equality) in mathematics, when the domain is restricted to some specific set of objects such as the set of natural numbers. Classical first-order logic with identity can be axiomatized as $K^{=}$, an extension of K (from Section 2.2) with (A6) and (A7).

[17] We leave it to the reader to *decide* for herself or himself whether the easy decidability or the arduous undecidability results are what amounts to *bad news* and to *good news*. :-)

(A6) $\forall x\, x = x$

(A7) $\forall x\, \forall y\, (x = y \supset (\mathcal{A}(x) \supset \mathcal{A}(y)))$

Axioms for identity may be added to LK, however, with that addition the cut theorem is no longer provable. In order to preserve the cut theorem, instead of axioms, rules can be added.[18]

The *sequent calculus* $LK_3^=$ is LK_3 extended by the following two rules.

$$\frac{x = x, \Gamma \vdash \Delta}{\Gamma \vdash \Delta} = \vdash \qquad \frac{P(y), x = y, P(x), \Gamma \vdash \Delta}{x = y, P(x), \Gamma \vdash \Delta} L\vdash^{\oslash}$$

The $^{\oslash}$ indicates a restriction (which is intended to be expressed by our notation too), namely, P must be a predicate; hence, $P(x)$ is an atomic formula. (L stands for Leibniz, as an allusion to (A7) (and $(L\vdash)$) being a weak version of the so-called Leibniz law.)

Exercise 9.2.1. Assuming the admissibility of the cut rule, prove that $K^=$ and $LK_3^=$ are equivalent.

Exercise 9.2.2. Construct proofs in $LK_3^=$ of formulas that express the symmetry and the transitivity of $=$.

Identity is an *order relation*. *Pre-orders* and *linear orders* are two frequently occurring types of order relations.

Definition 9.36. $\leq$ is a *pre-order* iff $\leq$ is *reflexive* and *transitive*. $\leq$ is a *linear pre-order* iff $\leq$ is a *pre-order*, and for any elements x, y of the underlying set, $x \leq y$ or $y \leq x$.

The theories of these two types of orders may be formalized by extending LK_3 with (pre_1), (pre_2), and then with (lin) too.

$$\frac{x \leq x, \Gamma \vdash \Delta}{\Gamma \vdash \Delta} \text{pre}_1 \qquad \frac{x \leq z, x \leq y, y \leq z, \Gamma \vdash \Delta}{x \leq y, y \leq z, \Gamma \vdash \Delta} \text{pre}_2$$

$$\frac{x \leq y, \Gamma \vdash \Delta \quad y \leq x, \Gamma \vdash \Delta}{\Gamma \vdash \Delta} \text{lin}$$

Exercise 9.2.3. Prove that given these rules, the wff's characterizing linear pre-orders are theorems.

The formalization of mathematical theories is a very promising area of investigation. A few more examples of mathematical theories, in particular, of various geometric theories, may be found in [144, Ch. 8].

---

[18]The extension of LK_3 with identity, as well as with $\leq$ is in [144].

9.3 Typed and labeled calculi

9.3.1 Simply typed calculi

Functions may be characterized—to some extent—by specifying what kind of inputs they take and what kind of output they return. This understanding of functions is well supported, for example, by the set-theoretic reduction of functions or by category theory. However, the theory of functions, which was formulated in the 1930s by Church, namely, λ-calculi, do not incorporate this idea at all. Similarly, combinators, which were invented by Schönfinkel in the early 1920s, are designed to take variables, predicates or combinators or their aggregates as arguments. In the 1940s, Curry noted the similarity between implicational formulas and the *input–output characterizations of functions*, specifically, of combinators.

Typed calculi and their meta-theory can easily provide enough content for a whole book, as [108] illustrates. We concern ourselves mainly with *simple types* and with *combinators*. Furthermore, we will work toward extracting combinators from certain consecution calculus proofs. We conceive this as a step toward extending the so-called *Curry–Howard correspondence* to classes of sequent calculus proofs and to combinators.

There are two main versions of typing combinatory or λ-terms, and they are often named after H. B. Curry and A. Church. *Church-typing* is also called rigid or monomorphic, because each term has exactly one type. *Curry-typing* is more flexible, and it permits some terms to have more than one type, hence, it is called polymorphic typing or a type-assignment.

Definition 5.4 described combinatory terms. We add here λ-terms, which will figure into a natural deduction calculus for $J_{\rightarrow}$ below.

Definition 9.37. (λ-terms) Let $\langle x_i \rangle_{i \in \mathbb{N}}$ be a sequence of (distinct) variables. The set of λ-*terms* is inductively defined by (1)–(3).

(1) If $x \in \langle x_i \rangle_{i \in \mathbb{N}}$, then x is a λ-term;

(2) if M and N are λ-terms, then (MN) is a λ-term;

(3) if $x \in \langle x_i \rangle_{i \in \mathbb{N}}$ and M is a λ-term, then $\lambda x.\, M$ is a λ-term.

The λ-operator's role is to *abstract out a variable*, and thereby create a function. Of course, this way of speaking is somewhat metaphorical, because the variable may not occur in M or may not occur free in M, in which case the abstraction is vacuous. Also, in the λ-calculus every term is interpreted as a function. Hence, by an application of λ, a function is created from a function.

A relation on terms, which is similar to weak reduction (that we mentioned in Section 5.2), is β-*conversion* (or β-*reduction*). $(\lambda x.\, M)N \rhd_{1,\beta} M'$, where

M' is obtained by substituting N for the free occurrences of x in M. β-reduction is the reflexive transitive closure of the one-step β-conversion relation, $\triangleright_{1,\beta}$. The λ-operator is a variable binding operator, which implies that provisions have to be made to avoid a "clash of variables" during substitution. The situation is similar to what happens with quantified wff's, where a term that is substituted may contain variables, which should not become bound as a result of the substitution. In the λ-calculus, the renaming of the bound variables is called *α-conversion*, and with α-conversion available, substitution may be defined rigorously (though somewhat tediously) so that no free variable of a substituted term becomes bound. (We will not go into those details here, because a somewhat informal understanding of β-reduction is sufficient for our purposes.[19])

Definition 9.38. (Simple types) Let $\langle p_i \rangle_{i\in\mathbb{N}}$ be a denumerable sequence of propositional variables. The set of *simple types* is inductively defined by (1)–(2), using the only type constructor $\rightarrow$.

(1) If $p \in \langle p_i \rangle_{i\in\mathbb{N}}$, then p is a simple type;

(2) if $\mathcal{A}$ and $\mathcal{B}$ are simple types, then $(\mathcal{A} \rightarrow \mathcal{B})$ is a simple type.

Simple types look like implicational formulas, and our choice of notation for them is intended to reinforce this idea. We will use conventions to omit parentheses. In terms, we may omit parentheses from left-associated ones, whereas in types, we may omit parentheses from right-associated ones. (These conventions intentionally match the earlier conventions for structures in consecution calculi and for implicational formulas. Despite appearances, these two conventions are also in harmony with each other.)

The *Church-typing* adds a type to each atomic term. Then the set of typed terms is defined.

Definition 9.39. (Typed terms) Let $\langle x_i \rangle_{i\in\mathbb{N}}$ be a denumerable sequence of variables, and let $\{ \mathcal{A}_i \}_{i\in\mathbb{N}}$ be the set of simple types generated from $\langle p_i \rangle_{i\in\mathbb{N}}$. The set of *typed terms* is inductively defined by (1)–(3).

(1) If $x \in \langle x \rangle_{i\in\mathbb{N}}$ and $\mathcal{A} \in \{ \mathcal{A}_i \}_{i\in\mathbb{N}}$, then $x^{\mathcal{A}}$ is a typed term;

(2) if $M^{\mathcal{A}\rightarrow\mathcal{B}}$ and $N^{\mathcal{A}}$ are typed terms, then $(MN)^{\mathcal{B}}$ is a typed term;

(3) if $x^{\mathcal{A}}$ is a typed variable and $M^{\mathcal{B}}$ is a typed term, then $(\lambda x.\, M)^{\mathcal{A}\rightarrow\mathcal{B}}$ is a typed term.

In the complex terms, we omitted repeating the types of the components to avoid clutter, but the component terms do retain their types.

[19]For more detailed accounts, see, e.g., [108] or [37].

Example 9.40. In the λ-calculus, combinators are closed λ-terms, that is, terms that do not contain free variables. Without types, let us say K is a λ-term that is unique up to α-conversion. K is $\lambda xy.\, x$. Once types are attached to terms, there are infinitely many versions of K, and some of them remain distinct even if both the α-conversion of terms and the substitution of types is permitted. To form a typed λ-term that is a version of KK, we need two λ-terms—one with the type of a function that can be applied to the type of the other term. Here are two suitable terms put together. (The types are explicitly indicated only for one occurrence of each variable and for the whole term.)

$$\left[(\lambda xy^{\mathcal{C}}.\, x^{\mathcal{A}\to\mathcal{B}\to\mathcal{A}})\, \lambda zv^{\mathcal{B}}.\, z^{\mathcal{A}}\right]^{\mathcal{C}\to\mathcal{A}\to\mathcal{B}\to\mathcal{A}}$$

KK is a well-formed and unproblematic term, whether in combinatory logic or in λ-calculus. However, $\mathsf{K}^{\mathcal{A}}\mathsf{K}^{\mathcal{A}}$ is not a typed term, because it cannot be formed according to the above definition. If $\mathcal{A}$ is $\mathcal{B} \to \mathcal{C}$, then $\mathcal{A}$ and $\mathcal{B}$ are not the same wff's. If $\mathcal{A}$ is p, then (2) from the above definition is not applicable at all.

Curry-typing preserves types, but adds some flexibility to what types can be assigned to closed λ-terms and combinators. In the case of λ-terms, flexibility appears along the lines of α-conversion and substitution, whereas in the case of combinatory terms, it is due to substitution. Variables unchangeably may have only one type. The more relaxed approach to typing is reflected in the expression of *a type being assigned to a term* rather than a type being *the type of a term*.

We introduce type-assignment for combinatory terms. The notation for a type $\mathcal{A}$ assigned to a term M is $M\colon \mathcal{A}$.

Definition 9.41. (Type-assignment) Let $\langle x_i \rangle_{i\in\mathbb{N}}$ be a sequence of variables, and let $\mathbb{C}$ be a *context*, that is, $\{\, x_i\colon \mathcal{A}_i \mid i \in I \wedge I \subseteq \mathbb{N} \,\}$, where for any x_i, $\mathcal{A}_i$ is unique. A *type-assignment* to combinatory terms relative to a context is defined by (1)–(4).

(1) Given a context $\mathbb{C}$, $x_i\colon \mathcal{A}_i$, if $x_i \in \pi_1(\mathbb{C})$ and $x_i\colon \mathcal{A}_i$; otherwise, x_i is not assigned a type relative to $\mathbb{C}$;[20]

(2) if $M\colon \mathcal{A} \to \mathcal{B}$ and $N\colon \mathcal{A}$ relative to $\mathbb{C}$, then $MN\colon \mathcal{B}$ relative to $\mathbb{C}$;

(3) $\mathsf{K}\colon \mathcal{A} \to \mathcal{B} \to \mathcal{A}$ relative to $\mathbb{C}$;

(4) $\mathsf{S}\colon (\mathcal{A} \to \mathcal{B} \to \mathcal{C}) \to (\mathcal{A} \to \mathcal{B}) \to \mathcal{A} \to \mathcal{B}$ relative to $\mathbb{C}$.

We consider only contexts that assign at most one type to a variable. (Sometimes these are called consistent contexts.) On the other hand, we do not require that a type is assigned to each variable, because the types assigned to

[20]We use π_1 for the first projection, and we think of type-assignments as ordered pairs. That is, $\pi_1(\mathbb{C})$ is the set of variables that are assigned a type relative to $\mathbb{C}$.

complex terms, obviously, do not depend on type-assignments to variables that do not occur in them.

Note that (3) and (4) ensure that the type assigned to $\mathsf{K}xy$ turns out to be the same as that assigned to x; similarly, $\mathsf{S}xyz$ gets the same type as $xz(yz)$. If $x\colon \mathcal{A}$ and $y\colon \mathcal{B}$, then $\mathsf{K}x\colon \mathcal{B} \to \mathcal{A}$ is possible. Similarly, if a context $\mathbb{C}$ includes $x\colon \mathcal{A} \to \mathcal{B} \to \mathcal{C}$, $y\colon \mathcal{A} \to \mathcal{B}$ and $z\colon \mathcal{A}$, then $xz\colon \mathcal{B} \to \mathcal{C}$ and $yz\colon \mathcal{B}$, by (2), and further, $xz(yz)\colon \mathcal{C}$. On the other hand, successively detaching the types of x, y and z from that of S, we get $\mathcal{C}$ too, that is, $\mathsf{S}xyz\colon \mathcal{C}$. It can be proved that for each typable combinator, there is a type *schema* $\mathcal{A}$ such that any other type that can be assigned to that combinator is a substitution instance of $\mathcal{A}$. These schemas are called the *principal type schemas*. This justifies clauses (3) and (4) above for K and S.

Example 9.42. It is straightforward to assign a type to SS, for instance. The principal type schemas are independent of contexts. Taking two suitable substitutions of S's principal type schema, SS can be assigned a type. Moreover, if we keep substitutions abstract and general (that is, if we substitute metavariables rather than propositional variables and we do not identify variables unless we have to), then the resulting type schema inherits principality.

$$\begin{aligned}
\mathsf{S}\colon & ((\mathcal{A} \to \mathcal{B} \to \mathcal{C}) \to (\mathcal{A} \to \mathcal{B}) \to \mathcal{A} \to \mathcal{C}) \to \\
& \qquad ((\mathcal{A} \to \mathcal{B} \to \mathcal{C}) \to \mathcal{A} \to \mathcal{B}) \to (\mathcal{A} \to \mathcal{B} \to \mathcal{C}) \to \mathcal{A} \to \mathcal{C} \\
\mathsf{S}\colon & (\mathcal{A} \to \mathcal{B} \to \mathcal{C}) \to (\mathcal{A} \to \mathcal{B}) \to \mathcal{A} \to \mathcal{C} \\
\mathsf{SS}\colon & ((\mathcal{A} \to \mathcal{B} \to \mathcal{C}) \to \mathcal{A} \to \mathcal{B}) \to (\mathcal{A} \to \mathcal{B} \to \mathcal{C}) \to \mathcal{A} \to \mathcal{C}
\end{aligned}$$

Self-application of functions may be somewhat counterintuitive nowadays, partly, because of our thinking of a (unary) function as a set of ordered tuples. In untyped combinatory logic or λ-calculus, self-application of any function is unproblematic. D. Scott showed that untyped calculi have models comprising functions on natural numbers (or on domains) in which all the terms—including those of the form MM—make sense in a mathematically rigorous way.

The two kinds of typing (with simple types) differ to some extent with respect to terms of the form MM. $x^{\mathcal{A}}x^{\mathcal{A}}$ is not a typed term in the sense of Church-typing. If $x\colon \mathcal{A}$ is in a context $\mathbb{C}$, then $x\colon \mathcal{A} \to \mathcal{B} \notin \mathbb{C}$, hence, $xx\colon \mathcal{B}$ is an impossible type-assignment. The typing systems agree on self-applied variables, but they disagree on the typability of self-applied closed λ-terms or self-applied combinators. $(\lambda xy^{\mathcal{B}}.\, x^{\mathcal{A}})^{\mathcal{A}\to\mathcal{B}\to\mathcal{A}}(\lambda xy^{\mathcal{B}}.\, x^{\mathcal{A}})^{\mathcal{A}\to\mathcal{B}\to\mathcal{A}}$ (or $\mathsf{K}^{\mathcal{A}\to\mathcal{B}\to\mathcal{A}}\mathsf{K}^{\mathcal{A}\to\mathcal{B}\to\mathcal{A}}$) is not a typed term in the Church-typing system. However, in the context $\{x\colon \mathcal{A} \to \mathcal{B} \to \mathcal{A}, y\colon \mathcal{C}\}$, x has a type assigned, and so in the context $\{x\colon \mathcal{A} \to \mathcal{B} \to \mathcal{A}\}$, $\lambda y.\, x\colon \mathcal{C} \to \mathcal{A} \to \mathcal{B} \to \mathcal{A}$. The next λ subtracts the type-assignment for x from the context and we obtain $\lambda xy.\, x\colon (\mathcal{A} \to \mathcal{B} \to \mathcal{A}) \to \mathcal{C} \to \mathcal{A} \to \mathcal{B} \to \mathcal{A}$. Similarly, we can get $\lambda xy.\, x\colon \mathcal{A} \to \mathcal{B} \to \mathcal{A}$. $(\lambda xy.\, x)\lambda xy.\, x\colon \mathcal{C} \to \mathcal{A} \to \mathcal{B} \to \mathcal{A}$ (or

$\mathsf{KK}\colon \mathcal{C} \to \mathcal{A} \to \mathcal{B} \to \mathcal{A})$ is a possible type-assignment in the Curry-typing system.

There are closed λ-terms and combinators that cannot be typed in either manner. Well-known examples include $\lambda x.\, xx$ or the combinator M (with axiom $\mathsf{M}x \rhd xx$).

Exercise 9.3.1. Find some other λ-term or combinator that cannot be typed, and prove that neither way of typing works for the term. [Hint: Notice that the terms given above have no free variables, thus it is the structure of the term that prevents typing.]

A complete combinatory base, for example, one containing S and K, is sufficient to define any function that can be defined via λ-abstraction. However, the simulated abstraction operator, often called the *bracket abstraction* in combinatory logic, is not as elegant on the surface as the λ by itself. Of course, one might contend that this is the price that must be paid for not needing variable binding, and it is quite worth the complication.

Type-assignment systems are usually formulated as *natural deduction systems.*[21] The introduction of $\to$ straightforwardly matches adding an argument to a function, that is, a λ-abstraction. As a result, combinatory type-assignment systems more closely resemble axiom systems than natural deduction systems. Although the abstraction of any variable (occurring arbitrarily many times and in arbitrary places in a term) is possible, but as a rule, not by a single S or K.

A *natural deduction-style type-assignment system* for λ-terms is defined by the following axiom and two rules. (We write the rules in the "horizontal" or "sequent-style" way, but it should be obvious that it is *not* a sequent calculus.)

$$\mathbb{C} \cup \{\, x\colon \mathcal{A} \,\} \vdash x\colon \mathcal{A} \ \ \text{id}$$

$$\frac{\mathbb{C} \vdash M\colon \mathcal{A} \to \mathcal{B} \qquad \mathbb{C} \vdash N\colon \mathcal{A}}{\mathbb{C} \vdash MN\colon \mathcal{B}}\ E{\to} \qquad\qquad \frac{\mathbb{C} \cup \{\, x\colon \mathcal{A} \,\} \vdash M\colon \mathcal{B}}{\mathbb{C} - \{\, x\colon \mathcal{A} \,\} \vdash \lambda x.\, M\colon \mathcal{A} \to \mathcal{B}}\ I{\to}$$

On the left-hand side of the turnstile, we have a *context*, which we defined to be consistent. The axiom is labeled with (id), because $x\colon \mathcal{A}$ appears on both sides of the $\vdash$. The $(I\to)$ rule drops the type-assignment for the variable that is bound by the newly introduced λ, which is expressed by subtracting a singleton set from $\mathbb{C}$.

A *natural deduction-style type-assignment system* for CL-terms (with the combinatory base $\{\, \mathsf{S}, \mathsf{K} \,\}$) is defined to have the following three axioms and the rule $(E\to)$.

$$\mathbb{C} \vdash \mathsf{S}\colon (\mathcal{A} \to \mathcal{B} \to \mathcal{C}) \to (\mathcal{A} \to \mathcal{B}) \to \mathcal{A} \to \mathcal{C}\ \ \mathsf{S} \qquad\qquad \mathbb{C} \vdash \mathsf{K}\colon \mathcal{A} \to \mathcal{B} \to \mathcal{A}\ \ \mathsf{K}$$

[21] See Section 8.1 for the natural deduction calculi NK and NJ.

$$\mathbb{C}\cup\{\,x\colon\mathcal{A}\,\}\vdash x\colon\mathcal{A}\ \text{id}\qquad\qquad \frac{\mathbb{C}\vdash M\colon\mathcal{A}\to\mathcal{B}\qquad \mathbb{C}\vdash N\colon\mathcal{A}}{\mathbb{C}\vdash MN\colon\mathcal{B}}\ E{\to}$$

Example 9.43. Let $\mathbb{C}$ be $\{\,x\colon\mathcal{A}\to\mathcal{A}\to\mathcal{B}, y\colon\mathcal{A}, z\colon(\mathcal{A}\to\mathcal{B})\to\mathcal{A}\to\mathcal{C}\,\}$.

$$\cfrac{\cfrac{\mathbb{C}\vdash z\colon(\mathcal{A}\to\mathcal{B})\to\mathcal{A}\to\mathcal{C}\qquad \cfrac{\mathbb{C}\vdash x\colon\mathcal{A}\to\mathcal{A}\to\mathcal{B}\qquad \mathbb{C}\vdash y\colon\mathcal{A}}{\mathbb{C}\vdash xy\colon\mathcal{A}\to\mathcal{B}}\ E{\to}}{\mathbb{C}\vdash z(xy)\colon\mathcal{A}\to\mathcal{C}}\ E{\to}}{\mathbb{C}-\{\,x\colon\mathcal{A}\to\mathcal{A}\to\mathcal{B}\,\}\vdash\lambda x.\,z(yx)\colon(\mathcal{A}\to\mathcal{A}\to\mathcal{B})\to\mathcal{A}\to\mathcal{C}}\ I{\to}$$

The next derivation illustrates the workings of the combinatory type-assignment system. Let $\mathbb{C}$ be $\{\,x\colon\mathcal{A}\to\mathcal{B}, y\colon\mathcal{E}\,\}$, and let $\mathcal{D}$ be the wff $(\mathcal{A}\to\mathcal{B}\to\mathcal{A})\to(\mathcal{A}\to\mathcal{B})\to\mathcal{A}\to\mathcal{A}$. We will use $\mathcal{D}$ as an abbreviation, just as $\mathbb{C}$ is a concise notation for the set of type-assignments to x and y.

$$\cfrac{\cfrac{\mathbb{C}\vdash \mathsf{S}\colon\mathcal{D}\qquad \mathbb{C}\vdash\mathsf{K}\colon\mathcal{A}\to\mathcal{B}\to\mathcal{A}}{\mathbb{C}\vdash\mathsf{SK}\colon(\mathcal{A}\to\mathcal{B})\to\mathcal{A}\to\mathcal{A}}\qquad \mathbb{C}\vdash x\colon\mathcal{A}\to\mathcal{B}}{\mathbb{C}\vdash\mathsf{SK}x\colon\mathcal{A}\to\mathcal{A}}$$

Here we chose $\mathbb{C}$ not to be minimal from the point of view of the final term, which has no occurrence of y. We also omitted annotating the steps, because each step is $(E{\to})$, since there are no other rules.

We will not pursue natural deduction-like type-assignment systems further. However, the above quick glimpse at them should explain why there has not been many type-assignment (or, typed, for that matter) sequent calculi proposed in the literature.[22]

There is a *sequent calculus*-like type-assignment system using λ-terms in [87]. Since a context is a set, this approach does not generalize in a straightforward way to non-classical logics beyond intuitionistic logic. The notation $M[N/x]$ denotes the substitution of the term N for the free occurrences of x in M (without changing the bindings in N). The *axiom* and *rules* for $LJ^{\lambda}_{\to}$ are as follows. (We adjust the notation to harmonize with ours.)

$$\mathbb{C}\cup\{\,x\colon\mathcal{A}\,\}\vdash x\colon\mathcal{A}\ \text{id}$$

$$\frac{\mathbb{C}\vdash N\colon\mathcal{A}\qquad \mathbb{C}\cup\{\,x\colon\mathcal{B}\,\}\vdash M\colon\mathcal{C}}{(\mathbb{C}-\{\,x\colon\mathcal{B}\,\})\cup\{\,y\colon\mathcal{A}\to\mathcal{B}\,\}\vdash M[yN/x]\colon\mathcal{C}}\ {\to}{\vdash}$$

$$\frac{\mathbb{C}\cup\{\,x\colon\mathcal{A}\,\}\vdash M\colon\mathcal{B}}{\mathbb{C}-\{\,x\colon\mathcal{A}\,\}\vdash\lambda x.\,M\colon\mathcal{A}\to\mathcal{B}}\ {\vdash}{\to}$$

[22] For a detailed exposition of typed and type-assignment systems utilizing simple types and (mainly) λ-terms, see [108].

The *cut rule* is changed too, so that it applies to type-assignments. Furthermore, one of the occurrences of the type, which is the analog of the cut formula, must be a type assigned to a variable, since the left-hand side of the $\vdash$ can contain only a context.

$$\frac{\mathbb{C} \vdash M\colon \mathcal{A} \qquad \mathbb{C} \cup \{x\colon \mathcal{A}\} \vdash N\colon \mathcal{B}}{\mathbb{C} - \{x\colon \mathcal{A}\} \vdash N[M/x]\colon \mathcal{B}}\ \text{cut}$$

The *cut rule is admissible* in $LJ^{\lambda}_{\rightarrow}$. It is proved in [15] via a complicated detour through the natural deduction-style type-assignment system and the simply typed λ-calculus (in which typed terms normalize). We do not present that proof here, nor we consider type-assignment systems with λ-terms any further.

Structurally free logics, which we briefly described in Section 5.2, provide an alternative approach to thinking about types and terms. First of all, structurally free logics can accommodate *combinators* in a sequent calculus, which seems cumbersome if not impossible in $LJ^{\lambda}_{\rightarrow}$. Second, combinators are taken to be formulas, that is, there is no separation between terms and formulas as in a type-assignment $M\colon \mathcal{A}$, where M must be a term and $\mathcal{A}$ must be a formula, and no term is a formula. Another crucial difference between $LJ^{\lambda}_{\rightarrow}$ and LC is that the former places sets of objects in the antecedent, whereas LC has structures there, which encode more information than sets do.

If we include (id), the rules for $\rightarrow$ and $\circ$ together with the combinatory rules ($\mathsf{S}\vdash$) and ($\mathsf{K}\vdash$), then it is easy to prove that the type-assignments for S and K, are provable in LC in the form $\mathsf{K} \vdash \mathcal{A} \rightarrow \mathcal{B} \rightarrow \mathcal{A}$ and $\mathsf{S} \vdash (\mathcal{A} \rightarrow \mathcal{B} \rightarrow \mathcal{C}) \rightarrow (\mathcal{A} \rightarrow \mathcal{B}) \rightarrow \mathcal{A} \rightarrow \mathcal{C}$. The fusion–implication fragment of LC permits proofs of fusion types, where by a fusion type we mean a formula built using $\circ$, such that the consecution matches the combinatory axiom for these combinators. That is, $\mathsf{S} \circ \mathcal{A} \circ \mathcal{B} \circ \mathcal{C} \vdash \mathcal{A} \circ \mathcal{C} \circ (\mathcal{B} \circ \mathcal{C})$ and $\mathsf{K} \circ \mathcal{A} \circ \mathcal{B} \vdash \mathcal{A}$ are provable.

The choice of a particular set of proper combinators for LC (or even the addition of a fixed point combinator) do not hamper the admissibility of the cut rule.[23]

Exercise 9.3.2. Verify the claims about provable consecutions and about the cut rule.

The type-assignment to $\mathsf{SK}x\colon \mathcal{A} \rightarrow \mathcal{A}$ may be replicated in LC as follows.

$$\cfrac{\cfrac{\cfrac{\mathcal{A} \vdash \mathcal{A}}{\mathsf{K}; \mathcal{A}; (\mathcal{B}; \mathcal{A}) \vdash \mathcal{A}}\ \mathsf{K}\vdash}{\mathsf{S}; \mathsf{K}; \mathcal{B}; \mathcal{A} \vdash \mathcal{A}}\ \mathsf{S}\vdash}{\mathsf{S}; \mathsf{K}; \mathcal{B} \vdash \mathcal{A} \rightarrow \mathcal{A}}\ \vdash\rightarrow$$

[23] The admissibility was originally proved in [81], but see also [30] and Section 7.3.

Here $\mathcal{B}$ plays the role of x, and the proof can be seen to follow one-step expansions. It is instructive to compare the above proof with the one below.

$$\dfrac{\dfrac{\dfrac{\mathcal{A} \vdash \mathcal{A} \qquad \mathcal{A} \vdash \mathcal{A}}{\mathcal{A}; \mathcal{A} \vdash \mathcal{A} \circ \mathcal{A}} {\scriptstyle \vdash\circ}}{\mathsf{I}; \mathcal{A}; \mathcal{A} \vdash \mathcal{A} \circ \mathcal{A}} {\scriptstyle \mathsf{I}\vdash}}{\mathsf{W}; \mathsf{I}; \mathcal{A} \vdash \mathcal{A} \circ \mathcal{A}} {\scriptstyle \mathsf{W}\vdash}$$

The introduction of fusion is necessary, because $\mathcal{A}; \mathcal{A}$ cannot be the result of $(\rightarrow\vdash)$, which is another way to say that an implicational formula can never be unified with its own antecedent. The $(\vdash\circ)$ rule is the only other connective rule that can add multiplicity on the left.

The typability of a combinator may be checked by simulating one-step weak reductions, and then determining whether the resulting structure can be obtained by applications of the $(\rightarrow\vdash)$ rule.

Example 9.44. The combinator B is definable as $\mathsf{S}(\mathsf{KS})\mathsf{K}$. The following schematic proof can be completed by replacing "..." by an atomic wff, which is the consequent of the formula appearing in the place of X. (X can be $\mathcal{A} \rightarrow \mathcal{B}$, and then Y is $\mathcal{C} \rightarrow \mathcal{A}$ and Z is $\mathcal{C}$.)

$$\dfrac{\dfrac{\dfrac{\dfrac{X; (Y; Z) \vdash \ldots}{\mathsf{K}; X; Z; (Y; Z) \vdash \ldots} {\scriptstyle \mathsf{K}\vdash}}{\mathsf{S}; (\mathsf{K}; X); Y; Z \vdash \ldots} {\scriptstyle \mathsf{S}\vdash}}{\mathsf{K}; \mathsf{S}; X; (\mathsf{K}; X); Y; Z \vdash \ldots} {\scriptstyle \mathsf{K}\vdash}}{\mathsf{S}; (\mathsf{K}; \mathsf{S}); \mathsf{K}; X; Y; Z \vdash \ldots} {\scriptstyle \mathsf{S}\vdash}$$

A clear diversion from the combinatory type-assignment system is the absence of the principal type schemas of the combinators (as axioms) appearing in the proof. *LC* can be straightforwardly extended with other connectives, such as $\wedge$ or even $\vee$, which have been added as type constructors in the so-called *intersection* and *union type disciplines*, respectively.

The insights from structurally free logics are very useful in *extracting inhabitants* from consecution calculus proofs. Specifically, we want to find a *concrete combinatory inhabitant* that has the theorem, which is proved in a consecution calculus proof, as its type. The motivation comes from results in the previous section, that is, from decidability. All the decidability results that we presented are based on sequent calculi. If a logic is decidable, then the corresponding problem of *inhabitation* is decidable. However, it can often be determined that a type is inhabited without an inhabitant being exhibited. For example, Gentzen's proof that propositional intuitionistic logic is decidable does not yield a combinator or a λ-term that inhabits a theorem—even for the implicational fragment.

Of course, we could take the following complicated route to find an inhabitant. First, we would determine using a sequent calculus whether a formula is a theorem or not. For a theorem, at least one sequent calculus proof exists.

Then we would proceed by turning to an axiomatic calculus or to a natural deduction calculus. In the axiomatic setting, we can run a semi-decision procedure without worrying about the need to bound the search, because we know from the start that the wff is a theorem. All the proofs can be systematically generated (and thereby, enumerated), and sooner or later, the theorem will show up as the last wff in a proof. Alternatively, we can try to translate the sequent calculus proof into a natural deduction proof, and further translate that proof into a type-assignment proof. All these moves are lengthy and overly complicated.

Combinatory terms *encode proofs* for implicational logics, in which each axiom is the principal type schema of a combinator. This excludes classical logic, but that is not a problem. There is a correspondence between *traversing proofs* in such an axiom system (with schemas) and combinatory terms.[24] The similar correspondence for $NJ_{\rightarrow}$ and λ-terms is often called the *Curry–Howard correspondence* or *Curry–Howard isomorphism*. The extraction of a concrete combinator from a consecution calculus proof is intended to be the first step toward extending the Curry–Howard correspondence to combinators and classes of consecution calculus proofs for implicational logics.

Section 9.1.3 described a proof of the decidability of $T_{\rightarrow}$, which motivated the extraction of combinators from proofs in the consecution calculus $LT^{t}_{\rightarrow}$. The same idea was adapted to the logic $R_{\rightarrow}$ using a consecution calculus—rather than the widely known sequent calculus $LR_{\rightarrow}$.[25] Here we deal with *pure entailment*.

The *consecution calculus* $LE^{t}_{\rightarrow 2}$ is $LE^{t}_{\rightarrow}$ with the $(\mathsf{B}\vdash)$ rule added. ($LE^{t}_{\rightarrow}$ and the rule are introduced in Section 5.3.) $LE^{t}_{\rightarrow}$ closely resembles an axiomatization of $E^{t}_{\rightarrow}$ by independent axioms. The addition of the $(\mathsf{B}\vdash)$ rule shortens some proofs, but it is harmless. The cut rule is admissible, however, from now on, we consider only cut-free proofs, in which $(\mathsf{W}\vdash)$ is applied to formulas. All theorems of $E_{\rightarrow}$ have such proofs.

$E_{\rightarrow}$ is decidable, and given a proof of a theorem of $E_{\rightarrow}$, there is a combinator over the base $\{\,\mathsf{B}, \mathsf{B}', \mathsf{4}, \mathsf{W}, \mathsf{I}\,\}$ that has the theorem as its principal type schema. The principal type schemas of the listed combinators may be taken as axioms with the rule detachment. $\mathsf{4}$ is CII, and it differs from the other combinators in the base by not being proper.

The rules of $LE^{t}_{\rightarrow 2}$ are well formulated, which yields the following result. (a is an atomic formula, that is, $\boldsymbol{t}$ or a propositional variable p.)

Lemma 9.45. *If* $\mathcal{A}_1 \rightarrow \ldots \rightarrow \mathcal{A}_n \rightarrow a$ *is a theorem of* $E^{t}_{\rightarrow}$, *then* there is a proof *of the consecution* $\boldsymbol{t} \vdash \mathcal{A}_1 \rightarrow \ldots \rightarrow \mathcal{A}_n \rightarrow a$ *in which the last* n *steps are applications of* $(\vdash\rightarrow)$ *following the consecution* $\boldsymbol{t}; \mathcal{A}_1; \ldots; \mathcal{A}_n \vdash a$.

[24]The notion of traversing proofs is defined in [32, §3.1] in the context of $T_{\rightarrow}$, but the definition is adaptable to other axiom systems, which have modus ponens as their sole rule.

[25]Those results may be found in a series of papers by J. M. Dunn and me, namely, [41], [42], [44] and [43].

Proof: The insight behind the claim is that $(\vdash\to)$ is the only rule that moves a formula from the antecedent into the succedent. If a left rule could be applied with fewer formulas in the antecedent, then it can be applied with the subaltern of the $(\vdash\to)$ rule in the antecedent. The latter means that an extra formula in the antecedent (toward the right) does not interfere with the applicability of the left rules. The proof is by induction on the height of the proof tree. What we prove is that the left rules and the $(\vdash\to)$ rule can be permuted. Of course, the lemma does not claim, and it would not be true that applications of the $(\vdash\to)$ rule can be switched with each other.

1. The case of (id) is obvious.

2. There are six left rules; any of them can be the one preceded by $(\vdash\to)$. We detail two subcases only.

2.1 If $(\vdash\to)$ is followed by $(\hat{t}_r\vdash)$, then the proof segment on the left is transformed into the one on the right.

$$\hat{t}_r\vdash\ \dfrac{{\vdash\to}\ \dfrac{\vdots \atop \mathfrak{A}[\mathfrak{B};t];\mathcal{A}\vdash\mathcal{B}}{\mathfrak{A}[\mathfrak{B};t]\vdash\mathcal{A}\to\mathcal{B}}}{\mathfrak{A}[\mathfrak{B}]\vdash\mathcal{A}\to\mathcal{B}} \qquad\qquad \dfrac{\dfrac{\vdots \atop \mathfrak{A}[\mathfrak{B};t];\mathcal{A}\vdash\mathcal{B}}{\mathfrak{A}[\mathfrak{B}];\mathcal{A}\vdash\mathcal{B}}\ \hat{t}_r\vdash}{\mathfrak{A}[\mathfrak{B}]\vdash\mathcal{A}\to\mathcal{B}}\ {\vdash\to}$$

2.2 If $(\to\vdash)$ is followed by $(\vdash\to)$, then we have the following two proof segments—one is given, the other is the modified one.

$${\to\vdash}\ \dfrac{\vdots \atop \mathfrak{A}\vdash\mathcal{A} \qquad \dfrac{\vdots \atop \mathfrak{B}[\mathcal{B}];\mathcal{C}\vdash\mathcal{D}}{\mathfrak{B}[\mathcal{B}]\vdash\mathcal{C}\to\mathcal{D}}\ {\vdash\to}}{\mathfrak{B}[\mathcal{A}\to\mathcal{B};\mathfrak{A}]\vdash\mathcal{C}\to\mathcal{D}} \qquad\qquad \dfrac{\dfrac{\vdots \atop \mathfrak{A}\vdash\mathcal{A} \qquad \vdots \atop \mathfrak{B}[\mathcal{B}];\mathcal{C}\vdash\mathcal{D}}{\mathfrak{B}[\mathcal{A}\to\mathcal{B};\mathfrak{A}];\mathcal{C}\vdash\mathcal{D}}\ {\to\vdash}}{\mathfrak{B}[\mathcal{A}\to\mathcal{B};\mathfrak{A}]\vdash\mathcal{C}\to\mathcal{D}}\ {\vdash\to}$$

qed

Exercise 9.3.3. Finish the proof of the lemma. [Hint: A similar lemma is true for $LT^t_\to$, which is proved in [44].]

We call the consecution $t;\mathcal{A}_1;\ldots;\mathcal{A}_n \vdash a$ the *source consecution* for the theorem $\mathcal{A}_1\to\ldots\to\mathcal{A}_n\to a$. If we consider a theorem of $E_\to$, then a is a propositional variable. The extraction of a combinatory inhabitant involves three steps: (1) *numbering*, (2) *insertion of combinators* and (3) BB'*-abstraction*. The whole procedure provides *exactly one combinator*, however, this does not mean that other combinators (even in weak normal form) cannot inhabit the same theorem. The uniqueness is a result of fixing a particular algorithm for (3) (as well as for (2) and (1)).

A consecution calculus can simulate detachment using cut (and some structural steps). However, consecution calculi are not designed to simulate *condensed detachment*, which is a version of detachment with a minimal amount of substitution built in, which can guarantee that only principal type schemas of combinators can be proved. In other words, the combinator that we find based on a proof in a consecution calculus is not guaranteed to be a combinator that has the theorem as its *principal* type schema.

(1) *Numbering* is a procedure that assigns a natural number to every formula occurrence in a proof ending in a source consecution. The procedure starts at the source consecution and proceeds upward, level by level, in the proof tree, that is, toward the leaves of the tree.

The formulas in the source consecution $\boldsymbol{t}; \mathcal{A}_1; \ldots; \mathcal{A}_n \vdash a$ are given their indices (if they have one) as their numbers; $\boldsymbol{t}$ is numbered by 0 and a is given the number $n+1$.

We specify the numbers for formulas in the upper consecutions of rules assuming that the formulas in the lower consecution have been numbered, and we know the set of numbers that has been used so far. (For example, if we move on a level from left to right, then we know which natural numbers have been used, and we know the numbers of the formulas in the lower consecution in the rule under consideration.)

If the rule is $(\check{\boldsymbol{t}}_l \vdash)$, then the numbers of the formulas are copied from the lower sequent—except for the $\boldsymbol{t}$ that is introduced by the rule. If the rule is $(\mathsf{B}\vdash)$ or $(\mathsf{B}'\vdash)$, then the numbers are transferred from the lower to the upper consecution via matching the formulas. If the rule is $(\mathsf{W}\vdash)$, then the numbers are copied again, and the formulas in the duplicated structure get the same numbers.

There are three rules that have some formulas in the upper consecutions that do not appear at all or do not appear as separate formulas in the lower consecution. If the rule is $(\hat{\boldsymbol{t}}_r \vdash)$, then we choose the least natural number not yet used, and assign that to the $\boldsymbol{t}$, which has disappeared as the result of the application of the rule. All the other formulas in the upper consecution get the same number that they have in the lower consecution. If the rule is $(\vdash\to)$ or $(\to\vdash)$, then we pick the least natural number not yet used for the consequent of the principal formula, and then the next least unused number for the antecedent of the principal formula. All the formulas except the principal formula appear in matching locations in an upper consecution, and they get the same number that they have in the lower consecution.

It may be useful to note that the procedure is not intended to (and does not) assign the same number to different occurrences of a wff. This does not cause any problems, because we start with a proof (not with a random collection of consecutions). The intention behind the numbering is to track how formulas *migrate* around in the proof. It is obvious (and provable by structural induction on the proof tree) that every occurrence of every formula is assigned exactly one natural number.

Lemma 9.46. *Every formula occurrence in a proof ending in a source consecution is assigned a* unique natural number.

Exercise 9.3.4. Prove the lemma. [Hint: The proof is an easy induction on the proof tree.]

Definition 9.47. (Predecessors) A formula occurrence in an upper consecution of a rule is a *predecessor* of a formula occurrence in the lower consecution iff they have the same number.

There are formulas with no predecessors, for instance, $\boldsymbol{t}$ of the $(\check{\boldsymbol{t}}_l \vdash)$ rule or the principal formulas of the $\to$ rules. There are formulas with two predecessors, namely, the formulas resulting by contraction. There are formulas that are not predecessors of a formula, for example, the subalterns in the $\to$ rules, and the $\boldsymbol{t}$ of the $(\hat{\boldsymbol{t}}_r \vdash)$ rule. However, if a formula is a predecessor of a formula, then it is the *predecessor of exactly one* wff.
(2) The *insertion of combinators* ensures that the structural rules are distilled into combinators, which serve as traces of or evidence for the application of those rules. Even with the details not yet provided, a similarity to structurally free logics may be suspected. We rush to note though that there is a crucial difference between how the combinators are inserted here and how they appear in structurally free logics. We treat combinators as enhancements for formulas that form *combinatorially augmented formulas.*

The structural rules are applied locally in a proof, but their effect impacts on all the consecutions below. Accordingly, we want to pass downward the combinators from upper consecutions to lower ones.

Definition 9.48. (Combinatorially augmented formulas) The set of *combinatorially augmented formulas* or caf's is inductively defined by (1) and (2).

(1) If $\mathcal{A}$ is a wff, then $\mathcal{A}$ is a caf;

(2) if X is a caf and Z is a combinator, then $\mathsf{Z}X$ is a caf.

We use $X, Y, Z, \ldots$ as meta-variables for caf's.

As the name for caf's suggests, they are practically formulas, but somewhat enhanced or augmented. In order not to overcomplicate things, we assume that the structures can be (and have been) defined anew from caf's. (This is a proper extension of the class of structures.) We keep $\mathfrak{A}, \mathfrak{B}, \mathfrak{C}, \ldots$ as meta-variables, and will continue to talk about structures without extra adjectives to refer to elements of this bigger set.

We note a fact from CL, which is useful in the process of the insertion of the combinators. It guarantees that combinators can be attached to atomic structures, that is, to caf's, and that any compound caf can be considered to be of the form $\mathsf{Z}\mathcal{A}$.

Let $\mathsf{Z}(M_1 \ldots M_n)$ be a CL-term. $\mathsf{B}(\mathsf{B} \ldots (\mathsf{B}\mathsf{Z})\ldots)M_1 \ldots M_n$, with n occurrences of B, is a CL-term, and the latter weakly reduces to the former. The combinator B is a regular associator, and Curry used the power of B to transpose the effects of various other combinators to more remote arguments. We will call this *the* $\mathsf{B}s$ *step* or $(\mathsf{B}s)$.

The insertion of the combinators proceeds from the top to the bottom of the proof tree, level by level, and let us say, from left to right. We need to make sure that when we consider a consecution obtained by a particular rule, at that time, the upper consecutions already have been subjected to the procedure and have their combinators (if any) inserted. The axiom does not trigger the insertion of a combinator, hence, we divide the insertion and propagation of combinators into steps according to the rule that has been applied with a general propagation step added.

1. *Default propagation step* (dps). If X is a caf of the form $Z\mathcal{A}$ in the upper consecution, where $\mathcal{A}$ is the predecessor of an alike formula in the lower consecution, then $\mathcal{A}$ is replaced by X in the lower consecution. (Dps) means applying this replacement to every atomic structure in the upper consecutions, which are not subalterns in the rule.

2. $(\check{t}_l \vdash)$. There is nothing combinatorially interesting happening in this rule (which is not to say that the rule is unimportant). (Dps) is applied.

3. $(\hat{t}_r \vdash)$. In the upper consecution, $\mathfrak{B}; \boldsymbol{t}$ occurs with this $\boldsymbol{t}$ disappearing in the lower consecution. If $\mathfrak{B}$ is the caf X, then $4X$ replaces $\mathfrak{B}$ in the lower consecution. If $\mathfrak{B}$ is a complex structure, let us say, $X; \mathfrak{C}_1; \ldots; \mathfrak{C}_n$, then 4 is added to X with (Bs). $\mathsf{B}(\ldots(\mathsf{B4})\ldots)X; \mathfrak{C}_1; \ldots; \mathfrak{C}_n$ replaces $\mathfrak{B}$ in the lower consecution. In both cases, (dps) is applied to the consecution to propagate the combinators that may occur outside of $\mathfrak{B}$.

4. $(\mathsf{B}\vdash)$. This rule is explicitly labeled by a combinator, and we insert that combinator into the lower consecution. If $\mathfrak{B}; (\mathfrak{C}; \mathfrak{D})$ is the substructure in the upper consecution that is affected by the rule, and $\mathfrak{B}$ is the caf X, then $\mathfrak{B}; \mathfrak{C}; \mathfrak{D}$ in the lower consecution is replaced by $\mathsf{B}X; \mathfrak{C}; \mathfrak{D}$ (where $\mathfrak{C}$ and $\mathfrak{D}$ are the structures in the upper consecution, which may contain complex caf's). If $\mathfrak{B}$ is of the form $X; \mathfrak{C}_1; \ldots; \mathfrak{C}_n$, then (Bs) is applied with B, and $\mathsf{B}(\ldots(\mathsf{BB})\ldots)X; \mathfrak{C}_1; \ldots; \mathfrak{C}_n; \mathfrak{C}; \mathfrak{D}$ replaces the $\mathfrak{B}; \mathfrak{C}; \mathfrak{D}$ in the lower consecution. Either way, the replacement is followed by (dps).

5. $(\mathsf{B}'\vdash)$. This step is very similar to the previous one except that B' is inserted instead of B. (Dps) is applied whether $\mathfrak{B}$ was atomic or complex.

6. $(\mathsf{W}\vdash)$. In the original proof, C (the formula that is contracted) is the same in both occurrences. However, the insertion of the combinators may result in two caf's containing C. There are four possibilities to consider.

If the two occurrences of C in the upper consecution remained formulas, then we want to add W to $\mathfrak{B}$. If $\mathfrak{B}$ is X, then $\mathsf{W}X; C$ replaces $\mathfrak{B}; C$ in the lower consecution. If $\mathfrak{B}$ is $X; \mathfrak{C}_1; \ldots; \mathfrak{C}_n$, then (Bs) is applied with W, and $\mathsf{B}(\ldots(\mathsf{BW})\ldots)X; \mathfrak{C}_1; \ldots; \mathfrak{C}_n; C$ replaces $\mathfrak{B}; C$ in the lower consecution. (Dps) follows in both cases.

If the two occurrences of C have become distinct caf's, then one or the other C became a caf with a combinator. We consider three cases. In the first case, $\mathfrak{B}; C; C$ turned into $\mathfrak{B}; ZC; C$ in the upper consecution. If $\mathfrak{B}$ is atomic, then we replace $\mathfrak{B}$ in the lower consecution by $\mathsf{BW}(\mathsf{B}'\mathsf{Z})\mathfrak{B}$. If $\mathfrak{B}$ is complex, then we apply (Bs) with this combinator.

If the upper consecution contains $\mathfrak{B}; C; ZC$, then we add the combinator $\mathsf{BW}(\mathsf{B}(\mathsf{B}'\mathsf{Z}))$—with or without a (Bs) step, depending on whether $\mathfrak{B}$ is complex or not.

In the third case, we have $\mathfrak{B}; \mathsf{Z}_1C; \mathsf{Z}_2C$ in the upper consecution. Then we add the combinator $\mathsf{BW}(\mathsf{B}(\mathsf{B}'\mathsf{Z}_2)(\mathsf{BB}'\mathsf{Z}_1))$ to $\mathfrak{B}$. Once again, if $\mathfrak{B}$ is complex, then (Bs) is applied too.

In the last three cases, (dps) is applied to all substructures in the antecedent in the consecution save $\mathfrak{B}$ and the caf's with $\mathcal{C}$'s.

7. $(\to\vdash)$. The antecedent of the principal formula of the rule is a formula, because no combinators are inserted into the consequent. However, the consequent may or may not be a formula in the upper consecution.

If the consequent is a wff, then (dps) is applied. If the consequent is X, which is of the form $Z\mathcal{C}$, then the principal wff $\mathcal{B}\to\mathcal{C}$ in the lower consecution is replaced by $\mathsf{B}Z\mathcal{B}\to\mathcal{C}$, which is followed by (dps).

8. $(\vdash\to)$. Now we have to look at the antecedent of the principal formula; the consequent is surely a wff. If the antecedent is a wff without any combinators, then we apply (dps). If $\mathcal{B}$ has become X and is of the form $Z\mathcal{B}$, then we want to transfer the combinator elsewhere in the consecution. In the consecution calculus $LE^{t}_{\to 2}$, the left-hand side of the $\vdash$ is never empty, hence, it is certain that the consecution is of the form $\mathfrak{A};Z\mathcal{B}\vdash\mathcal{C}$. $\mathfrak{A}$ may be atomic or complex. If the former, then $\mathsf{B}'Z$ is prefixed to $\mathfrak{A}$, and the resulting caf replaces $\mathfrak{A}$ in the lower consecution. If the latter, then (Bs) is applied with $\mathsf{B}'Z$ and the result is added to the leftmost caf in $\mathfrak{A}$, which replaces the leftmost wff in $\mathfrak{A}$ in the lower consecution. (Dps) is applied to the rest of the antecedent.

To make the numbering and the insertion of combinators more palatable, we give a detailed example.

Example 9.49. The restricted version of the permutation of the antecedents of a conditional is a prototypical theorem of $E_{\to}$. Here is a proof of the consecution for this theorem in $LE^{t}_{\to 2}$, where $(*)$ labels the source consecution.

$$\frac{\mathcal{B}\vdash\mathcal{B}}{t;\mathcal{B}\vdash\mathcal{B}}\ \check{t}_l\vdash$$

$$\frac{t;\mathcal{B}\vdash\mathcal{B}\qquad \mathcal{C}\vdash\mathcal{C}}{\mathcal{B}\to\mathcal{C};(t;\mathcal{B})\vdash\mathcal{C}}\ \to\vdash$$

$$\frac{\mathcal{B}\to\mathcal{C};(t;\mathcal{B})\vdash\mathcal{C}}{\mathcal{B}\to\mathcal{C};t;\mathcal{B}\vdash\mathcal{C}}\ \mathsf{B}\vdash$$

$$\frac{\mathcal{B}\to\mathcal{C};t;\mathcal{B}\vdash\mathcal{C}}{\mathcal{B}\to\mathcal{C};t\vdash\mathcal{B}\to\mathcal{C}}\ \vdash\to$$

$$\frac{\mathcal{B}\to\mathcal{C};t\vdash\mathcal{B}\to\mathcal{C}\qquad \mathcal{D}\vdash\mathcal{D}}{(\mathcal{B}\to\mathcal{C})\to\mathcal{D};(\mathcal{B}\to\mathcal{C};t)\vdash\mathcal{D}}\ \to\vdash$$

$$\frac{(\mathcal{B}\to\mathcal{C})\to\mathcal{D};(\mathcal{B}\to\mathcal{C};t)\vdash\mathcal{D}}{\mathcal{B}\to\mathcal{C};(\mathcal{B}\to\mathcal{C})\to\mathcal{D};t\vdash\mathcal{D}}\ \mathsf{B}'\vdash$$

$$\frac{\mathcal{B}\to\mathcal{C};(\mathcal{B}\to\mathcal{C})\to\mathcal{D};t\vdash\mathcal{D}}{\mathcal{B}\to\mathcal{C};(\mathcal{B}\to\mathcal{C})\to\mathcal{D}\vdash\mathcal{D}}\ \hat{t}_r\vdash$$

$$\frac{\mathcal{B}\to\mathcal{C};(\mathcal{B}\to\mathcal{C})\to\mathcal{D}\vdash\mathcal{D}}{\mathcal{B}\to\mathcal{C};(t;(\mathcal{B}\to\mathcal{C})\to\mathcal{D})\vdash\mathcal{D}}\ \check{t}_l\vdash$$

$$\frac{\mathcal{A}\vdash\mathcal{A}\qquad \mathcal{B}\to\mathcal{C};(t;(\mathcal{B}\to\mathcal{C})\to\mathcal{D})\vdash\mathcal{D}}{\mathcal{B}\to\mathcal{C};(t;(\mathcal{A}\to(\mathcal{B}\to\mathcal{C})\to\mathcal{D};\mathcal{A}))\vdash\mathcal{D}}\ \to\vdash$$

$$\frac{\mathcal{B}\to\mathcal{C};(t;(\mathcal{A}\to(\mathcal{B}\to\mathcal{C})\to\mathcal{D};\mathcal{A}))\vdash\mathcal{D}}{\mathcal{B}\to\mathcal{C};(t;\mathcal{A}\to(\mathcal{B}\to\mathcal{C})\to\mathcal{D};\mathcal{A})\vdash\mathcal{D}}\ \mathsf{B}\vdash$$

$$\frac{\mathcal{B}\to\mathcal{C};(t;\mathcal{A}\to(\mathcal{B}\to\mathcal{C})\to\mathcal{D};\mathcal{A})\vdash\mathcal{D}}{*\quad t;\mathcal{A}\to(\mathcal{B}\to\mathcal{C})\to\mathcal{D};\mathcal{B}\to\mathcal{C};\mathcal{A}\vdash\mathcal{D}}\ \mathsf{B}'\vdash$$

$$\frac{t;\mathcal{A}\to(\mathcal{B}\to\mathcal{C})\to\mathcal{D};\mathcal{B}\to\mathcal{C};\mathcal{A}\vdash\mathcal{D}}{t;\mathcal{A}\to(\mathcal{B}\to\mathcal{C})\to\mathcal{D};\mathcal{B}\to\mathcal{C}\vdash\mathcal{A}\to\mathcal{D}}\ \vdash\to$$

$$\frac{t;\mathcal{A}\to(\mathcal{B}\to\mathcal{C})\to\mathcal{D};\mathcal{B}\to\mathcal{C}\vdash\mathcal{A}\to\mathcal{D}}{t;\mathcal{A}\to(\mathcal{B}\to\mathcal{C})\to\mathcal{D}\vdash(\mathcal{B}\to\mathcal{C})\to\mathcal{A}\to\mathcal{D}}\ \vdash\to$$

$$\frac{t;\mathcal{A}\to(\mathcal{B}\to\mathcal{C})\to\mathcal{D}\vdash(\mathcal{B}\to\mathcal{C})\to\mathcal{A}\to\mathcal{D}}{t\vdash(\mathcal{A}\to(\mathcal{B}\to\mathcal{C})\to\mathcal{D})\to(\mathcal{B}\to\mathcal{C})\to\mathcal{A}\to\mathcal{D}}\ \vdash\to$$

Next we do the numbering. We display only the numbers that result from the procedure, but they can be thought of as subscripts on formula occurrences.

$$
\dfrac{\dfrac{\dfrac{\dfrac{\dfrac{\dfrac{\dfrac{\dfrac{\dfrac{\dfrac{11 \vdash 13}{7;11 \vdash 13} \quad 12 \vdash 10}{2;(7;11) \vdash 10}}{2;7;11 \vdash 10}}{2;7 \vdash 9} \quad 8 \vdash 4}{5;(2;7) \vdash 4}}{2;5;7 \vdash 4}}{2;5 \vdash 4}}{3 \vdash 6 \quad 2;(0;5) \vdash 4}}{2;(0;(1;3)) \vdash 4}}{\dfrac{2;(0;1;3) \vdash 4}{0;1;2;3 \vdash 4}}
$$

Finally, we insert the combinators.

$$
\dfrac{\dfrac{\dfrac{\dfrac{\dfrac{\dfrac{\dfrac{\dfrac{\dfrac{\dfrac{11 \vdash 13}{7;11 \vdash 13} \quad 12 \vdash 10}{2;(7;11) \vdash 10}}{\mathsf{B}2;7;11 \vdash 10}}{\mathsf{B}2;7 \vdash 9} \quad 8 \vdash 4}{5;(\mathsf{B}2;7) \vdash 4}}{\mathsf{B}'(\mathsf{B}2);5;7 \vdash 4}}{\mathsf{B}4(\mathsf{B}'(\mathsf{B}2));5 \vdash 4}}{3 \vdash 6 \quad \mathsf{B}4(\mathsf{B}'(\mathsf{B}2));(0;5) \vdash 4}}{\mathsf{B}4(\mathsf{B}'(\mathsf{B}2));(0;(1;3)) \vdash 4}}{\dfrac{\mathsf{B}4(\mathsf{B}'(\mathsf{B}2));(\mathsf{B}0;1;3) \vdash 4}{\mathsf{BB}'(\mathsf{B}0);1;\mathsf{B}4(\mathsf{B}'(\mathsf{B}2));3 \vdash 4}}
$$

Notice that instead of $\mathsf{B}'(\mathsf{B}2)$ we could have $\mathsf{BB}'\mathsf{B}2$; similarly, $\mathsf{BB}'(\mathsf{B}0)$ could be $\mathsf{B}(\mathsf{BB}')\mathsf{B}0$—by (Bs).

The combinators can be thought of as sets of types. Then $\mathsf{B}2$, that is, $\mathsf{B}\mathcal{B} \to \mathcal{C}$ can be $(\mathcal{B} \to \mathcal{B}) \to \mathcal{B} \to \mathcal{C}$. $\mathsf{B}'(\mathsf{B}2)$ is $((\mathcal{B} \to \mathcal{C}) \to \mathcal{D}) \to (\mathcal{B} \to \mathcal{B}) \to \mathcal{D}$, and $\mathsf{B}4(\mathsf{B}'(\mathsf{B}2))$ is $((\mathcal{B} \to \mathcal{C}) \to \mathcal{D}) \to \mathcal{D}$. The other formula in the source consecution that turned into a caf is $\boldsymbol{t}$. This $\boldsymbol{t}$ has the number 0, and we replace it with I. A suitable instance of the principal type schema of I is $((\mathcal{B} \to \mathcal{C}) \to \mathcal{D}) \to (\mathcal{B} \to \mathcal{C}) \to \mathcal{D}$. Then $\mathsf{B}0$ is $(\mathcal{A} \to (\mathcal{B} \to \mathcal{C}) \to \mathcal{D}) \to \mathcal{A} \to (\mathcal{B} \to \mathcal{C}) \to \mathcal{D}$. Finally, $\mathsf{BB}'(\mathsf{B}0)$ is $(\mathcal{A} \to (\mathcal{B} \to \mathcal{C}) \to \mathcal{D}) \to (((\mathcal{B} \to \mathcal{C}) \to \mathcal{D}) \to \mathcal{D}) \to \mathcal{A} \to \mathcal{D}$. The other occurrences of $\boldsymbol{t}$ are also taken to be instances of self-implication. In our example, 7 is $\boldsymbol{t}$, and it can be taken to be $\mathcal{B} \to \mathcal{B}$.

We want to interpret a consecution by selecting suitable types (of combinators that have been inserted, or of I when $\boldsymbol{t}$ occurs), and then to view ; as $\circ$. *Function application* in terms is *detachment* in types, and in the case of $E^t_{\to}$, $\to$ is the residual of $\circ$ (not of $\wedge$ as in $J_{\to}$). The source consecution, after steps (1) and (2), comprises $\mathsf{BB}'(\mathsf{B}0)$, from which 1, $\mathsf{B}4(\mathsf{B}'(\mathsf{B}2))$ and 3 can be detached to get $\mathcal{D}$.

Exercise 9.3.5. Go through the consecutions in the sample proof and verify that each consecution embodies a similar relationship between the antecedent and the consequent.

The interpretation of formulas, combinators and ; in the above example is how we want to think about consecutions after the numbering and the insertion of combinators. This intermediate interpretation step ensures that after the BB'-abstraction we will know that the theorem obtainable from the source consecution is a type of the combinator obtained by the BB'-abstraction.

Definition 9.50. (Structures as terms) The CL-*term* $\mathfrak{A}^\#$ that results from the structure $\mathfrak{A}$ in a consecution—once the numbering and the insertion of the combinators have been carried out—is inductively defined by (1)–(3).

(1) If $\mathcal{A}_n$ is $\boldsymbol{t}$, then $(\mathcal{A}_n)^\#$ is I; otherwise, $(\mathcal{A}_n)^\#$ is x_n;

(2) if X is a caf of the form $\mathsf{Z}\mathcal{A}$, then $X^\#$ is $(\mathsf{Z}\mathcal{A}^\#)$;

(3) if $\mathfrak{A}$ is $(\mathfrak{B};\mathfrak{C})$, then $\mathfrak{A}^\#$ is $(\mathfrak{B}^\#\mathfrak{C}^\#)$.

Exercise 9.3.6. Show that given a proof ending in a source consecution (with the numbering and the insertion of the combinators completed), the antecedent of each consecution is turned into a CL-term by an application of $^\#$.

Now we are ready to state and prove that the interpretation of consecutions ensures inhabitation. We assume that the set of variables in $\mathfrak{A}^\#$ can be determined, and $\mathfrak{A}$ determines not only $\mathfrak{A}^\#$ but also a context Δ in the sense of a type-assignment system. Namely, $\Delta = \{\, x_i\colon \mathcal{A}_i \mid x_i \in \mathrm{fv}(\mathfrak{A}^\#) \,\}$. In words, if a formula occurring in $\mathfrak{A}$ has been turned into a variable, then it itself is assigned to that variable in a type-assignment. Numbering is in harmony with the predecessor relation which guarantees that the context does not contain conflicting type-assignments (i.e., the resulting context is consistent in the usual sense).

Lemma 9.51. *Let a cut-free proof in $LE^t_{\to 2}$ end in the source sequent $\boldsymbol{t}; \mathcal{A}_1; \ldots; \mathcal{A}_n \vdash a$. For each consecution $\mathfrak{A} \vdash \mathcal{A}$ in the proof, $\Delta \Vdash \mathfrak{A}^\#\colon \mathcal{A}$, where Δ is the context determined by the antecedent of the consecution, with the proviso that if $\mathcal{A}$ itself is $\boldsymbol{t}$, then there is a suitable instance of the principal type schema of I.*

Proof: The proof is by induction on the height of the proof tree. As a preliminary remark, we note that the (Bs) and the BB$'$-abstraction steps *preserve types*, which is known from the type-assignment literature.

For instance, if $\mathsf{B}(xyz)\colon (\mathcal{C} \to \mathcal{A}) \to \mathcal{C} \to \mathcal{B}$, then if $y\colon \mathcal{D}$ and $z\colon \mathcal{E}$, then x must have the type $\mathcal{D} \to \mathcal{E} \to \mathcal{A} \to \mathcal{B}$. The (Bs) step yields $\mathsf{B}(\mathsf{BB})xyz$, and we choose the following types for the first two B's. $((\mathcal{E} \to \mathcal{A} \to \mathcal{B}) \to \mathcal{E} \to (\mathcal{C} \to \mathcal{A}) \to \mathcal{C} \to \mathcal{B}) \to (\mathcal{D} \to \mathcal{E} \to \mathcal{A} \to \mathcal{B}) \to \mathcal{D} \to \mathcal{E} \to (\mathcal{C} \to \mathcal{A}) \to \mathcal{C} \to \mathcal{B}$ and $((\mathcal{A} \to \mathcal{B}) \to (\mathcal{C} \to \mathcal{A}) \to \mathcal{C} \to \mathcal{B}) \to (\mathcal{E} \to \mathcal{A} \to \mathcal{B}) \to \mathcal{E} \to (\mathcal{C} \to \mathcal{A}) \to \mathcal{C} \to \mathcal{B}$. From the latter, $(\mathcal{A} \to \mathcal{B}) \to (\mathcal{C} \to \mathcal{A}) \to \mathcal{C} \to \mathcal{B}$ may be detached, and the resulting wff can be detached from the type of the first B giving $(\mathcal{D} \to \mathcal{E} \to \mathcal{A} \to \mathcal{B}) \to \mathcal{D} \to \mathcal{E} \to (\mathcal{C} \to \mathcal{A}) \to \mathcal{C} \to \mathcal{B}$. From this wff, the types of x, y and z can be successively detached resulting in $(\mathcal{C} \to \mathcal{A}) \to \mathcal{C} \to \mathcal{B}$.

1. The base case is when we have an instance of the axiom. If $\mathcal{A} \vdash \mathcal{A}$ is instantiated with $\boldsymbol{t}$, then $\Delta = \emptyset$, and we have $\boldsymbol{t}^{\#} \Vdash \boldsymbol{t}$. But $\boldsymbol{t}^{\#}$ is I, hence, an instance of the principal type schema of I is a suitable instance of self-implication. If $\mathcal{A}$ is not $\boldsymbol{t}$, then $\mathcal{A}^{\#}$ is x_n (for some n) and $\Delta = \{x_n\colon \mathcal{A}\}$. Obviously, $\Delta \Vdash x_n\colon \mathcal{A}$.

2.1 If the rule is $(\check{\boldsymbol{t}}_l \vdash)$, then by hypothesis, $\Delta \Vdash (\mathfrak{A}[\mathfrak{B}])^{\#}\colon \mathcal{A}$. In particular, $\Delta \Vdash (\mathfrak{B})^{\#}\colon \mathcal{B}$, for some $\mathcal{B}$. (The subterms of a typable term are typable, and the context that generates a type for a term, generates a type for each of its subterms.) $\boldsymbol{t}$ in $\mathfrak{A}[\boldsymbol{t};\mathfrak{B}]$ is I in $(\mathfrak{A}[\boldsymbol{t};\mathfrak{B}])^{\#}$. With $\mathsf{I}\colon \mathcal{B} \to \mathcal{B}$, $\Delta \Vdash (\boldsymbol{t};\mathfrak{B})^{\#}\colon \mathcal{B}$; hence, we have that $\Delta \Vdash (\mathfrak{A}[\boldsymbol{t};\mathfrak{B}])^{\#}\colon \mathcal{A}$.

2.2 If the rule is $(\hat{\boldsymbol{t}}_r \vdash)$, then we have that $\Delta \Vdash (\mathfrak{A}[\mathfrak{B};\boldsymbol{t}])^{\#}\colon \mathcal{A}$. By the definitions of $^{\#}$ and Δ, there is no variable in Δ that would have the index that $\boldsymbol{t}$ has, which means that Δ does not need to be changed. Since $(\mathfrak{B};\boldsymbol{t})^{\#}$ is typable, we know that $\mathfrak{B}$ has a type of the form $(\mathcal{B} \to \mathcal{B}) \to \mathcal{C}$, given Δ. $\Delta \Vdash (\mathfrak{B};\boldsymbol{t})^{\#}\colon \mathcal{C}$ then, with I's type chosen as $\mathcal{B} \to \mathcal{B}$. But then $\Delta \Vdash (4\mathfrak{B})^{\#}\colon \mathcal{C}$, because $\Vdash 4\colon ((\mathcal{B} \to \mathcal{B}) \to \mathcal{C}) \to \mathcal{C}$. Inserting $4\mathfrak{B}$ back into the consecution, we have that $\Delta \Vdash (\mathfrak{A}[4\mathfrak{B}])^{\#}\colon \mathcal{A}$, as desired.

2.3 If the rule is $(\mathsf{B}\vdash)$, then we have that $\Delta \Vdash (\mathfrak{A}[\mathfrak{B};(\mathfrak{C};\mathfrak{D})])^{\#}\colon \mathcal{A}$. There is no change in Δ as a result of an application of this rule. If $\Delta \Vdash (\mathfrak{B};(\mathfrak{C};\mathfrak{D}))^{\#}\colon \mathcal{B}$, then $\Delta \Vdash \mathfrak{B}^{\#}\colon \mathcal{C} \to \mathcal{B}$, and $\Delta \Vdash \mathfrak{C}^{\#}\colon \mathcal{D} \to \mathcal{C}$ as well as $\Delta \Vdash \mathfrak{D}^{\#}\colon \mathcal{D}$. In view of our earlier remark about the (Bs), we consider $\mathsf{B}\mathfrak{B}$ without further ado. We choose as an instance of the principal type schema of B the wff $(\mathcal{C} \to \mathcal{B}) \to (\mathcal{D} \to \mathcal{C}) \to \mathcal{D} \to \mathcal{B}$. Then $\Delta \Vdash (\mathsf{B}\mathfrak{B};\mathfrak{C};\mathfrak{D})^{\#}\colon \mathcal{B}$, hence, $\Delta \Vdash (\mathfrak{A}[\mathsf{B};\mathfrak{B};\mathfrak{C};\mathfrak{D}])^{\#}\colon \mathcal{A}$.

2.4 If the rule is $(\mathsf{B}'\vdash)$, then we proceed similarly, but with the insertion of B'. Assuming the same types for $\mathfrak{B}$, $\mathfrak{C}$ and $\mathfrak{D}$ in the upper consecution, as in the previous case, we choose the type of B' to be $(\mathcal{D} \to \mathcal{C}) \to (\mathcal{C} \to \mathcal{B}) \to \mathcal{D} \to \mathcal{B}$. Then $\Delta \Vdash (\mathsf{B}'\mathfrak{C};\mathfrak{B};\mathfrak{D})^{\#}\colon \mathcal{B}$, hence, $\Delta \Vdash (\mathfrak{A}[\mathsf{B}'\mathfrak{C};\mathfrak{B};\mathfrak{D}])^{\#}\colon \mathcal{A}$.

2.5 If the rule is $(\mathsf{W}\vdash)$, then we distinguish four subcases, according to whether any combinators have been or have not been attached to the two occurrences of $\mathcal{C}$.

2.5.1 In the simplest case, the $\mathcal{C}$'s have not been differentiated, and by hypothesis, we have $\Delta \Vdash (\mathfrak{A}[\mathfrak{B};\mathcal{C};\mathcal{C}])^{\#}\colon \mathcal{A}$. Then $\mathfrak{B}^{\#}$ must have a type $\mathcal{C} \to \mathcal{C} \to \mathcal{B}$, whereas $\mathcal{C}^{\#}$'s type is $\mathcal{C}$. Thus, $\Delta \Vdash (\mathfrak{B};\mathcal{C};\mathcal{C})^{\#}\colon \mathcal{B}$. The insertion of W and the deletion of one of the occurrences of $\mathcal{C}$ results in the consecution $\mathfrak{A}[\mathsf{W}\mathfrak{B};\mathcal{C}] \vdash \mathcal{A}$. We may choose $(\mathcal{C} \to \mathcal{C} \to \mathcal{B}) \to \mathcal{C} \to \mathcal{B}$ as the type of W, and then we have $\Delta \Vdash (\mathsf{W}\mathfrak{B};\mathcal{C})^{\#}\colon \mathcal{B}$. Δ remains the same in the lower consecution as it is in the upper consecution, because there is no new variable in the term in the lower consecution, and there is no loss of a variable either. To put everything together, $\Delta \Vdash (\mathfrak{A}[\mathsf{W}\mathfrak{B};\mathcal{C}])^{\#}\colon \mathcal{A}$.
2.5.2 If the upper consecution is $\mathfrak{A}[\mathfrak{B};\mathsf{Z}\mathcal{C};\mathcal{C}] \vdash \mathcal{A}$, then Z's type is $\mathcal{C} \to \mathcal{D}$ and $\mathfrak{B}^{\#}$'s type must be $\mathcal{D} \to \mathcal{C} \to \mathcal{B}$. For the type of B and W, we choose wff's so that BW has type $((\mathcal{D} \to \mathcal{C} \to \mathcal{B}) \to \mathcal{C} \to \mathcal{C} \to \mathcal{B}) \to (\mathcal{D} \to \mathcal{C} \to \mathcal{B}) \to \mathcal{C} \to \mathcal{B}$. Z's type is suffixed with $\mathcal{C} \to \mathcal{B}$, thus, $\mathsf{BW}(\mathsf{B}'\mathsf{Z})$ has type $(\mathcal{D} \to \mathcal{C} \to \mathcal{B}) \to \mathcal{C} \to \mathcal{B}$. For the new substructure, we have that $\Delta \Vdash (\mathsf{BW}(\mathsf{B}'\mathsf{Z})\mathfrak{B};\mathcal{C})^{\#}\colon \mathcal{B}$, hence, $\Delta \Vdash (\mathfrak{A}[\mathsf{BW}(\mathsf{B}'\mathsf{Z})\mathfrak{B};\mathcal{C}])^{\#}\colon \mathcal{A}$.
2.5.3 –2.5.4 These cases are similar, and we leave them for an exercise.
2.6 If the rule is $(\to\vdash)$, then we have $\Delta_1 \Vdash \mathfrak{A}^{\#}\colon \mathcal{A}$ and $\Delta_2 \Vdash (\mathfrak{B}[\mathcal{B}])^{\#}\colon \mathcal{C}$. The definitions of the numbering and of $^{\#}$ ensure that $\Delta_1 \cup \Delta_2$ is consistent. If the variable x_n, which corresponds to $\mathcal{B}$, does not have other occurrences in $\mathfrak{B}$, then the type-assignment for that variable may be omitted; otherwise, $\Delta = \Delta_1 \cup \Delta_2$. Clearly, $\Delta \Vdash (\mathcal{A} \to \mathcal{B};\mathfrak{A})^{\#}\colon \mathcal{B}$, and after inserting the structure into $\mathfrak{B}$, we obtain $\Delta \Vdash (\mathfrak{B}[\mathcal{A} \to \mathcal{B};\mathfrak{A}])^{\#}\colon \mathcal{C}$, as we wished to prove.

If $\mathcal{B}$ has become a caf X, which is of the form $\mathsf{Z}\mathcal{B}$, then Z's type is $\mathcal{B} \to \mathcal{D}$, and $(\mathsf{Z}\mathcal{B})^{\#}$'s type is $\mathcal{D}$. For B, we take the type $(\mathcal{B} \to \mathcal{D}) \to (\mathcal{A} \to \mathcal{B}) \to \mathcal{A} \to \mathcal{D}$. Then $(\mathsf{BZ}\mathcal{A} \to \mathcal{B})^{\#}$ yields $\mathcal{A} \to \mathcal{D}$, hence, $\Delta \Vdash (\mathsf{BZ}\mathcal{A} \to \mathcal{B};\mathfrak{A})^{\#}\colon \mathcal{D}$. Then, $\Delta \Vdash \mathfrak{B}[\mathsf{BZ}\mathcal{A} \to \mathcal{B};\mathfrak{A}])^{\#}\colon \mathcal{C}$, as desired.
2.7 Lastly, if the rule is $(\vdash\to)$, then $\Delta \Vdash (\mathfrak{A};\mathcal{A})^{\#}\colon \mathcal{B}$, by hypothesis. Δ is shrunk for the lower consecution, if $\mathcal{A}$'s variable has no other occurrence than what results from the explicit $\mathcal{A}$. Let Δ' be Δ or the new set of type-assignments. $\mathfrak{A}^{\#}$ must have the type $\mathcal{A} \to \mathcal{B}$, because of the form of the antecedent, but then the correctness of the typing of the lower consecution is immediate. $\Delta' \Vdash \mathfrak{A}^{\#}\colon \mathcal{A} \to \mathcal{B}$.

If $\mathcal{A}$ is no longer a wff, but a caf with a combinator, then it is of the form $\mathsf{Z}\mathcal{A}$, with some Z. The type of Z has to be $\mathcal{A} \to \mathcal{D}$, where $\mathfrak{A}^{\#}$'s type is $\mathcal{D} \to \mathcal{B}$. We select $(\mathcal{A} \to \mathcal{D}) \to (\mathcal{D} \to \mathcal{B}) \to \mathcal{A} \to \mathcal{B}$ as the type for B'. Then, $(\mathsf{B}'\mathsf{Z}\mathfrak{A})^{\#}$ has type $\mathcal{A} \to \mathcal{B}$, hence, $\Delta' \Vdash (\mathsf{B}'\mathsf{Z}\mathfrak{A})^{\#}\colon \mathcal{A} \to \mathcal{B}$. ($\Delta'$ is either Δ or Δ without the type-assignment for $\mathcal{A}^{\#}$, as before.) qed

Exercise 9.3.7. We did not include the details of Cases **2.5.3** and **2.5.4**. Complete the proof by working out the details.

(3) The BB'-*abstraction* is the final step in the extraction procedure. Before we started the numbering and the combinatory insertion, the source consecution was of the form $\boldsymbol{t};\mathcal{A}_1;\ldots;\mathcal{A}_n \vdash a$. If we are considering a theorem of $E_{\to}$, then we know that none of the $\mathcal{A}$'s is $\boldsymbol{t}$, moreover, $\boldsymbol{t}$ does not occur in them.

Now, it has the form $\mathsf{Z}_0\boldsymbol{t};\mathsf{Z}_1\mathcal{A}_1;\ldots;\mathsf{Z}_n\mathcal{A}_n \vdash a$, where any of the Z's may be missing. The term resulting from the antecedent is $\mathsf{Z}_0\mathsf{I}(\mathsf{Z}_1x_1)\ldots(\mathsf{Z}_nx_n)$. Each variable has a different index, and occurs exactly once. Moreover, if we scan the term from left to right, then the indices are monotone increasing. This term has a very special form in the broader class of terms that are called *hereditary right maximal terms*. There are some variations on how the label "hereditary right maximal term" is used, but the common motive is to permit a subterm like x_nx_m if $n \leq m$, and if this condition is satisfied, then give the index m to (x_nx_m) itself, as if the compound term *inherited* the right-hand side index. A prominent variation includes restricting our attention to variables with subscripts less than some fixed n, and then we overlook variables with a higher index. However, we only need to abstract the Z's from the term $\mathsf{Z}_0\mathsf{I}(\mathsf{Z}_1x_1)\ldots(\mathsf{Z}_nx_n)$, hence, we are not going to define the various subclasses of hereditary right maximal terms, or a full $\mathsf{BB'I}$- or a $\mathsf{BB'IW}$-abstraction.

Definition 9.52. (Restricted $\mathsf{BB'}$-abstraction)
The *restricted* $\mathsf{BB'}$*-abstraction* is defined by applying (1) followed by (2).
(1) Given a term of the form $\mathsf{Z}_0\mathsf{I}(\mathsf{Z}_1x_1)(\mathsf{Z}_2x_2)\ldots(\mathsf{Z}_nx_n)$, we apply $\mathsf{B'}$ expansions from right to left to obtain $\mathsf{B'Z}_n(\mathsf{B'Z}_{n-1}\ldots(\mathsf{B'Z}_1(\mathsf{Z}_0\mathsf{I})x_1)\ldots x_{n-1})x_n$. (If Z_m is missing then $\mathsf{B'Z}_m$ does not appear in the term.)
(2) Each $\mathsf{B'Z}_m$ (where $0 < m \leq n$) is replaced by $\mathsf{B}(\mathsf{B}\ldots(\mathsf{B'Z}_m)...)$, where $m-1$ B's are prefixed to $(\mathsf{B'Z}_m)$.

The combinator that we obtained this way has a special form. In general, we get a combinator (followed by n variables) in the following shape.

$$\mathsf{B}(\mathsf{B}\ldots(\mathsf{B'Z}_n)...)(\mathsf{B}(\mathsf{B}\ldots(\mathsf{B'Z}_{n-1})...)\ldots(\mathsf{B}(\mathsf{B'Z}_2)(\mathsf{B'Z}_1(\mathsf{Z}_0\mathsf{I})))...)x_1x_2\ldots x_{n-1}x_n$$

If any of the Z's is missing, then the matching part of this term is void. Typability is preserved by B and $\mathsf{B'}$ expansion steps. The abstraction could have been carried out in one step, of course. However, we separated the B and $\mathsf{B'}$ steps for clarity. The application of step (1) ensures that the order of the variables with respect to each other is unchanged. The $\mathsf{B'}$'s also guarantee that no combinator remains between the variables. The second step is like (Bs), that is, it simply disassociates the combinators from the variables.

Exercise 9.3.8. Finish the example that we used to illustrate the numbering and the insertion of combinators by applying the restricted $\mathsf{BB'}$-abstraction. Is the theorem $(\mathcal{A} \to (\mathcal{B} \to \mathcal{C}) \to \mathcal{D}) \to (\mathcal{B} \to \mathcal{C}) \to \mathcal{A} \to \mathcal{D}$ the principal type schema of the combinator you obtained?

To conclude this subsection we articulate three slogans. The well-known *"formulas are types"* view characterizes typed and type-assignment systems. Structurally free logics capture the *"combinators are formulas"* idea. Whereas our extraction procedure promulgates the *"formulas are* CL*-terms"* picture.

9.3.2 Categorial grammars and labeled calculi

Linguistics was a young and rapidly developing discipline in the middle of the 20th century. In the 1950s, several people seem to have hit upon the idea of applying formal, mathematical tools to languages. *Grammars* are traditionally one of the main concerns of those studying natural languages. The grammaticality or well-formedness of sequences of words might seem to be an easier task to tackle than capturing regularities of *meaning*, which is often elusive. Some linguists contend that to provide a systematic account of the meaning of natural language expressions, first, strings of expressions must be parsed into suitable syntactic categories. This also leads us back to the problem of creating or discovering grammars.

The Lambek calculi (outlined in Sections 4.1 and 5.1) were introduced based on linguistic motivations. The idea of parsing sentences into *functions* and their *arguments* was prominent in the Polish school of logic at the beginning of the 20th century—as shown by the work of Ajdukiewicz, who introduced *semantic categories*. Since its inception, categorial grammars have made big strides—both in terms of the developments of *formalisms*, as well as in terms of the size of natural language fragments and the number of linguistic phenomena that they can describe. There is a vast amount of literature on categorial grammars—even if we would consider only those that deal with English. Further references and sample categorial grammars may be found, for instance, in [57], [58], [139] and [140].

Concrete *lexical items* can be thought to replace λ-*terms* and CL-*terms* in type-assignments. A general pattern persists in that objects of different sorts are paired up with operations acting on them, and possibly, effecting changes in both. This idea has been applied to various areas in logic. For instance, calculi can be constructed where entities combine *syntactic* and *semantic information* such as a formula and a possible world (in the sense of relational semantics). $w\colon \mathcal{A}$ may be thought of as the statement that $\mathcal{A}$ is true at the possible world w (or as $w \vDash \mathcal{A}$). Examples of such calculi may be found in [127] and [144]. This approach appears to have great prospects in applications. It also seems that there are a lot of potential questions to be answered about such calculi. For example, in [144], substructural logics are formalized using labeled sequent calculi. However, it is unknown whether standard results obtained in other sequent calculus formalizations of the same logics can be replicated in some way in the labeled systems, or perhaps, new metatheorems can be proved, using the labeled approach, about those logics.

Appendix A

Some supplementary concepts

Modern logic uses certain mathematical tools and techniques—mainly from the area of *discrete mathematics*. There are comprehensive books on discrete mathematics, and some elements of discrete mathematics are included in books on theoretical computer science, on formal philosophy, on informatics, on linguistics, etc. For instance, [169] provides useful background concepts, which are related to computation, [106] is a comprehensive volume on graphs and [45] is an encyclopedic treatment of lattices.

The aim of this short appendix is to introduce some concepts that are paramount to the theory of sequent calculi, including those that we deal with in this book. These notions are typically not dealt with elsewhere, because they specifically pertain to issues in meta-logic.

A.1 Trees

Structures that we call *trees* play an important role everywhere in this book. Our trees always have a *root*, and they "grow" from their root. In other words, what we call a tree is sometimes called a *directed tree with a root* or a *rooted tree* in graph theory.[1] We have no use for connected acyclic graphs per se, and so our usage should not cause any confusion given the definitions below. Furthermore, we restrict out considerations to trees with no more than $\aleph_0$-many elements, because the cardinality of sequents and consecutions is $\aleph_0$.

We first give an *inductive definition* of trees.

Definition A.1. (Trees, 1) We define *trees* as certain ordered four-tuples $\langle V, E, r, L\rangle$, where V is the set of *vertices*, E is the set of *edges*, r is the *root* and L is the set of *leaves*.

(1) $\mathcal{T} = \langle \{ v \}, \emptyset, v, \{ v \} \rangle$ is a tree.

(2) If $L = \{ l_i \colon i \in I \}$ is an indexed set of points such that $r \notin L$, and $E = \{ \langle r, l_i \rangle \colon i \in I \}$, then $\mathcal{T} = \langle L \cup \{ r \}, E, r, L \rangle$ is a tree.

[1] Gross and Yellen [106] is a comprehensive volume on definitions and results in graph theory.

(3) Let $V = \{ v_i : i \in I \}$ be an indexed set that has order-type n (for $n \in \mathbb{N}^+$) or ω under a strict linear order relation ϱ. Let r be the ϱ-least element of V, and $E = \{ \langle v_k, v_j \rangle : \langle v_k, v_j \rangle \in \varrho \}$. Then $\mathfrak{T} = \langle V, E, r, L \rangle$ is a tree, where $L = \emptyset$ if $\mathrm{card}(V) = \aleph_0$, otherwise, $L = \{ l \}$, where l is the ϱ-maximal element of V.

(4) Let $\mathfrak{T} = \langle V, E, r, L \rangle$ be a tree; let $J = \{ l_i \in L : i \in I \}$ be an indexed subset of the leaves of $\mathfrak{T}$ and let $K = \{ w_i : \langle w_i, l_i \rangle \in E \wedge i \in I \}$, the set of predecessors of the elements of J. If $\mathfrak{T}_{i \in I} = \langle V_i, E_i, r_i, L_i \rangle$ is an indexed set of trees, then $\mathfrak{T}' = \langle V', E', r', L' \rangle$ is a tree, where the components are as follows.

(i) $V' = \bigcup_{i \in I} V_i \cup (V - J)$

(ii) $E' = \bigcup_{i \in I} E_i \cup (E - \{ \langle w_i, l_i \rangle : w_i \in K \wedge l_i \in J \}) \cup \{ \langle w_i, r_i \rangle : w_i \in K \wedge i \in I \}$

(iii) $r' = r$

(iv) $L' = \bigcup_{i \in I} L_i \cup (L - J)$

We tacitly assumed that all the trees in (4) have *pairwise disjoint sets* of vertices. Disjointness can always be guaranteed by indexing the sets and their elements. We simply make disjointness an assumption so as not to overcomplicate our notation.

Trees can be (and often are) defined without induction.

Definition A.2. (Trees, 2) The four-tuple $\langle V, E, r, L \rangle$ is a *tree* if and only if (1)–(4) hold.

(1) $V \neq \emptyset, \quad E \subseteq V \times V, \quad r \in V, \quad L \subseteq V$

(2) $\forall v \in V\, (v \neq r \Rightarrow \exists! w \in V\, \langle w, v \rangle \in E)$ and $\neg \exists v \in V\, \langle v, r \rangle \in E$

(3) $\forall v \in V\, (v \neq r \Rightarrow \exists! \langle w_0, \ldots, w_n \rangle\, (r = w_0 \wedge \forall i\, (0 \leq i \leq n-1 \Rightarrow \langle w_i, w_{i+1} \rangle \in E) \wedge w_n = v))$

(4) $l \in L$ iff $\neg \exists w \in V\, \langle l, w \rangle \in E$

Clause (2) means that for every vertex that is distinct from the root there is a unique vertex from which the former can be reached in one step. Such a vertex is called a *parent*, and the root has no parent at all. Clause (3) states the uniqueness of a *path* from the root to any other vertex.

In graph theory, the points on which a graph is defined are usually called *vertices* and the elements of the binary relation are *edges*—the terms we have been using so far. However, we will interchangeably use vertex and *node* or even simply *point*.

The *indegree* and the *outdegree* of a node is the number of incoming and outgoing edges. The root, which is also called the *origin* of the tree, is the only node with *indegree zero*, whereas nodes with *outdegree zero* are the *leaves*. Any node that has a positive outdegree is an *internal* or *intermediate* node.

The nodes of a tree are often categorized with respect to their absolute position in the tree as well as their relative positions to each other. If a node v_2 has an incoming edge from v_1, then v_2 is a *child* or *successor* of v_1, and v_1 is the parent or *predecessor* of v_2. The transitive closure of the successor relation is the *descendant* relation, with its converse being the *ancestor* relation. In other words, v_2 is a descendant of v_1 when there is a path from v_1 to v_2, which is necessarily directed and unique, because of the definition of trees. v_1 is an ancestor of v_2 in this situation. Nodes that have a common parent are called *siblings*.

A third definition of trees, which we adapt from [170], uses the function ℓ instead of singling out vertices that are a root and leaves.

Definition A.3. (Trees, 3) The three-tuple $\langle V, E, \ell \rangle$ is a *tree*, when (1)–(4) hold.

(1) $V \neq \emptyset, \quad E \subseteq V \times V, \quad \ell \colon V \longrightarrow \mathbb{Z}^+$

(2) $\exists! v \in V\, \ell(v) = 1,$ this node is called o, the origin;

(3) $\forall v' \in V (v' \neq o \Rightarrow \exists! v \in V\, \langle v, v' \rangle \in E)$

(4) $\forall v \in V\, \forall v' \in V (\langle v, v' \rangle \in E \Rightarrow \ell(v') = \ell(v) + 1)$

The existence of a distinguished node o, which is the root of the tree, is guaranteed by (2). ℓ is a total function that was not explicit in the previous definitions; the positive integer that ℓ assigns to a node is the *level* or *depth* at which the node is located in the tree. The level can be thought of as 1 added to the number of edges that have to be traversed from the root to the node in question. (It is convenient to think of the root as being at level 1.) The last condition, (4) ensures that ℓ can be given its intended meaning as the level where a node is.

Paths that start at the root and end in a leaf are of special importance. A path that is *maximal* in the sense that it is not contained in any other path is called a *branch*. The set of vertices that belong to a branch may comprise a *finite* or an *infinite* set. Finite branches end in a leaf. If all the branches of a tree are finite, then it has the *finite branch* property. The outdegree of a node is sometimes called its *branching factor*. If the branching factor of each node in a tree is a natural number, then we say that the tree is *finitely forking* or that it has the *finite fork* property.

Figure A.1 shows a tree (on the left) and two directed graphs that are not trees. The pictured tree "grows" from the top to the bottom, but trees can "grow" from the bottom (where their root is) to the top, or sideways. Sequent calculi proofs have their root at the bottom, but they expand downward, because their roots are replaced in each proof step.

Trees can be *finite* or *infinite* depending on the cardinality of the set V. Proofs in sequent calculi, as well as in some other proof systems are finite

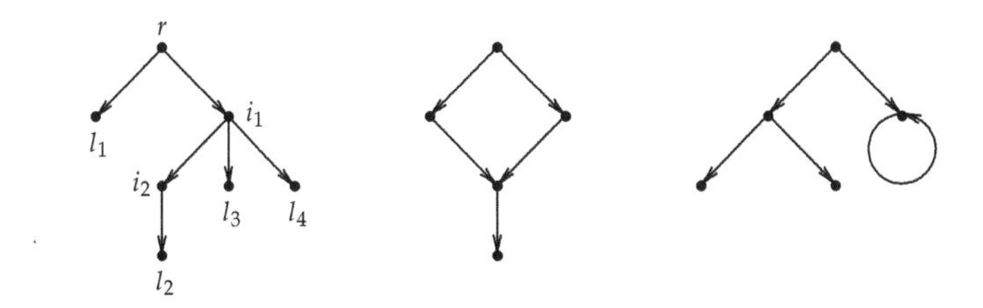

Figure A.1. A tree and two "non-trees."

trees. Often, a search for a proof may be constructed as a tree too, in which case it is important to know if a tree is finite or infinite. Figure A.2 illustrates the ways in which a tree can be infinite.

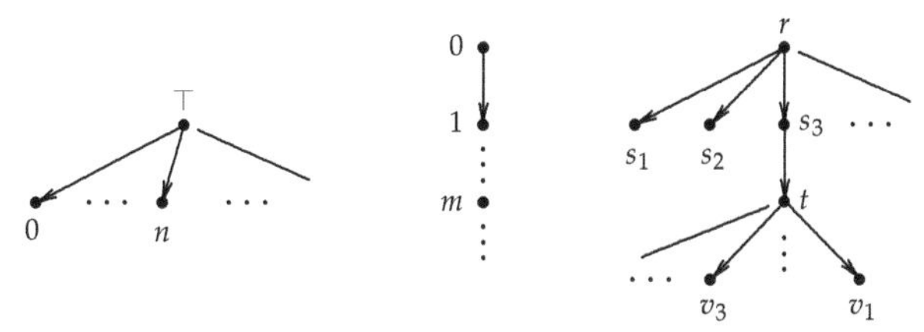

Figure A.2. Infinite trees.

Lemma A.4. (König's lemma) *A tree that has the finite fork and the finite branch properties is* finite.

Proof: Let the tree $\mathcal{T}$ be infinite, but have the finite fork property. Let us call a node infinite if it has infinitely many descendants; otherwise, we call the node finite. The root is infinite. Each node has finitely many children, because of the finite fork property. If all the children of a node are finite, then it is itself finite. Therefore, the root must have at least one infinite child, which in turn must have at least one infinite child. Selecting at each level an infinite child (of the selected infinite node), we obtain an infinite branch, which shows that the tree has at least one infinite branch; hence, it does not have the finite branch property. qed

The trees that occur in this book—such as proofs—are typically finite and have the finite fork property usually with their branching factor not exceeding 2. Occasionally, we will consider trees—such proof-search trees—with a higher (but still finite) branching factor, and we will aim at proving that all the branches are finite.

All the proof trees that we consider are *unordered*, that is, there is no linear ordering on the children of the nodes. Accordingly, even if we tend to place a node on the left or on the right (when, for instance, the branching factor is 2), this is only a coincidence and should not be conceived as an ordering.

A.2 Multiple inductive proofs

Inductive proofs are crucial in various areas—beyond formal theories of arithmetic. The principle of *weak mathematical induction* is, probably, the most widely known support for inductive proofs. Inductions are ubiquitous in meta-logic. However, it is often easier to use *complete induction* or *structural induction* than weak mathematical induction in proofs in meta-logic. A proof by complete induction allows a less cumbersome formulation of some claims. Furthermore, a proof by structural induction makes unnecessary the introduction of a function that maps into the natural numbers, such as the degree of a formula. In other words, the steps in a proof by structural induction follow the clauses of the inductive definition (which yields the class of objects that are dealt with).[2]

Assuming a language that includes at least a name constant for zero, let us say 0, and a unary function symbol for the successor function, let us say s, the *principle of weak induction* may be stated as (wi).

$$\mathcal{A}(0) \supset (\forall x\, (\mathcal{A}(x) \supset \mathcal{A}(s(x))) \supset \forall x\, \mathcal{A}(x)) \qquad \text{(wi)}$$

In a first-order theory, $\mathcal{A}(x)$ is a (possibly complex) formula with one free variable, whereas in a second-order theory, $\mathcal{A}(x)$ also could be a one-place predicate variable, which can be tacitly or explicitly universally quantified. Informally, the formula says that if $\mathcal{A}$ holds of zero, and $\mathcal{A}$ is passed along the successor function, then $\mathcal{A}$ holds of everything (i.e., $\mathcal{A}$ holds of all natural numbers).

If there is a binary function symbol in the language for addition, let us say $+$, then the two-place predicate $<$ is definable (or $<$ may be a primitive binary predicate). Then the *principle of complete induction* can be stated as (ci).

$$\forall x\, (\forall y\, (y < x \supset \mathcal{A}(y)) \supset \mathcal{A}(x)) \supset \forall x\, \mathcal{A}(x) \qquad \text{(ci)}$$

Here $\mathcal{A}(x)$ is exactly like in (wi). The wff may be read as "if $\mathcal{A}(x)$ holds, when $\mathcal{A}$ is true for all y that are strictly less than x, then $\mathcal{A}$ holds for all natural numbers."

These two principles (given that they are stated in a common language) are provably equivalent. To prove that (wi) implies (ci), a key step is to prove that $\forall x\, (\forall y\, (y < x \supset \mathcal{A}(y)) \supset \mathcal{A}(x))$ implies $\forall y\, (y \leq 0 \supset \mathcal{A}(0))$ as well as $\forall x\, (\forall z\, (z \leq x \supset \mathcal{A}(x)) \supset \forall z\, (z \leq s(x) \supset \mathcal{A}(s(x))))$ (where $x \leq y$ means $x < y \vee x = y$). The other direction is straightforward, given that $\mathcal{A}(0)$, and that $s(x) = y$ and $x < y \supset \mathcal{A}(x)$ imply $\mathcal{A}(y)$, hence, by (ci) $\forall x\, \mathcal{A}(x)$.

Inductions may be composed together in various ways. An induction can follow another one, or an induction may be carried out in the base case or

[2] An introduction to inductive definitions emphasizing a set-theoretical point of view is [2].

in the inductive case of another induction. These possibilities may be useful or even necessary, but from an abstract point of view they just stack several instances of a proof method on the top of each. A decisive observation is that the inductions can be separated from each other if desired.

We make a distinction between *two inductions* and a *double induction* as well as *three inductions* and a *triple induction*. A double induction is not two inductions and the inductions cannot be completely separated from each other. Similarly, for a triple induction.

In the concrete triple induction from Chapter 2, the non-separability stems from the interaction between the rank of the cut, the degree of the cut formula and the contraction measure of the cut. A reduction in the degree may increase the rank and the contraction measure of the cut. A reduction in the contraction measure, however, may increase only the rank. Finally, a decrease in rank never increases the degree of the cut formula or the contraction measure of the cut.

In Chapter 2, we gave a detailed proof of the admissibility of the single cut rule. The proof used a *triple induction* on three parameters, ϱ, δ and μ. The latter two range over non-negative integers, whereas ϱ is an integer and always at least 2. Triple inductions are not very common, hence, we first provide a graph that shows how the transformations in the various cases lead to other cases or completely eliminate a cut (which is indicated by $\emptyset$).

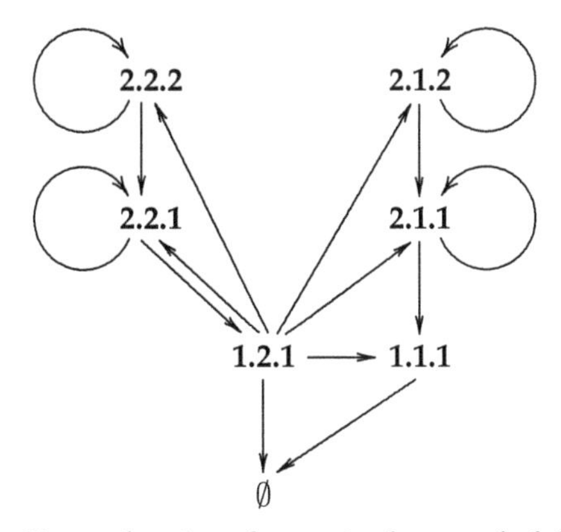

Figure A.3. Dependencies of cases in the proof of the cut theorem.

The graph in the diagram is obviously *cyclic*. However, this should not cause any concern, because the cases are sorted only with respect to a parameter having its least value or a greater value. For instance, in Case **2.2.1**, a reduction of ϱ by 1, let us say, from 14, produces a $\varrho' = 13$, and $13 > 2$. That is, the reduction in rank starting from this case does not lead in *one step* to another case.

The diagram does not show another aspect of the proof of the cut theorem for *LK* either. When δ decreases, because the cut formula is the principal formula of a rule for a binary connective, the cut is replaced by *two cuts*. Similarly, when μ decreases, because the cut formula is the principal formula of a contraction, the cut is replaced by *two cuts*. However, in both cases, each of the new cuts is eliminable by the hypothesis of the triple induction. In *LK*, the number of cuts cannot snowball out of control, because the duplication is accompanied with a decrease in a parameter.

Another way to see that the successive transformations of a proof with an application of the cut rule terminate in a cut-free proof is to depict the effect of the transformations in three-dimensional space.[3] The origin is at $\langle 2, 0, 0\rangle$ in each coordinate system in Figure A.4. The $\circ$ instead of $\bullet$ means that some transformations in that case eliminate the cut altogether.

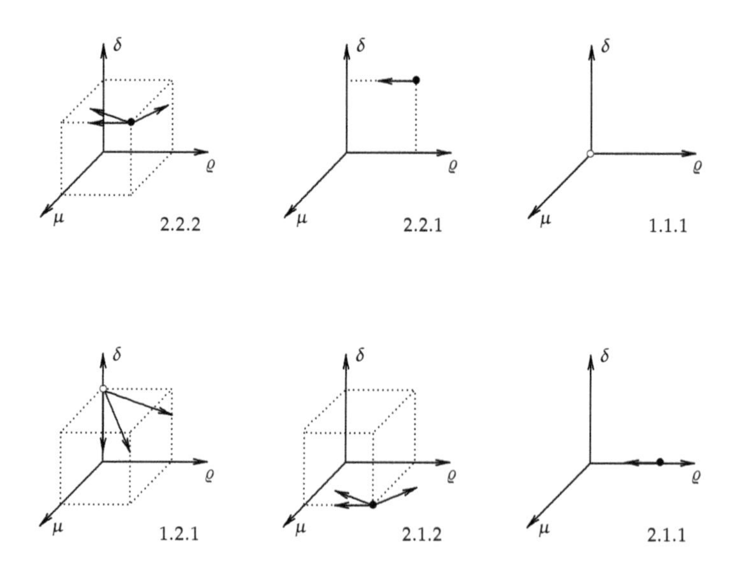

Figure A.4. Triple induction in three-dimensional space.

The three arrows in Cases **2.2.2** and **2.1.2** are in the same plane (which is a side of the virtual cube in the diagram). The three arrows in Case **1.2.1** lead to points that have a coordinate $\delta_1 < \delta_0$, where δ_0 is the coordinate of the $\circ$ on the vertical axis.

The triple induction principle may be formalized by a formula that we give below. We use D as a three-place predicate in the meta-language. It has a similar meaning as E had in Section 7.1, however, we add a third parameter.

[3]The idea behind this visualization comes from an illustration of Kripke's lemma in [77, §3.6].

$D(\delta, \mu, \varrho)$ is informally interpreted as "a cut with the cut formula of degree δ, with the contraction measure of the cut μ and with the rank of the cut ϱ is eliminable." In LK and LJ, contraction is one formula at a time, that is, reducing the number of the occurrences of a formula by 1. Moreover, the premise of an application of a contraction rule has two occurrences of a formula as subalterns. In the proof of the cut theorem that we presented in Chapter 2, the swap of a cut and of a contraction yields two cuts, each with lower contraction measure. However, in general, a contraction can reduce more than one occurrence of a formula (or of a structure), in which case, the similar transformation would yield several cuts (always finitely many though) with higher rank, but with lower contraction measure. (This situation can be easily illustrated in structurally free logics that we presented in Section 5.2.) A reduction in the degree of the cut formula may increase both the rank and the contraction measure. The next induction principle describes the triple induction. We use $<$ and $=$ to mean "strictly less than" and "equal," as usual.

$$\forall\varrho\,\forall\delta\,\forall\mu \Big[\big(\forall d\,(d < \delta \Rightarrow \forall m\,\forall k\,D(d,m,k)) \wedge \forall m\,(m < \mu \Rightarrow \forall k\,D(\delta,m,k)) \wedge \forall k\,(k < \varrho \Rightarrow D(\delta,\mu,k))\big) \Rightarrow D(\delta,\mu,\varrho)\Big] \Rightarrow \forall\varrho\,\forall\delta\,\forall\mu\,D(\delta,\mu,\varrho)$$

A.3 Curry's paradox

Haskell B. Curry was interested in combining combinatory logic with logic (in the sense of a calculus that has propositions, connectives, and perhaps, predicates and quantifiers). However, he soon discovered that some quite basic components of classical logic lead to inconsistency in the sense of absolute inconsistency. This is *Curry's paradox,* which he later slightly simplified. Nevertheless, it may be useful to emphasize that Curry's paradox—in either of its versions—concerns an *illative combinatory system* based on weak equality ($=_w$) enriched with a new binary constant ($\rightarrow$) that is characterized by axioms and a rule.[4] These axioms and the rule can be thought of as the positive implicational logic of Hilbert. Starting in the 1950s, following A. N. Prior's usage, a related, but essentially *semantical* paradox, has been labeled as "Curry's paradox" too. We are not concerned with this latter semantical paradox here, or with its relation to Curry's original paradox or to Löb's theorem.

We present very briefly the second variant that Curry gave. *Contraction* is a theorem of classical and intuitionistic (as well as of many other) logics. Instead of its usual form $(\mathcal{A} \rightarrow \mathcal{A} \rightarrow \mathcal{B}) \rightarrow \mathcal{A} \rightarrow \mathcal{B}$, we write it as

[4]We do not use Curry's notation for $\rightarrow$, which would be P, for the sake of transparency.

$\to(\to M(\to MN))(\to MN)$, where M and N are arbitrary CL-terms, including those that are intended to be thought of as formulas. The latter expression has a similarity to the contraction formula in prefix notation where C stands for $\to$ (i.e., for the conditional) $CC\mathcal{A}C\mathcal{A}\mathcal{B}C\mathcal{A}\mathcal{B}$—with the exception that in the prefix notation all the parentheses are omitted. The rule of detachment takes the form $\to MN$ and M imply N.

From the theory of equality in combinatory logic we need a form of *replacement*, namely, if $M =_{\mathrm{w}} N$, then $P[M/N] =_{\mathrm{w}} P$, where $P[M/N]$ is a notation for the term P in which an occurrence of the term M has been replaced by the term N, which weakly equals M, by assumption. Furthermore, there is a fixed point combinator, denoted by Y, for which $\mathsf{Y}M =_{\mathrm{w}} M(\mathsf{Y}M)$ holds, for any M. Lastly, there is a CL-term, that is, a combinator Z such that $\mathsf{Z}xyz \rhd_{\mathrm{w}} xz(xzy)$.

$\mathsf{Z}\!\to$ is a CL-term, and $\mathsf{Z}\!\to\! yx \rhd_{\mathrm{w}} \to x(\to xy)$. (The latter term looks like the antecedent of the contraction formula with x in place of M and y in place of N. $\mathsf{Z}\!\to\! y$ is also a CL-term, and so is $\mathsf{Y}(\mathsf{Z}\!\to\! y)$. The latter yields in one step $\mathsf{Z}\!\to\! y(\mathsf{Y}(\mathsf{Z}\!\to\! y))$, and further, $\to(\mathsf{Y}(\mathsf{Z}\!\to\! y))(\to(\mathsf{Y}(\mathsf{Z}\!\to\! y))y)$. Let us consider the following sequence of terms and weak equality statements.

1. $\mathsf{Y}(\mathsf{Z}\!\to\! y) =_{\mathrm{w}} \to(\mathsf{Y}(\mathsf{Z}\!\to\! y))(\to(\mathsf{Y}(\mathsf{Z}\!\to\! y))y)$
2. $\to(\to(\mathsf{Y}(\mathsf{Z}\!\to\! y))(\to(\mathsf{Y}(\mathsf{Z}\!\to\! y))y))(\to(\mathsf{Y}(\mathsf{Z}\!\to\! y))y)$
3. $\to(\mathsf{Y}(\mathsf{Z}\!\to\! y))(\to(\mathsf{Y}(\mathsf{Z}\!\to\! y))y)$
4. $\mathsf{Y}(\mathsf{Z}\!\to\! y)$
5. $\to(\mathsf{Y}(\mathsf{Z}\!\to\! y))y$
6. y

The first line is justified by the preceding reduction sequence. 2 is an instance of contraction—we simply chose $\mathsf{Y}(\mathsf{Z}\!\to\! y)$ for M and y for N. Then by replacement, we got 3 from 2 and 1, and by another replacement gave 4, from 3 and 1. However, 4 can be detached from 3 twice, yielding y in two steps.

For our purposes in Chapter 5, it is important to note that we justified the equation on line 1 by successive *reductions* of the term on the left of the $=_{\mathrm{w}}$ to the term on the right. (This is a perfectly good justification for a weak equality.) However, then we used the equation in the reverse direction, that is, we performed replacements according to weak *expansion* steps.

The first version of Curry's paradox assumes the provability of the self-implication $\mathcal{A} \to \mathcal{A}$ too. Let us fix a propositional variable, for instance, r. We have the following one-step reduction ($\rhd_1$) sequence. $\mathsf{Y}(\mathsf{C}\!\to\! r) \rhd_1 \mathsf{C}\!\to\! r(\mathsf{Y}(\mathsf{C}\!\to\! r)) \rhd_1 \to(\mathsf{Y}(\mathsf{C}\!\to\! r))r$. Again, we construct a series of terms ending with an arbitrary term, this time r.

1. $\to(\to(\mathsf{Y}(\mathsf{C}\!\to\! r))\underline{(\to(\mathsf{Y}(\mathsf{C}\!\to\! r))r)})(\to(\mathsf{Y}(\mathsf{C}\!\to\! r))r)$

2. $\to(\to(\mathsf{Y}(\mathsf{C}{\to}r))(\mathsf{Y}(\mathsf{C}{\to}r)))(\to(\mathsf{Y}(\mathsf{C}{\to}r))r)$

3. $\to(\mathsf{Y}(\mathsf{C}{\to}r))(\mathsf{Y}(\mathsf{C}{\to}r))$

4. $\to(\mathsf{Y}(\mathsf{C}{\to}r))r$

5. $\mathsf{Y}(\mathsf{C}{\to}r)$

6. r

1 is an instance of contraction in which we expanded the underlined term to obtain 2. 3 is an instance of self-implication, that we may detach from 2 to get the term on line 4. Then 4 expands to 5 yielding its own antecedent, which is detached resulting in r. Expansion steps are unproblematic in the theory of equality; however, it is important to note that, in the proof above, they cannot be replaced by reduction steps.

Bibliography

1. Wilhelm Ackermann. Begründung einer strengen Implikation. *Journal of Symbolic Logic*, 21:113–128, 1956.
2. Peter Aczel. An introduction to inductive definitions. In J. Barwise, editor, *Handbook of Mathematical Logic*, pages 739–782. North-Holland, Amsterdam, 1977.
3. Kazimierz Ajdukiewicz. Die syntaktische Konnexität. *Studia Philosophica*, 1:1–27, 1935.
4. Gerard Allwein and J. Michael Dunn. Kripke models for linear logic. *Journal of Symbolic Logic*, 58:514–545, 1993.
5. Alan R. Anderson. Entailment shorn of modality, (abstract). *Journal of Symbolic Logic*, 25:388, 1960.
6. Alan R. Anderson. Some open problems concerning the system E of entailment. *Acta Philosophica Fennica*, 16:9–18, 1963.
7. Alan R. Anderson and Nuel D. Belnap. *Entailment: The Logic of Relevance and Necessity*, volume I. Princeton University Press, Princeton, NJ, 1975.
8. Alan R. Anderson, Nuel D. Belnap, and J. Michael Dunn. *Entailment: The Logic of Relevance and Necessity*, volume II. Princeton University Press, Princeton, NJ, 1992.
9. Jeremy Avigad. Forcing in proof theory. *Bulletin of Symbolic Logic*, 10:305–333, 2004.
10. Arnon Avron. The semantics and proof theory of linear logic. *Theoretical Computer Science*, 57:161–184, 1988.
11. Arnon Avron. Combining classical logic, paraconsistency and relevance. *Journal of Applied Logic*, 3:133–160, 2005.
12. Matthias Baaz and Stefan Hetzel. On the non-confluence of cut-elimination. *Journal of Symbolic Logic*, 76:313–340, 2011.
13. Matthias Baaz and Alexander Leitsch. *Methods of Cut-Elimination*, volume 34 of *Trends in Logic*. Springer, Dordrecht, 2011.
14. Franco Barbanera and Stefano Berardi. A symmetric lambda calculus for classical program extraction. *Information and Computation*, 125:103–117, 1996.
15. Henk Barendregt and Silvia Ghilezan. Lambda terms for natural deduction, sequent calculus and cut elimination. *Journal of Functional Programming*, 10:121–134, 2000.
16. Jon Barwise, editor. *Handbook of Mathematical Logic*, volume 90 of *Studies in Logic and the Foundations of Mathematics*. Elsevier, Amsterdam, 1999.
17. J. C. Beall and G. Restall. *Logical Pluralism*. Oxford University Press, Oxford, UK, 2006.

18. Nuel D. Belnap. Intensional models for first degree formulas. *Journal of Symbolic Logic*, 32:1–22, 1967.

19. Nuel D. Belnap. Display logic. *Journal of Philosophical Logic*, 11:375–417, 1982.

20. Nuel D. Belnap. Linear logic displayed. *Notre Dame Journal of Formal Logic*, 31: 14–25, 1990.

21. Nuel D. Belnap. Life in the undistributed middle. In K. Došen and P. Schroeder-Heister, editors, *Substructural Logics*, pages 31–41. Clarendon, Oxford, UK, 1993.

22. Nuel D. Belnap and John R. Wallace. A decision procedure for the system $E_{\overline{I}}$ of entailment with negation. Technical Report 11, Contract No. SAR/609 (16), Office of Naval Research, New Haven, 1961.

23. Nuel D. Belnap and John R. Wallace. A decision procedure for the system $E_{\overline{I}}$ of entailment with negation. *Zeitschrift für mathematische Logik und Grundlagen der Mathematik*, 11:277–289, 1965.

24. Nuel D. Belnap, Anil Gupta, and J. Michael Dunn. A consecution calculus for positive implication with necessity. *Journal of Philosophical Logic*, 9:343–362, 1980.

25. Paul Bernays. Review of O. Ketonen, *Untersuchungen zum Prädikatenkalkul*. *Journal of Symbolic Logic*, 10:127–130, 1945.

26. Katalin Bimbó. Investigation into combinatory systems with dual combinators. *Studia Logica*, 66:285–296, 2000.

27. Katalin Bimbó. The Church–Rosser property in dual combinatory logic. *Journal of Symbolic Logic*, 68:132–152, 2003.

28. Katalin Bimbó. Semantics for dual and symmetric combinatory calculi. *Journal of Philosophical Logic*, 33:125–153, 2004.

29. Katalin Bimbó. The Church–Rosser property in symmetric combinatory logic. *Journal of Symbolic Logic*, 70:536–556, 2005.

30. Katalin Bimbó. Types of I-free hereditary right maximal terms. *Journal of Philosophical Logic*, 34:607–620, 2005.

31. Katalin Bimbó. Admissibility of cut in *LC* with fixed point combinator. *Studia Logica*, 81:399–423, 2005.

32. Katalin Bimbó. Relevance logics. In D. Jacquette, editor, *Philosophy of Logic*, volume 5 of *Handbook of the Philosophy of Science* (D. Gabbay, P. Thagard and J. Woods, eds.), pages 723–789. Elsevier (North-Holland), Amsterdam, 2007.

33. Katalin Bimbó. $LE^{t}_{\rightarrow}$, $LR^{\circ}_{\overset{\wedge}{\sim}}$, *LK* and cutfree proofs. *Journal of Philosophical Logic*, 36:557–570, 2007.

34. Katalin Bimbó. Dual gaggle semantics for entailment. *Notre Dame Journal of Formal Logic*, 50:23–41, 2009.

35. Katalin Bimbó. Schönfinkel-type operators for classical logic. *Studia Logica*, 95: 355–378, 2010.

36. Katalin Bimbó. *Combinatory Logic: Pure, Applied and Typed*. Discrete Mathematics and its Applications. CRC Press, Boca Raton, FL, 2012.

37. Katalin Bimbó. Combinatory logic. In E. Zalta, editor, *Stanford Encyclopedia of Philosophy*, pages 1–63. CSLI, Stanford, CA, URL: `plato.stanford.edu/entries/logic-combinatory`, 2nd, revised and expanded edition, 2012.

38. Katalin Bimbó and J. Michael Dunn. *Generalized Galois Logics. Relational Semantics of Nonclassical Logical Calculi*, volume 188 of *CSLI Lecture Notes*. CSLI Publications, Stanford, CA, 2008.

39. Katalin Bimbó and J. Michael Dunn. Symmetric generalized Galois logics. *Logica Universalis*, 3:125–152, 2009.

40. Katalin Bimbó and J. Michael Dunn. Calculi for symmetric generalized Galois logics. In J. van Benthem and M. Moortgat, editors, *Festschrift for Joachim Lambek*, volume 36 of *Linguistic Analysis*, pages 307–343. Linguistic Analysis, Vashon, WA, 2010.

41. Katalin Bimbó and J. Michael Dunn. New consecution calculi for $R^{t}_{\rightarrow}$. *Notre Dame Journal of Formal Logic*, 53:491–509, 2012.

42. Katalin Bimbó and J. Michael Dunn. On the decidability of implicational ticket entailment. *Journal of Symbolic Logic*, 78:214–236, 2013.

43. Katalin Bimbó and J. Michael Dunn. Combinatory inhabitants of $R_{\rightarrow}$ theorems extracted from sequent calculus proofs. *Bulletin of Symbolic Logic*, (to appear), 2015.

44. Katalin Bimbó and J. Michael Dunn. Extracting BB′IW inhabitants of simple types from proofs in the sequent calculus $LT^{t}_{\rightarrow}$ for implicational ticket entailment. *Logica Universalis*, 8:141–164, 2014.

45. Garrett Birkhoff. *Lattice Theory*, volume 25 of *AMS Colloquium Publications*. American Mathematical Society, Providence, RI, 3rd edition, 1967.

46. Patrick Blackburn, Maarten de Rijke, and Yde Venema. *Modal Logic*, volume 53 of *Cambridge Tracts in Theoretical Computer Science*. Cambridge University Press, Cambridge, UK, 2001.

47. Patrick Blackburn, Johan van Benthem, and Frank Wolter, editors. *Handbook of Modal Logic*, volume 3 of *Studies in Logic and Practical Reasoning*. Elsevier, Amsterdam, 2007.

48. Willem J. Blok and Don Pigozzi. *Algebraizable Logics*. Number 396 in Memoirs of the American Mathematical Society. American Mathematical Society, Providence, RI, 1989.

49. George Boolos. Don't eliminate cut. In *Logic, Logic, and Logic*, pages 365–369. Harvard University Press, Cambridge, MA, 1998.

50. Mirjana Borisavljević. Normal derivations and sequent derivations. *Journal of Philosophical Logic*, 37:521–548, 2008.

51. Ross T. Brady. Gentzenizations of relevant logics without distribution. I. *Journal of Symbolic Logic*, 61:353–378, 1996.

52. Ross T. Brady. Gentzenizations of relevant logics without distribution. II. *Journal of Symbolic Logic*, 61:379–401, 1996.

53. Ross T. Brady. Gentzenizations of relevant logics with distribution. *Journal of Symbolic Logic*, 61:402–420, 1996.

54. Ross T. Brady, editor. *Relevant Logics and Their Rivals. A Continuation of the Work of R. Sylvan, R. Meyer, V. Plumwood and R. Brady*, volume II. Ashgate, Burlington, VT, 2003.

55. Torben Braüner. A cut-free Gentzen formulation of the modal logic **S5**. *Logic Journal of IGPL*, 8:629–643, 2000.

56. Sabine Broda, Luís Damas, Marcelo Finger, and Paulo Silva de Silva. The decidability of a fragment of BB'IW-logic. *Theoretical Computer Science*, 318:373–408, 2004.

57. Wojciech Buszkowski. Linguistics and proof theory. In J. van Benthem and A. ter Meulen, editors, *Handbook of Logic and Language*, pages 683–736. Elsevier, Amsterdam, 1997.

58. Wojciech Buszkowski. Lambek calculus and substructural logics. In J. van Benthem and M. Moortgat, editors, *Festschrift for Joachim Lambek*, volume 36 of *Linguistic Analysis*, pages 15–48. Linguistic Analysis, Vashon, WA, 2010.

59. Brian F. Chellas. *Modal Logic: An Introduction*. Cambridge University Press, Cambridge, UK, 1995.

60. Alonzo Church. A note on the Entscheidungsproblem. *Journal of Symbolic Logic*, 1:40–41, 1936.

61. Alonzo Church. A formulation of the simple theory of types. *Journal of Symbolic Logic*, 5:56–68, 1940.

62. Alonzo Church. *The Calculi of Lambda-Conversion*. Princeton University Press, Princeton, NJ, 1st edition, 1941.

63. Alonzo Church. The weak theory of implication. In A. Menne, A. Wilhelmy, and H. Angsil, editors, *Kontrolliertes Denken, Untersuchungen zum Logikkalkül und zur Logik der Einzelwissenschaften*, pages 22–37. Komissions-Verlag Karl Alber, Munich, 1951.

64. Alonzo Church. *Introduction to Mathematical Logic*. Princeton University Press, Princeton, NJ, revised and enlarged edition, 1996.

65. William Craig. Linear reasoning. A new form of the Herbrand-Gentzen theorem. *Journal of Symbolic Logic*, 22:250–268, 1957.

66. William Craig. Three uses of the Herbrand–Gentzen theorem in relating model theory and proof theory. *Journal of Symbolic Logic*, 22:269–285, 1957.

67. Haskell B. Curry. The combinatory foundations of mathematical logic. *Journal of Symbolic Logic*, 7:49–64, 1942.

68. Haskell B. Curry. *Foundations of Mathematical Logic*. McGraw-Hill Book Company, New York, NY, 1963. (Dover, New York, NY, 1977).

69. Haskell B. Curry and Robert Feys. *Combinatory Logic, vol. I*. Studies in Logic and the Foundations of Mathematics. North-Holland, Amsterdam, 1st edition, 1958.

70. Haskell B. Curry, J. Roger Hindley, and Jonathan P. Seldin. *Combinatory Logic, vol. II*. Studies in Logic and the Foundations of Mathematics. North-Holland, Amsterdam, 1972.

71. Michael Detlefsen. Proof: Its nature and significance. In B. Gold and R. A. Simons, editors, *Proof and Other Dilemmas. Mathematics and Philosophy*, Spectrum Series, pages 3–32. Mathematical Association of America, Washington, DC, 2008.

72. Kosta Došen. Sequent-systems and groupoid models. I. *Studia Logica*, 47:353–385, 1988.

73. Kosta Došen and Peter Schroeder-Heister, editors. *Substructural Logics*. Clarendon, Oxford, UK, 1993.

74. Albert G. Dragalin. *Mathematical Intuitionism: Introduction to Proof Theory*, volume 67 of *Translations of Mathematical Monographs*. American Mathematical Society, Providence, RI, 1988.

75. J. Michael Dunn. A "Gentzen system" for positive relevant implication, (abstract). *Journal of Symbolic Logic*, 38:356–357, 1973.

76. J. Michael Dunn. Intuitive semantics for first-degree entailment and "coupled trees." *Philosophical Studies*, 29:149–168, 1976.

77. J. Michael Dunn. Relevance logic and entailment. In D. Gabbay and F. Guenthner, editors, *Handbook of Philosophical Logic*, volume 3, pages 117–224. D. Reidel, Dordrecht, 1st edition, 1986.

78. J. Michael Dunn. Gaggle theory: An abstraction of Galois connections and residuation with applications to negation, implication, and various logical operators. In J. van Eijck, editor, *Logics in AI: European Workshop JELIA '90*, number 478 in Lecture Notes in Computer Science, pages 31–51. Springer, Berlin, 1991.

79. J. Michael Dunn and Gary M. Hardegree. *Algebraic Methods in Philosophical Logic*, volume 41 of *Oxford Logic Guides*. Oxford University Press, Oxford, UK, 2001.

80. J. Michael Dunn and Robert K. Meyer. Gentzen's cut and Ackermann's gamma. In J. Norman and R. Sylvan, editors, *Directions in Relevant Logic*, pages 229–240. Kluwer, Dordrecht, 1989.

81. J. Michael Dunn and Robert K. Meyer. Combinators and structurally free logic. *Logic Journal of the IGPL*, 5:505–537, 1997.

82. Roy Dyckhoff and Sara Negri. Admissibility of structural rules for contraction-free systems of intuitionistic logic. *Journal of Symbolic Logic*, 65:1499–1518, 2000.

83. Melvin Fitting. Modal proof theory. In P. Blackburn, J. van Benthem, and F. Wolter, editors, *Handbook of Modal Logic*, volume 3 of *Studies in Logic and Practical Reasoning*, pages 85–138. Elsevier, Amsterdam, 2007.

84. Michael Gabbay. A proof-theoretic treatment of λ-reduction with cut-elimination: λ-calculus as a logic programming language. *Journal of Symbolic Logic*, 76:673–699, 2011.

85. Nikolaos Galatos and Hiroakira Ono. Cut elimination and strong separation for substructural logics: An algebraic approach. *Annals of Pure and Applied Logic*, 161:1097–1133, 2010.

86. Nikolaos Galatos, Peter Jipsen, Tomasz Kowalski, and Hiroakira Ono. *Residuated Lattices: An Algebraic Glimpse as Substructural Logics*. Elsevier, Amsterdam, 2007.

87. Jean Gallier. Constructive logics. Part I. *Theoretical Computer Science*, 110(2): 249–339, 1993.

88. Ian P. Gent. A sequent- or tableau-style system for Lewis's counterfactual logic VC. *Notre Dame Journal of Formal Logic*, 33:369–382, 1992.

89. Gerhard Gentzen. Untersuchungen über das logische Schließen. *Mathematische Zeitschrift*, 39:176–210, 1935.

90. Gerhard Gentzen. Die Widerspruchsfreiheit der reinen Zahlentheorie. *Mathematische Annalen*, 112:493–565, 1936.

91. Gerhard Gentzen. Investigations into logical deduction. *American Philosophical Quarterly*, 1:288–306, 1964.

92. Gerhard Gentzen. Investigations into logical deduction: II. *American Philosophical Quarterly*, 2:204–218, 1965.

93. Gerhard Gentzen. On the relation between intuitionist and classical arithmetic. In M. E. Szabo, editor, *The Collected Papers of Gerhard Gentzen*, Studies in Logic and the Foundations of Mathematics, pages 53–67. North-Holland, Amsterdam, 1969.

94. Steve Giambrone. *Gentzen Systems and Decision Procedures for Relevant Logics*. PhD thesis, Australian National University, Canberra, ACT, Australia, 1983.

95. Steve Giambrone. TW_+ and RW_+ are decidable. *Journal of Philosophical Logic*, 14:235–254, 1985.

96. Jean-Yves Girard. Linear logic. *Theoretical Computer Science*, 50:1–102, 1987.

97. Jean-Yves Girard. *The Blind Spot. Lectures on Logic*. European Mathematical Society Publishing House, Zürich, 2011.

98. Valeriĭ I. Glivenko. Sur quelques points de la logique de M. Brouwer. *Académie Royale de Belgique, Bulletin de la classe des sciences*, 15:183–188, 1929.

99. Louis F. Goble. Gentzen systems for modal logic. *Notre Dame Journal of Formal Logic*, 15:455–461, 1974.

100. Kurt Gödel. Zum intuitionistischen Aussagenkalkül. In S. Feferman, editor, *Collected Works*, volume I, pages 222–225. Oxford University Press and Clarendon Press, New York, NY and Oxford, UK, 1986.

101. Kurt Gödel. Eine Interpretation des intuitionistischen Aussagenkalküls. In S. Feferman, editor, *Collected Works*, volume I, pages 300–303. Oxford University Press and Clarendon Press, New York, NY and Oxford, UK, 1986.

102. Kurt Gödel. Zur intuitionistischen Arithmetik und Zahlentheorie. In S. Feferman, editor, *Collected Works*, volume I, pages 286–295. Oxford University Press and Clarendon Press, New York, NY and Oxford, UK, 1986.

103. Bonnie Gold and Roger A. Simons, editors. *Proof and Other Dilemmas. Mathematics and Philosophy*. Spectrum Series. Mathematical Association of America, Washington, DC, 2008.

104. Rajeev Goré and Ramanayake Revantha. Valentini's cut-elimination for provability logic resolved. *Review of Symbolic Logic*, 5:212–238, 2012.

105. Jean Goubault-Larrecq and Ian Mackie. *Proof Theory and Automated Deduction*, volume 6 of *Applied Logic Series*. Kluwer Academic, Boston, MA, 1997.

106. Jonathan Gross and Jay Yellen, editors. *Handbook of Graph Theory*. Discrete Mathematics and its Applications. CRC Press, Boca Raton, FL, 2004.

107. David S. Gunderson. *Handbook of Mathematical Induction: Theory and Applications*. Discrete Mathematics and Its Applications. CRC Press, Boca Raton, FL, 2011.

108. J. Roger Hindley. *Basic Simple Type Theory*, volume 42 of *Cambridge Tracts in Theoretical Computer Science*. Cambridge University Press, Cambridge, UK, 1997.

109. J. Roger Hindley and Jonathan P. Seldin. *Lambda-Calculus and Combinators: An Introduction*. Cambridge University Press, Cambridge, UK, 2008.

110. William A. Howard. The formulae-as-types notion of construction. In J. R. Hindley and J. P. Seldin, editors, *To H. B. Curry*, pages 479–490. Academic Press, London, UK, 1980.

111. Norihiro Kamide. Gentzen-type methods for bilattice negation. *Studia Logica*, 80:265–289, 2005.

112. Norihiro Kamide and Heinrich Wansing. Sequent calculi for some trilattice logics. *Review of Symbolic Logic*, 2:374–395, 2009.

113. Stephen C. Kleene. *Introduction to Metamathematics*. P. Van Nostrand Company, Inc., Princeton, NJ, 1952.

114. Alexei P. Kopylov. Decidability of linear affine logic. In A. R. Meyer, editor, *Special issue: LICS 1995*, volume 164 of *Information and Computation*, pages 173–198. IEEE, 2001.

115. Saul A. Kripke. The problem of entailment, (abstract). *Journal of Symbolic Logic*, 24:324, 1959.

116. Saul A. Kripke. Semantical analysis of modal logic I. Normal modal propositional calculi. *Zeitschrift für mathematische Logik und Grundlagen der Mathematik*, pages 67–96, 1963.

117. Saul A. Kripke. Semantical analysis of modal logic II. Non-normal propositional calculi. In J. W. Addison, L. Henkin, and A. Tarski, editors, *The Theory of Models*, pages 206–220. North-Holland, Amsterdam, 1965.

118. Saul A. Kripke. Semantical analysis of intuitionistic logic I. In J. N. Crossley and M. A. E. Dummett, editors, *Formal Systems and Recursive Functions. Proceedings of the Eighth Logic Colloquium*, pages 92–130, North-Holland, Amsterdam, 1965.

119. Hidenori Kurokawa. Tableaux and hypersequents for justification logics. *Annals of Pure and Applied Logic*, 163:831–853, 2012.

120. Joachim Lambek. The mathematics of sentence structure. *American Mathematical Monthly*, 65:154–169, 1958.

121. Joachim Lambek. On the calculus of syntactic types. In R. Jacobson, editor, *Structure of Language and Its Mathematical Aspects*, pages 166–178. American Mathematical Society, Providence, RI, 1961.

122. Joachim Lambek. From categorial grammar to bilinear logic. In K. Došen and Schroeder-Heister P., editors, *Substructural Logics*, pages 207–237. Clarendon, Oxford, UK, 1993.

123. Alexander Leitsch. *The Resolution Calculus*. Texts in Theoretical Computer Science. An EATCS Series. Springer-Verlag, Berlin, 1997.

124. Daniel Leivant. Proof theoretic methodology for propositional dynamic logic. In J. Díaz and I. Ramos, editors, *Formalization of Programming Concepts*, number 107 in Lecture Notes in Computer Science, pages 356–373, Berlin, 1981. Springer Verlag.

125. Daniel Leivant, editor. *Logic and Computational Complexity: International Workshop, LCC'94, Indianapolis, IN, USA, October 13-16, 1994: Selected Papers*. Number 960 in Lecture Notes in Computer Science. Springer, New York, NY, 1995.

126. Patrick Lincoln, John Mitchell, Andre Scedrov, and Natarajan Shankar. Decision problems for linear logic. *Annals of Pure and Applied Logic*, 56:239–311, 1992.

127. Wendy MacCaull. Relational semantics and a relational proof system for full Lambek calculus. *Journal of Symbolic Logic*, 63:623–637, 1998.

128. Edwin D. Mares. Relevance and conjunction. *Journal of Logic and Computation*, 22:7–21, 2012.

129. Errol P. Martin and Robert K. Meyer. Solution to the P–W problem. *Journal of Symbolic Logic*, 47:869–887, 1983.

130. Michael A. McRobbie and Nuel D. Belnap. Relevant analytic tableaux. *Studia Logica*, 38:187–200, 1979.

131. Michael A. McRobbie, Robert K. Meyer, and Paul W. Thistlewaite. Computer-aided investigations into the decision problem for relevant logics: The search for a free associative connective. Preprint Series 4/82, University of Melbourne, Melbourne, 1982.

132. Elliott Mendelson. *Introduction to Mathematical Logic*. Discrete Mathematics and its Applications. CRC Press, Boca Raton, FL, 5th edition, 2010.

133. Carew A. Meredith and Arthur N. Prior. Notes on the axiomatics of the propositional calculus. *Notre Dame Journal of Formal Logic*, 4:171–187, 1963.

134. George Metcalfe, Nicola Olivetti, and Dov M. Gabbay. *Proof Theory for Fuzzy Logics*, volume 36 of *Applied Logic Series*. Springer, New York, NY, 2009.

135. Robert K. Meyer. Topics in modal and many-valued logic. PhD thesis, University of Pittsburgh, Ann Arbor (UMI), 1966.

136. Robert K. Meyer. Improved decision procedures for pure relevant logic. In C. A. Anderson and M. Zelëny, editors, *Logic, Meaning and Computation: Essays in Memory of Alonzo Church*, pages 191–217. Kluwer Academic Publishers, Dordrecht, 2001.

137. Robert K. Meyer and J. Michael Dunn. E, R and γ. *Journal of Symbolic Logic*, 34: 460–474, 1969.

138. Robert K. Meyer, Michael A. McRobbie, and Nuel D. Belnap. Linear analytic tableaux. In P. Baumgartner, R. Hähnle, and J. Posegga, editors, *Theorem Proving with Analytic Tableaux and Related Methods*, number 918 in Lecture Notes in Computer Science, pages 278–293, Berlin, 1995. Springer.

139. Michael Moorgat. Symmetric categorial grammar: Residuation and Galois connections. In J. van Benthem and M. Moortgat, editors, *Festschrift for Joachim Lambek*, volume 36 of *Linguistic Analysis*, pages 143–166. Linguistic Analysis, Vashon, WA, 2010.

140. Glyn Morrill, Oriol Valentín, and Mario Fadda. The displacement calculus. *Journal of Logic, Language and Information*, 20:1–48, 2011.

141. Sara Negri and Jan von Plato. Cut elimination in the presence of axioms. *Bulletin of Symbolic Logic*, 4:418–435, 1998.

142. Sara Negri and Jan von Plato. *Structural Proof Theory*. Cambridge University Press, Cambridge, UK, 2001.

143. Sara Negri and Jan von Plato. Sequent calculus in natural deduction style. *Journal of Symbolic Logic*, 66:1803–1816, 2001.

144. Sara Negri and Jan von Plato. *Proof Analysis: A Contribution to Hilbert's Last Problem*. Cambridge University Press, Cambridge, UK, 2011.

145. Masao Ohnishi and Kazuo Matsumoto. Gentzen method in modal calculi. *Osaka Mathematical Journal*, 9:113–130, 1957.

146. Masao Ohnishi and Kazuo Matsumoto. Gentzen method in modal calculi, II. *Osaka Mathematical Journal*, 11:115–120, 1959.

147. Hiroakira Ono. Substructural logics and residuated lattices: An introduction. In V. F. Hendricks and J. Malinowski, editors, *50 Years of Studia Logica*, volume 21 of *Trends in Logic*, pages 193–228. Kluwer, Amsterdam, 2003.

148. Vincent Padovani. Ticket entailment is decidable. *Mathematical Structures in Computer Science*, 23:568–607, 2013.

149. Ewa Palka. An infinitary sequent system for the equational theory of $*$-continuous lattices. *Fundamenta Informaticae*, 78:295–309, 2007.

150. Francesco Paoli. *Substructural Logics: A Primer*, volume 13 of *Trends in Logic*. Kluwer, Dordrecht, 2002.

151. Jan von Plato. Rereading Gentzen. *Synthese*, 137:195–209, 2003.

152. Jan von Plato. Proof theory of classical and intuitionistic logic. In L. Haaparanta, editor, *The Development of Modern Logic*, pages 499–515. Oxford University Press, Oxford, UK, 2009.

153. Francesca Poggiolesi. *Gentzen Calculi for Modal Propositional Logic*, volume 32 of *Trends in Logic*. Springer, Dordrecht, 2011.

154. Emil L. Post. Introduction to a general theory of elementary propositions. *American Journal of Mathematics*, 43:163–185, (Reprinted in J. van Heijenoort, (ed.), *From Frege to Gödel: A Source Book in Mathematical Logic*, Harvard University Press, Cambridge, MA, 1967, pp. 264–283.) 1921.

155. Arthur N. Prior. The runabout inference-ticket. *Analysis*, 21:38–39, 1960.

156. David J. Pym, Peter O'Hearn, and Hongseok Yang. Possible worlds and resources: The semantics of **BI**. *Theoretical Computer Science*, 315:257–305, 2004.

157. James G. Raftery. Correspondences between Gentzen and Hilbert systems. *Journal of Symbolic Logic*, 71:903–957, 2006.

158. Greg Restall. Displaying and deciding substructural logics 1. *Journal of Philosophical Logic*, 27:179–216, 1998.

159. Greg Restall. A cut-free sequent system for two-dimensional modal logic, and why it matters. *Annals of Pure and Applied Logic*, 163:1611–1623, 2012.

160. Jacques Riche and Robert K. Meyer. Kripke, Belnap, Urquhart and relevant decidability & complexity. In G. Gottlob, E. Grandjean, and K. Seyr, editors, *Computer Science Logic (Brno, 1998)*, number 1584 in Lecture Notes in Computer Science, pages 224–240, Berlin, 1999. Springer.

161. John A. Robinson. A machine oriented logic based on the resolution principle. *Journal of ACM*, 12:23–41, 1965.

162. Dirk Roorda. Lambek calculus and Boolean connectives. OTS Working Papers CL–92–004, Faculteit Letteren, Universiteit Utrecht, Utrecht, Netherlands, 1992.

163. Richard Routley, Robert K. Meyer, Val Plumwood, and Ross T. Brady. *Relevant Logics and Their Rivals*, volume I. Ridgeview Publishing Company, Atascadero, CA, 1982.

164. Masahiko Sato. A cut-free Gentzen-type system for the modal logic S5. *Journal of Symbolic Logic*, 45:67–84, 1980.

165. Moses Schönfinkel. On the building blocks of mathematical logic. In J. van Heijenoort, editor, *From Frege to Gödel: A Source Book in Mathematical Logic*, pages 355–366. Harvard University Press, Cambridge, MA, 1967.

166. Peter Schroeder-Heister. A natural extension of natural deduction. *Journal of Symbolic Logic*, 49:1284–1300, 1984.
167. Peter Schroeder-Heister. Resolution and the origins of structural reasoning: early proof-theoretic ideas of Hertz and Gentzen. *Bulletin of Symbolic Logic*, 8: 246–265, 2002.
168. Wilfried Sieg. Hilbert's program sixty years later. *Journal of Symbolic Logic*, 53: 338–348, 1988.
169. Michael Sipser. *Introduction to the Theory of Computation*. PWS Publishing Company, Boston, MA, 1997.
170. Raymond M. Smullyan. *First-Order Logic*. Springer-Verlag, New York, NY, 1968. (Dover, New York, NY, 1995).
171. Raymond M. Smullyan. Uniform Gentzen systems. *Journal of Symbolic Logic*, 33: 549–559, 1968.
172. Raymond M. Smullyan. Analytic cut. *Journal of Symbolic Logic*, 33:560–564, 1968.
173. Morten H. Sørensen and Pawel Urzyczyn. Strong cut-elimination in sequent calculus using Klop's ι-translation. *Journal of Symbolic Logic*, 73:919–932, 2008.
174. Bayu Surarso and Hiroakira Ono. Cut elimination in noncommutative substructural logics. *Reports on Mathematical Logic*, 30:13–29, 1996.
175. Manfred E. Szabo, editor. *The Collected Papers of Gerhard Gentzen*. Studies in Logic and the Foundations of Mathematics. North-Holland, Amsterdam, 1969.
176. Mitio Takano. Subformula property as a substitute for cut-elimination in modal propositional logic. *Mathematica Japonica*, 37:1129–1145, 1992.
177. Gaisi Takeuti. *Proof Theory*, volume 81 of *Studies in Logic and the Foundations of Mathematics*. North-Holland, Amsterdam, 2nd edition, 1987.
178. Kazushige Terui. Which structural rules admit cut elimination? An algebraic criterion. *Journal of Symbolic Logic*, 72:738–754, 2007.
179. Paul B. Thistlewaite, Michael A. McRobbie, and Robert K. Meyer. *Automated Theorem Proving in Non-classical Logics*. Pitman, London, UK, 1988.
180. Anne S. Troelstra. *Lectures on Linear Logic*, volume 29 of *CSLI Lecture Notes*. CSLI Publications, Stanford, CA, 1992.
181. Anne S. Troelstra and H. Schwichtenberg. *Basic Proof Theory*, volume 43 of *Cambridge Tracts in Theoretical Computer Science*. Cambridge University Press, Cambridge, UK, 2nd edition, 2000.
182. Anthony M. Ungar. *Normalization, Cut-elimination and the Theory of Proofs*, volume 28 of *CSLI Lecture Notes*. CSLI Publications, Stanford, CA, 1992.
183. Alasdair Urquhart. The undecidability of entailment and relevant implication. *Journal of Symbolic Logic*, 49:1059–1073, 1984.
184. Alasdair Urquhart. The relative complexity of resolution and cut-free Gentzen systems. *Annals of Mathematics and Artificial Intelligence*, 6:157–168, 1992.
185. Alasdair Urquhart. The complexity of decision procedures in relevance logic II. *Journal of Symbolic Logic*, 64:1774–1802, 1999.
186. Alasdair Urquhart. Beth's definability theorem in relevant logics. In E. Orlowska, editor, *Logic at Work: Essays Dedicated to the Memory of Helena Rasiova*, volume 24 of *Studies in Fuzziness and Soft Computing*, pages 229–234. Physica Verlag, Heidelberg, 1999.

187. Alasdair Urquhart. The complexity of linear logic with weakening. In S. R. Buss, P. Hájek, and P. Pudlák, editors, *Logic Colloquium '98*, volume 13 of *Lecture Notes in Logic*, pages 500–515, Natick, MA, 2000. A. K. Peters.

188. Alasdair Urquhart. Four variables suffice. *Australasian Journal of Logic*, 5:66–73, 2007.

189. Heinrich Wansing, editor. *Proof Theory of Modal Logic*. Applied Logic Series. Kluwer Academic Publishers, Dordrecht, 1996.

190. Heinrich Wansing. Sequent systems for modal logics. In D. M. Gabbay and F. Guenthner, editors, *Handbook of Philosophical Logic*, volume 8, pages 61–145. Kluwer Academic Publishers, Dordrecht, 2nd edition, 2002.

191. Wojciech Zielonka. Axiomatizability of Ajdukiewicz–Lambek calculus by means of cancellation schemes. *Zeitschrift für mathematische Logik und Grundlagen der Mathematik*, 27:215–224, 1981.

Index

For Product Safety Concerns and Information please contact our EU
representative GPSR@taylorandfrancis.com
Taylor & Francis Verlag GmbH, Kaufingerstraße 24, 80331 München, Germany

www.ingramcontent.com/pod-product-compliance
Ingram Content Group UK Ltd.
Pitfield, Milton Keynes, MK11 3LW, UK
UKHW021115180425
457613UK00005B/99

* 9 7 8 1 4 6 6 5 6 4 6 6 4 *